Plant Resins

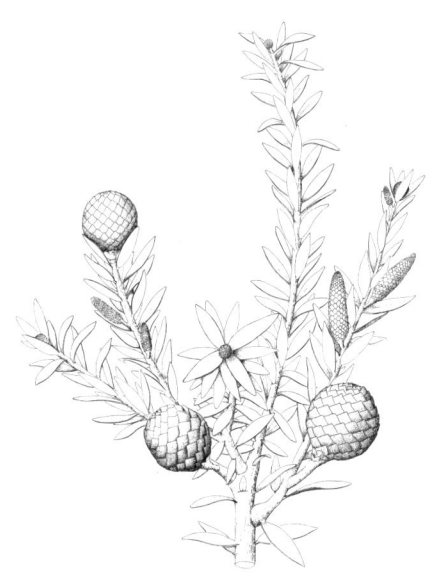

Plant Resins

Chemistry, Evolution, Ecology, and Ethnobotany

JEAN H. LANGENHEIM

Timber Press
Portland • Cambridge

. . . gum, the gum of the mountain spruce.
He showed me lumps of the scented stuff
Like uncut jewels, dull and rough.

—Robert Frost, "The Gum-Gatherer,"
from *Mountain Interval*, 1920

Page 1, *Agathis australis*, kauri; page 2, *Boswellia*, frankincense.

All drawings by Jesse Markman, maps by Gulla Thordarsen and Jesse Markman.

Published in 2003 by

Timber Press, Inc.
The Haseltine Building
133 S.W. Second Avenue, Suite 450
Portland, Oregon 97204, U.S.A.

Timber Press
2 Station Road
Swavesey
Cambridge CB4 5QJ, U.K.

Printed in Hong Kong

Library of Congress Cataloging-in-Publication Data

Langenheim, Jean H.
    Plant resins : chemistry, evolution, ecology, and ethnobotany / Jean H. Langenheim.
        p. cm.
    Includes bibliographical references (p.).
    ISBN 0-88192-574-8
        1. Gums and resins. 2. Gums and resins—Utilization. I. Title
SB289 .L36 2003
620.l'924—dc21

2002028941

Contents

5

Color plates follow page 432

Preface

When I was asked by Timber Press to write a new book on resins, including amber—Howes's 1949 *Vegetable Gums and Resins* was the most recent such effort—the breadth of interdisciplinary coverage seemed too ambitious for an individual person. There have been so many advances in resin research in the past half century, including the development of new fields of research such as chemical ecology, and the exploration of other interesting facets about resins made possible by new chemical, molecular, and microscopic techniques. With a little thought, however, I realized that my years of resin research had prepared me to accept the challenge enthusiastically, a challenge that has been stimulating and rewarding.

My interest in resins began with ambers formed over geologic time and proceeded rapidly to the evolutionary significance of the ecological roles resins play in plants. These were natural interests, arising from my training as an ecologist and paleobotanist. Later, my queries turned to how humans have used resins throughout history, and my interest in that intensified when I taught an undergraduate course, Plants and Human Affairs, and coauthored a textbook on the subject. I became convinced that resins are remarkable materials indeed, especially in their diversity and the length of time they have been such versatile substances in the lives of plants and humans. A university colleague, a philosopher, suggested that resin had created a "cosmos" for me because of the variety of topics I had been led to investigate: paleobotany, chemistry, systematics, ecology, anthropology, ethnobotany, art history, etc. There is no doubt, however, that I could only have delved into such wide-ranging topics with the collaboration and expertise of many individuals, which increased the value and enjoyment of the experience. Although most of the people associated with the development of my research were not directly involved in my writing *Plant Resins*, I want to acknowledge their contributions to the learning experiences that enabled me to accept the challenge. It also is interesting how serendipity played a role in the people I met or the events that took place, helping me as my research interests ramified.

My research into plant resins began as a member of a paleoecological expedition to study amber in Chiapas, Mexico, led by entomologists from the University of California, Berkeley. My role in this expedition was to determine which trees produced the resin in which a diversity of insects had been beautifully preserved and the kind of forest in which the trees and insects had lived. Previously, amber had not been analyzed chemically as a resin but rather had been described inorganically as a gemstone. My first hint of the botanical source of the Mexican amber was a chemical one—its use by the Maya as incense. The burning incense did not smell like burning pine resin, which had long been thought to be the source of the well-known Baltic amber and was assumed to be the source of Mexican amber. Thus I collected resins from all the kinds of resin-producing trees in Chiapas for chemical comparison with the amber, ushering me into the world of tropical resins and the forests in which the trees grew.

Fortunately, at this time I became a research fellow at Harvard University in the laboratory of the geochemist and paleobotanist Elso Barghoorn, who enthusiastically encouraged my exploration of the chemical criteria for determining the botanical sources of amber through geologic time. It was necessary to use solid-state analytic techniques, such as infrared spectroscopy, because the polymerization of amber precluded dissolving it for standard organic chemical analysis. I subsequently collaborated with spectral chemist Curt Beck of Vassar College, who I had serendipitously discovered was using infrared spectroscopy to determine the archeological provenance of European amber. My approach at that time established a new direction in the study of plant origins of ambers, by including chemosystematic data. Additionally, my approach had an even larger perspective, of integrating paleoecological data into the understanding of amber-producing plants. These chemical and paleoecological studies, together with my background as a plant ecologist, prepared me to be intrigued by the correlation that the greatest diversity of trees producing copious amounts of resins are tropical angiosperms (plants with true flowers). This interest coincided with the rapid advance of the field of biochemical ecology, and I was swept along with the tide of its development.

To understand tropical resin production, I decided to use the leguminous tree *Hymenaea* as a model, partly because I had determined it as the source of the amber in a number of large New World deposits. The genus has an amphi-Atlantic distribution, and the history of utilization of leguminous resins increased my growing interest in ethnobotany. Field investigation of *Hymenaea* led me from Mexico through Central America to South America and Africa. The formation of the Organization for Tropical Studies (OTS) coincided with my early studies of *Hymenaea* in Central America, and assistance from numerous OTS colleagues from various colleges and universities (too many to name) helped promote the ramifications of my overall investigation of *Hymenaea* and my interest in other resin-producing plants.

The center of distribution of *Hymenaea* is Amazonia and I had the good fortune to be introduced to the region by the late Richard E. Schultes, long-time Amazonian ethnobotanical researcher at the Harvard University Botanical Museum. He helped initiate my Amazonian research, which continued for many years, and importantly, further enhanced my interest in ethnobotany. Successful work on Brazilian Amazonian resin-producing plants also would not have been possible without the strong support and interest of Paulo Machado and Warwick Kerr, former directors of the Instituto Nacional de Pesquisas da Amazônia in Manaus; Paulo Cavalcante, Museu Goeldi in Belém; and others again too numerous to mention. Additionally, I had the unflagging interest and cooperation of Ghillean Prance, then director of research at the New York Botanical Garden, and later, director of the Royal Botanic Gardens, Kew, who was leading the research for *Flora Amazônica*.

Before I could investigate resin production throughout the geographic range of *Hymenaea*, revising the systematics of the genus was necessary since species had often been described from poor specimens collected during floristic surveys. This is a common situation for many of the plants belonging to tropical resin-producing families, a problem whose consequences are noted throughout *Plant Resins*. The *Hymenaea* revision, done in collaboration with a graduate student, Y. T. Lee, was approached as an interface between systematics and ecology, with amber providing the evolutionary context. During this revisionary work I interacted closely with tropical legume systematists such as the late Pat Brenan, Royal Botanic Gardens, Kew, and J. Léonard, Université de Bruxelles, a specialist on African copal producers. This opened my thinking on the important relationships of African and New World trees.

My interest in tropical resin-producing plants also expanded to discussion of taxonomic problems with specialists, including Douglas Daly, New York Botanical Garden (Burseraceae); T. C. Whitmore, Oxford University *(Agathis)*; and Peter Fritsch, California Academy of Sciences (Styracaceae), among others.

As my resin studies progressed, I had to learn more about the constituents of present-day rather than fossil resins. Thus I embarked on a determination of the components of *Hymenaea* resins with doctoral graduate students Susan Martin and Allan Cunningham, with assistance from chemists E. Zavarin, Forest Products Laboratory, University of California, Berkeley; George Hammond, University of California, Santa Cruz; A. C. Oehschlager, Simon Fraser University; and Duane Zinkel, Forest Products Laboratory, University of Wisconsin, Madison.

How resin is secreted into storage structures is significant to both plant defense and human use of resins. So another door to learning opened. In exploring the anatomy of secretory structures in *Hymenaea*, I was aided by the late Ralph Wetmore and I. W. Bailey as well as Margaret McCully, at Harvard University at the time, all of whom enthusiastically supplied the needed expertise. Lynn Hoefert, U.S. Department of Agriculture, Salinas, California, also assisted a graduate student, Gail Fail, with ultrastructural studies of resin secretion in *Hymenaea*. I increased my knowledge of resin secretory structures through contact with other researchers, too, including A. Fahn, Hebrew University of Jerusalem, and B. Dell and A. J. McComb, University of Western Australia, who studied secretory systems in a variety of resin-producing plants.

A major interest in the chemical ecology of *Hymenaea* was followed by comparison with the related legume, *Copaifera*. These investigations involved collaboration with another group of graduate students (Will Stubblebine, David Lincoln, José Carlos Nascimento, Matthew Ross, Craig Foster, Robert McGinley, Cynthia Macedo, Eric Feibert, and Susanne Arrhenius) on plant interactions with insects and fungi. Other avenues to understanding resin production were opened by graduate students (George Hall, Francisco Espinosa-García, and Wendy Peer) who worked on the chemical ecology of redwoods *(Sequoia)*. I also enjoyed numerous stimulating discussions on defensive mechanisms of other resin-producing plants with colleagues, including Karen Sturgeon, then at the University of Colorado; Kenneth Raffa,

University of Wisconsin, Madison; Marc Snyder, Colorado College; John Bryant, University of Alaska; and numerous others.

Archeological and anthropological studies of resin and amber were carried out in Angola in collaboration with Desmond Clarke, University of California, Berkeley. By serving on doctoral dissertation committees at Yale University and the University of Texas, Austin, I learned about the use of resin by the Semelai in peninsular Malaysia (with Rosemary Gianno) and by the Maya in Mexico and Central America (with Kirsten Tripplett). Moreover, these kinds of studies provided opportunities to observe art objects made from amber, and contacts with museums around the world. And who would not avail themselves of the opportunities to collect and enjoy amber jewelry!

Thus, from my varied experiences in research on resin and amber, I saw the need for an up-to-date book because so much disparate information is scattered throughout the literature. I decided that the book should tell the whole story of these fascinating plant substances. Despite the importance of a multidisciplinary approach, and my hope of raising awareness of that, I divided the book into three parts to make it easier to use by readers with diverse backgrounds, interests, and goals, who I knew might turn to such a volume for information. These parts may be read in any order, depending on the reader's interest. A glossary is also provided. The three chapters in Part I, The Production of Resin by Plants, provide biochemical, developmental, and systematic information. However, this information is repeatedly projected toward discussion of the value of resins to plants and humans in Parts II and III. Central to understanding the remainder of the book is my operational definition of resin, presented in Chapter 1. This definition comes from my struggle with the confused and vague usage of the term resin that has persisted through the years. I hope that my definition provides rigor and clarity by distinguishing resins from other materials with which they are commonly confused (e.g., gums and mucilage) based on three criteria: chemistry, secretory structures, and ecological roles in the plant. Part I also includes a discussion of more recent major breakthroughs in the understanding of terpenoid biosynthesis and the ultrastructural evidence for its compartmentation, and how this new information solves mysteries encountered in ecological studies of resins. The secretory structures are characterized, and the importance of understanding their functions in ecological interactions and human use is discussed. Furthermore, I introduce the reader to the distribution of resin-producing plants throughout the plant kingdom and for the first time present evolutionary convergences in different aspects of resin production.

Part II, The Geologic History and Ecology of Resins, includes topics that have been at the heart of much of my own research. The two chapters have a phytocentric approach whereas other publications covering these subjects are more insect-oriented. Questions on when resins first evolved and on which groups of resin producers have a geologic record are addressed in Chapter 4. Chapter 5 addresses the question of whether resin production is primarily a defense against herbivores and pathogens, and presents ecological and evolutionary data that support this view.

Part III, The Ethnobotany of Resins, presents in six chapters the substantial roles that different kinds of resins have played in most cultures of the world throughout human history. In Chapter 11, I consider whether the importance of resin to humans will become a historical remnant as they are replaced by petrochemicals and other alternatives, or whether new technologies as well as policies that preserve plant resources, particularly in the tropics, will enable change in uses of resins and an important future for them. *Plant Resins* only provides a progress report on our current knowledge—I hope this synthesis of the many facets of resins will stimulate future research on these remarkable plant products.

Acknowledgments

For *Plant Resins* specifically, I am grateful to friends, colleagues, and organizations who have contributed photographs as well as to those who provided comments that greatly improved the clarity of the chapters.

Numerous colleagues who shared photographs from their own resin research include Scott Armbruster, Norwegian University of Science and Technology; John Lokvam, and John Bryant, University of Alaska, Fairbanks; Ben LePage, University of Pennsylvania; A. Fahn, Hebrew University of Jerusalem; Duncan Porter, Virginia Polytechnic University; T. C. Whitmore, Oxford University; Robert Clarke, International Hemp Association; J. J. Hoffmann, S. P. McLaughlin, and D. L. Venable, University of Arizona; Robert Adams, Baylor University; Manuel Lerdau, State University of New York, Stonybrook; Jason Greenlee, Fire Research Institute, Fairfield, Washington; Hanna Czeczott, Museum Ziemi, Warsaw, Poland; Adam Messer,

University of Georgia; David Rhoades, Seattle, Washington; William Gittlin, Berkeley, California; Douglas Daly, New York Botanical Garden; John Dransfield, Royal Botanic Gardens, Kew; Rosemary Gianno, Keene State College, New Hampshire; M. Pennacchio, University of Technology, Western Australia; Bill Thomson, University of California, Riverside; J. G. Martínez-Avalas, Universidad Autónoma de Tamaulipas; Rudolf Becking, Humboldt State University, California; S. P. Lapinjoki, Kuppio University, Finland; William Crepet, Cornell University; Margaret McCully, Carleton University, Canada; Kennedy Warne, *New Zealand Geographic* magazine; Robert Wheeler, U.S. Forest Service, Fairbanks, Alaska; and Vito Polito, University of California, Davis. I owe special thanks to David Grimaldi, American Museum of Natural History, who so generously provided numerous amber photographs from his research and from his book, *Amber, Window to the Past*. I also gratefully acknowledge the following organizations for providing photographs: Royal Botanic Gardens, Kew; Danish National Museum, Copenhagen; and the National Library of New Zealand, Wellington.

I also express my gratitude to those who critically reviewed various drafts of different chapters: Ken Anderson, Argonne National Laboratory; Elizabeth Bell, Santa Clara University; Laurel Fox, University of California, Santa Cruz; Peter Fritsch, California Academy of Sciences; Jonathan Gershenzon, Max Planck Institute for Chemical Ecology; Cheryl Gomez, UCSC; David Grimaldi, American Museum of Natural History; Karen Holl and Ingrid Parker, UCSC; Campbell Plowden, Penn State University; Kirsten Tripplett, University of California, Berkeley; and Duane Zinkel, Forest Products Laboratory, Madison, Wisconsin. Again, I extend special thanks to Susan Martin, U.S. Department of Agriculture Research Laboratory, Ft. Collins, Colorado, and Marc Los Huertos and Thomas Hofstra , UCSC, for their particular care and thoughtfulness in reviewing numerous chapters. I also appreciate the generosity of the time given by classical historian Gary Miles, UCSC, and anthropologist Rosemary Joyce, UC Berkeley, to discuss details of the Chapter 6 time line.

I greatly appreciate the efforts of Gulla Thordarsen in drafting maps. Jesse Markman's contributions are special in that he did all drawings of plants, most maps, and generally shared in most aspects of the book's development. Jesse and I are grateful to Ann Caudle, Science Communications Program, University of California, Santa Cruz, for her assistance and critical comments on the plant drawings. The diligent help of the UCSC reference librarians was invaluable, and the cheerful persistence of the interlibrary loan librarians was essential in obtaining literature unavailable in our library. I am also grateful for the conscientious efforts of my editor at Timber Press to see that *Plant Resins* is as error-free and as understandable to a broad audience as possible. Finally, the book would not have been possible without Dorothy Hollinger's tireless word processing of the numerous drafts.

PART I

The Production of Resin by Plants

Overleaf: branch of *Pseudotsuga menziesii*, Douglas fir (see also Plate 4), with cone, and the structure of α-pinene, a common monoterpene in the resin.

CHAPTER 1

What Plant Resins Are and Are Not

The literature on resins, although relatively abundant, is not very precise as far as exact use of terms is concerned. . . . confusion which, in a way, reflects the complexity of the world of resins.

—Jost et al. 1989

To understand the many topics covered in *Plant Resins*, it is necessary to have a clear idea of what plant resin is and how it differs from other substances that have been called resins. Different readers doubtless have different concepts of what resin is. Some may be surprised at the number of plants that produce resin (Chapter 2) and consequently at the breadth and depth of the influence that resins have had throughout history (Chapter 6). The characterization of resins has changed greatly with the development of chemical, molecular, and microscopic technologies to analyze them. Associated with these technological breakthroughs have been advances in evolutionary and ecological concepts regarding the functions of resins in plants.

Definitions of Resin

Resin is sometimes referred to in a general manner, such as sap or exudate, both of which include numerous substances from plants. Throughout written history there has been a tendency to characterize resin vaguely as any sticky plant exudate. In some dictionaries, this definition has been extended to include substances that are mainly insoluble in water and that ultimately harden when exposed to air. Nevertheless, the vagueness of even this amended definition has led to continued confusion with other plant exudates, including gums, mucilages, oils, waxes, and latex. Some terms such as gum have often been used synonymously with resin; in fact, one prominent forest products researcher has referred to the use of these terms as "haphazard" (Hillis 1987). A better definition of resin, however, has awaited more knowledge about their chemistry, secretory structures, and functions in the plant.

Interest in the chemistry of resins and the secretory structures in which they are synthesized and stored began in the later 19th century in Germany. A pioneering book, *Die Harze und die Harzebehälter*, resins and resin-containing structures, was published by Tschirch and his students in 1906. Recognition that detailed chemical knowledge of plant exudates would be valuable, perhaps essential, for their commercial utilization led to the voluminous publications in the 1930s by Tschirch and Stock (1933–36) and others (e.g., Barry 1932). Nonetheless, only with the advent of various kinds of chromatography and spectroscopy in the 1940s and 1950s was real progress made in identifying the chemical constituents of resins and quantifying their composition. All the exudates that have been confused with resin in the past can now be distinguished from resin in their pure form by chemical composition and by the biosynthetic pathways through which they are formed. Information about resin secretory structures has become available through advances in plant anatomy, including electron microscopy (Chapter 3), and from ecological studies regarding the survival roles played by resins (Chapter 5). Together, these data provide criteria for a definition to minimize the confusion surrounding the term resin.

Thus in *Plant Resins*, plant resin is defined operationally as primarily a lipid-soluble mixture of volatile and nonvolatile terpenoid and/or phenolic secondary compounds that are (1) usually secreted in specialized structures located either internally or on the surface of the plant and (2) of potential significance in ecological interactions. Note that resins consist primarily of secondary metabolites or compounds, those that apparently play no role in the primary or fundamental physiology of the plant. In addition to being preformed and stored in secretory structures, resins sometimes may be induced at the site of an injury without forming in a specialized secretory structure. Moreover, resin occurs predominantly in woody seed plants. Amber is fossilized resin (Chapter 4).

Although terpenoid resins constitute the majority of copious internally produced resins that have been used commercially, some important resins are phenolic. Phenolic resin components occurring on the surfaces of plant organs have been used, particularly in medicines, and may be useful as a biomass source of fuel; however, their overall significance is probably greater as protection for vulnerable plant surfaces. Resin components are derived from photosynthetically produced carbohydrates that are broken down to produce simpler compounds (pyruvate products); terpenoid and phenolic compounds are then synthesized via different metabolic pathways (Figure 1-1).

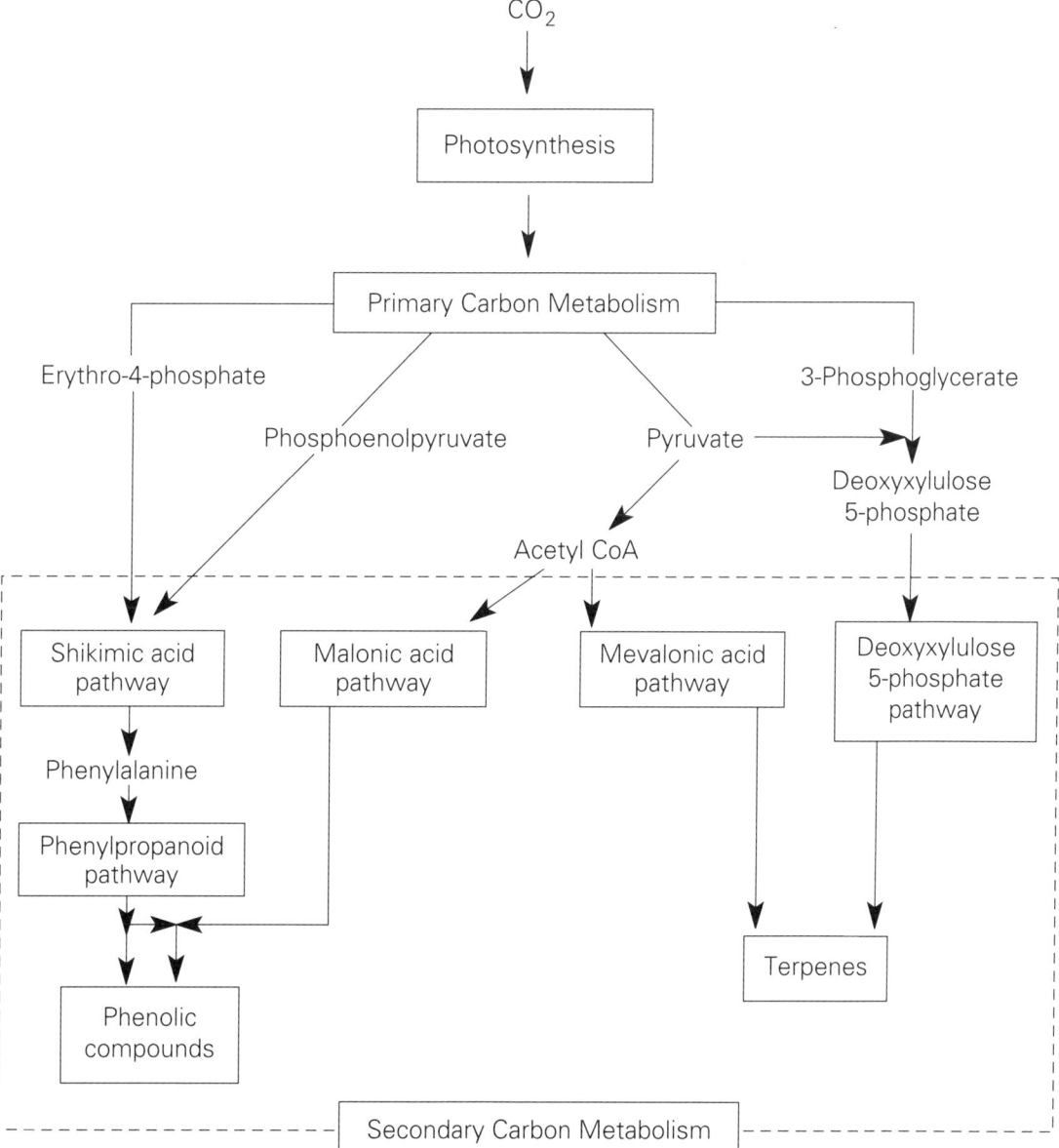

Figure 1-1. Generalized outline of biosynthesis of terpenes and phenolic secondary compounds constituting resins, showing interconnection with relevant primary compounds and processes.

I briefly discuss the biosynthesis of terpenoid and phenolic resins, as well as some of their characteristic components, as a basis for understanding the ecological interactions that are important to the plant and to the human use of its resin. Next, I review the substances that have been confused with resin or that occur intermixed with it. The distinguishing characteristics may be difficult to keep clear in some cases because the chemistry, kind of secretory tissue, and ecological roles of the material in question may be poorly known; therefore, *resin* may be ambiguously or dubiously designated in the literature on plant exudates. There may be a quandary as to whether the material is really a resin or not, a situation that is unavoidable.

I discuss resins in various categories (e.g., oleoresins, balsams, and copals) in Part III because these terms have gained prominence through human use of those resins. Again, there is confusion from varied use of these terms, especially in regard to the plants producing the resins. A list of common names of resins, their botanical sources, and major uses is provided in Appendix 5.

Terpenoid Resins

Terpenoid Synthesis

Terpenoids occur in all living organisms but attain their greatest structural and functional diversity in plants. In fact, terpenoids constitute the largest and most diverse class of plant compounds. The term terpenoid or terpene is derived from the German word for turpentine, *Terpentin*, from which the first members of this group of chemicals were isolated and their structures determined (Croteau 1998). Through continual development of chemical technology, especially gas chromatography–mass spectrometry (GC-MS) and nuclear magnetic resonance spectroscopy (NMR), the structures of approximately 30,000 terpenoids have been elucidated, but many more doubtless will be discovered. Although terpenoids exhibit enormous structural diversity and chemical complexity, they are united by a common biosynthetic origin that enables them to be grouped in useful categories by linkage of five-carbon (C_5H_8) isoprene structural elements. Consideration of these units help in the visualization of a terpenoid's biosynthetic assembly, although extensive metabolic rearrangement may complicate the picture. Terpenoids (referred to interchangeably as terpenes) are sometimes called isoprenoids.

Pathways and Cellular Compartmentation

Two biosynthetic pathways lead to formation of the basic structural unit of terpenoid synthesis (Figure 1-1). In the well-studied classical mevalonic acid (MVA) or mevalonate pathway, three molecules of acetyl coenzyme A are linked, (pyro)phosphorylated, decarboxylated, and dehydrated to yield isopentenyl diphosphate (IPP), traditionally called isopentenyl pyrophosphate. It has been discovered that IPP can be formed via a different pathway (Lichtenthaler et al. 1997). Although details of the complete pathway remain to be elucidated, 3-phosphoglycerate (3-PGA) and two carbon atoms derived from pyruvate apparently combine to generate a first intermediate, 1-deoxy-D-xylulose 5-phosphate (DOXP or DXP), then 2-C-methyl-D-erythritol 4-phosphate (MEP), which eventually is converted to IPP (Lange et al. 1998, 2000; Lichtenthaler 1999). This alternate pathway is referred to as the DOXP or DXP pathway, the MEP pathway, and sometimes as the nonmevalonic or mevalonate-independent pathway.

IPP and its isomer, dimethylallyl diphosphate (DMAPP), are the actual five-carbon building blocks for the formation of larger terpenoid molecules (Figure 1-2). DMAPP serves as a primer to which IPP units can be added in sequential chain-elongation steps. These reactions, catalyzed by prenyltransferase enzymes, connect isoprene units to one another. Thus IPP and DMAPP combine to form a C_{10} precursor (geranyl diphosphate, GPP) for all 10-carbon compounds, called monoterpenes. Addition of another molecule of IPP yields a C_{15} precursor (farnesyl diphosphate, FPP) for all 15-carbon isoprenoids, called sesquiterpenes. The structural diversity of sesquiterpenes greatly exceeds that of monoterpenes because many more types of cyclization can occur in a precursor with five additional carbon atoms. This diversity is evident in many resins, and some structures may polymerize, which can result in the formation of large deposits of fossil resin (Chapter 4). Mono- and sesquiterpenes generally are volatile, giving fluidity to the resin as well as acting as plasticizers for the more viscous components. When only the volatile mono- and sesquiterpenes occur, they often are called essential oils. This designation, however, is misleading because these terpenoids are neither essential to plant metabolism nor are they true oils; *essential* refers to their essence or fragrance, and *oil* to their feel. Essential oils as the only terpenoid fraction occur in a few trees, for example, those in the Lauraceae (e.g., *Laurus*, bay trees), but are found predominantly in herbaceous or shrubby plants, especially those in

Mediterranean climates (Ross and Sombrero 1991). Occasionally in the resin literature, the volatile fraction of resin as defined in *Plant Resins* is referred to as essential oil and only the nonvolatile fraction is called resin.

Addition of three molecules of IPP to DMAPP gives the C_{20} precursor (geranylgeranyl diphosphate, GGPP) of the diterpenes. More than 3000 diterpene structures have been defined, usually bearing a variety of oxygen-containing functional groups. Diterpene acids are particularly important in resin. Doubling (dimerization) of the C_{15} FPP leads to C_{30} compounds, the triterpenes. Triterpenes include a wide variety of structurally diverse substances, some of which have been so modified that they no longer contain the full complement of 30 carbon atoms. Numerous skeletal types occur in resin,

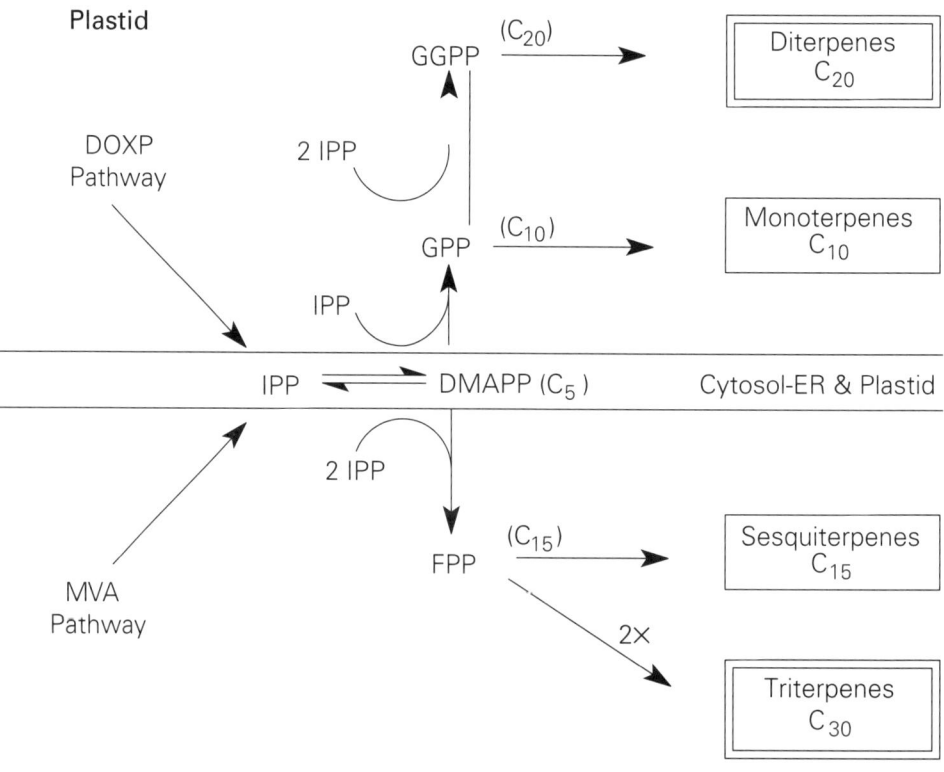

Figure 1-2. Biosynthesis of terpenoids presented according to the compartmentation of the two pathways. Terpene components of resins are indicated in boxes; volatile components have single outline and nonvolatile components have double outline. DMAPP, dimethylallyl diphosphate; DOXP, 1-deoxy-D-xylulose 5-phosphate pathway; FPP, farnesyl diphosphate; GGPP, geranylgeranyl diphosphate; GPP, geranyl diphosphate; IPP, isopentenyl diphosphate; MVA, mevalonic acid or mevalonate pathway.

characterizing some angiosperm families. Di- and triterpenes are nonvolatile components of resin (Figure 1-2). Resins in most plants generally do not contain both di- and triterpenes in the nonvolatile fraction, however, whereas all conifers and some angiosperms contain both mono- and sesquiterpenes in the volatile fraction. Both di- and triterpenes occur in a few genera of the large tropical rain forest angiosperm family Burseraceae (Chapter 8).

Dimerization of the C_{20} precursor (GGPP) leads to C_{40} compounds, the tetraterpenes, and addition of n C_5 isoprene units ($n > 8$) results in polyterpenes, well known as rubber and gutta percha. Tetraterpenes and polyterpenes are not known to be constituents of resin, although resin components may occur along with polyterpenes in several plants (Chapter 9).

Plant metabolism is extensively compartmentalized at the cellular level. Compartmentation is significant in regulating terpenoid synthesis because it allows independent control of different branches of the pathway at different sites in the cell. Within a compartment, metabolic dynamics depend on the kinds of enzymes (synthases) present and the permeability of intracellular membranes to precursors, intermediates, and products.

The two different pathways to IPP appear to be compartmentalized (Newman and Chappell 1999), with plastids, mitochondria, and cytosol–endoplasmic reticulum the compartments in which IPP is converted to various terpenoids (Kleinig 1989). Each compartment produces different products. For resins, the mevalonate pathway operates in the cytosol-ER compartment to produce sesqui- and triterpenes whereas the alternative DXP pathway operates in plastids to produce mono- and diterpenes (Figure 1-2). The synthases that produce terpenes differ for constitutive resin (preformed resin stored in secretory structures) and induced resin (that synthesized at the site of an injury). In fact, specialized secretory structures for many plant mono-, sesqui-, and diterpenes, such as those found in resin, are apparently required for synthesis of constitutive compounds (Gershenzon and Croteau 1990, 1993; Chapter 3). Therefore, the differentiation of such structures may provide another form of control over terpenoid production before any events are induced by injury.

Primary Versus Secondary Terpenoids

Although the basic pathways of terpenoid biosynthesis are present in all plants, relatively few terpenoids are known to play vital roles in plant growth and development in all plants; these few are considered primary metabolites.

For example, a sesquiterpene is a precursor to abscisic acid, and a diterpene is an intermediate in the synthesis of the gibberellic acids; abscisic and gibberellic acids are important plant growth regulators. Steroids are triterpene derivatives that are essential components of cell membranes. The red, orange, and yellow carotenoids (tetraterpenes) function as accessory pigments in photosynthesis and further serve to protect photosynthetic tissues from deleterious photooxidative effects. Terpene-derived side chains (e.g., the phytol side chain of chlorophyll) also help anchor certain molecules in membranes.

The vast majority of the 30,000 known terpenoids are secondary compounds, lacking any apparent role in primary physiological or metabolic processes in the plant. Thus terpenes are shown as part of secondary carbon metabolism in Figure 1-1 because most that occur in resin are considered to be secondary. Additionally, their formation originates from just a few intermediates of primary metabolism, including acetyl coenzyme A, mevalonic acid, and 3-phosphoglycerate. Many biochemists once thought that these terpenoids merely represented ways of disposing of excess acetate, that they were waste products. With the subsequent rapid development of the field of chemical ecology, terpenoid secondary compounds have been shown to play major defensive roles in the survival of the plant and in various interactions in ecosystems (Chapter 5). Chemists in the 19th and early 20th centuries also recognized the value of such compounds to humans and began to call them natural products to distinguish them from chemical compounds synthesized by humans. The term natural products is commonly used and preferred to the term secondary chemicals by some chemical ecologists (Romeo et al. 1996).

Enzymology

More than 30 genes have been isolated that encode terpene synthases (often called cyclases because the reaction products are frequently cyclic), the enzymes that catalyze formation of the basic skeleton of terpenoids. Sequence comparison and phylogenetic analyses show that all known terpene synthases share a common evolutionary origin (Mitchell-Olds et al. 1998, Phillips and Croteau 1999, Trapp and Croteau 2001). However, Bohlmann et al. (1998b) have suggested that there are some distinctive differences between gymnosperm and angiosperm synthases, indicating a bifurcation in primary metabolism from a common ancestor. They assumed that this bifurcation implies an independent functional specialization following separation of the gymnosperm and angiosperm lineages. This evolutionary comparison of synthases was made using resin-producing gymnosperms and terpenoid-producing (but not resin-producing) angiosperms. For example, they found that functional limonene synthases probably evolved separately in grand fir *(Abies grandis)* and mints (Lamiaceae, or Labiatae). Moreover, Savage et al. (1994) found that synthases producing similar terpenes are immunologically distinct in lodgepole pine *(Pinus contorta)* and grand fir. They hypothesized that there are different ways in which terpene synthase proteins catalyze allylic diphosphates to cyclic products even between genera in the same family (Pinaceae in this case).

Terpenoid synthases are not very similar to other enzymes except the mechanistically similar prenyltransferases. Isotopically sensitive branching experiments and cDNA cloning demonstrate that some of these enzymes have the unusual ability to synthesize multiple terpene products, which may represent a way to maximize diversity of compounds using minimal genetic machinery. Furthermore, the compounds may be produced in fixed ratios, thus contributing to the regulation of the relative proportions of compounds in the resin, that is, its composition. These ratios underlie chemical adaptation in individual trees and chemical diversity in populations (Katoh and Croteau 1998). Limonene synthase is an example of such a synthase, producing multiple monoterpenes that commonly occur in conifers, for example, (–)-limonene, myrcene, and α- and β-pinene. Likewise, phellandrene synthase in *Pinus contorta* produces α- and β-pinene as well as 3-carene and β-phellandrene (Savage et al. 1994) whereas (–)-pinene synthase in *Abies grandis* produces (–)-α- and (–)-β-pinene in a fixed ratio, 2:3 (Bohlmann et al. 1998b). Monoterpenes such as these commonly occur in various conifer and angiosperm resins (Figure 1-3).

Expression of cDNAs for several *Abies grandis* monoterpene synthases provides evidence that the complex mixture characterizing the resin forms through a family of both single- and multiproduct enzymes encoded by closely related genes (Bohlmann et al. 1997). Moreover, constitutive sesquiterpenes apparently are produced by even more remarkable synthases (Steele et al. 1998a). The longer chain length and additional double bond of the farnesyl substrate allow greater mechanistic flexibility in the construction of different carbon skeletons. In *A. grandis,* δ-selinene synthase and γ-humulene synthase yield 34 and 52 sesquiterpenes, respectively, from farnesyl diphosphate. These two constitutive sesquiterpene synthases are among the most complex terpenoid synthases. In contrast to constitutive (preformed) sesqui-terpenes, sesquiterpenes induced by injury, such as α-bisabolene and δ-cadi-nene, are produced by single-product synthases. Thus there are differences between synthases of constitutive and induced components of resins as well as between those of primary and secondary terpenoids. Trapp and Croteau (2001) have reviewed in detail the status of regulation, molecular genetics of protein-based genetics, genomic intron and exon organization in conifer resin biosynthesis, and important future challenges in identifying and isolating genes in resin pathways.

Monoterpenes

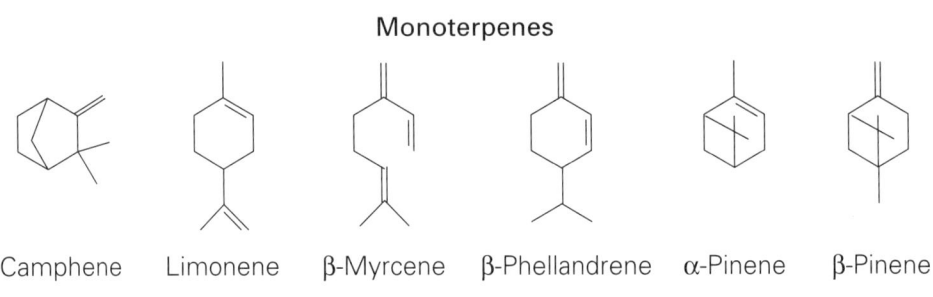

Camphene   Limonene   β-Myrcene   β-Phellandrene   α-Pinene   β-Pinene

Sesquiterpenes

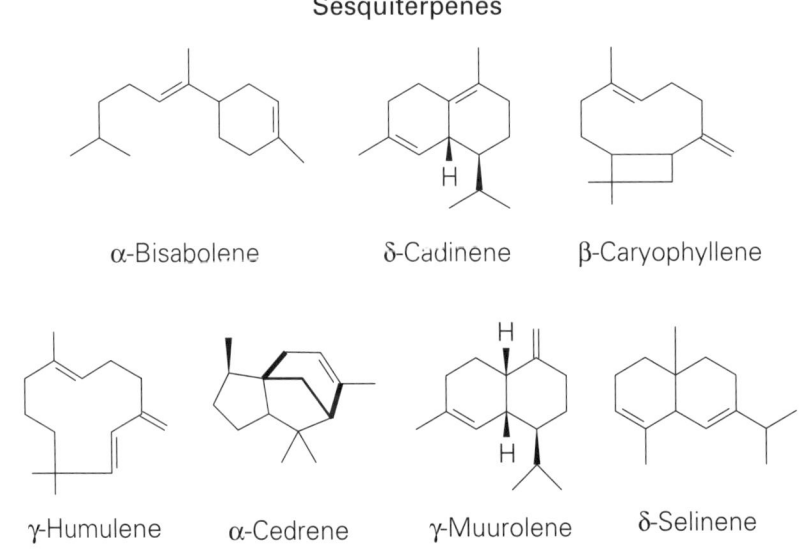

α-Bisabolene   δ-Cadinene   β-Caryophyllene

γ-Humulene   α-Cedrene   γ-Muurolene   δ-Selinene

Figure 1-3. Structures of some common mono- and sesquiterpenoids constituting conifer and angiosperm resins that are used commercially (Chapters 7–10). They include some constituents produced by multiproduct synthases.

Terpenoid Loss

A balance between the rate of synthesis and rate of loss controls the accumulation of any compound. In the plant's economy, it is important whether there is rapid metabolic turnover of secondary compounds, whether some may be catabolized to primary metabolites late in the life of a plant organ, or whether some may be released into the environment. A rapid rate of turnover (including biosynthetic interconversion, conjugation reactions, and polymerization) could increase the cost of maintaining a given concentration of defensive compounds and is often thought to be a major component of the cost of plant chemical defense (Gershenzon 1994a, b). Traditionally, terpenoid secondary chemicals have been viewed as stable end products of metabolism, although Burbott and Loomis (1969) demonstrated rapid turnover of monoterpenes in leaves of detached stem cuttings of peppermint *(Mentha)*. This report was much cited, especially by ecologists. On the other hand, when Mihaliak et al. (1991) repeated the experiments using rooted, intact plants, either low rates or no turnover was detected, suggesting that short-term turnover of monoterpenes does not occur normally in mint leaves but is an artifact seen only in cuttings. In further experiments to test various parameters that could affect turnover in intact plants, Gershenzon et al. (1993) were unable to detect significant turnover in developing leaves of species from a range of taxonomically distant terpene-accumulating families that synthesize mono-, sesqui-, and diterpenes and that store the products in various kinds of secretory structures.

In contrast to the lack of evidence for rapid or short-term turnover of monoterpenes, it is well known that various mono-, sesqui-, di-, and triterpenes (some of which occur in resin) may be lost from leaves late in their development. Some monoterpenes in mature leaves of several mint species are mobilized prior to senescence, when they no longer serve defensive roles (Gershenzon 1994a, b). These terpenes can be catabolized to water-soluble glycosides, which apparently are exported to the root and oxidatively degraded to acetyl coenzyme A (Croteau and Martinkus 1979, Croteau and Sood 1985, Croteau 1988). Thus, apparently, the fixed carbon of some terpenes can be recycled into usable primary metabolites for biosynthesis of new materials (Figure 1-1). Evidence further suggests that synthesis, storage, and catabolism of terpenes may be partially controlled by a balance of photosynthesis and use of the photosynthate through growth and differentiation into various structures and compounds (Loomis and Croteau 1973, Gershenzon

1994b). Although Gershenzon et al. (2000) found no evidence for monoterpene catabolism in peppermint, they suggested that large variances in monoterpene incorporation after pulse labeling may have prevented its detection. Alternatively, they hypothesized that the degradation enzyme in mints may detoxify monoterpenes that have come into contact with living cells following damage to the secretory structures. Although catabolism of terpenes may have considerable physiological and ecological significance, the data are fragmentary and little is known about the process or even if such catabolism occurs in the complex mixture constituting resin.

Studies of *Mentha* have shown that the rate of monoterpene biosynthesis, determined by $^{14}CO_2$ incorporation, closely correlates with monoterpene accumulation and appears to be the principal factor controlling the monoterpene level of peppermint leaves (Gershenzon et al. 2000). In addition to lack of detection of catabolic losses through leaf development, volatilization occurred at a low rate, which on a monthly basis represented less than 1% of the total pool of stored monoterpenes. Composition of the volatilized monoterpenes was sufficiently different from the total plant monoterpene pool that Gershenzon and coworkers suggested that the volatilized products may arise from a separate secretory system, as inferred from previous studies using other plant species (Chapter 3). It is not known if monoterpenes in a resin respond differently when they are formed in different secretory structures, especially with the evidence of terpenoid volatilization from conifers and its role in tropospheric chemistry (Chapter 5).

Characteristic Components

Secondary compounds such as those constituting resin differ from primary metabolites in having a restricted distribution in the plant kingdom. Usually, they occur only in particular groups of related plants. Terpenoid resin occurs in most conifer families but is widely scattered among the major evolutionary lineages of angiosperms (Chapter 2). Specific terpenoid skeletal types, however, often characterize taxa such as particular families and genera; thus it has been assumed that the evolutionary history of various taxa can be significant to the understanding of the taxonomic distribution of some of these chemicals (Gershenzon and Mabry 1983).

I introduce a few skeletal structures in this chapter to exemplify components of resins in important conifer and angiosperm plant families, discussed as of value either to the plants themselves or to humans in later chapters. Conifers only produce internally secreted terpenoid resin whereas angiosperms produce both terpenoid and phenolic resins, which may be secreted internally or on the surface of the plant. This is discussed in detail in Chapter 3.

In addition to the skeletal structure of the compounds, the complexity of the mixture of compounds constituting a resin is important for ecological interactions and human use. In general, among the 20–50 or more compounds that constitute a resin, only a few occur in high concentration. The relative proportions of the compounds in the mixture are called its composition, which may differ in constitutive and induced resins. Because this mixture involves volatile and nonvolatile fractions, the composition of either fraction (or just part of it) or both fractions may be analyzed and compared.

The volatile fraction, which has been most intensively studied, usually consists of mono- and/or sesquiterpene hydrocarbons with some oxygenated forms and, occasionally, diterpene hydrocarbons. The nonvolatile fraction of resin is primarily composed of di- or triterpene acids with some alcohols, aldehydes, and esters in addition to amorphous, neutral substances. The relative proportion of volatile to nonvolatile compounds, which can vary even between species of the same genus, determines a resin's fluidity, viscosity, and polymerization rate. These in turn influence its ecological properties (Chapter 5) as well as the methods used by humans to collect it (Chapters 7–10).

Conifer Resins

Conifer resins, such as those of the pine family (Pinaceae), are characterized by a large volatile fraction (20–50%) with monoterpenes predominating over sesquiterpenes. Both classes most commonly occur as hydrocarbons with a few oxidized forms, often as trace components. Under natural conditions, monoterpenes volatilize with varying degrees of rapidity, providing, for example, the fragrant aromas in conifer forests during warm weather and those from indoor Christmas trees. In fact, monoterpene hydrocarbons from these resins may reach significant proportions in our atmosphere and become troublesome as pollutants. In the soil, monoterpenes from resin may play a role in the nitrogen cycle in conifer forests by inhibiting nitrification. On the other hand, some may supply an energy source for forest soil microbes (Shukla et al. 1968), and others washed from conifer forest soils into estuaries may provide energy for marine microbes (Button 1984). These volatile components of terpenoid resin (both mono- and sesquiterpenes) play a major defensive role against insects and pathogens in amazingly intricate ways (Chapter 5). In commercial use in the naval stores industry, the volatile mono- and·sesquiterpenes of pine resin produce turpentine, a product used worldwide in solvents and as a feedstock for the flavor and fragrance industries (Chapter 7). Sesquiterpenes (e.g., cedrene) are used as cedarwood oil, again particularly in the aroma industry. Structures of some of the most common volatile mono- and sesquiterpenes in various conifer resins are shown in Figure 1-3. Note that the abundant monoterpenes are often the ones produced by multiproduct synthases.

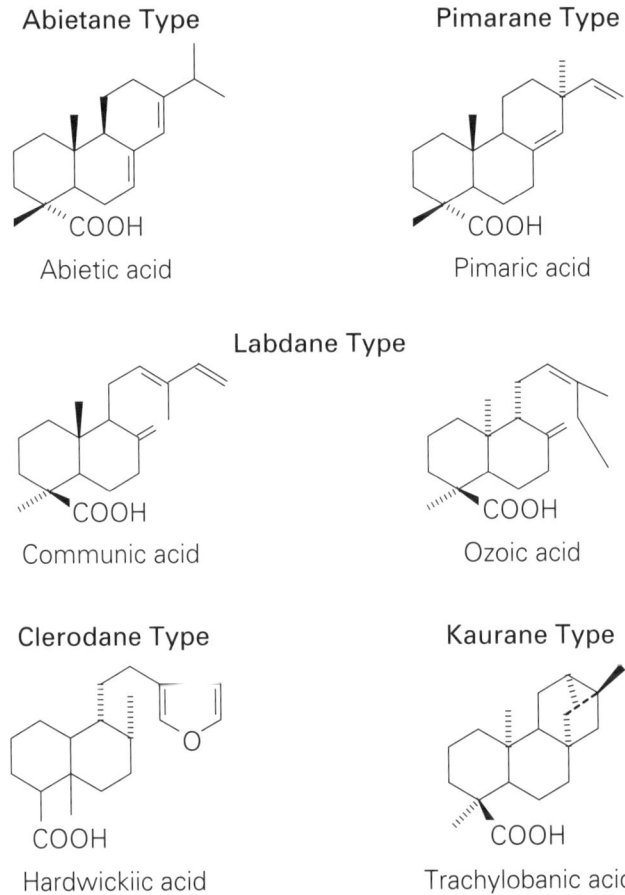

Figure 1-4. Some common diterpene resin acids. Those with abietane and pimarane structural types characterize conifer resins whereas those with labdane structural types occur commonly in both conifers and angiosperms. The conjugated diene in communic acid in conifers and ozoic acid in angiosperms enables polymerization and, hence, formation of amber (Chapter 4).

Nonvolatile terpenes in conifers are primarily diterpene acids. In pines, these diterpenes constitute what is known commercially as rosin, which has numerous uses but especially as a source of intermediate chemicals in various industries (Chapter 7). The nonvolatile fraction increases the viscosity of the resin, which can enhance the possibility of engulfing herbivores and other organisms visiting the tree. Such trapped organisms can be beautifully preserved in fossilized resin. That is, certain terpenoids polymerize and, hence, are able to withstand degradation under certain depositional conditions, forming amber (Chapter 4). Extensive accumulations of fossilized resin are significant components of some coals and even petroleum deposits (Chapter 9).

Diterpenes in conifer resins are characterized by three main skeletal types (abietane, pimarane, and labdane) that vary quantitatively in different conifer families (Chapter 2). Abietane- and pimarane-type diterpenic acids, for example, abietic and pimaric acids (Figure 1-4), are most abundant in resins of Pinaceae, remaining relatively soft and unpolymerized. However, resins with abietane-type compounds may sometimes become relatively solid with a hard surface, probably as a result of an abietadiene precursor that is prone to polymerization. On the other hand, labdane-type acids, such as communic and agathic acids, may contain conjugated diene compounds that readily polymerize. Labdane-type compounds are the primary diterpene constituents in the cedar family (Cupressaceae). All three skeletal types occur in resins of the araucarian family (Araucariaceae) although large quantities of labdanes in *Agathis* result in the production of very hard copals as well as amber (Chapters 4 and 9). In the Podocarpaceae and Cupressaceae s.l. (Chapter 2), an oxidation rearrangement leads to the formation of phenolic diterpenes such as ferruginol and totarol (Thomas 1990).

Angiosperm Resins

Although monoterpenes predominate in the volatile fraction of the resin of the chemically best known conifers, such as Pinaceae, sesquiterpenes generally dominate the volatile composition in most, but not all, flowering plants. For example, the volatile fraction in numerous genera of tropical trees in the legume family (Fabaceae, or Leguminosae, Chapter 2) consists of sesquiterpenes that most often occur as hydrocarbons (Figure 1-3). Caryophyllene is an example of a sesquiterpene that commonly occurs in angiosperm resins. The volatile fraction of resins from the large tropical family Dipterocarpaceae also is composed of sesquiterpenes, similar to those in leguminous resins. In both families there are genera in which the volatile fraction predominates, thus producing a more fluid resin that has been used medicinally and for fuel oil (Chapter 7), whereas in other genera the nonvolatile fraction predominates, resulting in a more viscous resin used for varnishes (Chapter 9).

On the other hand, the volatile fraction of resins in the large tropical family Burseraceae is much more diverse than that of resins of legumes and dipterocarps. It contains large proportions of both mono- and sesquiterpenes, giving it the characteristic high degree of fragrance when used for incense (Chapter 8). Monoterpenes that commonly occur in conifer resins are important in burseraceous resins, along with numerous sesquiterpenes with diverse skeletal frameworks (Figures 1-3 and 1-5). Aregullin et al. (2002) found a sesquiterpene lactone (8-β-hydroxasterolide) in *Trattinnickia* resin. This is the first report of a sesquiterpene lactone, so common in the Asteraceae, in Burseraceae.

Diterpenes are the dominant components in the nonvolatile fraction of leguminous resins. They form the very hard copals used for varnishes (Chapter 9) because of the presence of labdadiene-type acids (or alcohols) such as ozoic acid (Figure 1-4) or zanzibaric acid, which are enantiomers of communic acid. These components also can lead to fossilization of the resin in the legume *Hymenaea,* as they do in the conifer *Agathis* (Chapter 4). Leguminous resins also contain numerous other diterpenoids that do not polymerize, such as the clerodane-type hardwickiic acid.

In some angiosperm families, triterpenes rather than diterpenes dominate the nonvolatile composition of the resin. For example, triterpenes primarily with tetra- or pentacyclic skeletons (Figure 1-5) characterize resins from the large tropical families Burseraceae, Dipterocarpaceae, and Anacardiaceae. Resins from Burseraceae typically have tetracyclic euphane / tirucallane, and pentacyclic lupane, ursane, and oleanane triterpene skeletal types (Khalid 1985). Other structural types have been found in species of the chemically complex myrrh-producing genus *Commiphora* (Waterman and Ampoto 1985), however, emphasizing the great structural diversity of triterpenoids in Burseraceae. They have been much used medicinally (Chapter 8). Although α- and β-amyrins (Figure 1-5) occur in other plants, they are known to be components of resins only in the Burseraceae, where they are common. Interestingly, in *Bursera,* diterpenes occasionally occur along with triterpenes (Becerra et al. 2001).

Although the nonvolatile fraction of dammar resins from the Dipterocarpaceae also consists largely of triterpenes, the skeletal types are different from those of Burseraceae; the nonvolatile fraction of dipterocarps consists primarily of the tetracyclic dammarane series (Figure 1-5). The volatile fraction is composed of sesquiterpenes; cadinenes in some taxa may polymerize to form bicadinenes, structurally considered as triterpenoids (Chapter 4). Resins from certain genera of Anacardiaceae have some triterpene components in common with those of Dipterocarpaceae, but they are generally more numerous and have not been completely characterized (Mills and White 1994).

The structures of more than 200 terpene compounds elucidated by Ghisalberti (1994) from the Australian resin-producing shrub family Myoporaceae demonstrate the complexity that can occur in one family of only three genera. *Myoporum*, a small genus, is characterized by furanoid sesquiter-

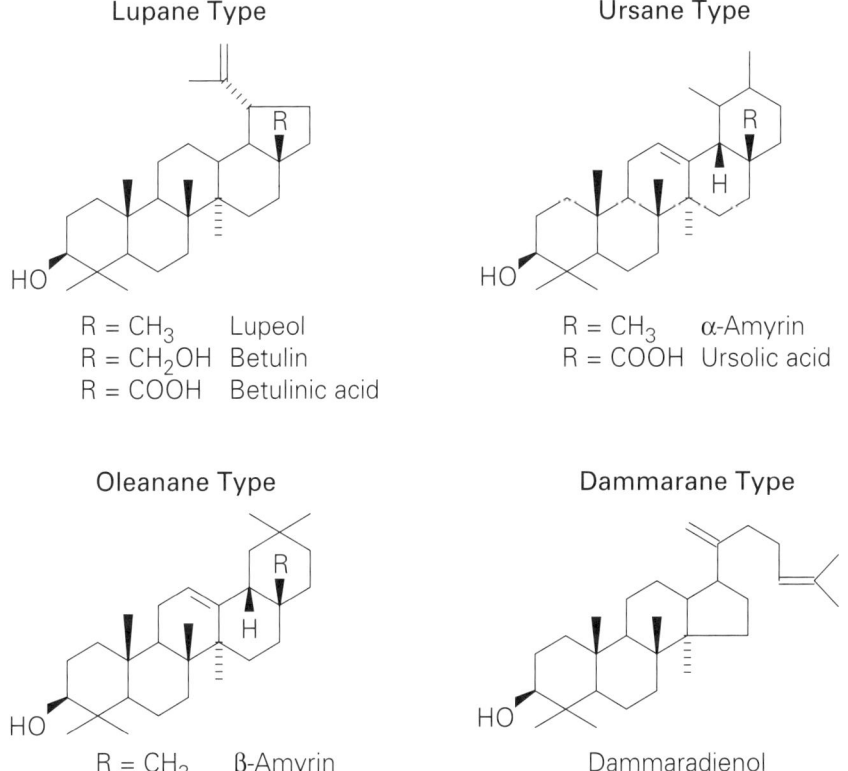

Figure 1-5. Examples of some structural types common in triterpenoid resin components in the large tropical families Burseraceae, Dipterocarpaceae, and Anacardiaceae (Chapters 8–10).

penes (Figure 1-6); (–)-ngaione is the best known because it is toxic to live-stock, but freelingyne was the first acetylenic terpene isolated from natural sources. In contrast, species of the large genus *Eremophila* (Plate 21) accumulate quantities of diterpenes that are all structurally and stereochemically unique. These diterpenes exhibit configurational differences from those of conifers and angiosperms (particularly legumes), with labdane, abietane, pimarane, and kaurane skeletons (Figure 1-4) that arise along the pathway to the physiologically necessary gibberellins (Figure 1-2). Thus Richmond and Ghisalberti (1994) suggested the possibility that diterpenes in *Eremophila* are synthesized by processes different from those observed in most terrestrial plants. Among the novel diterpenes, *Eremophila* generates numerous structural types (e.g., bisabolane, serrulatane, cedrane, and eremane) that bear resemblance to sesquiterpenes (Figure 1-6). *Eremophila* resin is an example of the amazing diversity of structural types that can occur even within one genus; other such cases may become evident as more resins are analyzed in detail.

Phenolic Resins

Phenolic compounds, those that include an aromatic ring plus at least one hydroxyl group (OH), are a dazzlingly diverse group of plant products. They are equally diverse in function, which includes structural support (lignin), pigmentation of flowers and other organs, protection from antioxidants and ultraviolet light, signaling between plants and animals or microbes, and plant

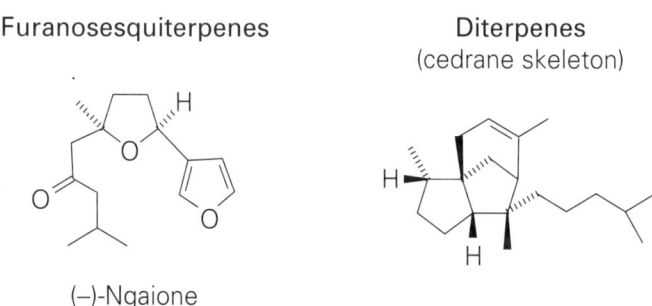

Furanosesquiterpenes

Diterpenes
(cedrane skeleton)

(–)-Ngaione

Figure 1-6. Novel sesquiterpenes and diterpenes characterize the resins of the Australian desert shrub *Eremophila* (Myoporaceae), including furanosesquiterpenes and diterpenes with skeletons that bear a resemblance to sesquiterpenes, sometimes called isoprenologues of sesquiterpenes.

defense. Such compounds frequently have several hydroxyl substituents, and some of the hydroxyl groups are often further substituted to form ethers, esters, or glycosyl (sugar) derivatives. Flavonoids, compounds with a C_{15} skeleton consisting of two benzene rings connected by a three-carbon bridge, are one of the most bioactive phenolic groups. Highly hydroxylated or sugar-substituted flavonoids are water soluble but many flavonoids are poorly water soluble or lipophilic, as are some simpler phenolics. Such lipophilic compounds are constituents of plant phenolic resins. Surprisingly, although water-soluble phenolic compounds are abundant in conifers, phenolic resin does not occur (Chapter 2). Internally produced phenolic resins occur only sporadically in flowering plants. More commonly, lipophilic compounds are intermixed with terpenoids in angiosperm resins, particularly those covering the surface of young organs.

Synthesis and Characteristic Components

Several metabolic pathways are involved in the synthesis of the mixture of components constituting phenolic resin. The shikimic acid pathway is the source of aromatic amino acids such as phenylalanine (Figures 1-1 and 1-7). A key step toward the formation of many components of phenolic resin is enzymatic conversion of phenylalanine to cinnamic acid, a reaction catalyzed by the important regulatory enzyme phenylalanine ammonia-lyase (PAL). Cinnamic acid is a simple C_9 phenolic compound called a phenylpropane because it contains a six-carbon (phenyl) benzene ring and a three-carbon (propyl) side chain (C_6–C_3). Phenylpropanoids derived from cinnamic acid are building blocks for many other phenolic compounds produced by the phenylpropanoid biosynthetic pathway. For example, benzoic acid derivatives, with a skeleton of a six-carbon benzene ring with a one-carbon substituent (C_6–C_1), are formed from phenylpropanoids by cleavage of a two-carbon fragment from the side chain. Cinnamic acid, benzoic acid, and its derivative, benzaldehyde (Figure 1-8), occur in the internally produced phenolic resins of such different plants as *Myroxylon,* leguminous trees that yield Peru balsam (Figure 8-2), and *Xanthorrhoea,* the Australian grass tree (Plate 12 and Chapter 9). Eugenol, found in numerous angiosperm resins, is synthesized in some cases from a phenylpropane derivative. Lignans (e.g., nordihydroguaiaretic acid, NDGA, Figure 1-8) are relatively common dimeric phenylpropanes that occur in surface leaf resins in desert shrubs such as *Larrea,* creosote bush (Plate 14).

Flavonoid components of phenolic resin are based on a structure of two benzene rings connected by a C_3 bridge (C_6–C_3–C_6), which is synthesized from components of two distinct pathways. One benzene ring and the C_3 bridge arise from the shikimic acid–phenylpropanoid pathway whereas the other benzene ring is formed from acetate units via the malonic acid pathway (Figures 1-1 and 1-7). Often, the C_3 chain cyclizes with an adjacent hydroxyl

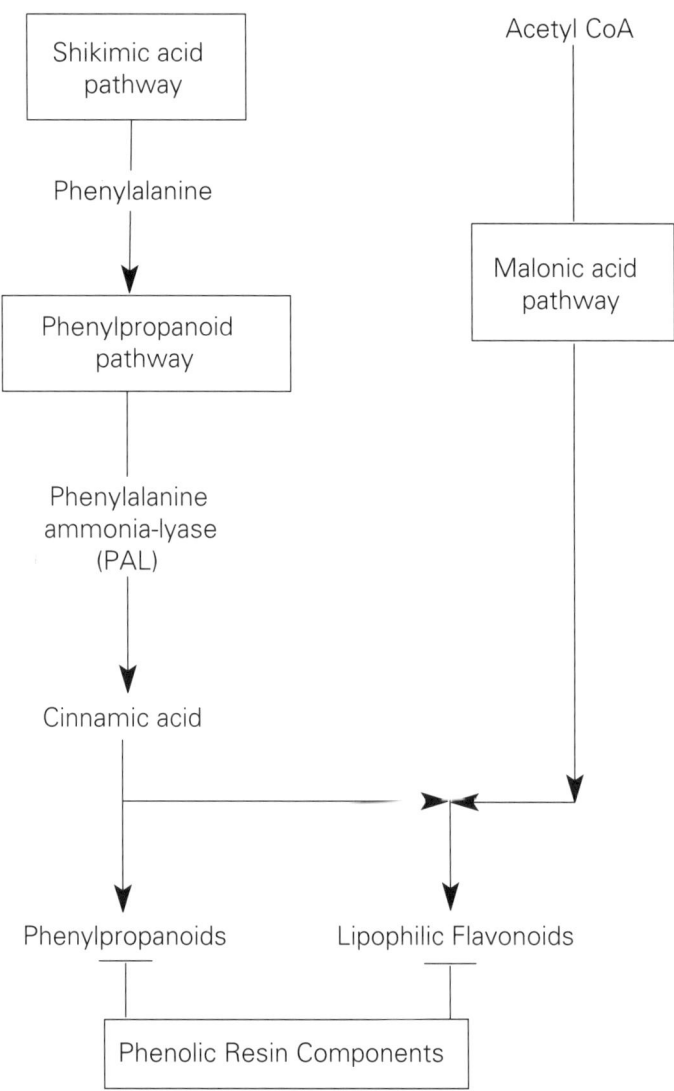

Figure 1-7. Generalized biosynthetic outline of phenylpropanoids and lipophilic flavonoids, two major groups of compounds characterizing phenolic resins.

group to form an oxygen-heterocyclic ring, and various classes of flavonoids are distinguished by the oxidation state of the C_3 chain or heterocyclic ring.

Many flavonoids occur as water-soluble glycosides. Those lacking a sugar substituent, called aglycones, often are lipophilic components of phenolic resin; these components frequently are members of flavonoid structural subclasses called flavones, flavanones, or flavonols. Flavonoid aglycones bearing fewer than five hydroxyl groups are hardly soluble in water. O-Methylation or methylenedioxy ring formation are also ways to mask reactivity of phenolic groups, at the same time increasing lipid solubility and volatility (Harborne 1980). Significant chemical features of the phenolics constituting resins are a reduced number of hydroxyl substituents (Figure 1-8) and a variable number of phenolic groups that are O-methyl substituted (methoxylated). Such lipophilic flavonoids commonly co-occur with terpenoids, particularly sesqui- and triterpenes (Wollenweber and Dietz 1981, Wollenweber and Jay 1988). Lipophilic flavonoids are less well studied than water-soluble ones and are best characterized in bud exudates from northern temperate zone angiosperm trees such as birches (*Betula*, Plate 33) and poplars *(Populus)* and in leaf resins of arid-zone shrubs such as monkey flower (*Mimulus*, Plate 22) and yerba santa (*Eriodictyon*, Figure 2-15). Diplacone and diplacol, lipophilic flavonoids from *Mimulus*, are shown in Figure 1-8.

Another way of introducing lipid solubility into phenolic compounds is to attach a hydrophobic side chain. The allergenic phenolic (catechol) compounds found in resins of Anacardiaceae are of this type (Figure 1-8; Chapters 9 and 10). Alternatively, one or more terpene (prenyl) residues may be attached to phenolic compounds to form prenylated phenolics (e.g., tetrahydrocannabinol, Figure 1-8). Most frequently, the terpenoid substituent is attached directly to the benzene ring, but it also can be attached to a phenolic group. Paseshnichenko (1995) summarized numerous prenylated phenolics, pointing out that they occur in nearly every phenolic structural class. He suggested that terpenoid and phenolic metabolism are linked through a control mechanism that regulates the distribution of precursors such as acetate, required for biosynthesis of both terpenoids and phenolics. Phenylpropanoid and flavonoid biosynthetic enzymes appear to form assemblies or complexes that cluster at the ER (Hrazdina and Wagner 1985, Winkel-Shirley 1999). Because the ER also appears to be the site of synthesis of most sesqui- and triterpenes, the opportunity exists for close localization of the synthesis of flavonoids and terpenoids and for their subsequent transport to a common storage site. These findings provide a plausible biochemical explanation for Wollenweber and Dietz's (1981) and Wollenweber and Jay's (1988) observation that lipophilic flavonoids frequently co-occur with sesqui- or triterpenes, as well as explaining the existence of so many prenylated phenolics. Much remains to be learned about the biosynthesis and co-occurrence of phenolics with terpenoids, and doubtless many mixtures of terpenoids and phenolics in resins will be reported.

Phenylpropanoids

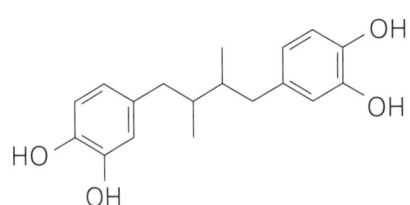

Cinnamic acid   Benzoic acid   Benzaldehyde

Dimeric Phenylpropanoid   Prenylated Phenol

Nordihydroguaiaretic acid (NDGA)   Tetrahydrocannabinol (Δ^1-THC)

Prenylated Flavonoid   Phenolic Allergens

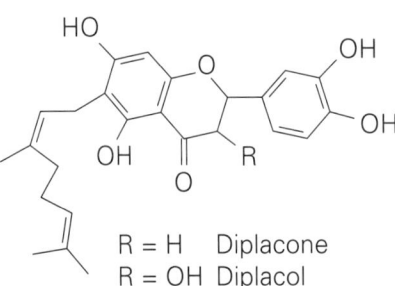

R = H   Diplacone
R = OH Diplacol

R = $C_{15}H_{23}$ Pentadecylcatechol
R = $C_{17}H_{23}$ Heptadecylcatechol

Figure 1-8. Structures of some of the kinds of phenolic compounds characterizing phenolic resins.

Substances Confused or Intermixed with Resin

Gums

Gums are often confused with resins, and terpenoid resins are frequently called gums commercially. In fact, only relatively recently have importers and exporters distinguished gum from resin in the naval stores industry (Chapter 7). Gum has not only been confused with resin but with latex-containing polyterpenes, for example, rubber (gum boots) and chicle (chewing gum). Clouding the issue further, a general definition of gum as "any plant substance that is both sticky and elastic as well as any glue used to bond surfaces" can be found in some dictionaries. Such definitions are not based on properties that clearly distinguish the various sticky exudates such as gum, resin, and latex (Table 1-1). Chemically, true gums are complex chains of hydrophilic polysaccharides (complex sugars) derived from monosaccharide (simple sugar) moieties such as galactose, arabinose, and rhamnose, and hence, are neither terpenoid nor phenolic in origin. Whistler (1993) described in detail the very complex structure of exudate gums.

Table 1-1

Characteristics of resins, gums, mucilages, oils, waxes, and latex. Secretory tissues are discussed in detail in Chapter 3

	PRIMARY COMPONENTS	SOLUBILITY	SECRETORY TISSUE
Resins	Terpenoids; phenolic compounds	Lipid soluble	Canals, pockets, cavities, trichomes, epidermal cells
Gums	Polysaccharides	Water soluble	Cavities
Mucilages	Polysaccharides	Water soluble	Idioblasts, epidermal cells, trichomes, ducts, cavities
Oils (fats)	Fatty acids and glycerol	Lipid soluble	None
Waxes	Fatty acids esterified with long-chain alcohols	Lipid soluble	Unspecialized epidermal cells
Latex	Complex mixture, may include terpenoids, phenolic compounds, proteins, carbohydrates, etc.	Lipid soluble	Laticifers

Gum is also formed differently from resin. Gummosis, as formation of gum is called, results primarily from metamorphosis of celluloses and hemicelluloses in the plant cell wall into unorganized amorphous substances. Starch also may be a source of gum formation. Gums usually are thought to be caused traumatically, induced by microbial infection as well as by insect attack and mechanical injury (Fahn 1979). Therefore, Whistler (1993) considered gum an induced natural defense mechanism. However, some long-time gum researchers do not agree that trauma is essential to the formation of gum (D. M. W. Anderson, pers. comm.).

Fahn (1979) showed that the process of gummosis begins after the formation of special groups of undifferentiated parenchyma cells, instead of normal wood elements, in some plant families. The rose family (Rosaceae), including fruit trees such as cherry, peach, and plum, is a good example. Disintegration of cell walls (Figure 1-9) leads to formation of a cavity, which fills with the polysaccharides that constitute gum. Vessels that carry water in the wood also may be filled with gum, formed only from the lamella of the secondary cell wall; gum is also formed in bark tissue, however, as in *Acacia* species that produce gum arabic. When this exuded gum material hardens, it can produce masses that resemble resin. Where the gum touches the bark, it

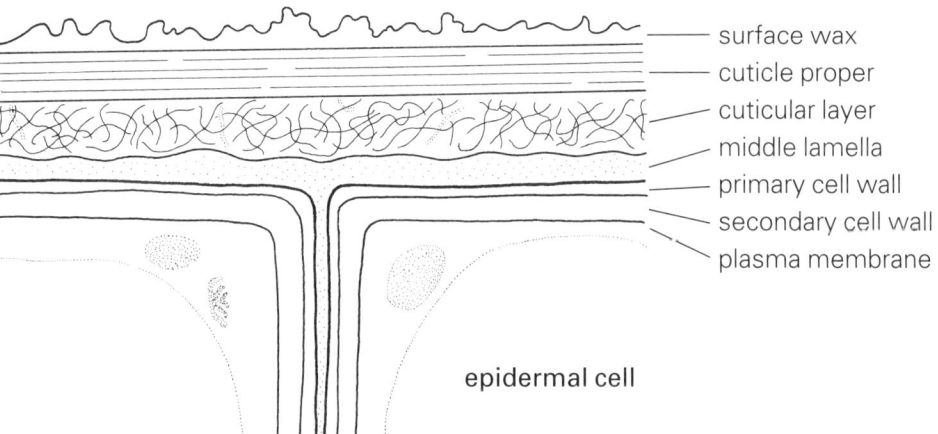

surface wax
cuticle proper
cuticular layer
middle lamella
primary cell wall
secondary cell wall
plasma membrane

epidermal cell

Figure 1-9. Schematic drawing of wax on a leaf surface, showing the epidermal cells, their plasma membranes, primary and secondary walls (with middle lamella above), covered by a cuticular layer (formed by cutin, wax, and cell wall carbohydrates), the cuticle proper, consisting of cutin embedded in wax, and a top coating of surface wax. Resin exuded from specialized epidermal cells and glands can burst through the cuticle onto the leaf surface and become intermixed with wax (Chapter 3).

may absorb tannins and become yellow. Confusion of gum with resin seems to occur particularly in exudations from leguminous trees such as *Acacia* and mesquite *(Prosopis)*; these gums are yellow, similar to the color of some resins, even amber.

Kino, which forms in the wood of plants such as *Eucalyptus,* is a type of gum that contains polyphenols. Kino veins develop in the zone of traumatic parenchyma cells produced by the cambium following injury. Groups of cells accumulate large quantities of red phenolic compounds; then these cells break down to form ducts into which the phenolics are released (Fahn 1979, Hillis 1987). The term kino is adapted from a West African word for gum from a leguminous tree, African rosewood *(Pterocarpus),* which was used medicinally in Europe (Burkill 1966).

Some plants produce both gum and resin. Cell walls surrounding the lumen of resin secretory structures in plants such as *Commiphora* and mango *(Mangifera)* slough off during duct development, thus incorporating carbohydrate material into their resin (Bhatt 1987). Halpine (1995) reported traces of amino acids in resins of artists' materials. She indicated that the hydroxyproline found in some resins may result from breakdown of cell walls because the main cell wall protein, extensin, contains as much as 25% 4-hydroxyproline. Sometimes when resin is induced by injury, gummosis occurs as cavities are created during breakdown of the cells in which terpenoid resin is synthesized. Again, in this manner, cell wall carbohydrates and, possibly, amino acids can become incorporated into the resin (Chapter 3). These resins are frequently referred to as gum resins because they may be slightly hydrophilic before they harden or swell somewhat when first put into solvent. A number of resins from both conifers and angiosperms belong to this category, such as those from spruce *(Picea)* and *Araucaria* among conifers and from the tropical leguminous *Hymenaea* and burseraceous *Commiphora* among angiosperms (Chapter 2).

Mucilages

The term mucilage is sometimes used colloquially to describe an aqueous solution of gum because of certain similarities between the two substances. However, mucilage and gum are clearly distinguishable by several criteria (Table 1-1). Plant mucilages are water-soluble complex acidic or neutral polysaccharide polymers of high molecular weight; as with gums, some components are closely related to cell wall compounds (e.g., galactose, arabinose, xylose, rhamnose, and galacturonic acid). In fact, their occurrence involves only a few enzymatic changes from these compounds. Differing from gum , however, mucilage occurs in various structures such as single secretory cells, canals and cavities, epidermal cells, and trichomes. This variability is reflected in the diversity of function and wide distribution of mucilages. For example, they serve multiple roles in higher plants, such as reserve food sources, water retention in succulents, lubrication for growing tips, regulation of germination in seeds, and seed dispersal.

Mucilage can become mixed with resin in surface coatings. In some plants, such as alders (*Alnus,* Betulaceae), the buds and developing leaves or their stipules, which envelop the shoot apex and leaf primordia, may bear glandular trichomes that secrete sticky substances containing either mucilage or resin, or both (Chapter 3). In *Populus* (Salicaceae), palisade-like epidermal cells under the cuticle may secrete both resin and mucilage. Mucilage also occurs commonly with phenolic leaf resin in leaves of the sticky monkey flower (*Mimulus,* Scrophulariaceae). Apparently, the relative proportion of mucilage to flavonoids varies between *Mimulus* species, which grow as herbs in moist habitats and as shrubs in dry areas.

Oils and Fats

Many components of resin are referred to as oils. For example, mono- and sesquiterpenes are called essential oils, cupressaceous resins with large quantities of sesquiterpenes are called cedarwood oil, and some leguminous resins are known as copaiba oil (Chapter 7). Oils and fats (and waxes), however, are distinguished chemically from terpenes in being alcoholic esters of fatty acids. Oils and fats are formed by synthesis of fatty acids from carbohydrates, followed by the combination of these fatty acids through enzymatic action with glycerol to form esters (triglycerides). Fatty acids are long hydrocarbon chains that carry a terminal carboxyl group, giving them the characteristics of a weak acid. Glycerol forms a link with the carboxyl groups, serving as a carrier for fatty acids.

Plant fats differ only slightly from oils in having fatty acid constituents that are more or less solid rather than liquid at ordinary temperatures. Fatty acids do occur in some conifer resins. For example, fatty acids (predominantly C_{18} oleic and linoleic) are co-isolated with rosin as by-products from pine resins in the kraft pulping process (Chapter 7). They also occur in abundance in the resin of glandular trichomes on the leaves of some Australian shrubs (Chapter 3). The utility of oils and fats to the plant is a bit obscure, although those in the seed constitute an abundant reservoir of stored energy for early growth of the seedling.

Waxes

Waxes are not macromolecules but, rather, complex lipid-soluble mixtures whose common components are alcoholic esters of fatty acids and straight-chain alkanes, for example, $CH_3(CH_2)_{24}CH_2OH$ and $CH_3(CH_2)_{27}CH_3$. Long-chain ketones and aldehydes, and free fatty acids, also occur in waxes. Wax occurs in almost all vascular plants as an important constituent of the cuticle, which acts as a protective coating on the epidermis of leaves, stems, and fruits, reducing desiccation or abrasion, or resisting pest attack (Figure 1-9). Cuticular waxes are synthesized by epidermal cells; they leave these cells as droplets, passing through pores in the cell wall into the cuticle. Wax, as the top coating of the cuticle, often crystallizes into an intricate pattern of platelets or rods, giving the surface a whitish bloom.

Other compounds may occur in waxes; hence, wax may differ chemically in different organs and with age, season, and local conditions. Components of resin such as various triterpenoids, for example, become intermingled in cuticular wax (Wollenweber et al. 1998). In fact, they can become a major constituent in some leaf waxes; for example, oleanolic acid forms 70% of grape *(Vitis)* leaf wax. Even in glaucous species of the succulent genus *Dudleya,* β-amyrin in one species, taxarone in another, constitute about 40% of the cuticular wax (Manheim and Mulroy 1978). In addition, resin produced in epidermal glands may burst through secretory structures to mix with wax on the leaf surface (Chapter 3).

Resin Compounds in Latex

Latex typically is a thick, creamy white, milky (as the Latin root *lac* indicates) emulsion. Sometimes, it may be a thin, clear, yellow or orange, aqueous suspension. In either case, latex comprises a complex of substances such as terpenoids—mono-, sesqui-, di-, tri-, and polyterpenes (rubber)—proteins, acids, carbohydrates, tannins, alkaloids, and minerals (Table 1-1; Fahn 1979). Often, resin is listed as a component of latex in members of some families (e.g., Euphorbiaceae and Convolvulaceae) although exactly what is meant is not clear in many cases. Latex is produced in vessels or special cells called laticifers, which are quite different from the internal secretory tissues (e.g., pockets, cavities, or canals) in which most resin is produced (Chapter 3). These latex-bearing structures vary in origin, anatomy, and distribution. The role that latex plays in the plant is not as well known as that of resin, but T. Lewinsohn (1991) presented evidence that it can be protective. Part of this protection may result from the terpenoid and phenolic components in the latex.

Miscellaneous Intermixed Compounds

Compounds from the epithelial cells of the secretory structure may get incorporated into the resin either in the breakdown of cell walls during lysigeny (Chapter 3) or during the wounding process. Cell wall polysaccharides and the amino acid hydroxyproline have been discussed for gum resins. Halpine (1995) found trace amounts of other amino acids (e.g., aspartic and glutaminic acids, serine, glycine, alanine, and valine) in resins used by artists. Wang et al. (1995) found a similar content of amino acids, averaging about 5 parts per million, in the matrix of resins, which did not systematically vary with ages from less than 100 million to 130 million years. The authors concluded that these amino acids had probably been derived either from plant cells or microorganisms that were trapped when the resin solidified on the outside of the plant.

Tannins and calcium carbonate from the bark have also been reported in resin, as well as particles of soil in resin that has fallen to the ground. Waxes and other cuticular compounds (Figure 1-9) get mixed with resins that are secreted directly to the epidermis of the stem or leaf (Chapter 3). The small amounts of miscellaneous compounds that get enclosed in resin need further analysis; they are of particular interest in analyzing the use of resins by animals (Chapter 5), including humans (Chapters 7–11).

CHAPTER 2

Resin-Producing Plants

Grouping resin-producing plants into genera, families, and orders is helpful. Although species are sometimes discussed in this chapter, they receive more attention in Chapters 6–10, dealing with particular aspects of resin production. Conifers have been put into one order, Coniferales; hence, the family becomes the focus for discussion of this group. The family has been a central unit in the systematics of flowering plants, or angiosperms; orders, however, are increasingly useful in discussing the relationships of families as the phylogeny of angiosperms is revealed in detail. The large sizes of some angiosperm families as well as a couple of conifer families have led to their division into smaller, more usable categories of subfamilies, and sometimes of those into tribes.

A phylogenetic framework facilitates insights into the evolutionary relationships of resin-producing plants. Phylogenies are used as road maps to navigate through the diversity of taxa and are simply models or hypotheses, best estimates, about the history of plants. Some estimates, of course, are better than others, especially in light of new information. Despite more than a century of research into conifers, there is no widely accepted phylogenetic framework, often referred to as a tree, for their families. Therefore, following a discussion of controversial views regarding their phylogeny, conifers are presented in accordance with their importance in resin production. Families, subfamilies, and genera of conifers are listed alphabetically in Appendix 1. On the other hand, angiosperm resin-producing plants are discussed in the phylogenetic framework presented by the Angiosperm Phylogeny Group (APG) in 1998 and listed as such in Appendix 2.

In Parts I and II, I compare production of conifer resin in a general way to that of angiosperms. Various aspects of resin production such as biochemistry, anatomy of secretory tissue, and ecology have received more intensive analysis in conifers than angiosperms, partly because of their prominence in temperate zone forests and their economic value. Further study of the diverse resin-producing angiosperms, especially in the tropics and arid areas, may open new horizons. This chapter thus provides a catalog of resin-producing conifers and angiosperms, including discussion of evolutionary topics for future consideration.

Resin-Producing Conifers

Many people, including some scientists, in Western countries tend to think first of conifers as resin producers among the plants living today because of their prominence in temperate zone forests, with which they are most familiar. At least in the northern hemisphere, the odor of the volatile terpenes in conifers (figure on Part I page) evoke the idea of resin for many people, for example, the smell of a conifer forest or a Christmas tree. Conifers and their progenitors are also the plants that left the earliest fossil record of resin (Chapter 4) and today occur in 80% of the world's kinds of habitats. Although five of the seven conifer families synthesize resinous terpenoids, only members of Pinaceae and Araucariaceae produce copious amounts. As a result, these taxa have been most studied, either for utilization of the resin or because of the resin's protection of valuable timber (Chapters 5, 7–10). Resin from some species of Cupressaceae s.l. and a few of Podocarpaceae, however, have been used by humans or have been analyzed in chemosystematic or chemical ecological research and are discussed where appropriate.

Classification

Conifers are the only gymnosperms that produce resin as defined here, although cycads produce mucilage and *Ginkgo biloba* produces terpenes that sometimes are referred to as resin. Conifers display such diverse characteristics that it is difficult to define them by a single feature, such as the flower for angiosperms (Miller 1999). The female cone (hence, the name conifer) is considered the primary character in defining the families, but not all taxa have a woody or leathery multiseeded cone, often considered by nonbotanists as characteristic of conifers. Some species are multiseeded but their cones have more or less juicy scales and appear berry-like (e.g., junipers, *Juniperus*), or others have cones that are greatly reduced, with highly modified, juicy, brightly colored scales and just one seed (e.g., yews, *Taxus*). Furthermore, evolutionary relationships of conifers have remained controversial despite much work on their morphology, anatomy, embryology, cytology, phytochemistry, and paleobotany. This controversy has ranged from questions about the origin of conifers to their ordinal and familial relationships.

Since the seminal research of Florin (1951) on Paleozoic fossils, conifers have been associated with the fossil genus *Cordaites* (Chapter 4). Some researchers, however, have considered *Cordaites* and conifers to be in separate evolutionary lines. According to Doyle (1998), modern conifers are linked with *Cordaites* and ginkgos in analyses that include all extant groups of conifers. The earliest fossils attributable directly to conifers occur in the late Carboniferous (about 310 Ma), and conifers were diverse and floristically important by the end of the Paleozoic era (290 Ma). In a study reexamining the phylogenetic relationships of fossil and living conifers, Miller (1999) pointed out that since the 1950s, modern conifer families were thought to have evolved from late Paleozoic ancestors via early Mesozoic transition conifers. In the 1990s, however, expanding knowledge of Paleozoic conifers led to the view that modern conifers diverged directly from Paleozoic families, particularly the Majonicaceae. The exception is the yew family (Taxaceae), derived from a different ancestral family than other modern conifers, a finding that supports the often held idea of placing Taxaceae in its own order, Taxales.

In contrast to paleobotanical support for two orders, Taxales and Coniferales, morphological and molecular cladistic analyses support the monophyly of conifers, that is, that conifer taxa are all descended from one ancestral population (e.g., Chaw et al. 1997, Stefanoviac et al. 1998). In this interpretation, Taxaceae are included in the Coniferales. Stefanoviac et al. (1998) thought that Taxaceae are closely related to the Cephalotaxaceae, which have sometimes been considered to be members of the Taxodiaceae, the redwoods and their relatives. The analysis by Stefanoviac and coworkers, as well as other morphological and molecular studies, support the merger of Taxodiaceae into Cupressaceae although there are some strong opponents to this merger (e.g., Page 1990). In *Plant Resins*, Taxodiaceae are included in Cupressaceae and the combined group is referred to as Cupressaceae sensu lato (s.l.). In some cases where the Taxodiaceae are not included in the group, the Cupressaceae are denoted as sensu stricto (s.s.). Stefanoviac and coworkers have put the so-called umbrella pine *(Sciadopitys)* into its own family, Sciadopityaceae; they connected Araucariaceae to Podocarpaceae, thereby linking the two families with current southern hemispheric distribution and suggesting that they had a common ancestor. They also thought that the Pinaceae are the sister group to all other conifers, a view supported by others, including Chaw et al. (1997). Based on fossil cones, Miller (1999) changed his 1988 view that Araucariaceae and Pinaceae were sister groups in a basal clade to one in which Pinaceae branch from a basal position to all other conifer families except Taxaceae.

Only the Pinaceae, Araucariaceae, Cupressaceae s.l., and Podocarpaceae unequivocally produce resin in secretory structures, although resin synthesis can be induced by injury without a secretory structure in some cases (Chapter 3). The Cephalotaxaceae have resin ducts but there is little evidence of their producing resin today, and *Sciadopitys* has not been recorded as producing resin. Taxaceae produce terpene synthases similar to those of the Pinaceae (Chapter 1); there are no secretory structures, although Bierhorst (1971) considered that they have been lost during evolution. Thus copious resin producers and little- or non-resin-producing families appear to be dispersed in the order Coniferales.

Pinaceae

Although essentially restricted to the northern hemisphere, the Pinaceae form the largest conifer family with approximately 200 species (about 35% of all conifers); 80–100 of the species, depending on one's taxonomic perspective, are pines *(Pinus;* Richardson 1998). There are nine other genera; the next largest are the true firs *(Abies)* and spruces *(Picea),* with hemlocks *(Tsuga)* and larches *(Larix)* following. Douglas firs *(Pseudotsuga),* true cedars *(Cedrus),* golden larches *(Pseudolarix), Cathaya,* and *Keteleeria* each has only a few or one species (Appendix 1). The genera have been put into either three or four subfamilies over the years (e.g., Melchior and Werdermann 1954–64, Frankis 1989). Four subfamilies are most commonly recognized today. Phylogenetic studies, using molecular analyses based on chloroplast DNA and supported by fossil evidence, generally indicate that the Abietinoideae—*Abies, Cedrus, Keteleeria, Pseudolarix,* and *Tsuga* (including *Nothotsuga)*—are the basal group that appeared during the Jurassic. The Pinoideae *(Pinus)* and Piceoideae *(Picea)* follow during the early Cretaceous, and the Laricoideae *(Larix, Pseudotsuga,* and *Cathaya)* in the late Cretaceous (Labandeira et al. 2001).

Leaves of all Pinaceae contain terpenoid resin but in some genera resin may be produced only traumatically in the trunk of the tree (e.g., *Cedrus, Pseudotsuga, Pseudolarix,* and *Tsuga*). Wu and Hu (1997) thought the distribution of resin ducts in stems and leaves such a "striking morphological character" that it should be used in evaluating relationships of genera in Pinaceae. They thus divided the genera into three groups based on the presence or absence of resin ducts in the wood (Chapter 3). The importance of differences in the kind of secretory structure and quantity of resin produced is elucidated in the contexts of the defensive properties of resin in Chapter 5, and human use in Chapters 7–10.

Paleobotanists such as Miller (1976, 1977) thought that the early evolution of Pinaceae was centered on *Pinus*. Miller (pers. comm.) has concluded more recently, however, that there are three evolving lines within the Pinaceae and not a central pine group. *Pinus* today is not only the largest genus in the family but is particularly noted for its great ecological diversity and adaptability. Pines occupy a wide range of habitats from the arid plateaus of western North America, the mountains of Mexico and Central America, to the tropical lowlands of the Caribbean. They also cover large areas across Eurasia into the mountains and tropical lowlands of eastern Asia (Mirov 1967, Richardson 1998). Furthermore, pines are the most widely planted exotic trees because of their adaptability and their large-scale use for timber, paper, and resin. Thus they are found out of their native range in South America, Africa, and Australasia.

Despite fragmentation of large pine forests during the past few hundred years through land clearance and pest introduction, pines continue to be conspicuous components of forests in many parts of the northern hemisphere. For example, *Pinus sylvestris* (Plate 1), known as Scots or Scotch pine in England and common pine on the European continent, is a complex of many intergrading varieties or subspecies with a range encompassing more than 10,800 km^2 (Nikolov and Helmisaari 1992), the widest distribution of any pine. It spans a wide range of environments from the Scottish highlands along the Atlantic to the Pacific coast of eastern Siberia to northwestern Asia, occurring with most of the boreal tree species of Europe and Asia; it also has scattered populations throughout the Mediterranean region (Mirov 1967). The resin has been used by Europeans throughout history (Chapter 7).

Ponderosa pine *(Pinus ponderosa)* is the widest ranging pine in North America. A drought-tolerant species, it persists in dry environments. Two varieties are recognized, partly related to the difference in the drought seasons in California and the Pacific Northwest versus those in the southwestern United States. In the driest environments, ponderosa pine forms savannas, but in more mesic environments it occurs in mixed stands with various other conifers. For example, it commonly occurs with Douglas fir *(Pseudotsuga menziesii)* in the Rocky Mountains and with *Abies grandis, Calocedrus decurrens,* and *Pinus lambertiana* in the Sierra Nevada. Ponderosa pine and Douglas fir are the two most important western North American timber trees, and their resins have been intensively studied for the protection they provide to valuable timber (Chapter 5).

Lodgepole pine *(Pinus contorta)* is another pine widely distributed across western North America, with four varieties. Some varieties have cones that are more serotinous (closed, often aided by the gluing effect of resin, until opening because of high temperatures such as those caused by fire) than others. This characteristic is partly responsible for the various ecological roles played by lodgepole pines in diverse ecosystems. In the Rocky Mountains, *P. contorta* var. *latifolia* is primarily a pioneer or early forest successional tree in the *Picea engelmannii–Abies lasiocarpa* forest zone (Plate 3), where infrequent but high-intensity fires occur. Those trees with serotinous cones have little or no seed dispersal in the absence of fire but typically generate dense, even-aged stands following fire. There are other areas, however, where lodgepole pine is the only tree species that can reproduce with or without fire, and it can become dominant in both early and late forest succession. Most of the late-succession or mature forest stands have lodgepole pines with nonserotinous cones, allowing regeneration after disturbances other than fire. The successional status of resin-producing conifers such as *Pinus contorta* var. *latifolia* has been shown to be related to the characteristics of the resin defense and long-term health of the forest (Chapter 5).

Pines of the coastal plain of southeastern North America (e.g., *Pinus elliottii,* Figure 7-2; *P. palustris;* and *P. taeda*), occurring in widespread savannas, and those around the rim of the Mediterranean Basin (e.g., *P. halepensis*) have been some of the most heavily used for resin. Today, however, pines native to the Asian tropics (e.g., *P. kesiya, P. massoniana,* and *P. merkusii*) are among the most productive for the naval stores industry (Chapter 7). Numerous pines also cover midelevations of mountains in Mexico (e.g., *P. montezumae,* a variable complex of subspecies; Figure 7-6) into Central America (e.g., *P. oocarpa*) and much of the Caribbean (e.g., the *P. caribaea* complex). These pines have a long history of human use.

In many cases, species of different genera of Pinaceae either codominate or are prominent components over vast expanses of boreal and high-elevation forests in mountains. It is assumed that resins probably provide defensive properties for these important forest trees, some of which have been heavily used by humans (Chapter 5). Examples include the boreal spruce-fir forests across North America (with *Picea glauca* and *Abies balsamea*, Plate 2 and Figure 2-1), the subalpine spruce-fir zone in the Rocky Mountains (dominated by *P. engelmannii* and *A. lasiocarpa*; Plate 3), and the ponderosa pine–Douglas fir zone in the Rocky Mountains. In the Sierra Nevada, Douglas fir occurs in a zone with *Pinus jeffreyi* (Plate 4), a close relative of *P. ponderosa*. Large areas in the arid southwestern United States and northern Mexico have a low open forest or woodland characterized by mixed species of pinyon pine (e.g., *P. edulis* and *P. monophylla*) and juniper (e.g., *Juniperus monosperma* and *J. scopulorum*, Plate 5). The resins from these trees have long been used by Native Americans (Chapter 10), and pinyon pines were once tapped for their resin on a small scale. Desert junipers also could be used as a basis for small cedarwood oil industries (Chapter 7).

The true cedars *(Cedrus)* consist of only three or four species that occur naturally in the western Mediterranean in North Africa to the western Hima-

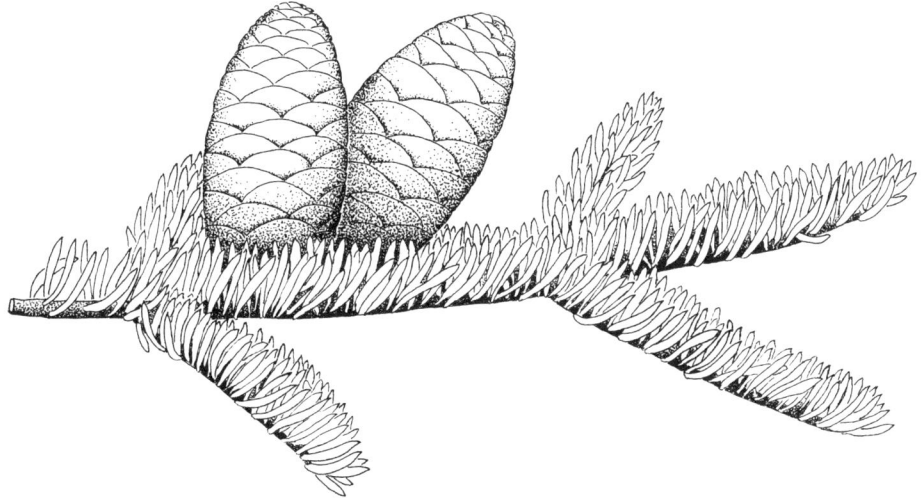

Figure 2-1. *Abies balsamea,* balsam fir, an important balsam-producing tree of North American boreal forests (see also Plate 2).

laya. Differences between the species are slight, and Melchior and Werdermann (1954–64) considered their nearest relatives to be the deciduous larches *(Larix)*. The name *Cedrus* is from *kedros*, used by the ancient Greeks to designate a resinous tree. General use of this name by the Greeks may have led to trees other than *Cedrus* being called cedars, as discussed under Cupressaceae. The leaves of true cedars are resinous, and resin exudes from young cones (Plate 6). Resin that accumulates in the heartwood gives the wood its fragrant odor and is distilled as cedarwood oil (Chapter 7). Cedars of Lebanon *(C. libani)* are highly resistant to pests and also were the most massive trees known to the ancient Israelites, who used them to build the temple and palace of Solomon. Other ancient potentates exploited *C. libani* so intensively that only small remnant populations remain.

Larix, a cool northern hemisphere tree, is the most abundant deciduous conifer and occurs in some of the most extreme environments, above the northern and elevational limits of most conifer growth (Gower and Richards 1990). The American larch or tamarack *(L. laricina)*, a tree of the cooler northern hemisphere, produces an economically valuable resin, as does the European larch *(L. decidua)*, that is similar to Canada balsam from *Abies balsamea* (Chapter 8). *Pseudolarix*, with only one species, *P. amabilis* (Plate 7), is indigenous to a very limited area in the coastal mountains of eastern China. Considered to be closely related to *Larix* by Melchior and Werdermann (1954–64), *Pseudolarix* is also deciduous. Before dropping, the leaves change to a rich golden yellow, hence its common name, golden larch. On the other hand, Frankis (1989) considered *Pseudolarix* more closely related to *Abies* and *Cedrus* than to *Larix*. Despite its restricted distribution today, *Pseudolarix* fossils indicate the genus was widely distributed in mid- to high latitudes of Asia, North America, and Europe during the Cretaceous and Tertiary (Le Page and Basinger 1995). *Pseudolarix* resin forms amber and may be related to the producer of Baltic amber because it is one of the rare conifers with resin that contains succinic acid (Chapter 4).

Members of the North American and Asiatic genus *Tsuga* are called hemlock in North America and hemlock-spruce in the British Isles. Despite the name, they are not relatives of the poison hemlock (*Conium maculatum*, Apiaceae, or Umbelliferae) used by the ancient Greeks to poison Socrates among others. Species of *Tsuga* are either small or large trees in northern coastal or subalpine forests. Hemlocks are considered by Melchior and Werdermann

(1954–64) to be close kin of firs *(Abies)* and spruces *(Picea)* and often are confused with them. Hemlock leaves are resinous, but like *Abies, Cedrus,* and *Pseudolarix, Tsuga* only produces resin in the wood following injury (Chapter 3).

Since species of Pinaceae tend to occur in high density, dominating forests that cover vast areas, there are probably more individuals of these northern temperate zone dwellers than other resin-producing trees; however, resin-producing angiosperm trees of the tropics have a far greater diversity of taxa.

Araucariaceae

Araucariaceae have been reported from the Triassic about 200 Ma (Chapter 4). In fact, possible araucarian wood *(Araucarioxylon)* is thought to be related to that of *Dadoxylon,* a cordaitalean tree abundant during the Carboniferous period, 300 Ma. Today, Araucariaceae are a small family with only three genera and about 40 species (Appendix 1). *Araucaria* (19 species) is the most diversified genus and is disjunctly distributed throughout the warm southern temperate zone. However, araucarian fossils have been discovered widely in both the northern and southern hemispheres. Beautifully preserved fossil cones of *Araucaria* have been identified from as long ago as the Triassic and Jurassic periods (Miller 1977, 1988; Stockey 1982). Although some *Araucaria* species today produce abundant resin, often associated with insect damage, *Araucaria* has not been studied from an ecological perspective. The resin has a complex composition, often admixed with gum. There is relatively little variability in the gum fraction of resin from different species of *Araucaria* (D. Anderson and Munro 1969); the terpenoid fraction, however, has received little study.

The taxonomy of *Agathis* remains confused, with about 20 species recognized (Whitmore 1980a, de Laubenfels 1989). The genus may have evolved from *Araucaria* in the late Jurassic (Miller 1977). It contains by far the most important resin-producing species of Araucariaceae, some of which produce copious quantities of resin (Thomas 1969, Whitmore 1980a). *Agathis* is the genus of conifers most nearly confined to the tropics; in fact, *A. robusta* in southern Queensland, Australia (Plate 8), and *A. australis* in northern New Zealand (Figures 2-2 and 2-3) are the only species of the genus that extend beyond the tropics into the subtropics. Within the tropics, *Agathis* occurs in a wide range of habitats, for example, those with dry seasons varying in length from one to a few months, lowlands on a diversity of soil types, and at elevations up to 2000–2500 m. Both the timber and resin are of sufficient commercial value that rapidly growing species have been considered for enrichment planting, particularly in Irian Jaya (or West New Guinea), Indonesia (Chapter 11). The very hard (highly polymerized) copal-type resin has an extensive fossil record; researchers have chemically identified *Agathis*-type resin from the Triassic, 200 Ma (Chapter 4).

Figure 2-2. *Agathis australis,* kauri. The broad evergreen leaves, globose female cones, and elongated male cones are characteristic of the genus.

Figure 2-3. *Agathis australis,* towering monarch and copal-producing tree of northern New Zealand. With permission of the Alexander Turnbull Library, National Library of New Zealand, Te Puna Mātauranga o Aotearoa (timber industry, logging, reference 6266 1/1).

Agathis australis is the most famous species in the genus, and its Maori vernacular, kauri, has become a common name for the genus in many European languages. It is also commonly referred to as kauri pine, an obvious misnomer since it is not a pine. About 1850, kauri occurred over millions of hectares, but now only about 7000 hectares of virgin forest remain on the northern tip of New Zealand's North Island (Whitmore 1977). *Agathis australis* is one of the world's forest giants, and trees can form a canopy 40–50 m or more tall. The trees also have vast girth, more than 10 m being common, with clear boles of 10–12 m before the crowns spread (Figure 2-3). Trees self-prune their lower branches as they grow, leaving massive limbs sometimes a meter in diameter in the crowns of older trees. Although the big trees are thought to be 500–800 years old, with a few approaching 1000 years, their precise age is difficult to determine because the trees are often hollow. These giant trees contain a large volume of millable timber and produce enormous quantities of resin. The copal-type resin, used for fine, very durable varnishes, was sufficiently important commercially that it played a role in the establishment of the economy of northern New Zealand (Chapters 6 and 9).

Wollemia nobilis has been discovered relatively recently in a sheltered gorge of temperate rain forest in New South Wales, Australia (Jones et al. 1995). It occurs as an emergent tree in this one population. Morphological characteristics of *Wollemia* are intermediate between those of *Araucaria* and *Agathis;* similar characteristics are known only in fossil taxa such as *Podozamites,* which occurred from the Jurassic to Tertiary periods. Therefore, Setoguchi et al. (1998) called *Wollemia* a living fossil. The chemistry of the resin of *Wollemia* has been studied from several perspectives such as carbon isotope biogeochemistry (Murray et al. 1998) and NMR (Lambert et al. 1999). NMR spectra of *Wollemia* and *Agathis* are similar.

Podocarpaceae

The Podocarpaceae are the second largest conifer family with 18 genera and 178 species (Appendix 1). Despite being the most diverse family of conifers morphologically and ecologically, Kelch (1998) has shown the family to be monophyletic. Most genera contain only a few species, but *Podocarpus* has more than 90 species spanning warm temperate, subtropical, and tropical regions. *Dacrydium* is the only other large genus, with 25 species. Although there are some fossil records of Podocarpaceae in the northern hemisphere beyond those areas where they still occur, the family has an overwhelmingly southern hemispheric distribution, past and present (Enright and Hill 1995). Despite the fact that *Podocarpus* and other genera of Podocarpaceae produce terpenoid resin, the family has not been analyzed chemically in detail. Resin occurs notably in leaves but not in sufficient quantity in stem tissues to have been used commercially, nor has it been studied ecologically.

Cupressaceae

Eckenwalder (1976) proposed merging the Cupressaceae and Taxodiaceae into one family based on morphological characters. This merger has subsequently been supported by various kinds of evidence such as cladistic studies of morphology (Hart 1987), paleobotany (Miller 1977, 1988), and more recently, molecular data (Brunsfeld et al. 1994, Stefanoviac et al. 1998, Kusumi et al. 2000). On the other hand, Miller (1999) changed his mind because of evidence from fossil cones, suggesting that Taxodiaceae had a biphyletic origin. Some genera are definitely related to members of the Cupressaceae, but others seemed to Miller to have diverged from the Araucariaceae. Molecular data suggest that the major lineages of Taxodiaceae originated before those of the Cupressaceae s.s. and that these latter lineages were derived from within the Taxodiaceae. Taxonomically, however, Taxodiaceae were subsumed into the Cupressaceae s.l. because the latter name had priority. On the other hand, some researchers dissent from this merger. Page (1990) argued that the two families "differ in many fundamental aspects of vegetative and reproductive morphology" whereas Chaw et al. (1997) indicated that the familial relationships cannot be resolved by nuclear 18s rRNA sequences. Moreover, other researchers have preferred to retain the name Taxodiaceae "by convention" (e.g., Enright and Hill 1995); this has also been true in discussion of Cretaceous conifers, among which taxodiaceous taxa are so prominent (e.g., Graham 1999, Grimaldi et al. 2000, 2001; Chapter 4).

Cupressaceae s.l. is the largest and most widespread conifer family, extending through the northern and southern hemispheres. The family has the greatest number of conifer genera: 30. Gadek et al. (2000) proposed an infrafamilial classification of Cupressaceae from combined morphological and molecular approaches in which seven subfamilies are recognized (Appendix 1). Junipers *(Juniperus),* with a primarily northern hemispheric distribution, is by far the largest genus with 50 or more species, followed by cypresses

(Cupressus) with about 13 species in the Mediterranean and Sahara as well as North America and Asia.

The name cedar is often applied to several different cupressaceous plants, for example, white cedar *(Chamaecyparis thyoides),* incense cedar *(Calocedrus decurrens),* red cedar *(Juniperus virginiana),* and western red cedar *(Thuja plicata).* To prevent confusion, some taxonomists have suggested that without a qualifying adjective, the name should be reserved for the true cedars *(Cedrus,* Pinaceae). Although most cupressaceous trees are resinous, especially the leaves, which give off a strong aroma if rubbed or bruised, the accumulation of volatile terpenes in the heartwood of *Juniperus* and *Cupressus* has been the most exploited, as cedarwood oil (Chapter 7). A genus, *Xanthocyparis,* has been described from northern Vietnam (Farjon et al. 2002). The species *X. vietnamensis* is called the golden Vietnamese cypress. Its closest relative is *Chamaecyparis nootkatensis* (yellow spruce or Nootka false cypress) from the U.S. Pacific Northwest. In fact, Farjon and coworkers renamed that species *X. nootkatensis.* Resins distilled from *Xanthocyparis* wood as well as from other cupressaceous genera such as *Fokenia* in this region are used for incense.

Cupressaceous genera in the southern hemisphere, such as *Callitris,* occupy habitats ranging from rain forest to desert. *Callitris* (Plate 9) has 14 species, 12 of which are small trees in Australia (where it is known as cypress pine), and 2 in New Caledonia. *Callitris* is a characteristic tree across Australian ecosystems but is a major feature in arid zones. Intermediate forms resulting from hybridization have led to the description of numerous species and consequent taxonomic confusion (Bowman and Harris 1995). *Tetraclinis,* once placed in *Callitris,* is now a genus of a single species extending from Spain to North Africa. Li (1953) subsequently included *Tetraclinis* as the only northern hemispheric genus in the otherwise southern hemispheric subfamily Callitroideae. Gadek et al. (2000) placed *Tetraclinis* in the northern hemispheric Cupressoideae, however, based on morphological and molecular data. The resin of *Callitris* and *Tetraclinis* is known commercially as sandarac, of which similar nonvolatile terpenoid components have been used for specialized varnishes (Chapter 9).

Ten genera traditionally and still commonly placed in Taxodiaceae are now considered to be members of several subfamilies of Cupressaceae s.l. (Gadek 2000; Appendix 1). The Taxodiaceae, more than any other group of conifers, had diverse and extensive representation in the geologic past but today comprise only a few species restricted to small areas. Some also left a fossil record of amber (Chapter 4). Taxodiaceous species generally have abundant leaf resins, some of which have been studied ecologically (Chapter 5). They only produce resin in the trunk traumatically, however, often in relatively small amounts unless the trunk has been severely injured (Chapter 3).

Genera of Cupressaceae subfamily Sequoioideae—*Sequoia*, coastal redwood; *Metasequoia*, dawn redwood; and *Sequoiadendron*, Sierran redwood or big tree—are particularly interesting in terms of their past and present distributions. Each genus now has only a single species of very restricted distribution. *Sequoia sempervirens* is essentially restricted to the coast of northern and central California, and *Sequoiadendron giganteum* to the central Sierra Nevada of California. The native distribution of *M. glyptostroboides* is even more limited, only a small remote area in central China (Plate 11). The fossil ancestors of these redwood species, however, occurred in vast forests. *Metasequoia* was the most abundant conifer in western and arctic North America from the late Cretaceous period to the Miocene (i.e., through most of the Tertiary period). *Sequoia* and *Sequoiadendron* were also widespread, especially during the Tertiary when the climate was much warmer than at present. Although amber has been found in association with remains from *Sequoia*, most chemically authenticated taxodiaceous amber was produced by *Metasequoia* (Chapter 4). Cooling of the global climate resulted in southern migration and restriction of these trees to their present localized areas.

Even though it is now known that *Metasequoia* was extremely widespread and abundant in the geologic past, the genus was only formally described as separate from *Sequoia* following study of fossils in Japan in 1941. It had been recognized that some fossils have leafy twigs with needles arranged oppositely whereas *Sequoia* has alternating needles (Figure 2-4). Furthermore, the cones are attached to long, naked peduncles (stems) whereas cones of *Sequoia* do not have such stems. These characteristics appeared to be intermediate between *Sequoia* and *Taxodium*. But only later was the fossil officially named *Metasequoia*, appropriately indicating its intermediate characteristics. Incredibly, a Chinese forester, also in 1941, found small stands of an unknown species that he called water pine along the Chang (Yangtse) River in central China. A short time later, Chinese botanists made the connection that this Chinese tree is identical to the fossil *Metasequoia* (Gittlen 1998a). Thus in China, *Metasequoia* is a living fossil.

Unlike *Sequoia* or *Sequoiadendron,* which grow in groves, *Metasequoia* trees grow in fog-enshrouded valleys in a diverse primary forest (Plate 11), similar to the association of trees that constituted the Arcto-Tertiary forests. Paleobotanists on the first expedition to collect *Metasequoia* in 1941 estimated that about 1000 *Metasequoia* trees remained and made a plea to the Chinese government to preserve them (Gittlen 1998a, b). Although most of the native forests in which the trees grew have been destroyed through human population pressure, about 4000 protected mature dawn redwoods currently live in those isolated foggy valleys in central China. The deciduous trees have been planted in 130 countries around the world and grow especially well in the southeastern United States. Neither as tall nor as long-lived as *Sequoia, Metasequoia* attains heights of 33 m and lives to 600 years; *Sequoia* can exceed more than 110 m and live 2000–3000 years.

Apparently, *Metasequoia* used to occupy swamps in some areas of North America similar to those of the present-day swamp or bald cypress *(Taxodium distichum),* a deciduous tree that grows in swamps of the southeastern United States (Plate 10). *Taxodium* has been placed in Cupressaceae subfamily Taxodioideae rather than Sequoioideae (Appendix 1). *Taxodium mucronatum* is the only other species of the genus; it is the national tree of Mexico and has one of the largest girths of any tree. Its resin was used by the Maya.

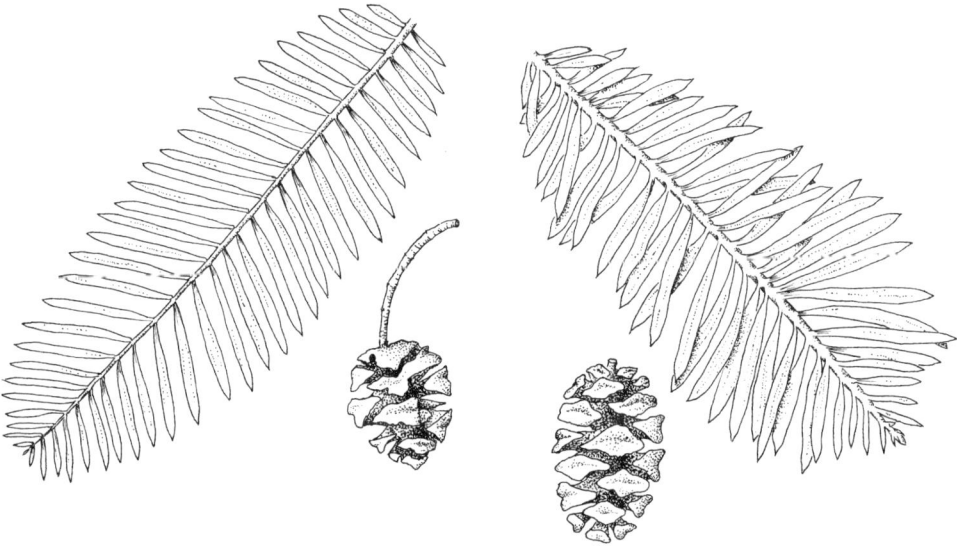

Figure 2-4. Differences in needles and cones led to description of the living fossil *Metasequoia glyptostroboides* (dawn redwood, Plate 11), left, as distinct from *Sequoia sempervirens* (coastal redwood), right.

Resin-Producing Angiosperms

Classification

Substantial agreement between the classification systems of angiosperms (plants with true flowers and fruits) by Cronquist (1981) and Takhtajan (1980, 1997) led to their wide acceptance, but these schemes have been superseded by the accumulation of molecular evidence. These classification systems are still in frequent use, however; hence, I allude to them. The classification of flowering plants by the Angiosperm Phylogeny Group (APG; Figure 2-5) has been presented in the more modern form of clades, whose relationships in a cladogram serve the same purpose as a "tree." The APG classification recognizes a selected number of monophyletic suprafamilial groups, that is, clades supported by at least one and often several lines of evidence. Although APG employs the ranks of order and family, it is emphasized that categories of the same rank are not evolutionarily comparable units unless they are sister groups, that is, close relatives.

In general, APG has adopted a broad circumscription of orders compared to earlier classification systems. Their ordinal classification has been based on the principle of monophyly, with well-established and familiar taxonomic entities maintained to preserve some stability in nomenclature. In a phylogenetic classification, names are given only to groups that are monophyletic. APG (1998) thus recognized 462 families in 40 orders of flowering plants whereas Cronquist (1981) had 321 families in 64 orders, and Takhtajan (1997), 589 families in 232 orders. These differences in numbers of orders and families clearly point out the somewhat arbitrary circumscription of families and orders. It may seem shocking to nonsystematists that there is no objective way to recognize such levels of taxa, that such decisions result partly from tradition, utility, and taste. The decision by APG, that classification is most useful as a reference tool with a limited number of orders, is accepted here.

For centuries, flowering plants have been divided into monocotyledons and dicotyledons (hereafter referred to as monocots and dicots) on the basis of several morphological characteristics, usually at the rank of class. The simplistic division of angiosperms into monocots and dicots, however, does not accurately reflect phylogenetic history. The monophyly of only two major angiosperm clades—monocots and eudicots—is supported. The remaining families form a largely unresolved complex. Nearly every one of these fami-

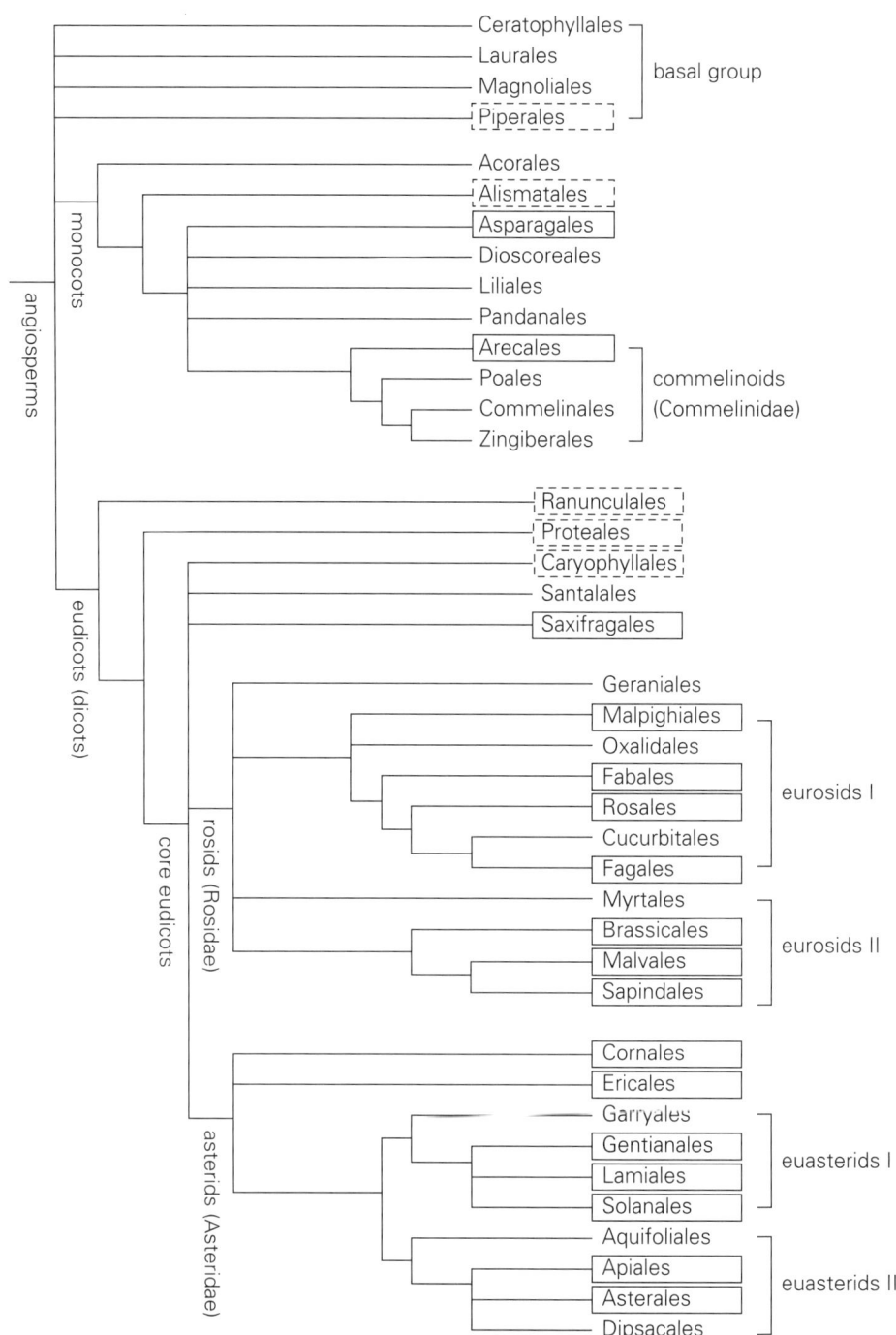

Figure 2-5. Ordinal classification of angiosperms, adapted from the Angiosperm Phylogeny Group (1998). Resin-producing orders are in boxes with solid borders; those described as resin producing, but unsupported by adequate chemical data, are in boxes with dashed borders. Also see Appendix 2.

lies has been suggested, at one time or another, as the most primitive extant angiosperm. Because of incomplete knowledge of the phylogenetic relationships of angiosperms, the early divergent clades continue to be analyzed and discussed. This group is referred to as the basal group or the nonmonocot paleoherbs. APG did not name monocots and eudicots formally because it could not be determined at which rank these two taxa should be recognized (Figure 2-5). Nor were commelinoids (monocots) or rosids and asterids (core eudicot subgroups) formally named. These groups, however, are commonly known as the subclasses Commelinidae, Rosidae, and Asteridae, respectively, and I occasionally refer to them in this manner. Within the eudicots, there is increasing support for the large subgroup, the core eudicots, which in turn is divided into rosids and asterids, each of which is increasingly considered monophyletic. There is further division into eurosids I and II and euasterids I and II, with one rosid and two asterid orders placed outside these groups (Figure 2-5). Some families are recognized as belonging to one of the major groups but their ordinal position remains uncertain.

Resin production is scattered throughout the flowering plants, which is generally true for the various kinds of secondary chemicals. The distribution of resin-producing plants among the 40 orders recognized by APG is shown in Figure 2-5. This dispersion is even clearer in the listing of genera within families (Appendix 2).

Resins are produced primarily by woody angiosperms, that is, flowering trees and shrubs, but they also occur in a few herbs. The most copious quantities are produced in the trunks of trees, particularly those in subtropical and tropical climates (Langenheim 1969, 1990). However, the greatest amount of resin produced by organ weight may be that coating leaves of shrubs in semiarid regions (Dell and McComb 1978). Not all resins of species within a genus listed in Appendix 2 have been reported as used by humans or as playing an ecological role in the plant. Only species for which the significance of resin to plants or humans has been recorded in the literature are discussed in the following chapters.

Basal Group

Among the four orders in the basal group of plants recognized by APG (Figure 2-5), resin has been reported to occur only in the Piperales. Moreover, only a few of the 2000 species of the tropical genus *Piper* (Piperaceae) have been recorded as producing flavonoid "resins" from the root and unripe fruit. One large shrub in particular, *P. methysticum*, known as kava, is widely cultivated, the resin used as a drug by the native people of Oceania (Chapter 10). It is not known how many other species of *Piper* produce this resin, although it is assumed that others do but have not been reported. There are numerous taxa in the Laurales that produce large amounts of volatile essential oils but without the nonvolatile fraction that characterizes resin.

Monocots

Only a few genera in four monocot families in three orders produce resin (Figure 2-5). Two families, Convallariaceae and Xanthorrhoeaceae, are in the order Asparagales. The Arecaceae (or Palmae) are the sole family in the commelinoid order Arecales with a resin producer. Araceae (order Alismatales) produce a chemical substance that has not been substantiated as a resin. The resin of all producing monocot species, except in the Araceae, is phenolic based.

Species of the large tropical genus *Dracaena* are put by APG in the Convallariaceae (Liliaceae s.l.). However, they are often placed in Dracaenaceae (Agavaceae in the sense of Cronquist 1981). Evergreen shrubby *Dracaena* species are familiar to many as indoor and greenhouse ornamentals and are grown outdoors in the tropics and subtropics. Most species are native to warm arid areas and may occur as trees. The name, from the Greek *drakoina*, female dragon, alludes to the massive trunks and branches of the dragon tree *(D. draco)*; the red phenolic resin produced is known as dragon's blood. *Dracaena draco* is the most massive species of the genus, often growing 12–19 m high. A notable specimen (Figure 2-6) growing in Tenerife, Canary Islands, was 22 m high, had a trunk girth of nearly 14 m, and was thought to be the oldest tree in the world by ancient peoples (Everett 1981). It was alleged to be 6000 years old in 1868, but in 1991 no plant more than 365 years was found alive (Mabberley 1997). The resin, collected from incisions of the stems of several other *Dracaena* species from Somalia and the island of Socotra, was used pharmaceutically and for coloring varnishes (Chapter 10).

In the family Xanthorrhoeaceae (Rudall and Chase 1996), the single Australian endemic genus *Xanthorrhoea* produces a yellow or red phenolic acaroid resin. In fact, the name *Xanthorrhoea* was derived from the Greek *xanthas* (yellow) and *rhoea* (flowing), referring to the yellow resin noted in the type species, *X. resinifera. Xanthorrhoea* trees are a characteristic feature of the vegetation in many parts of Australia. They grow slowly and usually attain a height of 2–3 m before branching into a tuft of grass-like leaves about a meter long. As the plants grow, the basal leaf bases remain attached to the fibrous trunk, making the plant look like a small palm at a distance. The tuft of grass-like leaves and the tree-like appearance led to the plant's common name, grass tree (Plate 12); it is sometimes called black boy because of a fire-blackened trunk. It is very resistant to fire, especially when young, because the bud is 12 cm below the ground; later buds are protected by old leaves. Some species such as *X. preissii* live at least 350 years, and flowering may be delayed until 200 years. Flowering is stimulated by fire, however, and some species growing near human habitations flower more frequently because more fires occur there. The resin, which accumulates in the old leaf bases of several species, has been used commercially for various purposes (Chapter 10).

The Arecales is the sole order of commelinoids in which resin is produced, and its occurrence is restricted to only one genus of the palm family (Are-

Figure 2-6. The very large, legendary dragon tree, *Dracaena draco,* in Tenerife, Canary Islands, drawn from a photograph of a mural painted by Charles Corwin in the Field Museum of Natural History, Chicago.

caceae). Although palm fruits are better known for their true oils, a distinct group of climbing species of the large Indomalesian rattan genus *Daemonorops* (Figure 2-7) produces a red resin on the surface of unripe fruits (Plate 47) that has been used medicinally and for colored varnish (Chapter 10). Like the red resin from *Dracaena,* it has been called dragon's blood and has been the most important commercial source of this kind of red resin.

The aroids (Araceae) *Philodendron* and *Monstera* produce a sticky material in the flowers that has been called resin, but the chemical components

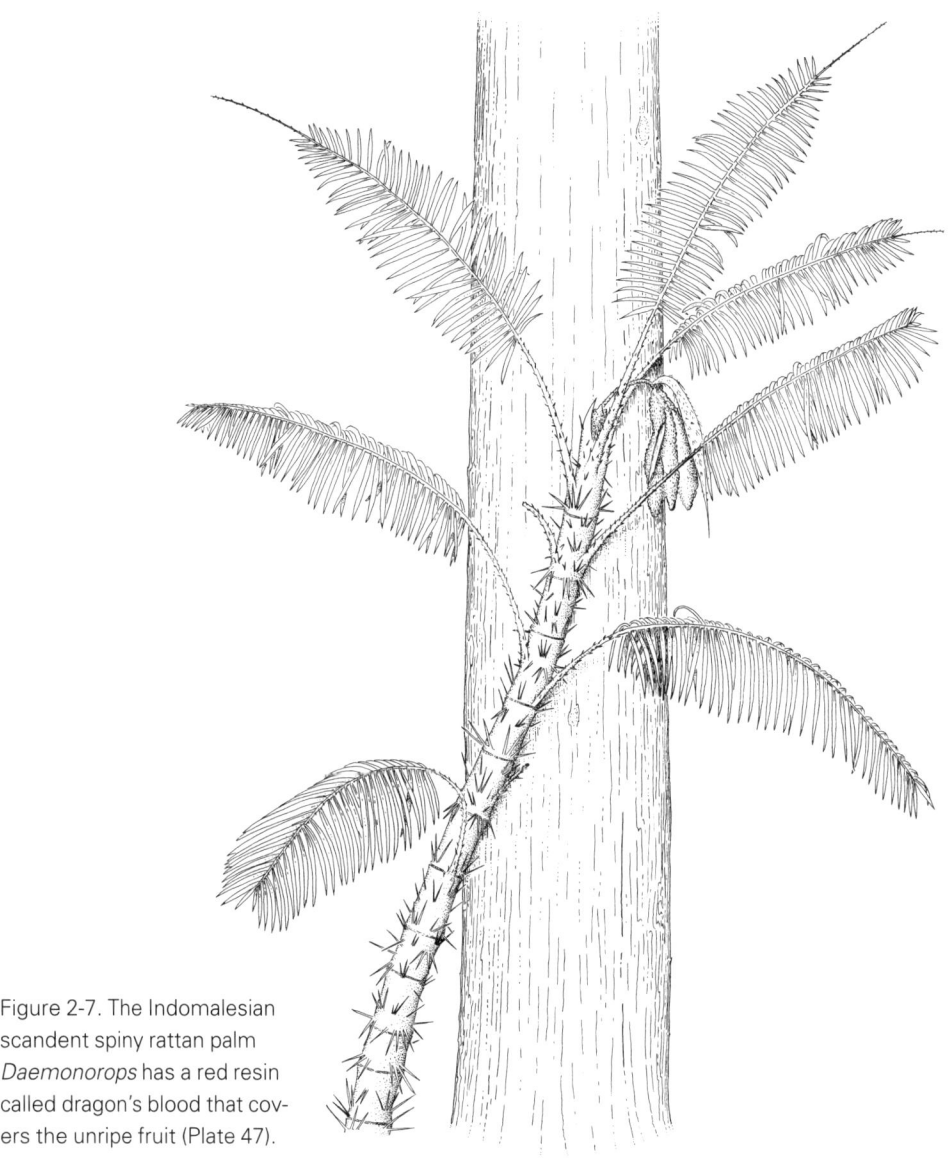

Figure 2-7. The Indomalesian scandent spiny rattan palm *Daemonorops* has a red resin called dragon's blood that covers the unripe fruit (Plate 47).

have not been analyzed. Grayum (1996) stated that this "resinous secretion" in the spathes of most species of *Philodendron* subgenus *Pteromischum* probably functions primarily in the adhesion of pollen to the exoskeleton of pollinating dynastine scarab beetles. Resin-gathering trigonid bees also collect "resin" from *Monstera* flowers for nest construction (Ramírez and Gómez 1978). However, Grayum (1996) was unaware of resin-gathering bees involved in pollination of *Philodendron* whereas such bees are involved in the pollination of *Clusia* (Clusiaceae) and *Dalechampia* (Euphorbiaceae) (Chapter 5). Better understanding of the chemical nature and role of this secretion awaits further study.

Eudicots

Two orders, Ranunculales and Proteales, are eudicots but are outside the core eudicots (Figure 2-5). Relatively little chemical information is available concerning the more primitive herbaceous dicots reported as producing resin. For example, the small perennial herb *Podophyllum* (Berberidaceae, Ranunculales) produces a "resinous substance" from a long rhizome. Podophylloresin yields a starting material for semisynthesis of effective anticancer drugs. Although most podophyllotoxin comes from the endangered Himalayan *P. emodi*, there is interest in obtaining the resin from leaves of the eastern North American mayapple, *P. peltatum* (Chapter 11).

Various families with reduced, wind-pollinated flowers, often aggregated into dangling inflorescences, have traditionally been considered as forming a group of related families called the Hamamelidae (in the sense of Cronquist 1981). However, Judd et al. (1999) showed that this assemblage is not monophyletic. Some families traditionally placed in Hamamelidae, such as Platanaceae, are considered relicts of an early floral reduction phase in plant evolution and have been put in the order Proteales. *Platanus,* the only genus of Platanaceae, is distributed primarily in the northern hemisphere, with some species in Southeast Asia, and produces a lipophilic bud exudate containing prenylated flavonols (Wollenweber and Dietz 1981).

Two orders, Caryophyllales and Saxifragales, part of the core eudicots but not belonging to either the rosid or asterid groups, contain resin-producing plants. In the small family Plumbaginaceae (Caryophyllales), only *Plumbago,* a genus of scandent shrubs of tropical and other warm regions, has been reported to produce resin. It is a lipophilic substance thought to be terpenoid resin (Rachmilevitz and Joel 1976). The resin, which occurs in a persistent calyx that is shed from the plant with the fruit, is believed to facilitate fruit dispersal by animals (Fahn and Werker 1972).

Resins of the tree genera *Liquidambar* and *Altingia* (Altingiaceae, Saxifragales) have been studied in more detail than the other early dicots. These genera are characterized by a mixture of primitive and advanced morphological and anatomical features that have led to varying interpretations of their relationships and phylogenetic position (Bogle 1986). Now in a separate family, Altingiaceae, they have traditionally been placed in the Hamamelidaceae (Cronquist 1981), which has sometimes been considered to be related to the Platanaceae in the Hamamelidae.

Liquidambar is a tall deciduous tree with palmately lobed leaves that people sometimes confuse with maple *(Acer)* leaves (Figure 2-8); the genus is disjunctly distributed in eastern Asia, Asia Minor, and North and Central America. Three of the five species are important producers of a phenolic resin from the trunk called storax that has been used for various purposes (Chapter 8). *Liquidambar styraciflua,* native to North and Central America, has several varieties grown widely as ornamental trees. The trees are noted for their fragrance as well as their brilliant foliage in the fall in warm temperate and subtropical areas (Plate 13). Amber chemically determined to be resin from *Liquidambar,* or a related progenitor, has been found in deposits at the Cretaceous–Tertiary boundary (about 65 Ma; Chapter 4), making its amber one of the first angiosperm ambers unequivocally identified. *Altingia* is a massive tropical tree, growing to 60 m high, with elliptic to oblong leaves; there are eight widely distributed species in Southeast Asia. Resin from *A. excelsa* is particularly used for scents, often as a replacement for *Styrax.*

Rosids

APG (1998) considered Zygophyllaceae to be a member of the rosids (subclass Rosidae) but did not assign the family to an order. Previously, Zygophyllaceae was considered as in either the order Sapindales (Cronquist 1981) or Zygophyllales (Takhtajan 1997). Prominent among the nine resin-producing genera in this family is the warm temperate North and South American shrub *Larrea.*

The five species of *Larrea* dominate the warm deserts of North and South America. Of the three North American deserts (Mojave, Sonoran, and Chihuahuan), the Mojave overall has the lowest diversity of perennials. Creosote bush *(L. tridentata)* by itself, or with a single associate, frequently dominates the Mojave's well-drained sandy flats, alluvial fans, and upland slopes, from below sea level in Death Valley to about 1500 m elsewhere (Plate 14). In sandy flats, it can occur in pure stands with as many as 6000 plants per hectare. Creosote bush maintains its dominance by monopolizing moisture and nutrients with a deep taproot that can extend 5 m into the soil, with lateral roots that can fan out to cover more than 40 m² of soil (Schultz and Floyd 1999). Suckering clones with a radius of 7–8 m occur that are estimated to be 11,700 years old (Mabberley 1997). Their sheer numerical abundance makes

Figure 2-8. Leaf and fruit of *Liquidambar styraciflua*, sweet gum, which is often confused with maples *(Acer)* because of leaf shape and red fall foliage (Plate 13).

creosote bushes a major influence not only on the desert plant communities but on associated animal and even human communities.

Larrea tridentata is thought to be derived from *L. divaricata,* one of the four species that occur in the deserts of Argentina. Argentina is considered to be the center of radiation not only for *Larrea* but the entire family Zygophyllaceae. *Larrea* is thought to have arrived in North America, possibly as seeds on the feathers of migrating birds during the last glacial episode 10,000–14,000 years ago (Schultz and Floyd 1999). Resin analysis of South and North American species also supports a recent origin of *L. tridentata* from a South American progenitor (Sakakibara et al. 1976). The creosote-like odor of the complex mixture of primarily phenolic resin (with some volatile terpenes) coating the leaves is responsible for the common name, creosote bush. The ecological roles of resins of the North and South American species (particularly *L. tridentata* and *L. cuneifolia*) have been studied intensively (Chapter 5), and medical uses are varied (Chapter 10).

A warm temperate to subtropical evergreen tree, *Guaiacum,* produces abundant resin that is used medicinally (Chapter 10). *Guaiacum* also has one of the hardest of commercial timbers (known as lignum vitae), used for bowling balls, pulleys, etc. Other zygophyllaceous resin-producing genera include *Fagonia,* 40 species of shrubs in warm dry areas; the arborescent *Balanites,* 25 species from Africa to Myanmar (Burma); and the shrub *Porlieria,* 6 species from Mexico and the Andes. Other genera have 1–3 species: *Metharme* and *Pintoa* from Chile, *Sericodes* from Mexico, and *Plectrocarpa* from temperate South America.

Four orders (Malpighiales, Fabales, Rosales, and Fagales) contain resin-producing species within the eurosids I. In the Malpighiales, resin producers are represented in three families from the tropics (Clusiaceae, Euphorbiaceae, and Humiriaceae) and one family from the temperate zones (Salicaceae). In the Clusiaceae (or Guttiferae or Hypericaceae), terpenoid resins are secreted in most tissues of the large, climbing, epiphytic or partially epiphytic *Clusia* (145 species in tropical America with some in Madagascar and New Caledonia) and the arborescent *Garcinia* (200 species in tropical Asia and Africa). The floral resins of *Clusia* (Plate 30) play an unusual role in pollination and other bee activities (Chapter 5). Fossil evidence from the late Cretaceous suggests that resin may have been available to bees in some of the earliest fossil flowers, including relatives of *Clusia* called *Paleoclusia* (Figure 4-8). Crepet and Nixon

(1998) stated that these flowers indicate that Clusiaceae were well differentiated and diverse by that time. *Garcinia* has numerous species in which the resin has been used by humans for varnish and medicine (Chapter 9). Other resin-producing trees in the Clusiaceae include the large pantropical *Calophyllum*, the Southeast Asian *Cratoxylum*, and *Symphonia* of Madagascar and tropical South America. *Moronobea*, a small genus in tropical America, produces an adhesive and inflammable resin known as hog gum within a latex.

In the large cosmopolitan family Euphorbiaceae, traditionally considered to be in the order Euphorbiales (in the sense of Cronquist 1981), "resins" have been reported as occurring in the latex secreted into laticifers in some tropical species (e.g., *Euphorbia resinifera*; Chapter 11). Also, "resins" in the latex of the large tropical genus *Croton* are used medicinally (Tripplett 1999); it is difficult to consider these euphorbs as primarily resin-producing plants, however. On the other hand, some euphorbiaceaous genera exude large quantities of terpenoid resin over the leaves from glandular trichomes, as in the shrubs *Beyeria* (turpentine bushes) and *Bertya,* which occur in the tropical deserts of Australia (Chapter 3). *Dalechampia* is a large tropical vine genus (Plate 31) that, like *Clusia,* produces resin in the flowers as well as in the leaves. The ecological role of these resins has been studied intensively in an evolutionary context (Chapter 5). One very small tropical American resin-producing genus, *Humiria* (Humiriaceae), produces a balsam-type phenolic resin in the trunk of the tree (e.g., *H. balsamifera*) similar to the resin in the leguminous *Myroxylon* (Chapter 8).

The Salicaceae, often put in the order Salicales (Cronquist 1981, Takhtajan 1997), includes the primarily northern temperate zone deciduous poplars and aspens *(Populus)* and willows *(Salix)*. Although *Salix* is sometimes listed as producing bud resin, it is minor compared to *Populus*. Species of *Populus* produce copious bud resins, differing chemically in the three sections of the genus (English et al. 1991), in specialized secretory structures (Figure 3-10). More than 30 different flavonoids have been identified in resinous bud secretions and used chemosystematically (Wollenweber 1975). Bud resin from species of section *Aigeiros,* such as *P. nigra,* characterized by a diversity of flavonoids, is collected by honeybees for use in nest construction (Chapter 5). Called propolis, the resin has a long history of pharmaceutical use for various purposes (Chapters 10 and 11). In contrast, species of section *Tacamahaca,* such as *P. balsamifera,* have bud resins composed of terpenoids and phenylpropanoids and not collected by honeybees. The role of these resins in defense against herbivores has been studied in detail (Chapter 5). The aromatic resinous *P. balsamifera* subsp. *trichocarpa*, known as western balsam poplar, is often grown as an ornamental tree. Many species of *Populus* characteristically grow along rivers or streams in North America, including the balsam poplars in boreal regions and *P. deltoides* and *P. fremontii* in more southerly regions. Others, however, such as quaking aspen *(P. tremuloides)*, occur over large areas of North America, extending throughout Alaska and Canada to the northern limit of forest, and throughout northeastern, northern central, and the Rocky Mountain regions of the United States.

Particularly noteworthy are the genera of tropical resin-producing leguminous trees (Fabaceae, or Leguminosae, order Fabales). Eleven genera (Chapter 9) have been reported to produce terpenoid resin in the tropical subfamily Caesalpinioideae, which is sometimes recognized as a separate family, Caesalpiniaceae (Langenheim 1973). The flowers of Caesalpinioideae are different from the butterfly-shaped flowers of subfamily Papilionoideae, so well known to people living in the temperate zones from peas and beans. The distinctiveness of Caesalpinioideae is exemplified by the copiously resin-producing tree genera *Hymenaea* (Figure 2-9) and *Copaifera* (Figure 5-5), which occur across a gradient of moist to arid ecosystems in Africa and the Americas. *Daniellia* (Plate 15), *Gossweilerodendron*, *Guibourtia*, *Oxystigma* (syn. *Pterygopodium*), and *Tessmannia* are important West African forest genera. However, *Daniellia* has two species in the Sudanian and Zambezian savannas. Additionally, *Guibourtia*, *Oxystigma* and *Tessmannia* have outlier species in coastal East Africa. *Colophospermum*, with a single species, is a small tree or shrub that forms widespread mopane woodlands in southern Africa. *Eperua* and *Prioria* occur in moist habitats in the New World, as does *Sindora*, the only significant Asian resin-producing legume genus (although it has one species in West Africa).

Systematists have long despaired in defining tribes in subfamily Caesalpinioideae. However, Léonard's (1957) insightful revision of the tribe Cynometreae, separating the tribes Detarieae and Amherstieae from it, greatly clarified the tribal placement of resin-producing genera. Although Cowan and Polhill (1981) sometimes referred to the grouping as Detarieae-Amherstieae, Polhill (1994) recognized a single tribe, Detarieae s.l., a view followed by Mabberley (1997) and further supported by molecular studies (Bruneau et al.

2001). Using chloroplast *trn*L sequences, Bruneau and coworkers nonetheless reported that a "prominent characteristic" of the Detarieae s.s. clade is that it groups together resin-producing genera—*Copaifera, Daniellia, Eperua, Guibourtia, Hymenaea, Sindora,* and *Tessmannia*—from various generic groups recognized by Cowan and Polhill (1981). On the other hand, *Colo-*

Figure 2-9. Flowering branch of the copal-producing tropical tree *Hymenaea courbaril* (see also Plate 16), showing caesalpinioid floral characteristics that differ from those of the papilionoids (e.g., peas and beans), common in legumes of the temperate zones. The pod structure of *Hymenaea* is shown in Figure 5-6.

phospermum, Gossweilerodendron, Oxystigma, and *Prioria* are put into the *Prioria* clade in the molecular analysis. Based on morphology, Breteler (1999) combined the African genera *Gossweilerodendron* and *Oxystigma,* along with the non-resin-producing Asian genus *Kingiodendron,* into the New World genus *Prioria.* Bruneau et al. (2001), however, indicated that more molecular analyses are needed to help resolve perpelexing questions of generic delimitation in the Detarieae. They further pointed out that species from Africa, Asia, and the New World occur together in these clades but that molecular relationships between them are insufficiently clear to indicate biogeographic relationships.

Many leguminous resins are dominated by diterpenoids that produce very hard copal-type resin and thus have been in particular demand for their toughness and durability for varnishes (Chapters 6 and 9). Exceptions include the New World species of *Eperua* and *Prioria* as well as *Copaifera,* which produce more fluid resins that often consist predominantly of sesquiterpene hydrocarbons and/or diterpenes that do not polymerize (Chapter 7). Similarly, the Asian *Sindora* produces an oleoresin.

Eperua has been revised taxonomically (Cowan 1975), *Prioria* has received recent, albeit controversial, revision (Breteler 1999), but the New World species of *Copaifera* are in need of revision. *Copaifera* comprises variable New World species, many of which have been distinguished only by leaf characteristics (Dwyer 1951). The Venezuelan species have been carefully assessed (Xena de Enrech et al. 1983) but most species are in Brazil. The economic importance of *Copaifera* is suggested by the abundance of common names for different species in the New World, the majority of which are Brazilian, of which the most common are *copaiba, copaiva, copahyba,* and *aceite.* In fact, it was *copei* or *copai,* the name used by the Amazon Tupi Indians, from which the genus acquired its name (Dwyer 1951). Tropical rain forest species such as *C. multijuga* produce the largest quantities of resin. *Copaifera langsdorfii* is one of the most widely ranging species, however, occurring in deciduous and savanna forests in Brazil, and its resin has been of commercial importance (Chapter 7).

In contrast to taxonomic problems with *Copaifera, Hymenaea* has been assessed using morphological criteria (Lee and Langenheim 1975) and leaf resin chemical characteristics (Martin et al. 1976). Numerous species previously based on leaf characteristics were revised, and the important African copal-producing *Trachylobium* was placed back into the genus *Hymenaea* (Langenheim and Lee 1974). Similar to *Copaifera* in the New World, *Hymenaea* species range across a wide range of forest types from rain forest (Plate 16) to desert thorn forest. Species of *Copaifera* and *Hymenaea* frequently co-occur (Chapter 5). *Hymenaea courbaril* is a particularly important species that occurs throughout the New World range of the genus, from Mexico to Brazil. It is such a useful tree that this single species has a different common name in most countries where it occurs: *guapinole, algarrobo, jatobá, jutaí,* West Indian locust, etc. Trees are smaller and produce less resin in drier sites. In rain forests, *Hymenaea* is very dispersed, like *Copaifera*, making commercial collection for some purposes difficult. In subdeciduous tropical forests, species of *Hymenaea* occur in relatively dense stands on sandy soils along rivers, where local people often tap them for resin (Plate 43). Because of the high degree of polymerization of labdane diterpenoids, *Hymenaea* is an important source of amber (Chapter 4).

Myroxylon is a small tropical American genus and the only unequivocal leguminous resin producer not in subfamily Caesalpinioideae; it is a tree of subfamily Papilionoideae (Figure 8-2). Once described as having two species, the genus now is considered to consist of a single species, *M. balsamum*, with two varieties, one in Central America and the other in South America. *Myroxylon* does not produce the hard copal-type resin characteristic of Caesalpinioideae but a softer, balsam-type resin that contains primarily phenolic compounds accompanied by a few terpenoids. These resins have a long history of human use (Chapter 8).

Dell (1977) reported that *Acacia*, the large genus of subfamily Mimosoideae, produces a resinous coating on young leaves, and Wollenweber and Dietz (1981) listed the genus as producing leaf aglycones. *Acacia glutinosissima*, in Western Australia, was specifically mentioned as producing leaf resin (Dell and McComb 1978). Although resin components may occur in the leaf coatings of some *Acacia* species, without additional information these are questionably resins, as defined here. *Acacia* is noted for its production of tree gums from stem tissue (Chapter 1).

Two resin-producing families, Cannabaceae and Moraceae, have been placed in the order Rosales (Urticales in the sense of Cronquist 1981) along with the rose family (Rosaceae). The annual herbs *Cannabis* and *Humulus* are the only genera of Cannabaceae. The family is so closely related to both

Urticaceae and Moraceae that Schultes (1970) considered Cannabaceae as intermediate between these two large families.

Cannabis has a long, complicated history, with estimates that humans have used it for more than 6000 years. Thus *Cannabis* is one of the oldest domesticated plants and the most widely distributed one. *Cannabis* is the ancient Greek name for hemp, a commonly used name for the plant. Hemp also refers to the fiber used from the stem. Marijuana, hashish, and cannabis (versus *Cannabis*) refer to use of the resin, although marijuana and cannabis are also commonly used names for the plant. The genus is now generally considered to comprise a single highly variable species, *C. sativa*, adapted to a variety of habitats throughout the world (Figure 10-3). Two subspecies are recognized: the more northern *sativa*, cultivated primarily for fiber, and the more tropical *indica*, grown principally for the psychoactive resin (either as marijuana or hashish). Some researchers (e.g., Clarke 1998), however, recognize subspecies *indica* as a separate species. *Cannabis* is related to, although clearly delimited morphologically from, the northern temperate zone genus *Humulus*, which includes three species of large climbing twiners (Figure 10-4). The most copious resin occurs in trichomes in female inflorescences of *C. sativa* and *H. lupulus* (the most commonly utilized *Humulus* species), but their resin chemistry is distinctively different, and accordingly, the resins are used for different purposes (Chapter 10). Resin of *Humulus* is used primarily to provide flavor and aroma to beer.

In the Moraceae, resins have been reported to occur in the latices of the tropical genus *Ficus* (800 species), the Southeast Asian *Artocarpus* (50 species), and the warm temperate and tropical African *Morus* (12 species). As with resins reported from latex of Euphorbiaceae, however, more chemical analysis is needed before definite conclusions can be made about the actual nature of these secretions.

Although *Prunus* is well known for the production of true gums in the stems, aglycone phenolics are produced in leaf buds of *Prunus* and *Sorbus* (Wollenweber and Dietz 1981). Furthermore, some members of the Rosaceae bear glandular trichomes and accumulate triterpenoids rather than the usual mono- and sesquiterpenes, along with the flavonoids.

The fourth resin-producing order in the eurosids I is Fagales, which includes the family Betulaceae. Chen et al. (1999), in a study using DNA, morphology, and paleobotany, characterized the Betulaceae as a well-defined family with six living genera. Most species are distributed in the northern temperate zone with a few species in other areas. Most researchers have supported divisions of the family into two clades, treated either as subfamilies or tribes, although other systematists have placed the clades in two families. Gene sequence data support the subfamilies Betuloideae and Coryloideae within Betulaceae. Betuloideae includes birches *(Betula)* and alders *(Alnus)*, two of the best-known genera producing bud resin that covers young leaves (Chapter 3). Chen and coworkers indicated that *Betula* appears in some respects as transitional between subfamilies Betuloideae and Coryloideae. One leaf-resin-producing genus, *Ostrya*, is in the Coryloideae, although other genera of this subfamily have resins that have not been analyzed.

Betula has 35 species that are prominent primarily in boreal and subarctic forests whereas the 25 species of *Alnus* arc commonly temperate zone riparian trees. Birches and alders produce resins composed primarily of terpenoids mixed with some phenolics that coat buds, developing leaves, and young stems. Terpenoids from *Betula* bud resin are used on Russian leather, giving it a characteristic odor. The resin secretory structures are particularly conspicuous on the young stems and leaf buds of some species such as European birch (*B. pendula*, Figures 2-10, 3-11 to 3-13). These resins provide defense against herbivory and, possibly, desiccation when the young organs are emerging in early spring in northern latitudes. The ecological roles of *Betula* resin have been studied in considerable detail in subarctic regions (Chapter 5).

Resin-producing plants occur in three (Brassicales, Malvales, and Sapindales) of the four orders of eurosids II, two of which contain important tropical resin producers. In the Brassicales, only the small genus *Didymotheca* (Gyrostemonaceae) has been reported as a resin producer. This shrub, distributed in Western Australia, is noted for its leaf-coating terpenoid resin.

On the other hand, the order Malvales includes such different families as the cosmopolitan Thymelaeaceae, the large tropical Dipterocarpaceae, and the small subtropical Cistaceae. Several species of the small Southeast Asian rain forest genus *Aquilaria* and one species of *Gonystylus*, whose resin-saturated aromatic wood has been used for centuries in Asian countries (Chapter 10), are the resin producers among the toxic trees and shrubs of the Thymelaeaceae.

Dayandan et al. (1999) pointed out that despite the high diversity of species and ecological dominance of Dipterocarpaceae in Asian rain forests, phylogenetic relationships within the family as well as relationships with other families are poorly defined. In their molecular phylogenetic analysis (based on nucleotide sequences of the chloroplast *rbc*L gene), Dayandan and coworkers recognize three subfamilies, including the Monotoideae of Africa and South

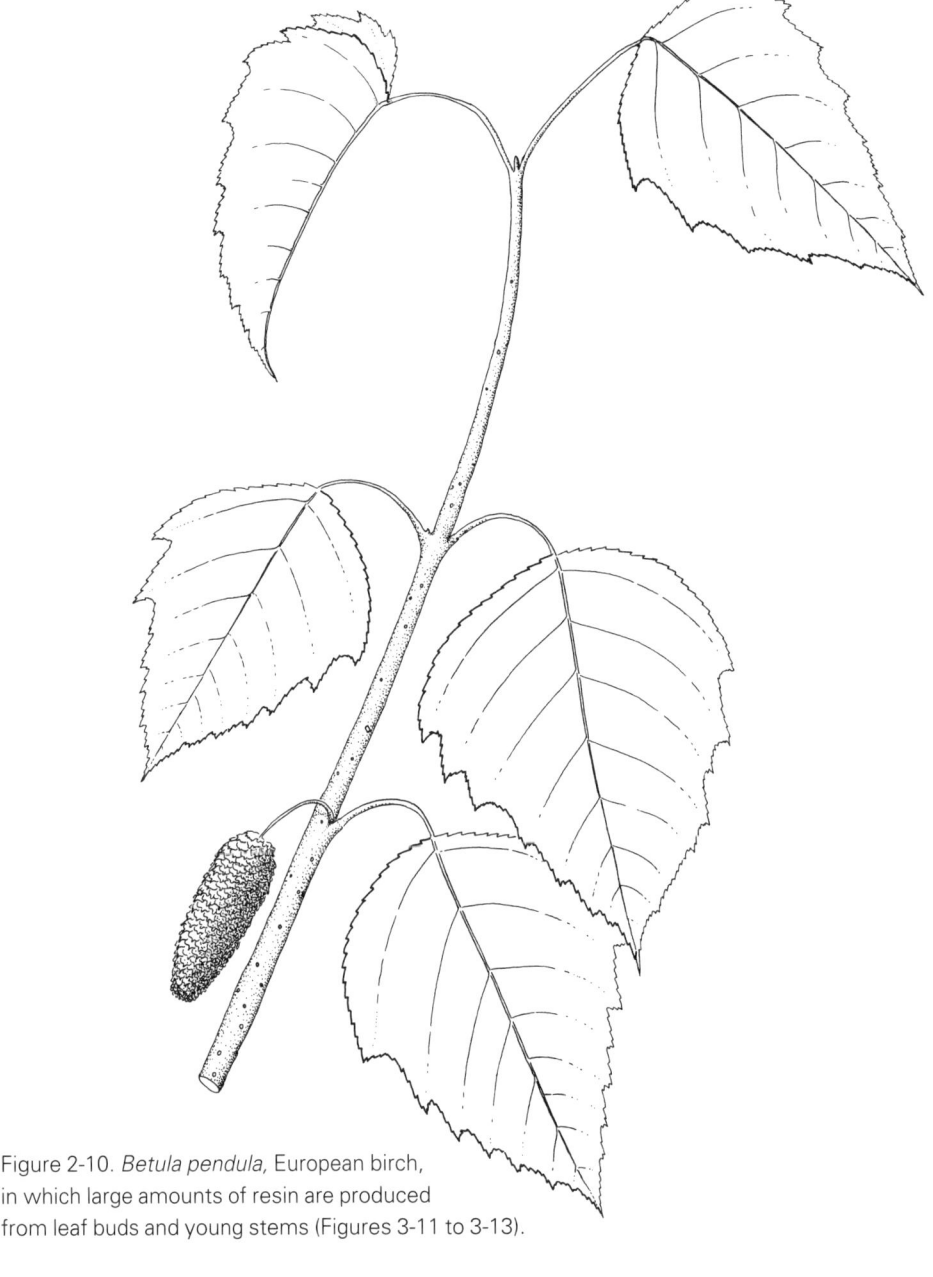

Figure 2-10. *Betula pendula,* European birch, in which large amounts of resin are produced from leaf buds and young stems (Figures 3-11 to 3-13).

America (30 species), Pakaraimoideae of South America (1 species), and the primary Dipterocarpoideae of Asia (470 species), which form a monophyletic group allied to the Malvales. Alverson et al. (1998) also had provided more inclusive molecular data supporting the malvalean affinity of Dipterocarpaceae.

Similar to Fabaceae in which resin production is essentially restricted to one subfamily, the resin producers of Dipterocarpaceae only occur in subfamily Dipterocarpoideae (Appendix 2). Dipterocarpaceae is the most important resin-producing family in forests of Southeast Asia, just as the Fabaceae are in tropical forests of Africa and the New World. Some of the most prominent resin-producing dipterocarps are *Shorea* (194 species; Plate 44 and Figure 9-1) and *Hopea* (102 species) in tribe Shoreae, and *Dipterocarpus* (69 species; Plate 40 and Figure 2-11), *Vatica* (65 species), and *Anisoptera* (11 species) in tribe Dipterocarpeae. These genera are prominent components of rain forests in Southeast Asia, India, and Sri Lanka (Ashton 1982). In fact, dipterocarps may constitute 80% of all the emergent trees that tower over the canopy in the rain forests as well as 40% of the understory trees. Under increasing seasonal drought (dry season lasting as long as 6 months), both forest strata become less dense and the representation of dipterocarps decreases except where gregarious species occur. There are areas, however, such as forests of New Guinea, where dipterocarps play a subsidiary role or where they tend to be endemic to particular islands. Dipterocarps are noted for their extremely irregular flowering and fruiting, which has numerous consequences for their utilization (Chapter 10).

Dipterocarp trees are valuable for their resins and, with their great girth, straight trunks, and resinous wood that prevents termite infestation, their timber. The protective properties of dipterocarp resins against insects and fungi have received study (Chapter 5). The resins, known as damar or dammar, consisting of sesqui- and triterpenes, have long been used for various purposes by the Malays and exported to be used as a highly lustrous varnish. In some genera, cadinene sesquiterpenes polymerize; hence, the resin can accumulate as large deposits of amber, some of which are processed for petroleum (Chapters 4 and 9).

The resin-producing rockroses *(Cistus)* in the small family Cistaceae (Malvales) have a Mediterranean and Transcaucasian distribution in dry scrub and open woodlands. Several species are well known as ornamental shrubs because of the large, handsome flowers (Figure 10-6). Leaf-coating resin from several of the 18 species has been used throughout history in various ways, particularly those involving the aroma of the volatile terpenes, as have the labdane diterpenes for their antimicrobial activity (Chapter 10).

The Sapindales is another important order of eurosids II with five tropical resin-producing families. One of the large and most copiously resin-pro-

Figure 2-11. *Dipterocarpus vernicifluus* with winged fruit characteristic of the copiously wood-oil-producing genus (see also Plate 40 for *D. kerrii*, another wood-oil-producing dipterocarp).

ducing families is Burseraceae, which has been placed by others in the order Rutales (Waterman and Grandan 1985) or Burserales (Takhtajan 1997). Often called the balsam or torchwood family, Burseraceae are widely distributed in tropical and subtropical regions worldwide and are divided into three subfamilies: Bursereae, Canarieae, and Protieae (Appendix 2). Resin-producing trees of several genera are prominent components of rain forests, including the African and Asian *Canarium* (77 species) and pantropical *Dacryodes* (40 species; Plate 17) of subfamily Canarieae. Although *Canarium* occurs in Africa, the center of its distribution is Southeast Asia, where the species are mainly tall primary and secondary forest trees growing at low elevations (Leenhouts 1959), whereas *Dacryodes* has numerous species in Africa and Asia and a few in America. *Santiria* (24 species), also in the Canarieae, occurs in West Africa and Southeast Asia. On the other hand, *Protium* (more than 120 species; Plate 41), primarily in tropical America with a few species in Madagascar and Malesia, belongs to subfamily Protieae along with other prominent tropical American resin-producing genera, *Tetragastris* and *Trattinnickia*. All these genera have species noted for their copious resin production and are used by humans (Chapter 8). Two other small Asian genera in the Protieae, *Scutinanthe* and *Triomma*, also produce resin.

Except for *Bursera* and *Protium*, the distribution of Burseraceae is primarily either African or Asian. Most *Protium* species occur in the New World; many are prominent resin producers, especially in Brazilian rain forests. *Protium copal* resin also was an important part of the Mayan cultures in southern Mexico and Guatemala (Chapters 6 and 8). The copiously resin-producing genus *Bursera* (about 100 species), on the other hand, is restricted to tropical America from the southwestern United States to Peru but with species reaching their maximum diversity on the Pacific slopes of Mexico. In these coastal areas, 70 of the approximately 80 species are endemic (Becerra and Venable 1999b). They range from small shrubs to large trees but typically are low to medium-sized trees. Many trees have succulent trunks colored yellow, green, red, or purple. The outer bark of some species exfoliates in papery flakes or sheets of bright colors often different from that of the trunk (Plate 18).

Because *Bursera* is one of the dominant components of Mexican tropical dry forests, the genus has been studied taxonomically there since the end of the 19th century. Nevertheless, its classification remains problematic and its phylogenetic relationships are not well understood (Rzedowski and Kruse

1979). *Bursera, Boswellia,* and *Commiphora* have been placed in subfamily Bursereae. Some systematists have included the species of *Bursera* in *Commiphora* (Gillett 1980) whereas others have considered *Bursera* closer to *Boswellia* (Lam 1932). Through molecular studies, however, Becerra and Venable (1999b) supported the view that *Bursera* is monophyletic and more closely related to *Commiphora* than *Boswellia. Bursera* produces balsam-type terpenoid resin (comprising mono-, sesqui-, and triterpenoids as well as diterpenes rarely) from canals in the bark (Chapter 3). Resins from numerous species have been used by humans for various purposes, particularly because of the fragrant volatile monoterpenes (Chapter 8).

Commiphora (190 species; Plate 19) and *Boswellia* (about 20 species—the nomenclatural confusion regarding species of *Boswellia* is discussed in Chapter 8; see Plate 42) are the sources of the classical incenses, myrrh and frankincense, respectively. These genera occur commonly in tropical dry forests or shrubby vegetation of northeastern Africa, Arabia, and parts of Iran, Pakistan, and India. *Commiphora,* however, is pantropical with an additional few species in Mexico and South America (Figure 2-12). Numerous species of *Commiphora* can grow in the same area. Waterman (1986) showed that 10 species of *Commiphora* growing at the same locality differed markedly in the chemistry of resin as well as the quantity and the manner in which it was produced. The resins from *Commiphora* and *Boswellia* are significant not only historically because of their use as incense and in traditional folk medicine, but also for continued interest in their myriad medical uses today (Chapters 8 and 11). As a result, resins from various species (e.g., *C. wightii, C. opobalsamum,* and *B. serrata*) are being studied experimentally from the perspective of Western medicine.

The Anacardiaceae are another important resin-producing family in the Sapindales. Molecular analyses by Gadek et al. (1996) show that Burseraceae and Anacardiaceae are related. Anacard resin occurs in internal canals, trichomes on leaves and young stems, or as part of the latex that characterizes many members of the family. The resins may be predominantly terpenoid with some similarities to those from various members of the closely related Burseraceae (Chapter 8). One of the prominent terpenoid resin-bearing anacards is *Pistacia* (9 species), native from the Mediterranean to southern Asia, the Canary Islands, and Central and North America. The word turpentine may trace its origin to the turpentine tree *(P. terebinthus).* The most familiar species of the genus is pistachio *(P. vera)*, used for its nut, but *P. lentiscus* is best known for its resin. Called mastics, the resins have been used as a masticatory, for fine varnishes, and in medicines since ancient time (Chapter 9). *Schinus* (27 species), an entirely tropical American genus, is called the American mastic tree (Plate 20). Terpenoid resins from *Pistacia* and *Schinus* are generally not allergenic, in contrast to phenolic resins of other anacard genera.

Figure 2-12. Spiny branch of *Commiphora africana,* one of the sources of myrrh, in fruit. See also Plate 19.

The predominantly phenolic resins of other Anacardiaceae are allergenic to varying degrees (Chapter 10), including those of cashew tree *(Anacardium)*, Burmese lacquer tree *(Gluta)*, mango tree *(Mangifera)*, poison wood *(Metopium)*, tar tree *(Semecarpus)*, *Tapirira*, and poison ivy, poison oaks, and poison sumac *(Toxicodendron)*. There is considerable controversy over

Figure 2-13. Comparison of fruiting western poison oak (*Toxicodendron diversilobum*, below; Figure 10-5) and flowering poison ivy (*T. radicans*, above), showing similarities and differences in leaves and inflorescences or infructescences.

the nomenclature of the genus *Toxicodendron*. Traditionally, poison ivy and poison oak, well known to North American inhabitants (Figure 2-13), have been put in the genus *Rhus*. Furthermore, the toxic effects of the urushiol complex of these plants are generally referred to as rhus dermatitis in the medical literature. Gillis (1971), however, placed the toxic plants in *Toxicodendron* and the nontoxic in *Rhus*. The genus *Toxicodendron* thereby includes 30 species. Resins from some species of *Toxicodendron*, such as *T. vernicifluum*, are used for lacquers (Chapter 9).

Other resin-producing families in the Sapindales include the tropical families Simaroubaceae, Rutaceae, and Sapindaceae. *Ailanthus* (5 species, Simaroubaceae) is a resin-producing genus of Asiatic and Australian deciduous trees. By far the best known species is *A. altissima*, often called the tree of heaven in Asia, where the resin is used for incense in Hindu temples. Although a native of China, where it is rare, it has freely naturalized in North America, Europe, and Asia. It even occurs as a weed in urban areas, where it flourishes in poor soil, and was made famous by the novel *A Tree Grows in Brooklyn*. It is frequently used for soil conservation in many temperate zone regions. It can reach 30 m in height and has leaves as long as 0.5 m. In India and Sri Lanka, the resin of *A. malabarica* is collected for medical purposes as well as for incense.

Rutaceae are cosmopolitan in distribution. Tropical members of the family are especially noted for producing highly methylated flavonoids, especially flavones and flavonols. They also are known for producing essential oils in lysigenous oil cavities, as in *Citrus*. Phenylpropanes as well as di- and triterpenes also occur (Wollenweber and Dietz 1981). Although the combination of lipophilic flavonoids and terpenes in other plants has been called resin, these substances in the rind of citrus fruits generally have not been so named. On the other hand, the Neotropical genus *Amyris* is well known for its balsam-type resin, called elemi, which exudes from the trunk of the tree (Chapter 8).

Buckeyes *(Aesculus)* are deciduous trees. The genus had been placed in the Hippocastanaceae (in the sense of Cronquist 1981) but APG (1998) put it in Sapindaceae. *Aesculus* (Figure 2-14) has 13 species with a wide distribution from warm temperate Europe and North America to the Asian tropics. Some species secrete a sticky resinous coating over buds, young leaves, and stems, as do birches and poplars. Other genera of Sapindaceae include the resinous *Diplopeltis* and *Dodonaea*. Resin in these Australian desert shrubs is exuded onto young leaves and stems, playing an ecological role in deterring herbivores or acting as antidesiccants (Chapters 3 and 5).

Asterids

Although there are a few resin-producing trees among the asterids (subclass Asteridae), resin-producing shrubs are common, occurring primarily in warm arid areas. Two orders, Cornales and Ericales, were placed within the asterids by APG (1998) but not in either the euasterids I or II (Figure 2-5). The

Figure 2-14. *Aesculus californica,* a common buckeye tree in California oak woodlands. Resin coats buds, young leaves, and stems of various *Aesculus* species.

important resin-producing plants in these orders are tropical trees. The Indo-malesian tropical genus *Mastixia* (13 species) is placed in the family Corna-ceae (Cornales); other systematists have put it in a separate family, Mastixi-aceae. It tends to occur scattered in mixed dipterocarp forests and produces a very fragrant terpenoid resin from the bark. *Mastixia* is also known to have occurred in numerous areas with tropical vegetation in the northern hemi-sphere during early Tertiary time; amber in *Mastixia* fruit has been reported from Eocene deposits in Europe (Chapter 4). Some species of the temperate zone dogwoods *(Cornus)* have "resin" in the bark (Duke 1985).

In the order Ericales (Ebenales in the sense of Cronquist 1981 and Takh-tajan 1980), the tree and shrub genus *Styrax* (Styracaceae) has more than a dozen species of about 130 in which the internally produced phenolic resin has long been prized by humans (Chapter 8). The name *Styrax* is derived from the Arabic *astirax*, meaning a sweet-smelling exudation. There are two sections of the genus, one evergreen and the other deciduous (Fritsch 1999, 2001). Resin-producing tree species in the evergreen section are almost exclu-sively tropical, with six Neotropical (American) and four Paleotropical (Asian) species. All Paleotropical species are in need of taxonomic revision; hence, there is some nomenclatural confusion (van Steenis 1932, Fritsch 1999). Three of the resin-producing species in the deciduous section also occur in the Asian tropics with another (shrub) species in the Mediterranean subtropics. Species such as *S. benzoin* (Figure 8-3) produce resin in sufficient quantity to have been an important export for centuries and today are often grown in swidden agriculture for this purpose (Chapters 8 and 11). Although less resin is produced in the American than the Asian species, the resin still has been used medicinally by indigenous people.

Among euasterids I there are three orders (Gentianales, Lamiales, and Solanales) and one family (Boraginaceae) not yet placed in an order that have resin-producing plants (Figure 2-5). Included in the Gentianales are the large tropical families Apocynaceae and Rubiaceae. The Apocynaceae (including Asclepiadaceae in the sense of Cronquist 1981) are noted for their production of latex, but resin has only been reported from the small Madagascan genus *Cryptostegia* (which also produces rubber) and several tropical American species of *Plumeria*.

The very large, cosmopolitan but primarily tropical family Rubiaceae has several genera that produce surface resin. The Old World *Gardenia* and New

World *Carphalea* secrete resin from structures similar to those found on leaf and stem buds of some other angiosperms (Chapter 3). Quinine *(Cinchona)* and coffee *(Coffea),* native to Africa, also are known to produce resin. *Elaeagia* is a montane South American tree in which the surface phenolic resin has been used since precolonial times to make a unique lacquer for covering wooden and leather objects (Chapter 9). *Coutarea* resin has been used as an incense in Mesoamerica. Mora-Osejo (1977) predicted that other rubiaceous plants probably produce resin, but none has been reported. From an evolutionary perspective, rubiaceous plants appear to be exceptions in that most tropical trees produce resin in internal ducts rather than on the surface of young shoots (Chapter 3).

The order Lamiales includes three families with resin-producing shrub species from warm dry areas: Lamiaceae (or Labiatae), Myoporaceae, and Scrophulariaceae. In the Lamiaceae, there are the Australian desert shrubs *Cyanostegia* (often considered in a separate family, Dicrastylidaceae), *Newcastelia,* and *Prostanthera.* Myoporaceae include several noteworthy Australian genera that produce copious leaf-coating resins that play ecological roles in their arid habitats (Chapters 3 and 5). *Eremophila* (206 species; Plate 21) is one of the most significant desert genera in the biogeography of Australia (Richmond and Ghisalberti 1994). Secretory structures involved in production of the abundant leaf resins have been intensively studied in several species (Chapter 3). *Myoporum* (28 species) occurs throughout southwestern Australia, whereas *Bontia* (1 species) is endemic to the West Indies. A detailed chemical analysis of the resin of *Eremophila* reveals that it is composed primarily of terpenoids, often with very unusual or novel structures (Chapter 1; Ghisalberti 1994). Aborigines used the resins medicinally, and some of these compounds are being analyzed for their medicinal value (Chapters 10 and 11).

Diterpenes in the Myoporaceae suggest that the family is related to Scrophulariaceae. Some taxa of Scrophulariaceae, however, such as the Mediterranean shrub or subshrub *Diplacus* (subsumed in the large genus *Mimulus*), copiously produce phenolic leaf resin (Plate 22) that deters insect herbivores (Chapter 5). Further confusing the chemical picture in *Mimulus,* the secretions of most species appear to be primarily mucilages with some flavonoid components incorporated into them.

In the Solanales there are two resin-producing families, Convolvulaceae and Solanaceae. Convolvulaceae, including the cosmopolitan *Convolvulus*

(100 species) and the tropical and warm temperate *Ipomoea* (650 species), contain a complex mixture of resinous substances called resin glycosides by McNair (1930). These compounds remain an ill-defined chemical complex although they have been used medicinally for centuries (Chapter 10). There has been much confusion in the literature regarding the placement of species in either the genus *Convolvulus* or *Ipomoea*, and this problem is exemplified by the resin-producing species. For example, *I. purga* (Figure 10-2), long used for the purgative action of its resin, has been listed as either *I. purga* or *C. purga*. Apparently, the resin produced in the tuberous roots of only several species in these large genera has been used. It is not known whether other species also produce resins that have not been used for some reason. *Anthocercis* (9 species) of the Australian desert is the only known resin producer in the very large, nearly cosmopolitan family Solanaceae.

There is one resin-producing family not placed in an order of the euasterid I group, the Boraginaceae. This family includes the shrubby *Halgania* (18 species) of the Australian desert and the evergreen shrubby *Eriodictyon* (8 species, previously placed in Hydrophyllaceae) of southwestern North America (Figure 2-15). The ecological roles of phenolic leaf resin in yerba santa *(E. californicum)* have been studied and a suggestion made regarding convergent evolution of its resin with that of other chaparral species (Chapter 5). Numerous North American arid shrubs and herbs in the Boraginaceae contain sticky substances that could be resins but as yet have not been analyzed chemically.

In the euasterids II, important resin-producing families occur in two orders, Apiales and Asterales (Figure 2-5). Apiales contains two families, the Apiaceae (or Umbelliferae), which have internal resin ducts rather than external trichomes, and Araliaceae, in which resin is secreted internally. The Apiaceae include several relatively tall perennial herbaceous genera whose gum resins are used medicinally (Chapter 10). Resins are less commonly produced in herbs, where they sometimes are primarily synthesized in the roots, as they are in Apiaceae and Convolvulaceae. These root resins are mainly phenolic but volatile terpenoids often occur in the walls of the fruits (often referred to as seeds, e.g., celery seeds) and in the leaves of Apiaceae, where they are commonly used in temperate climates to flavor food (e.g., parsley, dill, fennel, and caraway).

The Apiaceae form a large cosmopolitan family (446 genera) with most genera comprising only a small number of species. Despite its large size, widespread distribution (particularly in the northern hemisphere), and economic importance, there is no widely accepted modern classification of the family (Downie et al. 2000). The resin-producing *Ferula* (Plate 46), one of the largest genera (172 species), occurs from the Mediterranean to central Asia (Figure 10-1). The genus is morphologically variable and Downie et al. (2000) have questioned its monophyletic status (Pimenov and Leonov 1993). Numerous

Figure 2-15. *Eriodictyon californicum,* yerba santa, a common shrub in California chaparral with leaf resin that defends against beetle herbivory (Chapter 5). Lower leaves are typically blackened by a sooty epiphytic fungus.

species of *Ferula* produce gum resins known as asafoetida or galbanum. *Dorema* (12 species), a prominent resin-producing genus, is distributed throughout central and southwestern Asia. Resins called ammoniacum from both these perennial herbaceous genera have a long history of human use for various purposes. Although poorly known chemically, resin from *Opopanax* has been used medicinally and for scent (Chapter 10). Resins from *Laretia* and *Thapsia* are used for purposes other than medicine, such as plasters.

Species from three genera of Araliaceae yield resins that have been used as a readily photopolymerizable protective varnish for armor and other metalwork in Japan. Resins from *Acanthopanax sciadophylloides, Dendropanax trifidus,* and *Evodiopanax innovans* were used in Japan, the three variously known as *gonzetsu* or *koshiabura* (Mills and White 1994). Other species of *Dendropanax* yielded similar resins in Korea and China, known during the T'ang dynasty as *ushitsu* (golden varnish). Furthermore, although the chemistry of the exudates from the rhizomes and thick roots of *Aralia* and *Panax* is poorly known, their resins have been used medicinally. Some species of *Aralia* are known for sarsaparilla and those of *Panax* for ginseng.

In the order Asterales, the large family Asteraceae (or Compositae) has numerous genera that produce resinous secretions. They occur in the subfamily Asteroideae and tend to be concentrated in the tribe Astereae. The tremendous success of the Asteraceae "might well depend more on their chemical arsenal than on specialized floral structure" (Cronquist 1981). Resins are only one group of chemicals in the armory of this family, and the resin-producing plants tend to be shrubs that occur in warm arid areas. There are, however, a few perennial herbs and even annuals (e.g. *Holocarpha*, Figure 5-9) that persist long enough to grow in grasslands during the summer in dry Mediterranean climates. Certain resin-producing genera have many species, such as *Baccharis* (400), *Haplopappus* (150), *Brickellia* (110), and *Olearia* (100). Some, such as *Grindelia*, have been studied intensively for economic use of their resins (Chapter 10). Other genera of Asteraceae noted for resinous exudates that have been analyzed for potential use include *Balsamorhiza, Chrysothamnus, Flourensia, Gutierrezia, Helianthus, Madia, Tagetes,* and *Xanthocephalum*. Often, the common names of these plants refer to the resins, indicating how characteristic they are. For example, *Madia, Hemizonia,* and *Holocarpha* are called tarweeds; *Flourensia*, tar bush or varnish bush; *Calycadenia* and *Silphium*, rosin weed; and *Grindelia*, gum plant.

Parthenium illustrates how species can differ chemically in significant ways within one genus. The 15 species of *Parthenium* display extremes in life forms and habitats, ranging from desert shrubs, tropical deciduous small trees, submontane caespitose plants, to annual and perennial herbs in prairies and temperate forests. The desert shrub guayule (*P. argentatum*, Figure 10-9), however, is the only species that does not produce sesquiterpene lactones but, instead, resin and rubber (Rodriguez 1977). The resin, unique in its terpenoid constituents, has been studied in some detail because it can hinder the processing of the plant for its rich rubber content, or alternatively, can be extracted for separate use (Chapter 10).

The Goodeniaceae are another resin-producing family in the Asterales. The family includes the Australian desert shrub *Coopernookia* and the Australian lactiferous *Goodenia*.

Evolutionary Trends in Resin-Producing Plants

Available information limits discussion of the evolution of resin production to interesting occurrences in different plant groups and chemical similarities that occur in unrelated taxa. This raises questions of evolutionary convergence and parallelism, that is, the appearance of similar traits (in this case, resin production) in unrelated organisms. Although some evolutionists distinguish between convergence and parallelism, others use the terms interchangeably, the view I adopt here.

Taxonomic Distribution of Resin Producers

I summarize the taxonomic distribution of resin-producing plants in subclasses, orders, and families, with some reference to genera, from the general phylogenetic framework in which they have been presented. Comparison of the occurrence of resin production at these higher-level categories provides a picture of how resin-producing plants are distributed throughout the plant kingdom, highlighting phenomena common to various taxa that may be useful in stimulating research. Such comparisons, however, do not establish evolutionary relationships because groups at the same rank are not evolutionarily comparable unless the taxa constituting them are closely related.

Although it has usually been assumed that all conifers produce resin, the relatively recent inclusion of the families Taxaceae and Sciadopityaceae in

the order Coniferales raises the question of whether all conifer families produce resin as defined here. Only two coniferous families have members that
produce large quantities of trunk resin whereas resin occurs primarily in
leaves of the other families. Interestingly, the two most important resin-producing conifer families are distributed today principally in the northern hemisphere (Pinaceae) or southern hemisphere (Araucariaceae). In any event, a
proportionally smaller number of angiosperm families than conifer families
have resin-producing members, but the actual number of resin-producing
angiosperm taxa greatly exceeds that of conifers because flowering plants are
much more numerous and diverse than conifers. Since resin is characteristically synthesized in woody plants, the smaller percentage of angiosperms may
in part reflect their large number of herbaceous plants, whereas all conifers
are woody.

There are geographic and associated climatic differences that distinguish
the present distributions of conifer and angiosperm resin-producing plants.
Conifers occur primarily in the temperate zones with the most copious resin
production, however, in the tropics and subtropics (Appendix 1). On the
other hand, resin-producing angiosperm trees occur predominantly in tropical or subtropical regions, and shrubs primarily occupy warm dry areas (some
derived from tropical ancestors).

Even in the restricted number of orders (40) in the APG (1998) classification of angiosperms, the scattered occurrence of resin-producing taxa is obvious (Figure 2-5; Appendix 2). Chemically authenticated resin producers
occur in only 3 of the 10 monocot orders, of which the last is included in the
commelinoid subgroup (subclass Commelinidae). Resin-producing taxa
occur in 18 of the 26 orders (78%) of eudicots. Among the large core eudicot
grouping, 15 orders (60%) contain unequivocal resin-producing taxa, with
one order somewhat questionable. In the rosid subgroup (subclass Rosidae),
resin producers are scattered among 7 of 11 orders (63%) and, somewhat
similarly, in the asterid subgroup (subclass Asteridae), in 7 of 10 orders
(70%). It is interesting that only plants whose production of resin (as defined
here) is questionable occur in the basal group, and then they occur in only 1
of 4 orders.

In the APG system, 13% of eudicot families have resin-producing members, their families constituting 15% of the rosids and 13% of the asterids.
These percentages exclude families in the basal group and core eudicots that

are not members of the rosid and asterid subgroups. However, a higher percentage of families with resin-producing members characterize the eurosids II (19%) and euasterids I (21%). A total of 12% of dicot families recognized by Cronquist (1981) have resin-producing members; the largest number of resin-producing families occur within his advanced subclasses Rosidae and Dilleniidae, which APG mainly places within the rosids. These families have numerous genera (often with many species) that produce copious resins in tropical trees. In each of two of the large tropical families Fabaceae and Dipterocarpaceae, resin production is concentrated in one subfamily, Caesalpinioideae and Dipterocarpoideae, respectively. On the other hand, in another large tropical family, Burseraceae, resin producers are distributed more or less equally among three subfamilies. The highest percentage of families with resin-producing members, however, occur in Cronquist's subclass Asteridae, that is, plants that are predominantly shrubs, producing resin on the surface of leaves or stems. As in APG's asterids, the largest number of these taxa occur in the family Asteraceae, and one subfamily, the Asteroideae. Thus, despite different conceptual bases for the phylogenetic systems of Cronquist and APG, the general distribution of resin-producing angiosperms is similar.

Resin production is generally found in only some genera within a family and sometimes *apparently* in only some species within a genus. For example, only certain species of the large tropical genus *Styrax* have been recorded as producing resin. A question, however, arises as to whether all species within a resin-producing genus actually have the capacity to synthesize resin but may not have been reported as producing it. Resin production may be overlooked for several reasons. One is that synthesis may occur only when the plant is injured, or production of a noticeable quantity may depend on seasonal environmental conditions such as moisture and temperature. Another reason, especially in the large tropical families, is that observations often have not been recorded for plants in which the resin has not been used by humans —unless resin production has been studied in a scientific survey of the species. For example, a survey of the tropical leguminous genus *Hymenaea* showed that all species produce resin, but only those in the moist tropics produce it copiously (Chapters 5 and 9). Previously, the only species reported to produce resin were those in which the resin is sufficiently copious to be used commercially. Thus without such surveys it is risky to conclude that resin production is restricted to certain species within a genus, particularly tropical trees. On the other hand, surveys of *Betula* and *Eremophila* demonstrate that some species of each genus produce resin whereas others do not (Chapter 3). There also are cases, such as the genus *Parthenium,* where chemical surveys have shown that only one species in the genus produces resin whereas the other species produce related but different compounds (Chapter 10). Therefore, definite conclusions for any genus depend on detailed studies; these may have important ecological and evolutionary implications.

Convergence in Aspects of Resin Production

There are intriguing questions concerning convergence or parallel evolution in various aspects of the complex process of resin production. How is terpenoid versus phenolic resin distributed across the plant kingdom? What is the distribution among taxa of resins with both terpenoid and phenolic components? Is there a pattern to the relationship of these three chemically characterized resins, with the kind of secretory tissue in which they occur? These questions are discussed in more detail for specific groups in other chapters, but here I consider some of them in relation to patterns across plant groups.

All resin-producing conifers and most angiosperm trees synthesize terpenoid resin. Among the latter, 17 rosid families have terpenoid resin-producing trees, 13 of which comprise predominantly tropical evergreens; the 4 other families include deciduous trees with temperate zone distributions (Appendix 2). Frequently, these tropical families also have numerous resin-producing genera. Sometimes there are terpenoid-resin-producing shrubs as well as trees, in the Burseraceae and Zygophyllaceae, for example.

Although only a few genera are involved, the repeated independent occurrence of phenolic balsam-type resins in morphologically and phylogenetically distinct angiosperm taxa is striking. The dicots with these resins are all tropical trees that occur in four families. *Altingia* and *Liquidambar* (Altingiaceae) are among the core eudicots but are not included among the rosids or asterids (Figure 2-5; Appendix 2). *Myroxylon* (eurosids I, Fabaceae) is the only resin-producing representative of subfamily Papilionoideae. This is particularly interesting because numerous terpenoid-producing genera occur in subfamily Caesalpinioideae. *Humiria* (Humiriaceae) is also in the eurosids I whereas *Styrax* (Styracaceae) is in the euasterids I. Phenolic balsam-type resin occurs in monocots in subtropical to tropical habitats, including the Australian grass tree *Xanthorrhoea* (Xanthorrhoeaceae), *Dracaena* (Convallari-aceae), and the tropical vine *Daemonorops* (commelinoid, Arecaceae); phe-nolic resins from the latter two monocots are called dragon's blood. All these genera are further discussed in Chapters 8 and 10.

Another interesting question is why certain components of particular ter-penoid and phenolic resins occur repeatedly, but sporadically, across the plant kingdom. Repeated occurrence of groups of certain monoterpenes (e.g., α- and β-pinene, limonene, and myrcene) may be explained by multiproduct synthases, for example, limonene synthase (Chapter 1), that occur in differ-ent taxa. Moreover, these monoterpenes have been shown to serve various ecological roles (Chapter 5) that may have been advantageous in particular environments. Multiproduct terpene synthases may explain other common chemical co-occurrences such as that of the sesquiterpenes caryophyllene and various cadinenes (Chapter 1) in different taxa. The production of 52 com-pounds by a single enzyme, γ-humulene synthase in grand fir *(Abies grandis),* is truly remarkable. Some of these sesquiterpenes as well as certain diterpenes with labdane structures, which polymerize to form amber, occur in both conifers and angiosperms (Chapter 4).

There are also different characteristic combinations of volatile and non-volatile fractions of terpenoids that characterize resins in certain families, such as sesqui- and diterpenoids in Fabaceae, sesqui- and triterpenoids in Dipterocarpaceae, mono-, sesqui-, and diterpenoids in Pinaceae, and mono-, sesqui-, and triterpenoids in Burseraceae (Chapter 1). Moreover, some genera of legumes and dipterocarps produce a fluid resin composed primarily of sesquiterpenes (Chapter 7) whereas species in other genera in each family produce predominantly nonvolatile components (Chapter 9). There is also the striking independent occurrence or convergence of a particular terpenoid chemistry in tropical trees in the Dipterocarpaceae (eurosids II, Malvales) and those in the Cornaceae (euasterids I, Cornales). The polymer chemistry (polycadinenes) in these unrelated families is sufficiently similar to make the origin of some ambers ambiguous solely from chemical characteristics (Chap-ter 4). An occurrence of certain predominant diterpene labdanoid polymers in the coniferous *Metasequoia* and araucarian trees also leads to confusion in determining the source of ambers produced by these trees.

There seems to have been convergence in the production of terpenoid and phenolic resins in internal ducts in trees. Whether their compounds are produced in canals, pockets (cysts), or cavities (Chapter 3), however, differs considerably between taxa. Within the same family of conifers, such as Pinaceae, and angiosperms, such as Fabaceae, pocket or cyst secretory structures often occur in genera when related genera have canals. There has been discussion since the early 1900s about the evolutionary sequence of development of different kinds of secretory tissue in Pinaceae. Did the pocket or cyst precede the canal or vice versa? Molecular and paleobotanical studies of the phylogeny of Pinaceae support the subfamily Abietiodeae as ancestral to subfamilies Pinoideae, Piceoideae, and Laricoideae (Chaw et al. 1997, Labandeira et al. 2001) and, hence, the presence of pockets or cysts before the development of a canal system. Furthermore, there are important ecological consequences associated with different secretory systems. For example, some tree taxa with cysts may occur at different stages of forest succession than those with canals. This difference and its relation to resin response to injury have been shown to be significant in some but not all members of Pinaceae (Chapters 3 and 5), and a general relationship has not been carefully evaluated. In conifers that have cysts, an induced resin defense is essential because there is a less immediate resin defensive response than from constitutive resin flowing in canals (Chapters 1 and 5).

Mixtures of terpenoid and phenolic compounds (lipophilic flavonoids) are particularly prevalent in externally produced resins that coat leaves and young stems of shrubs in arid habitats and some deciduous flowering trees of temperate and boreal climates. Although terpenoids and flavonoids occur together, terpenoids predominate in some taxa, and flavonoids in others. Novel sesqui- and diterpenes are present in the resins of the shrubs (Chapter 1). Where there is a predominance of flavonoids, sesqui- and triterpenes are frequently the accompanying terpenoids. There has been convergence or parallel development of a basic kind of secretory structure—glandular trichome or epidermal cell—but with considerable variation in architecture (Chapter 3).

Status of Evolutionary Interpretation

It is important to identify the processes (e.g., genetic and developmental constraints, natural selection pressures, and chance) that result in evolutionary convergences or parallelisms. The complexity of the genetics controlling secondary compounds is staggering, and the information available for resin is limited despite advances resulting from the study of a few conifers, especially with respect to the evolution of terpene synthases (Chapter 1). The genes for synthesis of resin are likely to be many and tend to be clustered. Secondary chemical genes generally are highly regulated to respond to both specific and general exogenous signals, and many systems are highly plastic (Jarvis and Miller 1996). At all levels, there is evidence for self-organizing forces, for example, from binding to receptors; to development of cellular structures where the chemicals are synthesized, stored, and transported; to complex organism–organism interactions. Jarvis and Miller, and Maplestone et al. (1992), concluded that clustering of functionally related genes on a chromosome, as demonstrated for some secondary chemicals, implies that they have evolved as a unit because they conveyed a selectional advantage. For resins, clustering of genes seems apparent because of the close relationship between synthesis of constitutive resin and development of secretory structures (Chapters 1 and 3), and thence, to their ecological relationships with other organisms (Chapter 5). However, the development of different enzymes for induced resin synthesis adds yet another layer of complexity.

Natural selection, conferring some fitness advantage over time, seems the most likely scenario to account for much of the remarkable independent evolution of resin production in conifers and angiosperms. Identical compounds, but in different arrangements and proportions, are produced by multiple unrelated taxa. Random accumulation of structurally diverse compounds occurs, but this seems to augment the repeated occurrence of the more common compounds and the clusters of terpenoids resulting from multiproduct synthases (Chapter 1).

Some convergences may have evolved through natural selection in answer to similar defense problems that plants face (at different life stages or in the development of different organs in various plants) as well as resulting from the complex patterns of coevolution of plants with insects, microbes, and mammals (Chapter 5). Most antiherbivore and antimicrobial ecological studies have been done at the levels of species or population, where minor differences in the process of adaptation at different localities can lead to differences in resin chemistry. Resin that coats buds and young leaves could not only protect against herbivory but also against desiccation, high temperatures, and ultraviolet radiation. The interaction of chance and selection (exaptation in the sense of Gould and Vrba 1982, or preadaptation) has also been shown to be important in the origin of defense and pollination systems, and that each system has influenced the evolution of the other (Chapter 5).

It is impossible at present to move much beyond description of the repeated sporadic occurrences of resin production at the different levels of complexity, and it is hoped that future studies will meet the challenges of discovering the processes involved. Meanwhile, however, ecological studies at the levels of species and population help sort out the roles of genetics in controlling developmental differences and plasticity versus those of biotic selection pressures in creating and maintaining resin chemical diversity (Chapter 5). Furthermore, molecular techniques provide critical information regarding genes that encode terpene synthases and their unusual ability to synthesize multiple products (Chapter 1). Molecular techniques have also facilitated coevolutionary studies such as those of Pinaceae and bark beetles, of Burseraceae and leaf-eating chrysomelid beetles, and of floral resins in Euphorbiaceae or Clusiaceae and pollinating bees (Chapter 5). Additionally, biochemical and coevolutionary studies with an ecological perspective may provide further insights into the evolution of resin production in plants.

CHAPTER 3

How Plants Secrete and Store Resin

Resin secretory structures, where resin is synthesized and stored, are central to the ecological interactions and human use of resin. For example, how and in what kinds of structures resin is stored in the plant or develops as a result of injury, and how the resin exudes to the plant's surface, can determine its effectiveness against enemies as well as how it is collected for use.

A plethora of terms has been developed to describe the complexity associated with elimination of secondary compounds such as resins, from the cellular sites of synthesis into different kinds of structures, and thence, exudation to the outside of the plant, with or without injury. However, Fahn's (1979) view of using the overarching term secretion seems most useful for resins. Likewise, his handling of the concepts of internal (endogenous) versus external (exogenous) secretion is helpful in understanding the basic types of extracellular resinous secretion. Endogenous secreted material accumulates in various internal structures (generally referred to as glands or ducts), which essentially are intercellular spaces surrounded by secretory cells. Normally, such material only exudes from the plant when it is injured. Exogenous secretion, on the other hand, occurs in various types of epidermal secretory cells (often, glandular hairs) that may discharge the material to the outside of the organ either directly or first into a subcuticular space before further secretion.

The same type of secretory structure may be present in all organs or may be confined to specific organs within a plant. Conversely, different types of secretory structures may occur in different parts of the plant or at different ontogenetic stages of the plant.

Ultrastructural Features of Resin Secretory Structures

Biochemical evidence for the cellular sites of synthesis of components of terpenoid resins is presented in Chapter 1. Electron microscopy provides a means for better understanding of secretory structures and their function, helping in the evaluation of regulation of resin component biosynthesis, metabolic trafficking between organelle sites, and secretion into a storage area (intercellular lumen in endogenous structures, subcuticular spaces in exogenous ones).

Sites of Synthesis

Early ultrastructural studies were hampered because there were no definite stains for resin; hence, positive identification of the resin could not be made at the site of synthesis. Terpenoids could only be identified as lipoidal substances by their electron density after staining with osmium tetroxide. It was assumed that the lipoidal material in known resin-producing plants was resin, but there was uncertainty without biochemical support. Nonetheless, early studies of *Pinus* (e.g., Wooding and Northcote 1965, Vassilyev 1970, Fahn and Benayoun 1976) and the shrub *Eremophila* (Dell and McComb 1977) suggested that different organelles are involved in synthesis of resin components. These speculations were largely based on apparent sites of lipid accumulation during the secretory phase of development.

Within secretory cells, two prominent organelles, the plastid and smooth endoplasmic reticulum (smooth ER), have distinctive ultrastructural features. In *Pinus*, Gleizes et al. (1980a) reported evidence that sesquiterpenes are formed in the ER, and Gleizes et al. (1983) first showed the involvement of plastids in monoterpene biosynthesis. These plastids, called leucoplasts, are devoid of thylakoids and ribosomes, unlike chloroplasts. Their ratio of surface to volume is high, indicating an important increase in membrane surface and metabolic exchange during active secretion and thus representing a distinct class of organelle (Carde 1984). In a study of 45 species belonging to 23 angiosperm families, Cheniclet and Carde (1985) correlated ultrastructural data with extracted yields of mono- and sesquiterpenes, and phenylpropanoids. Their study included both endogenous and exogenous secretory tissue and supported the view that the leucoplast is the site of synthesis of monoterpenes but not sesquiterpenes or phenylpropanoids.

Sustained, multifaceted research on resin in *Pinus pinaster* by combined biochemical (^{14}C-labeled molecules) and ultrastructural approaches led to greater knowledge about the sites of synthesis (Bernard-Dagan et al. 1982). In 1988, Bernard-Dagan concluded that constitutive mono- and diterpenes are synthesized by large leucoplasts in needles of *P. pinaster* only during a short time when the secretory epithelial cells surrounding the canal lumen are young and active. Assimilates from photosynthesis are brought toward these cells from the phloem and are transformed by enzymes within the cytoplasm and leucoplasts. As the tissue in *P. pinaster* became older, the activity of the secretory cells progressively decreased. The ER appeared to be involved in the synthesis of sesquiterpenes, which supported the research of Gleizes et al. (1980a). In contrast to mono- and diterpene synthesis, sesquiterpene synthesis occurs throughout the year. Thus these studies not only pinpointed different sites of synthesis for mono- and diterpenes versus sesquiterpenes but suggested seasonal differences in synthesis of these components, which could influence the ecological roles of constitutive resin.

Early ultrastructural studies of exogenous resin secretory structures in Australian desert shrubs (Dell and McComb 1978) revealed similarities to the ultrastructural features of endogenous structures such as those from pine. Similarities include the extended, net-like tubular smooth ER and abundant leucoplasts with poorly developed internal structures commonly associated with the smooth ER. Most of the known components of the resins they studied were diterpenes, with some sesquiterpenes and numerous lipophilic flavonoids. However, chemical analysis of the components was not directly correlated with the ultrastructural study.

More recent ultrastructural work on structural transformations during terpenoid secretion has focused on monoterpenes in peppermint *(Mentha ×piperita)*. Peppermint has been developed as a model system to study the site of monoterpene biosynthesis and transport to the storage depot. Results from ultrastructural techniques such as cryofixation are related to previous biochemical analyses that substantiated the location of biosynthesis in the leucoplast (Turner et al. 2000). Characteristic ultrastructural transformations in the exogenous glands occur at the onset of secretion, correlated to the peak of monoterpene production and supporting the roles of both leucoplast and smooth ER in monoterpene synthesis. In fact, it appears that monoterpene biosynthesis is initiated in the leucoplast but completed (hydroxylation steps) in the smooth ER. The secretion of peppermint consists primarily of monoterpenes but contains some flavone aglycones. Thus Turner and co-workers suggested that some of the ultrastructural features they described may also represent specializations for the secretions of flavonoids. Various flavonoid and phenylpropanoid enzymes are clustered at the ER, implicating the ER in synthesis of these compounds (Winkel-Shirley 1999).

Stalk cells in peppermint glands undergo modification correlated with secretory activity (Turner et al. 2000). This includes development of distinctive plastids, numerous microbodies, and abundant mitochondria. Turner and coworkers suggested that some of these specializations relate to their essential role in supplying carbon substrates to the nonphotosynthetic secretory cells during the brief but intense period of secretion.

Turner et al. (2000) further discovered that significant changes occur in the composition of the monoterpenes when the secretory cells are no longer active in synthesis. Developmental changes in monoterpene accumulation are accompanied by changes in monoterpenoid composition (Gershenzon et al. 2000). The proportions of some components decrease whereas others increase substantially. Minor constituents tend to increase during leaf development. Although such compositional shifts have been reported in ecological studies, they have not been related to the activity of the secretory cells.

Most of the ultrastructural studies of mono- and sesquiterpenoids have assumed that similar secretory structures are involved in synthesis whether they are in essential oil- or resin-producing plants and whether in endo- or exogenous glands. However, such overall conclusions are less clear regarding the ultrastructure of resin-secreting cells for phenolic resins and those that also contain different kinds of terpenoids. For example, ultrastructural studies of the resin in exogenous secretory tissue in birch *(Betula),* composed of flavonoids and triterpenes, led to the assumption that resin components are synthesized initially in the ER (Raatikainen et al. 1992). In *B. pendula,* myelin-like deposits and osmiophilic substances accumulate in the columnar secretory cells (Figure 3-11). The authors suggested that the myelin-like material consists of triterpenes in concentric layers of cytoplasmic membranes, and that the osmiophilic substances are flavonoid compounds.

An ultrastructural study of the secretory structures in creosote bush *(Larrea tridentata),* which produces a predominantly phenolic resin, presents a complex picture with somewhat different secretory structures than those in plants producing primarily volatile terpenoid resins. The creosote bush, a dominant shrub in the deserts of the southwestern United States and northern Mexico (Plate 14) has single-celled glands on young leaves, petioles, and stipules (Thomson et al. 1979). The glands contain a resin consisting of more than 100 compounds, mostly phenolics but including some volatile terpenoids (Chapter 1). The exogenous secretory structures become isolated from other epidermal leaf cells by the development of a suberin layer internal to the primary wall of the structure. After attaining maturity, the gland senesces and collapses (Figure 5-8), extruding the material in the lumen through pores onto the surface of the leaves. Extrusion occurs primarily on young leaves, and Thomson and coworkers suggested that chemical modification of the resin occurs during and after extrusion. Although the dominant ultrastructural feature is an elaborate system of smooth and rough ER, other features of the secreting cell differ from those usually seen in terpenoid-secreting cells.

Export of Resin Components

A central role for the ER in transporting terpenoid resin components from intracellular sites of synthesis into the lumen of an endogenous secretory structure had long been assumed because ER can fuse with membranes of other organelles and form vesicles that move through the cell to the plasma membrane. In resin-producing epithelial cells, the resin components could be transported from the plastid or cytosol-ER by vesicles that are enveloped by the plasma membrane prior to discharge into the intercellular storage space. Carde et al. (1980) suggested this transport role for ER in *Pinus*, and their results were supported by Bernard-Dagan (1982). In *Parthenium*, resin masses enclosed in ER vesicles near the plasma membrane suggested this means of transport into the canal lumen (Gilliland et al. 1988). Resin-containing vesicles in *Hymenaea* epithelial cells are attached to the plasma membrane as well as occurring between it and the cell wall (Fail 1990). Also, a loose fibrillar mesh that characterizes the canal cell walls in *Commiphora* could facilitate easy transfer of resin to the lumen. The presence of bodies between the plasma membrane and cell wall lining the duct, as well as the absence of plasmodesmata, suggest that resin passes across the relatively porous cell wall via vesicles (Setia et al. 1977). A myelin-like structure along the epithelial cell wall in *Commiphora* may aid in the movement across the wall (Bhatt 1987).

Evidence points to the role of ER in transport of resin components to the storage areas of exogenous structures as well. Studies of secretory tissue in *Betula pendula* suggest that exported resin is formed by combining the triterpenoid and flavonoid components in the ER. The resin then accumulates in vesicles that fuse with larger ones, which are deposited in the periplasmatic area, whence they diffuse through the cell wall into the subcuticular space of the exogenous structures (Raatikainen et al. 1992). In peppermint, Turner et al. (2000) indicated that associations of smooth ER with both leucoplasts and the plasma membrane bordering the subcuticular storage area of the exogenous structure strongly suggest involvement of ER in transport of the secreted material.

Although it seems probable that ER vesicles play a role in transporting resin components from epithelial cell to the storage area, it is not clear how components synthesized in different organelles mix in specific proportions to produce constitutive compositions characteristic of mature organs in various taxa (Chapter 5). It is possible, as indicated for monoterpenes in peppermint, that important conversions occur after the fundamental components arrive in the storage area. Another intriguing question is how the chemicals in the storage areas move against a pressure gradient; it is assumed that they probably require a number of specialized lipid carriers and transfer proteins. For example, in pines, Lorio and Hodges (1968) showed that the contents of the endogenous canals in the trunk of the tree are under pressure (typically 3–10 bars but sometimes 0–30 bars). This pressure is very sensitive to water stress, which thus can also influence the rate of resin exudation to the outside of the tree. High moisture stress results in low resin exudation pressure, which is important both in tapping the tree to use the resin (Chapter 7) and in the tree's defense against attack by insects such as bark beetles (Chapter 5).

There is much to be learned about the structure of the cellular site of synthesis of terpenoid resin components, their transport to the lumen of the storage area, and their exudation to the outside of the plant. The most definitive information will come from correlated biochemical, molecular, and ultrastructural investigations. Little doubt exists that terpenoid components of resin are continuously compartmentalized, however, first within plastids or cytosol-ER during their synthesis, then in ER membrane-bound vesicles during their transfer and sequestering within the storage structures. The events involved in compartmentation and, ultimately, sequestration of endogenous phenolic resin have been less studied than those of terpenoid resin but are assumed to occur because they appear necessary for cell survival (Douce et al. 1978). Flavonoid aglycones and phenylpropanoids are thought to be synthesized in the ER, but there has been less study of them in relation to the ultrastructure of the organelles. Moreover, there is little information on the ultrastructure of cellular sites of synthesis and export of resins containing both terpenoids and flavonoids, especially when terpenoids occur in a predominantly flavonoid resin.

Internal Resin Secretory Structures

Resins synthesized in internal (endogenous) secretory structures or ducts are common in conifers and tropical angiosperm trees. Such structures typically are lined with parenchyma cells called epithelial cells, within which the resin components are synthesized.

Canals Versus Pockets or Cysts

When the ducts are elongated, they are called canals; when rounded, pockets or cysts (sometimes called blisters); and when enlarged and irregularly formed, cavities. Most anatomists (e.g., Esau 1977, Fahn 1979, Mauseth 1988) have reported that these containers, into which the resin is secreted, may arise by schizogeny, lysigeny, or a combination of both (schizolysigeny). Schizogeny involves the separation of parenchyma cells from each other. After several cell divisions, intracellular space is increased, producing a lumen, with the secretory cells then occurring in an epithelial layer or layers surrounding it (Figure 3-1). The epithelial cells of canals (e.g., *Pinus*) are thin-walled and remain active for a longer time than the epithelial cells of cysts (e.g., *Abies*), which may have thick lignified walls. Thus more resin is secreted initially from canal than cyst epithelial cells. Canals or pockets and cysts may occur in vegetative structures such as roots, stems, and leaves, or reproductive structures such as flowers and fruits. Lysigeny is a process involving breakdown of a mass of mature secretion-filled cells that release the secretion as they degenerate. Despite continued reports of lysigeny, Turner (1999) questioned whether lysigeny is a "false category of gland development caused by misinterpretation of artifacts." On the other hand, schizolysigeny, where a cyst or cavity is initiated schizogenously but increases in size with breakdown of the glandular cells as a result of trauma of some kind, seems likely. Although the location of resin cavities varies in different genera, they may occur in any part of the stem, root, or reproductive organs such as fruit, but are essentially unknown in leaves. The resin is contained within the various secretory structures internally (endogenously) until some injury allows exudation to the outside of the plant.

Canals or pockets seldom occur consistently in genera throughout a plant family, but there are families in which they are common. The large tropical rain forest family Dipterocarpaceae is exceptional in having canals in nearly all genera. The types of resin secretory systems in important resin-producing genera in many conifer and angiosperm families emphasized in later chapters are listed in Table 3-1.

There are two kinds of endogenously secreted terpenoid resin (Chapter 1): constitutive (preformed) and induced (reaction). Constitutive resins, stored in canals, pockets, or cysts in tree trunks, or resins induced by injury and devel-

Table 3-1

Types of secretory structures with examples in important resin-producing genera

SECRETORY STRUCTURES	GENERA
Endogenous canals	*Agathis, Anacardium, Bursera,* †*Canarium, Cathaya,* **Commiphora, Copaifera, Dipterocarpus, Eperua,* **Grindelia, Keteleeria, Larix,* †*Liquidambar, Parthenium, Picea, Pinus, Pistacia, Protium, Pseudotsuga, Rhus,* †*Sequoia,* *Toxicodendron
Endogenous cells and pockets (cysts); cavities often traumatically formed	*Abies, Boswellia,* †*Canarium, Cedrus,* **Commiphora, Cryptomeria, Cupressus, Hymenaea, Juniperus,* †*Liquidambar, Metasequoia, Myroxylon, Protium, Pseudolarix,* †*Sequoia, Styrax, Taxodium, Thuja, Tsuga*
Endogenous laticifers with resin compounds	*Calophyllum,* **Cannabis, Convolvulus, Euphorbia,* *Humulus
Exogenous glandular trichomes or epidermal cells	*Aesculus, Alnus, Anthocercis, Betula, Beyeria,* **Cannabis,* **Commiphora, Elaeagia, Eremophila, Fagonia,* **Grindelia,* **Humulus, Larrea, Mimulus, Newcastelia, Populus,* *Toxicodendron

*Endogenous and exogenous structures occur in different organs; †species may have either canals or pockets in different organs.

oping in traumatic canals or large cavities occur in such copious amounts that they have been used industrially. Examples include resins from Pinaceae, Burseraceae, Fabaceae (or Leguminosae), and Dipterocarpaceae (Chapters 7–10). Induced resin, synthesized in a localized area at the site of injury and accumulated without a secretory structure, benefits the tree defensively, although the saturated wood may be used in some cases. However, humans have induced resin by various tapping procedures and further stimulation by chemicals or fire when only small quantities of constitutive resin are synthesized in the secretory structures.

Endogenous secretory tissues are differentiated in the parenchyma tissue produced by the actively dividing meristematic region, the vascular cambium (Figure 3-1). During growth, cells are produced either inwardly into the xylem (wood) tissues or outwardly into the phloem (bark) tissues. These structures may be produced vertically (axially), tangentially to the axis, or horizontally (radially) in the ray parenchyma. Interconnected canals, often vertical and horizontal, may span both xylem and phloem (e.g., *Pinus*) whereas spheroidal or ovoid cysts are unconnected and much more localized in their occurrence in xylem or phloem (e.g., *Abies*). Pockets or cysts, however, may occur in tangential series (e.g., *Tsuga*). Both canals and pockets can occur at different stages of development of the organ, especially in the primary and secondary bodies of the stem. Because terpenoids are synthesized in epithelial cells lining an intercellular lumen into which they are moved, the resin is sequestered such that the plant's sensitive cellular machinery is protected from poisoning by the resin components.

Conifers

The literature presents a confusing picture regarding the developmental anatomy of secretory structures in some conifers, the terminology for the various degrees of development, and how much structure can be ascribed to wounding in the secondary body. Some of these problems may stem from variation in secretory structures within different taxa and attempts to generalize from inadequate samples. Furthermore, most descriptive anatomical studies were done in the early 1900s (Jeffrey 1905, Penhallow 1907, Hanes 1927, Bannan 1936). More recent studies have focused on ultrastructural analysis of resin-producing cells and were designed to understand the role of various organelles in the synthesis of resin and its transport into the secretory container.

Therefore, a gap often exists between the information available from gross anatomical studies and detailed ultrastructural analyses.

Although most conifers produce resin (Chapter 2), the secretory structures in Pinaceae have been studied most intensively. Despite consensus that all members of Pinaceae develop resin ducts, less is known about differences in

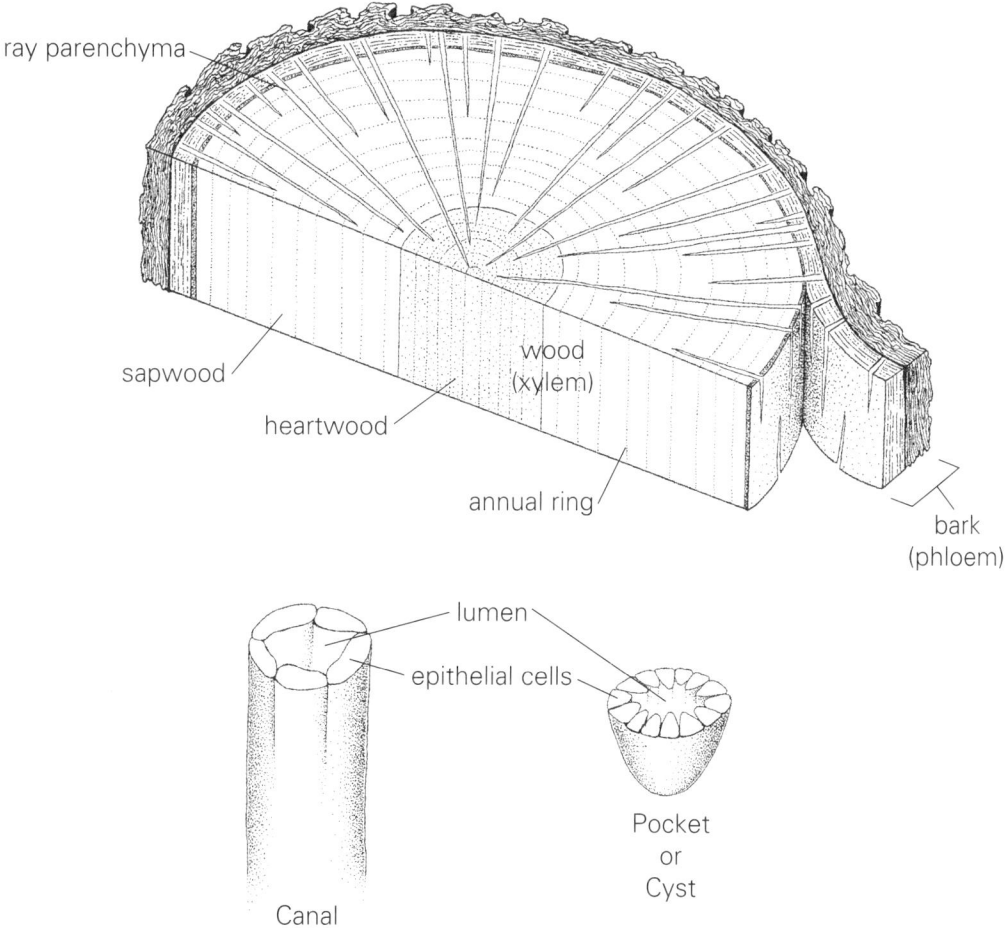

Figure 3-1. Stylized cross section of a tree trunk (secondary stem tissue), showing the vascular cambium, cells from which xylem (wood) and phloem (bark) tissue is differentiated. The inner phloem is constituted of living cells that transport the photosynthate (glucose) in the stem tissue, and the outer phloem is dead corky protective tissue, which is often referred to as bark. Annual rings, comprising spring and summer wood, characterize temperate zone trees. Sapwood is the light-colored layer most active in the transport of water; heartwood is the older, dark wood in the center of the trunk in which water conduction has ceased. Canals and pockets, shown greatly enlarged, occur in either the xylem or phloem, often in the ray parenchyma. Trees generally produce either canals or pockets (the latter also called cysts or blisters). Resin is synthesized in the epithelial cells of the canal or pocket, then transported into the lumen where it is stored.

the structures, how and when they develop in the ontogeny of the tree, the response to wounding, and differences in development between species of a genus. Wu and Hu (1997) provided a rare modern anatomical analysis of resin ducts in stems and leaves of 50 species from 10 genera of Pinaceae. They found differences in the structure and distribution of the duct between genera but no significant differences between species of the same genus, except in *Keteleeria*.

In Wu and Hu's comparative study, resin ducts were found in leaves and the cortex of the primary stem structure in all genera except *Larix* and *Pseudolarix*; resin ducts in the primary xylem only occurred in *Pinus* and *Pseudotsuga*, and vertical ducts were absent in the phloem. Normal, scattered, tube-like canals occurred in the secondary xylem of stems in *Cathaya*, *Larix*, *Picea*, *Pinus*, *Pseudotsuga*, and some *Keteleeria* species but were absent from *Abies*, *Cedrus*, *Pseudolarix*, and *Tsuga* except at sites of injuries. Sac-like cysts filled with constitutive resin in the bark of *Abies* are sufficiently prominent to be called blisters (Figure 8-1); it is this resin that is collected for commercial purposes (Chapter 8). Traumatically induced cysts in *Cedrus* and *Tsuga* occur in tangential series that are isolated or that branch and fuse with others in a series to form an anastomosing network of cavities (Bannan 1936). In fact, such cavities can be extensive in *Pseudotsuga* after attack by the Douglas fir bark beetle *(Dendroctonus pseudotsugae)*. Resin streaks also can develop through accumulation of resin in abnormal amounts in localized areas of injury. In these regions, the resin fills the lumina of cells and permeates the cell wall so that the wood takes on a resin-soaked appearance (Hillis 1987). The kind of duct system, the location of ducts within the tree trunk, and how ducts are influenced by wounding are important in choosing the procedures for maximum yield of resin from taxa that are used commercially.

Although canals generally occur in tangential series in *Larix*, *Picea*, and *Pseudotsuga*, they are neither as abundant nor as evenly distributed as in *Pinus* (Bannan 1936, Fahn 1979). Figure 3-2 shows a cross section of two canals in spring wood of *Pinus ponderosa*; in Figure 3-3 a cryo-scanning electron micrograph gives an enlarged view of the cross section of a canal in *P. strobus* filled with resin. Fast freezing the tissue (McCully et al. 2000) allows preservation of the resin in situ. New secretory tissue is formed from the vascular cambium, placing it close to the active phloem, the source of photosynthetic carbon substrate for resin synthesis. The difference in complexity of the canal system in secondary xylem of *Pinus* contrasted with that of *Larix*,

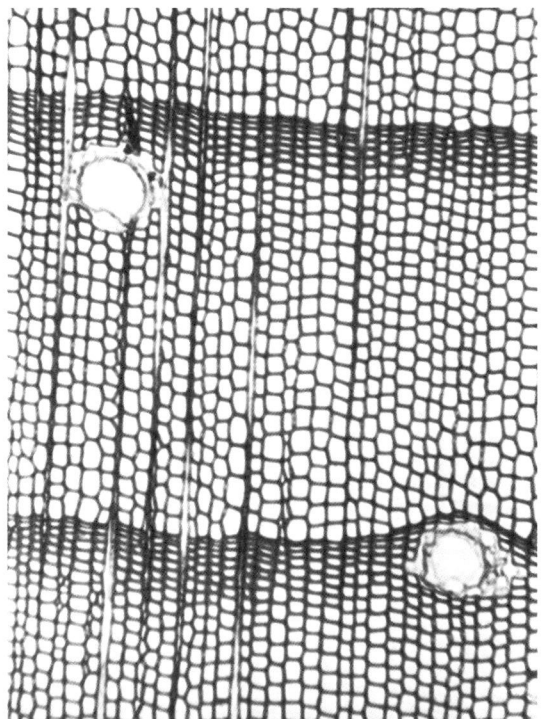

Figure 3-2. Light micrograph of cross section of *Pinus ponderosa* wood, showing normally formed resin canals near the border of early (spring) wood and late (summer) wood, ×120. This relationship is often more pronounced with increased production of vertical canals during the summer months in the temperate zone. From Fahn (1979) with permission of Academic Press.

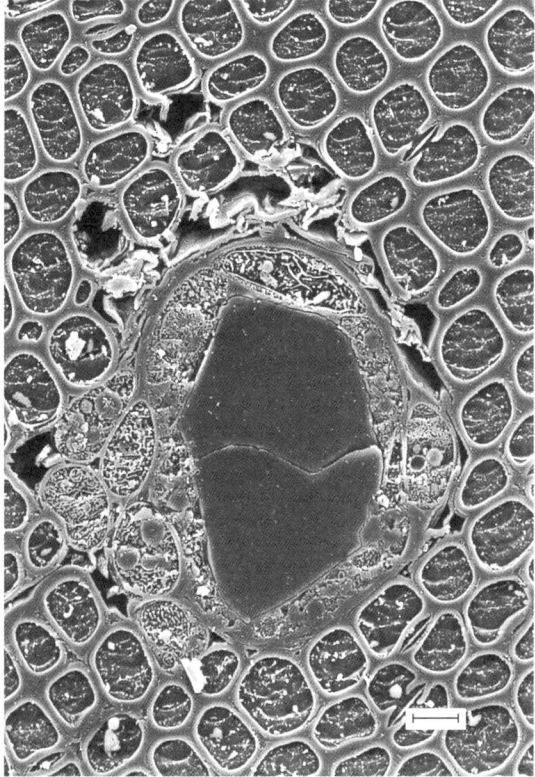

Figure 3-3. Cryo-scanning electron micrograph of a cross section of a canal in young branch tissue of *Pinus strobus,* showing the lumen filled with resin. Fast freezing allows retention of the resin. Scale, 10 μm. Micrograph by M. E. McCully.

Picea, and *Pseudotsuga* led Penhallow (1907) to consider the latter as transitional from a cyst-like structure, and he referred to them as resin passages. In this manner, Penhallow distinguished less developed canals in these genera from the more highly developed canals in *Pinus,* a useful distinction for evolutionary considerations.

Both horizontal and vertical canals occur in the xylem of *Pinus halepensis,* with the vertical canals connected by the horizontal (radial) ones (Werker and Fahn 1969). Only horizontal canals occur in the phloem of *P. halepensis;* numerous plasmodesmata (fine cytoplasmic interconnections) in the primary walls of the epithelial canal cells enable transport of terpene precursors from the phloem to all the resin-synthesizing epithelial cells. Connections do not exist between resin-producing tissues of different organs in either the primary or secondary bodies, or between the canal system of the primary and secondary bodies. This lack of interconnection between organs may be a factor in maintaining resin composition differences between organs, which is determined developmentally and/or by different biotic selection pressures (Chapter 5). McReynolds et al. (1989) summarized studies of the distribution of resin canals in the xylem of pines used for naval stores products. They pointed out that vertical and horizontal canals in the xylem of these pines are interconnected in a radial but not in a tangential direction. Radial canals may intersect vertical ones, but not all vertical canals have a connection with radial ones. Although there may be 2.5–4 times as many radial canals as vertical ones, the number of epithelial cells (where resin is synthesized) is greater and the lumen of the canal is 3 times larger in the vertical canals. Such information is important in planning tapping procedure to obtain the resin (Chapter 7).

There is much less information about resin secretory structures in the Cupressaceae than in the Pinaceae. Some conifer woods are characterized by numerous resin-producing wood parenchyma cells, which Penhallow (1907) called resin cells. These cells are particularly prominent because of the reddish color of the resin in cupressaceous trees, including *Cupressus, Sequoia,* and *Taxodium.* Although the resin comprises primarily terpenoids, the reddish color may result from inclusion of phenolic compounds such as those that produce the characteristic wood color of the coastal redwood *(S. sempervirens). Sequoia* is characterized by widely scattered resin cells that can be aggregated into groups, forming cysts from which red resin teardrops exude

at a site of injury to the trunk of the tree. Cell aggregation, which precedes formation of cysts in *Sequoia*, occurs in *Cedrus* and *Juniperus* following injury
(Figure 3-2). In fact, the trunk of cupressaceous trees may contain relatively
large accumulations of resin in cavities where severe injury has occurred, as
shown by Eocene amber in the trunk of *Metasequoia* (Plate 24). In *Calocedrus* and *Taxodium*, prominent resin cells occur in the wood in well-defined
zones concentric with the growth rings, but in *Cryptomeria* and *Thujopsis*,
resin cells are found scattered throughout a transverse section with little evidence that they form aggregates leading to the development of cysts.

Anatomy in Relation to Resin Production

Monoterpene cyclase activity was analyzed in relation to monoterpene production in constitutive resin in stems of saplings from a range of conifer species showing various degrees of anatomical complexity in their resin secretory
structures (E. Lewinsohn et al. 1991b). Species of *Sequoia* and *Thuja* (Cupressaceae) with less differentiated secretory structures were contrasted to species
of *Abies*, *Larix*, *Picea*, *Pinus*, and *Pseudotsuga* (Pinaceae) with a range of
complexity in the secretory tissue. Although *T. plicata* has only resin cells
whereas *S. sempervirens* also has cysts, both exhibited very low levels of
cyclase activity. On the other hand, *A. grandis*, which also has cysts, showed
increased levels of cyclase activity, and species with an even more developed
canal system (Penhallow's resin passages) such as *L. occidentalis*, *Picea pungens*, and *Pseudotsuga menziesii* had still higher levels. The highest levels of
monocyclase activity, however, occurred in the highly developed, interconnected canal system of *Pinus* species. Species with high rates of monoterpene
cyclase activity generally exhibited the highest levels of total monoterpenes.
Although cyclase activity was correlated with total monoterpene content,
this correlation was less pronounced than that between cyclase activity and
the degree of anatomical differentiation of the resin-secreting structures.
Lewinsohn and coworkers concluded that the general correlation of cyclase
activity and monoterpene content probably reflects the total number of epithelial cells associated with the different secretory structures. Thus the more
extensive duct systems (varying from resin passages to more sophisticated
canals) probably contain the greatest number of cells secreting resin.

To assess the induction of monoterpenes by wounding, the monoterpene
cyclase activity of wounded and unwounded conifer saplings was analyzed

(E. Lewinsohn et al. 1991a). *Abies grandis* and *Picea pungens*, with low to moderate levels of constitutive monoterpene cyclase activity, showed a 5- to 14-fold increase in cyclase activity 7 days after wounding. In contrast, species of *Pinus* with high levels of constitutive cyclase activity did not significantly respond to wounding by altering the level of cyclase activity. Detailed studies of *A. grandis* showed that wound response was localized, and both bark and wood tissues had increased cyclase activity at the wound site, the magnitude of which depended on wound severity. These results support those from earlier ecological studies, discussed next.

Secretory Structures in Relation to Ecology

Some pines (e.g., *Pinus contorta* var. *latifolia*) that occur mostly in early, pioneer stages of forest succession have an extensive canal system centered in the xylem that contains abundant constitutive resin. When wounded, these pines primarily transport constitutive resin from other parts of the canal system to the wound site, and a new suite of compounds is not induced by de novo synthesis (Raffa and Berryman 1982). *Abies grandis,* however, which occurs in mature, old-growth stages, has isolated blisters or cyst secretory tissue in the phloem (Figure 8-1). Upon wounding, the secretory area around the blister increases, or if away from a blister, de novo synthesis of additional monoterpenes commences at the site of wounding, as indicated in the previously discussed cyclase experiments. Raffa and Berryman hypothesized that constitutive resin produced copiously in healthy trees with a well-developed canal system may be adequate for plant defense except during epidemic outbreaks of herbivorous insects. Therefore, little resin may be induced, increased flow still consisting primarily of constitutive resin. Constitutive resin produced in isolated cysts by healthy trees may be inadequate for defense, however; hence, induced or reaction resin in these trees may be produced copiously following wounding by herbivores or infection by microbes. Furthermore, this resin is generally synthesized de novo and the composition differs from that of constitutive resin, often varying in response to the specific attacking organism (Figure 5-3). The consequences of different secretory structures to the defensive strategies of trees that occur in different stages of forest succession are discussed in more detail in Chapter 5.

Norway spruce *(Picea abies)* pretreated by wounding and fungal infection shows highly enhanced resistance to subsequent inoculations with pathogenic blue-stain fungus *(Ceratocystis polonica)* carried by *Ips* bark beetles (Christensen et al. 1999). All trees had a marked increase in the number of resin ducts in the year of inoculation compared to previous years, suggesting that formation of traumatic resin ducts plays an important role in the development and maintenance of enhanced resistance. Resistance of Norway spruce to blue-stain fungus correlates positively with both constitutive flow and the amount of resin induced near the site of injury (Horntveldt 1988; Chapter 5). Nagy et al. (2000) experimentally induced canal development by inoculating *P. abies* with *C. polonica* and characterized the spatial and temporal features of traumatic canal development. Following heavy fungal inoculation, there was massive destruction of the cambial zone in areas close to the wound. Typically, traumatic canals are associated with radially arranged rays (Figure 3-1), but Nagy and coworkers showed how axial traumatic ducts are induced. In contrast to normal axial canals, which are generally single and have a scattered distribution, by 18 days the mature traumatic canal system in the spruce formed a complex of interconnected networks of resin-filled cavities in the tangential plane of new sapwood.

Because the traumatic duct system is not normally formed by the vascular cambium, it requires coordinated reprogramming of cambial-zone cells. Nagy et al. (2000) indicated that initiation of traumatic canals implies both a predisposition to form them and tangential spread of an inductive stimulus for their differentiation. Ray tissue may play a role in transfer of inductive stimuli and nutrients from active phloem. The ultimate benefit of such a structural and metabolic investment is an enhanced defensive potential through massive accumulation of resin and increased resin flow at the injury site.

Polyphenol-rich parenchyma cells in phloem tissue also appear close to the developing traumatic duct system (Nagy et al. 2000). Presumably, this adds to the resin's defense against fungi, as found by Franceschi et al. (2000). Thus the traumatic canal system and surrounding polyphenolic cells may provide a potent protective structure ready to retard both longitudinal and inward spread of pathogenic fungi that attempt to invade following wounding by fungus-bearing bark beetles.

Other considerations of possible ecological importance regarding secretory structure relate to age of the tree and resin flow from radial and axial canals. De Angelis et al. (1986) found that radial resin canals are more abundant, thereby increasing resin flow, in juvenile rather than adult loblolly pine

(Pinus taeda). On the other hand, Blanche et al. (1992) demonstrated greater resin flow from axial than radial canals in adult loblolly pine. These studies demonstrate the need for more comparative analyses of how resin production relates to the development of the secretory structure at different stages of development of the tree. The importance of the type and development of secretory systems for defensive capacity in different conifer taxa is discussed in Chapter 5.

Angiosperms

Even though there are more numerous resin-producing angiosperm genera, their endogenous secretory structures generally are less well studied than those in conifers. Some tropical trees such as the Asian *Dipterocarpus* and New World *Copaifera* produce fluid (oily) resins somewhat similar to oleoresins in *Pinus* (Chapter 7) and also have a canal secretory system. In *Dipterocarpus* and most species of *Copaifera* (Moens 1955, Alencar 1982), axial canals are arranged in complete circles or may be confined to short sectors of the cross section, forming an interrupted ring. Where rings of canals are close together, they often anastomose (Metcalfe and Chalk 1983). The secretory structures of *Copaifera* and *Daniellia* are similar (Guignard 1982). On the other hand, in tropical leguminous trees that produce hard copal-type resins, there are pocket (cyst) secretory structures.

Secretory structures in *Hymenaea* have been studied in some detail and may serve as representative of numerous genera in Fabaceae subfamily Caesalpinioideae (Chapter 2) and of pocket development in other angiosperms. Schizogenously developed ovoid pockets appear early in the ontogeny of the plant, occurring one or two cells under the epidermis of young stem tissue (Langenheim 1967). Since bark tissue is initiated in the cell layer adjacent to the epidermis in the young stem, the secretory pockets formed early among the thin-walled parenchyma cells of the cortex are preserved as the stem grows. In the young stem of seedlings and saplings and in branches, the resin-filled pockets may form a continuous band (Figure 3-4), storing resin that can deter various agents attempting to enter the tissue.

The schizogenous pockets that develop in the cortex in the primary stem tissues, however, progressively disappear with secondary development of the stem (i.e., the woody trunk). If a cut is made into either the trunk or root of the mature tree, resin slowly exudes from undifferentiated xylem tissue in the cambial zone (Figures 3-5 and 3-6). Apparently, on further wounding, there is lysis of the secretory cells, and their subsequent breakdown forms cavities. Thus the exudation of large masses of resin from secondary tissue probably results from schizolysigenous development of secretory tissues. Figure 3-6 shows that during lysigeny, carbohydrates from the cell wall (Figure 1-9) can become mixed with the terpenoids to form a gum resin in *Hymenaea*.

Development of secretory tissue apparently occurs similarly in *Hymenaea* pods and in the secondary stem or root. Initial schizogenously developed pockets form a continuous layer around the seed as well as under the pod epidermis (Figure 5-6). In the African species *H. verrucosa*, the pockets form an exaggerated bumpy or varicose surface, which is different from all the New World species, represented by the widely distributed *H. courbaril*. Synthesis and exudation of the resin can be increased by breakdown of the secre-

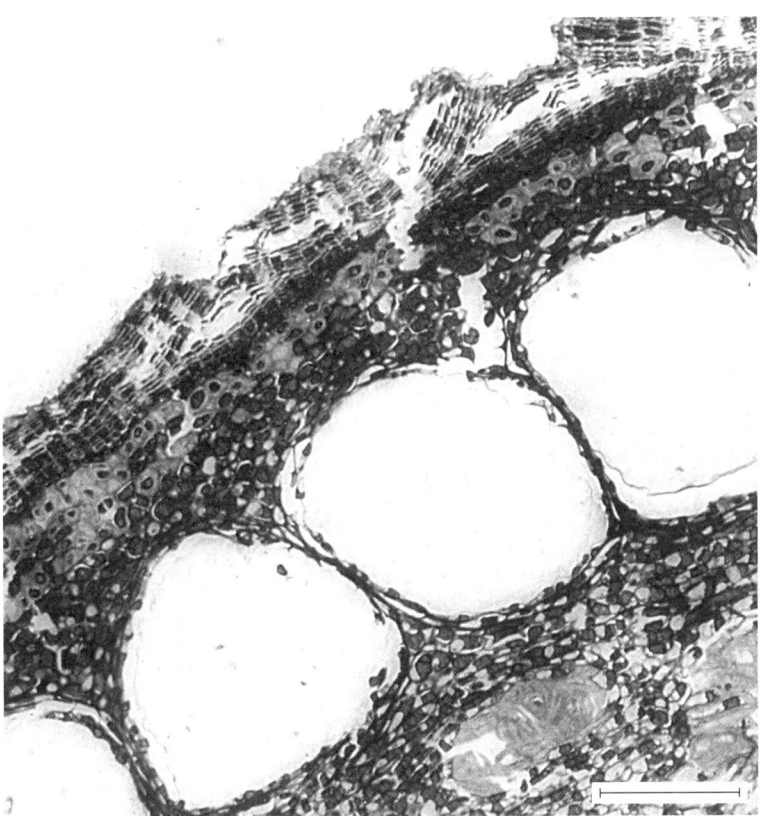

Figure 3-4. Continuous band of pockets in the cortex of a branch of *Hymenaea courbaril*, demonstrating how the location of these structures forms a defensive wall of resin under the developing layer of thin bark. Scale, 50 µm.

Figure 3-5. A cut through the bark of *Hymenaea courbaril* to the vascular cambial region, from which the resin exudes.

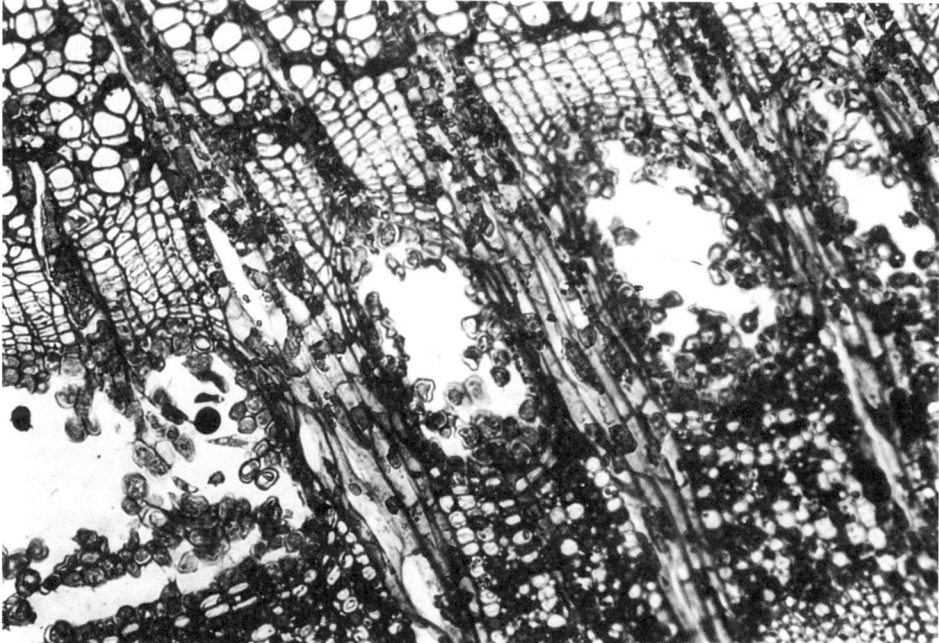

Figure 3-6. Light micrograph of a cross section taken from the cambial region of *Hymenaea courbaril,* showing pockets in the xylem side (lower left) with epithelial cells beginning to lyse to form large areas of secretory cavities. In this manner, cell wall components can be incorporated with the terpenoids to form gum resin.

tory cells following wounding in New World and African species (Langenheim et al. 1978). Fruits have been observed to exude such large masses of resin that they often encase the pod.

Resin pockets in *Hymenaea* leaves are schizogenously produced and cannot be increased in size by traumatic breakdown of the secretory tissue, contrary to the case in secondary stem and root tissue, and pods. Size and shape of the leaf blade pockets vary between species, as do their arrangement at the margin, dispersal through the leaf, and so on. These differences could result from various kinds of defensive selection pressures, such as different feeding patterns of foliovores (Langenheim et al. 1982). Difference in sesquiterpene composition occurs in different parts of the leaf blade as well as during development of the leaf, both of which have ecological implications (Chapter 5).

The secretory structures of Burseraceae have been most intensively studied in species of *Bursera* and *Commiphora*, in subfamily Bursereae, whereas most information regarding genera in subfamilies Protieae (e.g., *Protium* and *Tetragastris*) and Canarieae (e.g., *Canarium*) have been derived coincidentally from ecological or commercial studies. Evidence generally suggests that constitutive resin is produced in canals in both primary xylem and phloem tissue in young stems but occurs primarily in the secondary phloem (bark) in mature trees.

In *Bursera*, Ramos and Engleman (1982) found both vertical and radial canals that form a single network in the phloem of the trunk and branches of *B. copallifera* and *B. grandiflora*. They also reported schizogenously formed canals in the primary body (young plant) of both species, and in *B. copallifera* that radial canals in ray parenchyma of the xylem connect with those in the phloem. Because of the resin's varied uses (Chapter 8), secretory structures and sites of resin synthesis have been analyzed in *Commiphora wightii* (syn. *C. mukul*) in some detail. In the young stem, resin canals develop schizogenously in the phloem region; in the stem, with secondary growth they develop in the vascular cambial zone toward the phloem (Setia et al. 1977, Bhatt 1987). Canals in the secondary phloem are discontinuous, oriented parallel to the longitudinal axis of the trunk, and anastomose tangentially (Bhatt et al. 1989). *Commiphora wightii* is an interesting species in that the leaves and young stems also have glandular trichomes (Ahmed et al. 1969). Having both endogenous and exogenous secretory systems is rare in trees; it is not known whether their co-occurrence in *C. wightii* has any evolutionary or ecological significance.

Since some burseraceous plants produce gum resins, Bhatt (1987) discussed two hypotheses as to how gum (polysaccharides) enters the resins as part of an ultrastructural study of resin secretion in *Commiphora*. It is well established that gum is produced by dictyosomes and that these vesicles may deposit the material from the epithelial cell into the lumen of the resin canal by exocytosis. However, dictyosomes are not commonly seen in resin-producing cells. Alternatively, new cell wall layers are deposited while the outer wall layers of the epithelial cells facing the lumen of the resin canal slough off. The loose and dispersed microfibrillar texture of the outer walls of epithelial cells in *Commiphora* and its appearing to slough off into mature ducts support this latter hypothesis. Although more studies are necessary to fully understand the discharge of polysaccharides into the resin, it is inevitable that during lysis of epithelial cells, either normally or following injury, the degrading cell walls become incorporated into the resinous secretion, as previously suggested for *Hymenaea*.

Studies of the commercial use of some species of *Canarium* (e.g., *C. luzonicum*; Brown 1921) and *Protium* (*P. copal*; Neels 2000) indicate that most constitutive resin is produced in the bark as a result of wounding (Chapter 8). Research on wound closure in Panamanian species of *Protium* and *Tetragastris* demonstrates copious resin production from the cambial zone (Guariguata and Gilbert 1996), which the authors assumed was constitutive. Canals have only been reported in scattered multiseriate rays in the secondary xylem of *Protium*, but schizolysigenous vertical canals occur following wounding (Webber 1941). Thus it seems that most of the resin used commercially from these genera comes from synthesis induced by either pests or humans. More anatomical research is needed to understand the secretory structures and their possible variation between genera in the different burseraceous subfamilies, which are increasingly being utilized (Chapters 8 and 11).

The secretory tissue in guayule (*Parthenium argentatum*, Asteraceae), is very unusual because both resin and rubber (polyterpenes) are synthesized in the epithelial cells of schizogenously formed canals (Gilliland et al. 1988). Previously, it had been assumed that resin and rubber are synthesized in different secretory tissues in guayule because rubber generally occurs in laticifers in other plants. Although both resin and rubber accumulate in canals of leaves, stems, and roots, resin amounts are greater in leaves than stems, and in younger stems during summer compared to older ones in winter. In contrast, rubber production is greater in older stem tissue, where it also accumulates in droplets in the cytoplasm and vacuoles of parenchyma cells surrounding the canal, especially in winter (Schloman et al. 1986). Gilliland and coworkers found resin canals in the pith and cortex of the primary stem tissue, but in secondary tissue the canals occurred in concentric rings from derivatives of the vascular cambium.

Myroxylon, the only resin-producing genus in Fabaceae subfamily Papilionoideae, and *Styrax* (Styracaceae) secrete phenolic resin into schizolysigenous structures following injury. Again, wounding the trunk to induce synthesis is necessary in both genera to obtain sufficient resin for commercial use (Chapter 8). The secretory structures of these phenolic balsams, however, have not been studied in detail, nor has resin production been related to ecological interactions.

Resin in Laticifers

Laticifers are latex secretory structures; they are heterogeneous developmentally and structurally and occur only in angiosperms. Resin is often listed as a component of latex in some families, such as Euphorbiaceae and Convolvulaceae. Since the resin occurs in the mixture constituting the latex, the definition of resin here is not clear. Laticifers are typically classified into two basic types: (1) single-celled, nonarticulated, and (2) many-celled, articulated. The first type is called a laticiferous cell because it originates from a single cell and occurs primarily in young tissue (e.g., *Cannabis*). In fact, in *Cannabis* and *Humulus* (Cannabaceae) these nonarticulated laticifers occur in the outer portion of the primary phloem of the shoot; they are apparently absent in the roots and secondary tissue (Fahn 1979). However, the cannabinoid and hop resins occur in trichomes, discussed under External Resin Secretory Structures. The other type of laticifer, often called a laticiferous vessel, originates from a longitudinal chain of cells in which the walls separating the individual cells become perforated or are completely resorbed, thus forming a tube. In some cases the laticifer is nonanastomosing (e.g., in Convolvulaceae) whereas in others it is anastomosing (e.g., rubber, *Hevea*, in Euphorbiaceae).

Because of their economic potential, especially for rubber, laticiferous plants have been surveyed on a worldwide basis (Archer and Audley 1973). There are 40 families of latex-producing angiosperms, with more than

20,000 species, that have laticiferous structures of some kind (T. Lewinsohn 1991). How many of these laticifers contain resinous compounds has not been documented. Moreover, it is not known if the "resinous compounds" act as resin within the complex latex mixture or if they do so following exudation of the latex (Chapter 1).

There are more tropical than temperate latex-bearing families, and more tropical species both in absolute numbers and proportionally. T. Lewinsohn (1991) thought that the greater diversity of plants producing latex in the tropics may be related to the higher rates of herbivory there (Coley and Aide 1991). In fact, Dussourd et al. (1991) proposed that resin in canals and latex in laticifers may be similarly perceived by insects, each thus acting as a plant defense mechanism (Chapter 5).

Evolution of Internal Secretory Structures

Much remains to be learned about the complex development of resin secretory systems and their variations, let alone their evolution. Explanations for the evolution of different kinds of secretory tissue in Pinaceae created a controversy around the turn of the 20th century. Jeffrey (1905) thought that resin canals in the wood of *Pinus* were ancestral and that the resin cysts (blisters) in *Abies* and other conifers had been derived from *Pinus* by disappearance of the canals during the course of evolution. On the other hand, Penhallow (1907) considered the development of resin canals as a specialization from a primitive condition in which resin cells scattered throughout the wood aggregated to form resin cysts. Resin canals then developed from these cysts. Bannan (1936) supported Penhallow's ideas, suggesting that variation in form and activity of resin ducts imply a phylogenetic series of increasing sensitivity to injury. Jain (1976) agreed with Penhallow's views, based on study of wood from conifers in the Himalaya and fossil data. More recently, molecular studies of the Pinaceae indicate that the Abietinoideae are ancestral to the Pinoideae, thus supporting the view that resin cells and cysts preceded resin canals. Careful study of endogenous secretory structures preserved among now better known Paleozoic conifers (Chapters 2 and 4) could possibly add other dimensions to the understanding of the evolution of these basic types of resin secretory structures.

There is a similar occurrence of both canals and pockets in angiosperm genera within a family, for example, the Fabaceae. In most cases a genus is characterized by one or the other secretory type, but related and often ecologically co-occurring genera such as *Copaifera* and *Hymenaea* have different secretory structures in the tree trunk. Thus, as Penhallow thought for conifers, there may be an evolutionary gradation between the two distinct types that has not been analyzed. In fact, little thought has been given to the evolution of resinous secretory tissues in angiosperms, outside of considering the coevolution of tubular secretory systems (resin canals and laticifers) and specialized insects.

Despite anatomical differences between a resin canal and a latex laticifer, Dussourd and Denno (1991) argued that both form a complex network of tubes throughout the plant, with damage eliciting an abrupt release of chemicals, and thus both could be considered a single defensive syndrome against which some foliovores have adapted by vein cutting or trenching. Tubular systems containing either latex or resin apparently originated at least 40 times (Farrell et al. 1991). They concluded that the functional similarity of diverse canals is suggested by the host ranges of behaviorally adapted insects, some of which attack phylogenetically, morphologically, and chemically diverse plants that are united only by the occurrence of chemicals in tube-like canals. Leaf-feeding insects from 11 families and 3 orders exhibited convergent behavioral counteradaptations to secretory canals of host plants in at least 9 families (in the sense of Cronquist 1981). Farrell and coworkers further concluded that secretory canals are an independently defined, repeatedly evolved characteristic that functions to protect the plant. Furthermore, the canals could be expected to allow plant radiation in an adaptive zone of reduced herbivory. Farrell and coworkers quantified evidence for this hypothesis by comparing diversities of plant lineages that independently evolved canals to their sister groups for as many plant lineages that the taxonomic evidence permitted at the time of their analysis. They demonstrated that the canal-bearing plant lineages consistently had higher diversities than their sister groups and concluded that enhanced individual resistance to enemies is the most probable explanation for these increased diversity rates. There was no apparent trend with respect to age of origin of the lineage, suggesting that selective advantage of canals might persist through time. The studies of Farrell et al. (1991) and Dussourd and Denno (1991) are among the few in which a kind of resin secretory structure is related to groups of interacting organisms with consideration of the coevolutionary implications. Their research entices more research into resin secretory structures from an ecological and evolutionary perspective.

External Resin Secretory Structures

Plant trichomes, or hairs, on plant surfaces were among the first anatomical features recognized and depicted by the early microscopists of the 17th century. Since then, they have become one of the more intensively studied plant characters because of their ubiquity among flowering plants and their extreme morphological variation and function. From the perspective of resin secretion, it is necessary to distinguish those that produce resin as being glandular. The importance of distinguishing glandular from nonglandular trichomes is exemplified by a genus such as *Betula*. Some birch species produce resinous hairs on young stems whereas others in the same geographic region produce nonresinous ones as a pubescence (Taipale et al. 1994). Interestingly, glandular trichomes are generally much more diverse morphologically than nonglandular ones (McCaskill and Croteau 1999). These exogenous secretory structures, in contrast to endogenous ones, do not occur in conifers and are more common in angiosperm shrubs and herbs than trees.

Glandular Trichomes

Resins that are secreted externally from either glandular trichomes or epidermal cells usually coat the surfaces of stipules (which sheathe leaf buds), young leaves, young stems, and/or the floral calyx. Terpenoid and phenolic surface resins, or frequently a mixture of terpenes and lipophilic flavonoids, are produced on plants growing in diverse environments. They occur commonly on shrubs and some herbs in semiarid to arid regions (Table 3-1), especially on young leaves and stems of diverse shrub taxa in Western Australia (Dell and McComb 1978), in the deserts of the American Southwest (McLaughlin and Hoffmann 1982), and in the chaparral of California. Resin-producing plants may well occur in other desert or Mediterranean-climate areas but have not yet been reported. In European Mediterranean vegetation, the shrubs primarily produce essential oils rather than resins. In fact, 49% of essential-oil-producing genera (frequently in the families Lamiaceae and Asteraceae) occur in regions with a Mediterranean climate (Ross and Sombrero 1991). The surface-coating resins are a prominent feature of genera in numerous plant families, including Boraginaceae, Euphorbiaceae, Goodeniaceae, Lamiaceae, Myoporaceae, Sapindaceae, Scrophulariaceae, and Solanaceae in Australia. The Asteraceae in particular, but also Boraginaceae, Scrophulariaceae, and Zygophyllaceae, are noteworthy in North America. Despite the diversity of families, most of these shrubs belong in the asterid groups recognized by the Angiosperm Phylogeny Group (APG 1998; Figure 2-5 and Appendix 2).

Arid-zone shrubs of these plant families produce copious leaf surface resins that may constitute 17–30% of the dry weight of the leaf in Western Australian shrubs (Dell and McComb 1978). The density of the glandular trichomes varies with age of the leaf, with the greatest density generally occurring in early stages of leaf development, decreasing as the leaf expands to full size. Trichomes can reach a density of 30,000 per leaf (e.g., some species of *Beyeria*, Euphorbiaceae) or even double that number in other genera. Although the greatest quantity of resin thus is secreted onto the young leaf, trichomes in some plants continue to secrete significant amounts of resin as the leaf matures. Glandular trichomes occur in the flowers of some Australian shrubs but the secretion is far less significant than on the vegetative organs, whereas the reverse is true for some North American arid-zone shrubs and herbs.

In examining the secretory structures in representatives of many families that produce surface resin in arid Australia, Dell and McComb (1978) found that not all species in a genus produce resin. For example, 39 of the 90 or so species (43%) of *Eremophila* (Plate 21) have resinous leaves and stems. Those with nonresinous leaves usually have a dense pubescent covering of nonglandular trichomes; pubescence often is cited as an antidesiccant strategy of desert plants and would provide an alternative to the resin coating (Johnson 1975). The resin of *E. fraseri,* predominantly diterpenes accompanied by some highly methoxylated flavones, is produced in glandular trichomes that occur at a frequency of about 60,000 per leaf (Dell and McComb 1977). The glandular trichomes develop early during leaf expansion; hence, the young leaf becomes covered by a thick layer of resin. In other species, where the resin remains on the mature leaf, raised stomata protrude through the resin layer, thereby allowing gas exchange.

In several Australian shrubs from various plant families, the glandular trichomes typically consist of one to many cells. The number of head cells (those at the distal end of the stalk, where the resin is secreted) is not constant within a species, and there may be more than one type of trichome on a leaf. The stalk may be short, consisting of only a foot and supporting cell, or long, including one to several stalk cells, as shown by *Eremophila* (Figure 3-7). Long-stalked glandular trichomes are sometimes branched. The head may comprise more than one cell, as in *Anthocercis* (Solanaceae). Head cells in *Eremophila*

(Myoporaceae), *Beyeria* (Euphorbiaceae), and *Newcastelia* (Lamiaceae) are characterized by amoeboid leucoplastids and a proliferation of ER, which Dell and McComb (1974, 1975, 1977) thought are involved in synthesis and secretion of the resin components. Chloroplasts, which probably supply the carbon for terpenoid and flavonoid synthesis, occur in the stalk cells near the secretory cells. Further analysis of these stalk cells, using refined fixation techniques, may show additional features associated with provision of the carbon substrate, as found by Turner et al. (2000) in stalk cells of peppermint.

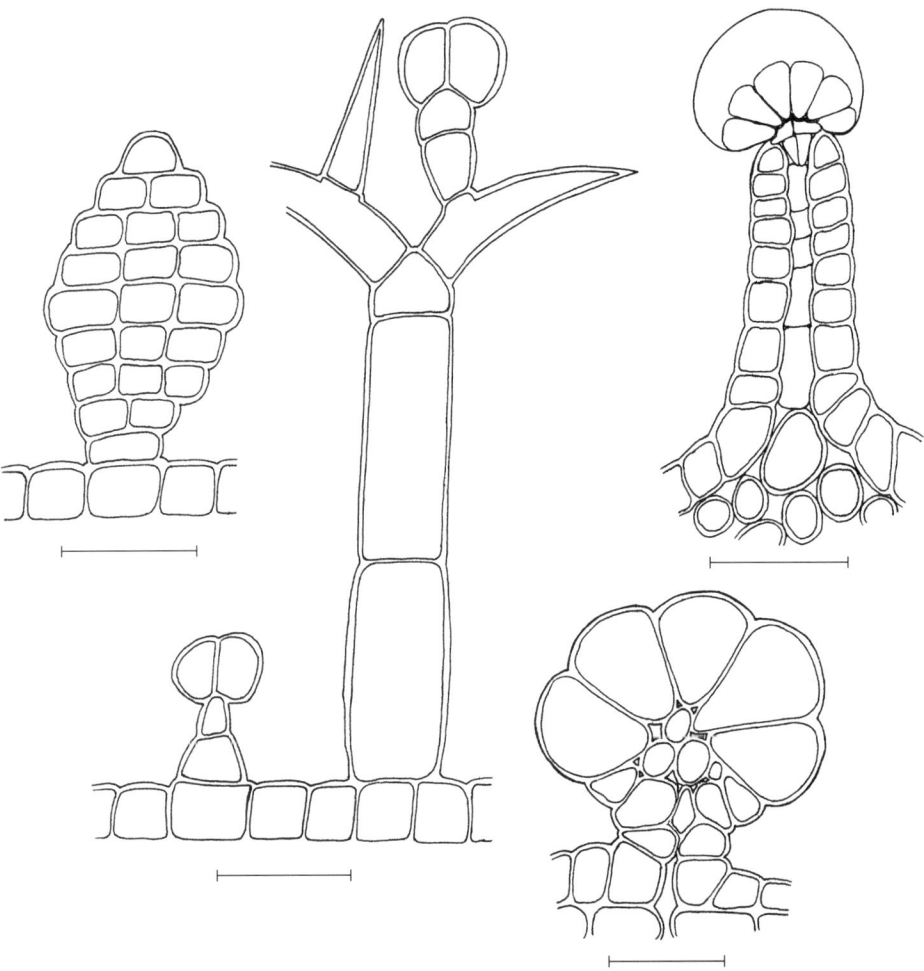

Figure 3-7. Diversity of glandular trichome structures, from shrubs and trees discussed in text. Left, *Anthocercis littorea* (scale, 50 μm); center, *Eremophila leucophylla* with two kinds of trichomes (scale, 50 μm); right above, *Cannabis sativa* (scale, 100 μm); right below, *Alnus incana* (scale, 1 μm). Redrawn from Dell (1977), Clarke (1998), and Wollenweber et al. (1971).

Numerous genera of Asteraceae tribe Astereae produce terpenoid and phenolic surface resins, for example, *Baccharis, Chrysothamnus, Grindelia, Gutierrezia, Haplopappus,* and *Xanthocephalum* (McLaughlin and Hoffmann 1982). *Grindelia,* commonly called gum weed because the plants are usually tacky or sticky to the touch, has been studied in detail. The tackiness is from the mixed terpenoid and flavonoid resin, not true gum as the implied by the name (Chapters 1 and 7). Apparently, *Grindelia* is characterized by a diversity of resin glands, and *G. camporum* has two types of resin-secreting tissues (Figure 3-8): (1) multicellular trichomes in shallow pits on the surface of stems, leaves, and phyllaries (bracts surrounding the flower heads) and (2) canals in the leaf mesophyll and stem cortex (Hoffmann et al. 1984). Resinous trichomes are most abundant on the involucres of the flower heads, densely distributed on leaves, but sparse on stems. Accordingly, resin constituted 20% of dry weight of flower heads, 14% of leaves, and only 2% of stems (Hoffmann and McLaughlin 1986). Furthermore, because resin was easily extracted from *Grindelia* flower heads compared to the stem, Hoffmann and McLaughlin concluded that most resin is produced by trichomes rather than in the few internal canals. Thus flower heads are the most important source of resin for use, with leaves secondary (Chapter 10).

Wollenweber and Dietz (1981) and Wollenweber and Jay (1988) emphasized the significance of accumulation of free flavonoid aglycones in special-

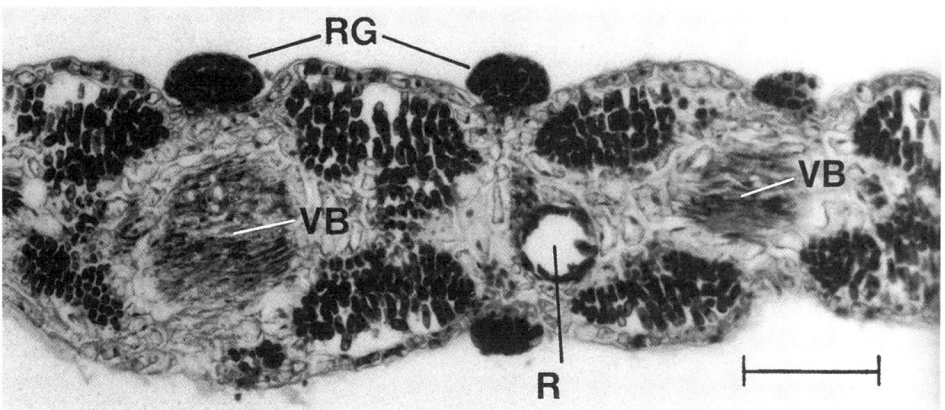

Figure 3-8. Light micrograph of cross section of leaf of *Grindelia camporum,* showing several resin-secreting trichomes (RG) and one resin canal (R); vascular bundles (VB) also indicated. Scale, 100 μm. From Hoffmann and McLaughlin (1986) with permission of the New York Botanical Garden Press.

ized exogenous secretory structures in Asteraceae. Although they particularly mentioned tribe Astereae, genera in other tribes produce aglycones along with terpenoids in epidermal structures. The ecological role of surface resin in *Holocarpha* (tribe Helenieae, subtribe Madiinae) is discussed in Chapter 5.

Both genera of Cannabaceae, *Cannabis* and *Humulus,* secrete resin from glandular trichomes. Although *C. sativa* synthesizes lipophilic material in single-celled laticifers, cannabinoid resins are synthesized in trichomes in vegetative and floral parts of the plant (Figure 10-3). Among the variety of trichomes that occur, Hammond and Mahlberg (1973) and Mahlberg et al. (1984) distinguished three primary types: small bulbous, capitate sessile, and capitate stalked, the latter shown in Figures 3-7 and 3-9. Hammond and Mahlberg pointed out that bulbous and capitate sessile forms occur on both the vegetative and floral axes whereas the highly evolved stalked form is only present on floral organs. Development is controlled independently in each trichome type, emphasizing the complexity of the situation. Figure 3-9 shows individual stalked capitate trichomes on the surface of a floral leaflet. The head cells secrete resin containing cannabinoids and numerous volatile mono-

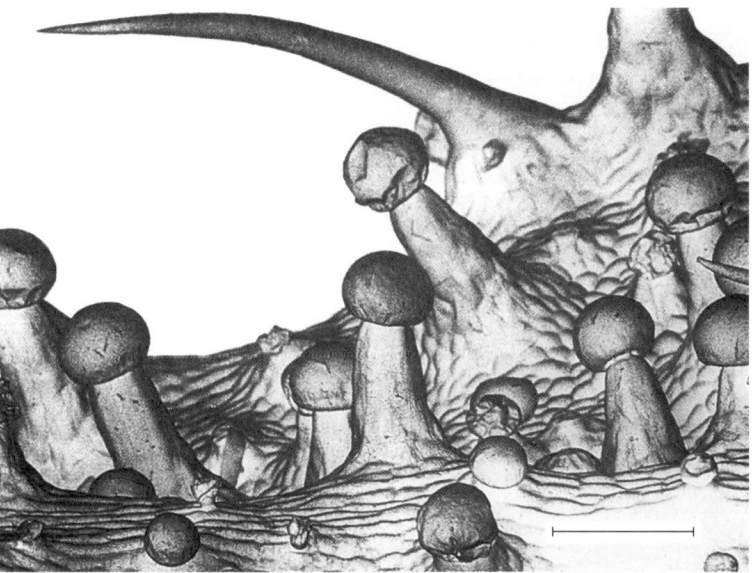

Figure 3-9. Scanning electron micrograph of stalked capitate trichomes on the surface of a leaflet subtending a female flower of *Cannabis sativa*. The head cells at the distal end of each trichome stalk secrete the resin. An elongated nonglandular trichome is at the top in the background. Scale, 100 μm. Courtesy of Robert C. Clarke and the International Hemp Association.

and sesquiterpenes (Chapters 1 and 10). The highest levels of resin (cannabinoids) occur in the capitate stalked trichomes in the flowers and the small bracts interspersed in the floral clusters of the female plant. Although resin in *Cannabis* remains in the glandular head until the structure is ruptured for some reason, mature gland heads detach easily when threshed for collection as resin powder.

In *Humulus,* glandular trichomes occur on female flowers and the bracts subtending the flowers as well as on young stems and leaves. The trichomes on leaves and stems contain smaller quantities of resin than those in the female inflorescences; hence, the floral parts are used for hops (Chapter 10). Because most plants grown commercially for hops develop fruit without fertilization, however, resins are only obtained from the cone-like floral bracts (Figure 10-4). Resins are secreted from two basic kinds of different trichomes: unicellular stinging and hooked hairs, and multicellular clubbed, disk- and cup-shaped, glandular trichomes that can also occur on young stems and leaves (Takahashi et al. 1993). The multicellular trichomes are sometimes referred to as lupulin glands, which can vary, with cup-shaped ones forming on the surface of bracts in the female inflorescence, and disk-shaped ones scattered over young stems and the underside of leaves.

The calyx of *Plumbago capensis* (Plumbaginaceae) has large trichomes that develop from epidermal emergences. Each trichome consists of a multicellular stalk of elongated parenchyma cells and a head with a mantle covered by a thick cuticle. In fully developed trichomes, the resin is secreted into the head, then to the outside by bursting through blisters in the cuticle at several points on the head (Rachmilevitz and Joel 1976). Because fruits are shed from the plant together with a persisting calyx, the sticky resin may facilitate fruit dispersal by animals (Fahn and Werker 1972).

Epidermal Cells and Bud Trichomes

Northern temperate zone trees belonging to the rosid groups in the APG system (e.g., Salicaceae and Betulaceae), produce resins in epidermal cells and bud trichomes of leaves and young stems (Table 3-1). Tropical trees, rare among asterids, also produce bud trichomes (Rubiaceae). The ecological roles of these kinds of resin-secreting structures have been considered.

In contrast to secretion from glandular trichomes, resin exudation in *Populus* (Salicaceae) occurs from epidermal cells, seen as secretory spots on the surface of stipules, and from teeth of young leaves (Curtis and Lersten 1974). The secretory tissue of *P. pyramidalis* and *P. alba* buds consists of palisade-like epidermal cells called prismatic cells. The secreted material is eliminated into a space formed between the outer walls of the prismatic cells and the cuticle covering them, forming a blister or spot. Later, the cuticle bursts and the resin collects between leaves and stipules of the bud (Figure 3-10). Charrière-Ladreix (1973) deduced from optical fluorescence microscopy of *P. nigra* that the resin is not only eliminated to the outside but is secreted intracellularly in the epidermis and extracellularly into the parenchyma below the prismatic epidermis.

The timing of resin secretion in *Populus* is important ecologically. In *P. deltoides*, for example, buds accumulate resin in late summer and secrete it over leaves the following spring, thus enhancing their resistance to leaf beetles

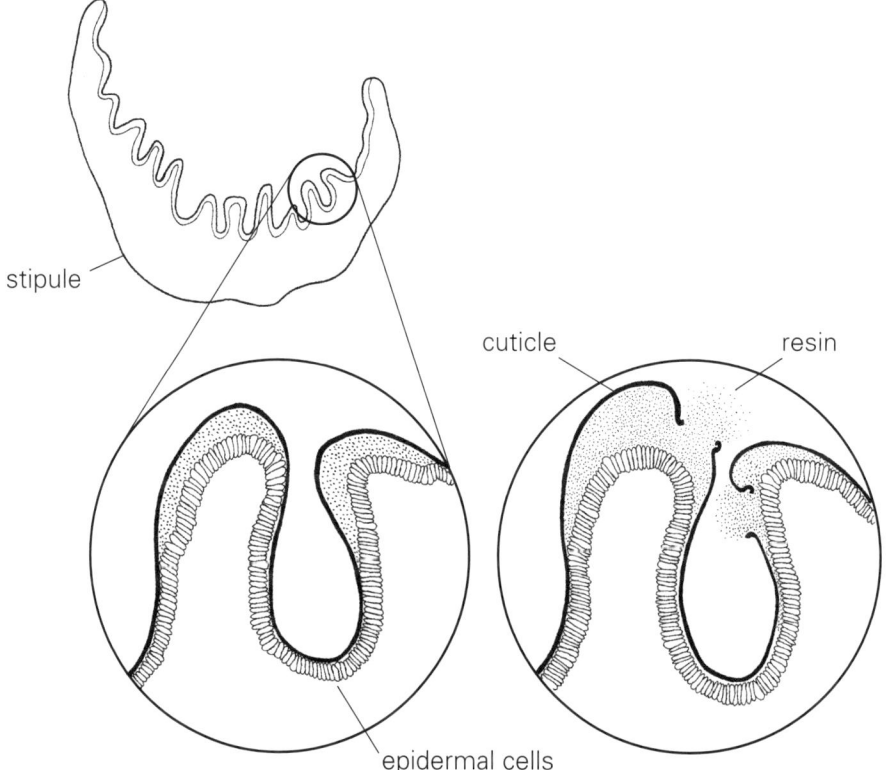

Figure 3-10. Elimination of resin from specialized epidermal cells of *Populus pyramidalis*. Cross section of a stipule, about ×140. Lower left, a portion of a stipule with the cuticle beginning to detach from the epidermal (prismatic) cells as resin collects in the subcuticular space. Lower right, as more resin collects in the space, the cuticle ruptures and the resin flows onto the leaf surface. Redrawn from Fahn (1979).

(Curtis and Lersten 1974) and gypsy moths (G. Meyer and Montgomery 1987). Resistance to gypsy moth declines rapidly as the leaves expand and the resin is diluted, volatilized, or weathered. Furthermore, the timing of the resin's availability to honeybees is important for their nest construction (Chapter 5) as well as to beekeepers who collect it commercially as propolis (Chapter 10).

Bud trichomes of Betulaceae are more similar to epidermal cells than to trichomes, so they are discussed here. Various classes of triterpenoids are major constituents of *Alnus, Betula*, and *Ostrya* resins. Species of these genera, along with some of *Aesculus* (Sapindaceae), also contain methylated flavonoid aglycones (flavones and abundant flavonols). Sometimes mixed with mucilage (Charrière-Ladreix 1973, 1975; Wollenweber and Dietz 1981; Palo 1984), the resins are generally secreted via multicellular glandular trichomes (often called colleters), which may not have stalk cells.

A comparison of bud trichomes and their chemical content on juvenile shoots of six *Betula* species from different geographic areas showed that morphological differences in the secretory structures are often related to chemical differences in the triterpenoid-dominated resins (Taipale et al. 1994). Apical shoots of *B. ermanii, B. pendula, B. platyphylla*, and *B. resinifera* are covered with resin droplets secreted by multicellular peltate glands (Figure 3-11)

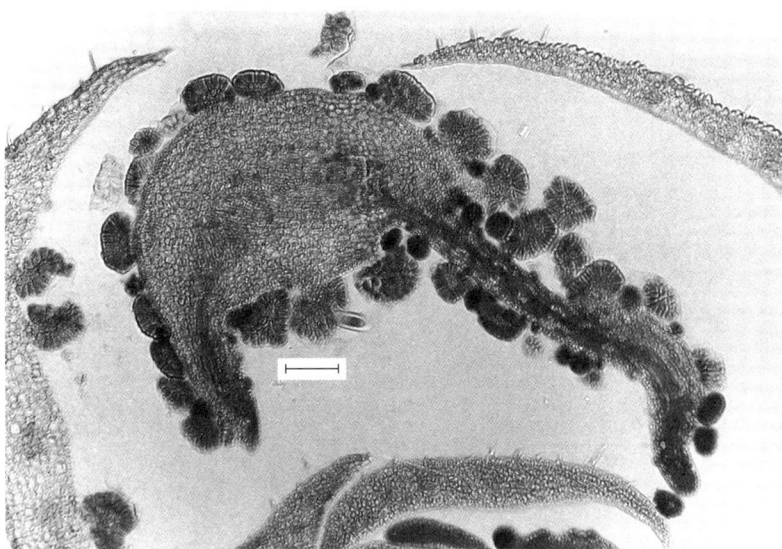

Figure 3-11. Light micrograph of cross section of an apical leaf bud of *Betula pendula*, showing peltate trichomes in which resin is secreted. Scale, 100 μm. From the archives of S. P. Lapinjoki.

whereas those of *B. papyrifera* and *B. pubescens* are covered with long non-glandular hairs that form a pubescence. Interestingly, both resinous and pubescent hairs can occur within a single species, such as *B. pendula*. Resin gland morphology and chemistry are similar in *B. pendula, B. platyphylla,* and *B. resinifera* but are different from those in *B. ermanii*.

Detailed study of the resin secretory tissue in European birch *(Betula pendula)* exemplifies the importance of understanding secretory activity to solve ecological problems arising in commercial use of the plant. Lapinjoki et al. (1991) analyzed resin secretory activity in *B. pendula* because juvenile plants of this species, important as a source of short fiber pulp and for reforestation in Scandinavia, are susceptible to many herbivores. Although resin in birch had been shown to deter hares (Tahvanainen et al. 1991) and other animals that browse on the juvenile plant (Chapter 5), there was marked intraplant variation in resistance (Roussi et al. 1989). Little was known, however, about the relationship between development of resin-secreting structures, changes in resin chemistry, and ontogeny of young organs of European birch.

Lapinjoki et al. (1991) reported that two phases of resin production in *Betula pendula* are related to different stages of growth of juvenile shoots and twigs as well as to defense of these tissues. They found that the resin trichomes, composed of a peltate gland of differentiated columnar epithelial cells (appearing as a palisade-like epidermis) radially surrounding a core of cells, are similar in the young leaf and young stem (Figures 3-11 and 3-12). They suggested that material stored from the preceding season is used for trichome formation and early secretion from densely arranged glands resulting from primary growth. The sticky resin in the buds and along the young stems may serve as an antidesiccant, protecting developing tissues from the harsh conditions of late winter and early spring, and also deter insects and birds when less food is available. Secondary growth in later seasons moves glands farther apart, and resin production depends on large trichomes that are gradually worn off. Large resin droplets on the stem (Figure 3-13), as well as particular triterpenes that lead to winter protection (Chapter 5), are produced only after the most intensive stage of primary growth. These observations helped clarify ecological confusion about differences in palatability for vertebrate herbivores between first-year shoots and twigs of older saplings (Roussi et al. 1989).

The peltate glandular trichomes on young leaves or buds of alder *(Alnus)* possess a short stalk of four or more cells, several intermediate cells, and a flattened head composed of a layer of relatively large secretory cells (Figure 3-7). In ultrastructural studies of the development of the glandular trichomes in *Alnus*, Wollenweber et al. (1971) discovered that some of the trichomes are active in fall, others in spring, a situation perhaps similar to what is found in *Betula*. At the stage of secretion in *Alnus*, the cuticle together with the cutic-

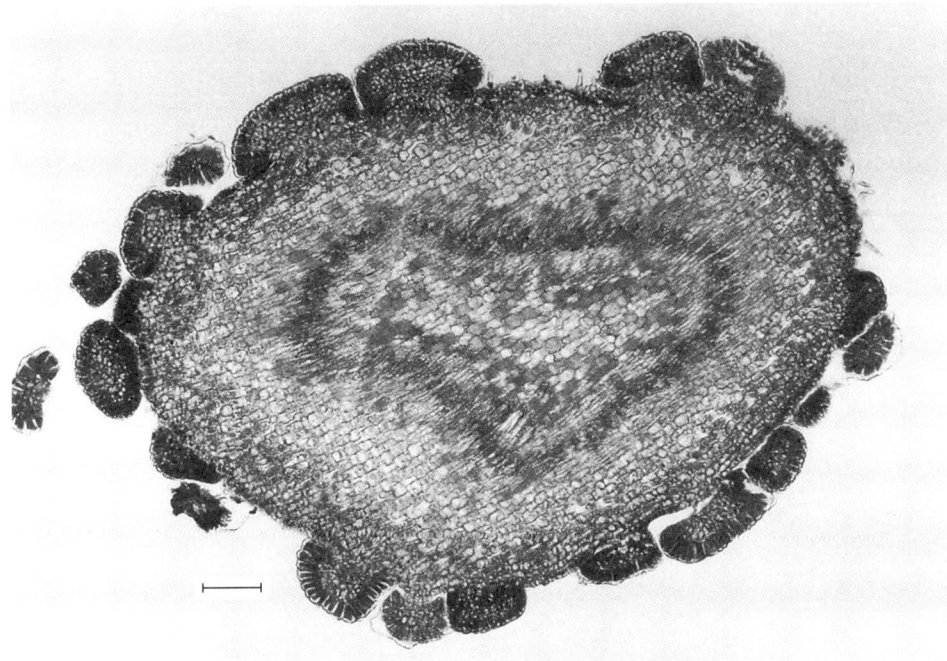

Figure 3-12. Light micrograph of cross section of newly grown internode of *Betula pendula* with the same peltate resin-secreting trichomes as in the leaf buds. The columnar epithelial cells that surround the core of the gland radially are more obvious in this section than in Figure 3-11. Scale, 100 µm. From the archives of S. P. Lapinjoki.

Figure 3-13. Close-up of resin droplets from trichomes along an adventitious shoot of *Betula pendula*. Scale, 1 mm. From the archives of S. P. Lapinjoki.

ular layer splits from the cell wall to form subcuticular spaces that become filled with resin. During the course of this exudation, subcuticular spaces increase and the resin passes through openings in the cuticle and fills the spaces between young leaves and bud scales.

Certain genera of Rubiaceae produce resin in multicellular colleters (Fahn 1979). In some cases the exudate is primarily phenolic resin whereas in others it may be mucilage or may contain both substances. Solereder (1908) described colleters in several species of *Burchellia, Carphalea* (syn. *Dirichletia*), and *Gardenia,* and Mora-Osejo (1977) found that this type of trichome in *Elaeagia* has a structure somewhat similar to that in other rubiaceous species. The resin exudes from this type of trichome at the base or edge of stipules and accumulates at the ends of shoots, forming a spherical cap that encloses the entire bud. The resin thus impregnates leaves and inflorescences, in fact, practically the entire aerial portion of the plant. Krause (1909) thought that the stipular resin decreases transpiration in species of *Gardenia,* observing that some species in moist habitats do not possess colleters. On the other hand, *Gardenia* species in Indian and eastern Asian moist forests produce resin in colleters, with the species producing the most resin growing in drier habitats. On the contrary, Mora-Osejo pointed out that Colombian species of *Elaeagia* in humid montane tropical forests produce the greatest amount of resin. Therefore, he proposed that resin in *E. pastoensis* protects against the insects and microorganisms abundant in these moist tropical habitats. The use of *Elaeagia* resin, as *barniz de Pasto,* is discussed in Chapter 9.

Resin is produced in thin-walled epidermal cells, rather than colleters, in young stems of *Elaeagia* (Mora-Osejo 1977). It also occurs in the interior regions of the secondary cortex, found as well in *Gardenia* (Solereder 1908). Scattered sclerified cells in this region in *Elaeagia* had resin, which Mora-Osejo reported as probably similar to that observed by Solereder in the stems of *Cinchona* and *Coffea.* Apparently, rubiaceous plants contain diverse secretory tissues that also may contain different kinds of chemicals.

PART II

The Geologic History and Ecology of Resins

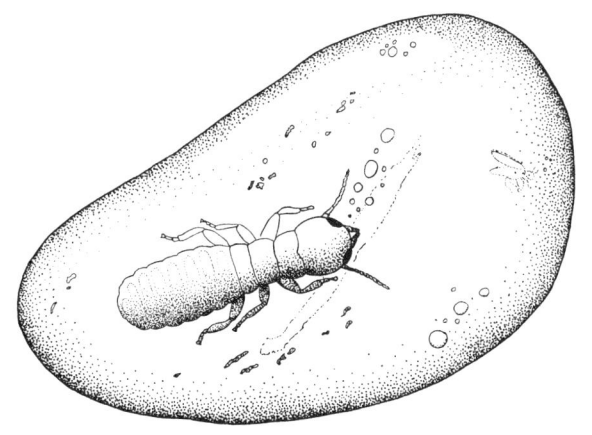

Overleaf: a termite enclosed in a piece of amber.

CHAPTER 4

Amber: Resins Through Geologic Time

A fascinating aspect of resins is the ability of some of them to fossilize and thus be preserved in the geologic record. This is indeed a rarity among secondary chemicals (Chapter 1), and in fact, fossil resin (amber) has even been called the archetype of a chemical fossil (Murray et al. 1998). Furthermore, among sedimentary organic materials, amber is unique in the exceptional preservation of small organisms or parts of larger ones embedded in the resin. These inclusions, together with the beauty of the material, have led to worldwide use of amber in both science and art (Chapter 6).

Varying use of the terms fossil resin, amber, and resinite has led to considerable confusion, unfortunately. In many cases the terms have been used interchangeably whereas in others they have been used with very restrictive connotations. In the older European literature, for example, amber was used in a narrow sense to refer specifically to succinite, the major mineralogically described variety of Baltic amber. In fact, this usage can still be found in the literature in all modern European languages (Beck 1986), probably because of the importance of the extensive Baltic deposits throughout European history. Today, on the other hand, amber is generally considered to be synonymous with fossil resin, with different ambers identified geographically. Resinite is a term also often used by coal geologists to include all fossil resins. To clarify this latter usage, K. Anderson (1997) suggested that amber and fossil resin should be used synonymously for macroscopic material whereas resinite should be reserved for microscopic material. This dichotomy appears to be useful, particularly for research focused on fossil resin as part of the sedimentary record.

How Is Resin Fossilized and When Is It Amber?

Fossilization of resin differs from that of plant parts. For example, the process is neither like the preservation of a film of carbon, as in a compression of fossilized leaves, nor like mineral replacement of the original cellular structure, as in petrified wood. Not all resins fossilize, but fossilization of those that produce large deposits begins with polymerization, that is, the combining of two or more resin monomers to form a complex compound of higher molecular weight. Most ambers are derived from components of terpenoid resins. For example, diterpenes with a labdanoid structure (Figure 1-4) polymerize to form some of the most abundant ambers (Figure 4-1 and Appendix 3). Initial polymerization occurs across terminal groups located on the side chain, resulting in formation of the general polymeric structure (K. Anderson et al. 1992). Because resin is a mixture of compounds, the polymeric structure of amber may involve different monomers, cross-links with nonpolymeric compounds (e.g., succinic acid), as well as occluded compounds (e.g., monoterpenes). Hence, amber is not based on pure or completely consistent polymers, but the predominant monomers in this example appear to be labdatriene acids and/or alcohols. This polymerization apparently is rapid via a free radical mechanism that is photoinitiated when the resin exudes from the plant and hardens as it is exposed to sunlight and air (Cunningham et al. 1987).

Other ambers that occur in large deposits are based on polymers of sesquiterpene hydrocarbons such as cadinene. Unlike diterpenoid ambers, these are substantially soluble in organic solvents and may include triterpenoids. Ambers from phenolic resins that have been analyzed are derived from polymers of styrene (Figure 4-1) and are less common than those with terpenoid polymers. Although most ambers result from polymerization of particular constituents in some resins, small amounts of other ambers result from preservation of nonpolymerizable terpenoids protected by occurrence in fossilized plant parts (Appendix 3; Otto et al. 2002 a, b).

Like other forms of organic matter, the structural characteristics of polymerized resin undergo regular changes over geologic time in response to various conditions (K. Anderson et al. 1992). These progressive changes, reflecting both age and burial history, are considered collectively as maturation. With increasing age, the maturity of any given resin will increase, but the rate at which it occurs depends on the prevailing geologic conditions as well as the composition of the resin. Therefore, maturity and age usually cannot be directly correlated. Changes appear to be a response primarily to geothermal stress since chemical change in the resin accelerates at higher temperatures. Because resins are deposited under very different geologic conditions, there is a continuum of *relative* maturity, beginning with recently deposited resin and extending back to the oldest fossil resin in late Carboniferous sediments.

When in the maturation process does resin become fossilized? This question is one of the most persistent and controversial. The confusion increases with the addition of terms such as young amber, subfossil amber, or subfossil resin, used by some investigators (as well as amber dealers). Reactions in the maturation process, such as cross-linking, isomerization, and cyclization, account for decreased solubility (Clifford and Hatcher 1995) but assertions that maturation to fossil resin takes millions of years (e.g., Poinar 1992a) are

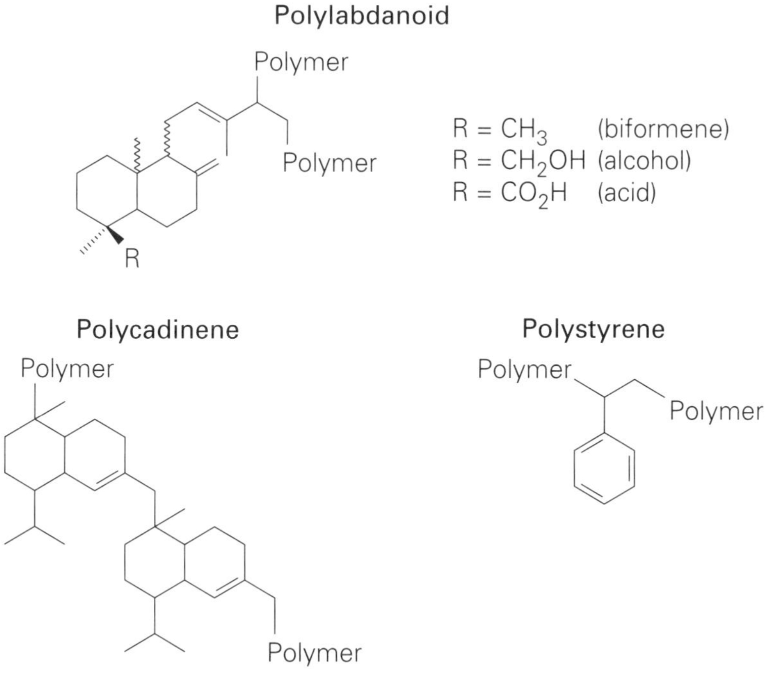

Figure 4-1. Skeletal structures of resin components that form polymers and, hence, fossilize. The macromolecular structures are those that characterize some of the most abundant ambers and provide the basis of a chemical classification of amber (Appendix 3). The polycadinene illustrated is the current model, but additional investigations are necessary to define the precise structural characteristics of these polymers (K. Anderson and Muntean 2000).

unsupported by data and do not contribute to understanding how resin should be distinguished as fossilized. Rice (1987) and Poinar (1992a), however, have made suggestions for evaluating physical characteristics such as hardness, specific gravity, melting point, and solubility, which provide a first estimate of the maturation state of resin as well as help in the determination of imitations of amber. In fact, these are some of the criteria that have been used in describing mineralogical varieties of amber. On the other hand, there is no objective chemical analysis for reliable determination of the relative maturity of fossil resin; hence, alternative criteria for distinguishing recent from fossil resin have been proposed.

To help clarify the terminology used to distinguish the age of resins, K. Anderson (1997) proposed a scale based on carbon-14 dating (Table 4-1), and this terminology is followed in *Plant Resins*. Resins older than 40,000 years are considered fossil. Dates for fossil resin older than 40,000 years would be based on stratigraphic dating of the sediment from which the resin was recovered because 40,000 years is the limit for carbon-14 dating. Inadequate amounts of potassium in the amber preclude using potassium–argon or other trace isotope dating methods for older materials. Gas chronology, however, involving the ratio of argon-40 to argon-39, may offer potential for dating (Landis and Snee 1991). A 40,000-year date for fossilization fits reasonably well within a paleontological framework as there is no firm age definition for the term fossil, and remains of mammoths and other extinct species considered to be fossils are also of the order of tens of thousands of years old. Between 40,000 and 5000 years the resin is considered subfossil. From 5000 to 250 years, it has the status of ancient resin, but less than 250 years, it is considered modern or recent. Although the distinctions are arbitrary and

Table 4-1

Modern versus fossil resin based on carbon-14 dating (K. Anderson 1997)

RADIOCARBON (^{14}C) AGES (YEARS)	TERMINOLOGY
0–250	Modern or recent resin
250–5000	Ancient resin
5000–40,000	Subfossil resin
>40,000	Fossil resin, amber, resinite

doubtless controversial, these ages can be directly and objectively determined based on a small sample, and the scale provides a consistent terminology that should minimize the problems caused by vaguely defined, even undefined, terms (K. Anderson 1997).

Distribution of Amber Deposits

Fossilization of resins allows us to trace the history of their production in trees through millions of years, which is not possible for most plant products. Since not all resins have the requisites for fossilization, amber provides only an incomplete geologic record of the kinds of trees that produced resin. Secretory structures preserved with amber give some idea of when plants began to sequester these secondary chemicals, probably to defend themselves against an increasing diversity of diseases and phytophagous insects. Although precise age determinations are relatively rare because stratigraphic and associated age data depend on the availability of reliable index fossils, amber has been documented from late Carboniferous to late Pleistocene sediments throughout the world, about 310 million to 40,000 years before the present (Figure 4-2 and Appendix 4).

Although hundreds of deposits or occurrences of amber have been found, with more continually being reported by geologists, in most cases the amber occurs in small amounts (Krumbiegel and Krumbiegel 1994). Just as special conditions are necessary for polymerization of resin, special conditions are necessary for accumulation and preservation of large quantities of amber. Therefore, only about 20 deposits are sufficiently rich that they have been mined, and these are of a wide range of ages.

Amber is generally preserved in sediments (sandstones, shales, or mudstones) formed in bays or estuaries fed by weak streams, in deltas, or in the mouths of rivers in continental coastal zones (Figure 4-3). The resin originally exudes from the tree and is deposited in the soil around it, then it gets washed into nearby streams. Because resin tends to be buoyant, it can be carried downriver along with logs from fallen amber-producing trees (which may contain unexuded resin) as well as with other logs from the forest. Over time, sediments gradually bury the wood and resin. The resin becomes amber, and the wood becomes lignite (an early stage in development of coal, still retaining a woody texture). Sometimes amber and resinite occur in later

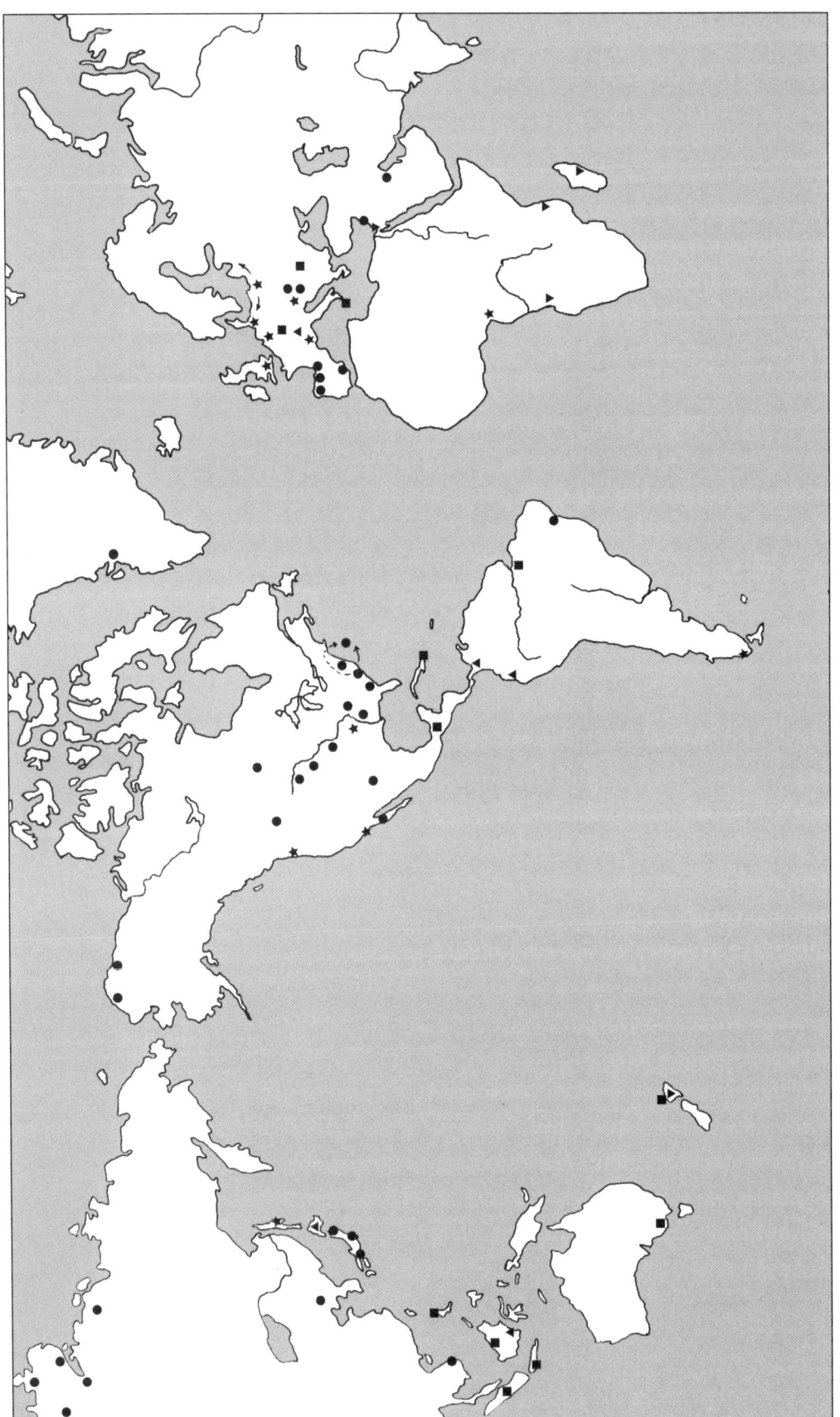

Figure 4-2. Distribution of sizable amber deposits. Geologic age: ▼, Pleistocene; ◀, undetermined Tertiary; ■, Miocene and Pliocene; ★, Eocene and Oligocene; ●, Cretaceous (early, late, and undetermined).

stages of coalification. A rich amber deposit is thus a fortuitous combination of factors, involving forests with trees that produce resin with the chemical requisites for fossilization growing relatively close to a deposition site where resin is collected and concentrated, and burial conditions devoid of oxygen, which would deteriorate the resin. Furthermore, in some cases, amber formed at one site may be washed down and redeposited in another, thereby confusing the age of the amber, based on the deposits in which it is found. The specific conditions determining either the deposition of some of the important amber deposits or their age are discussed later.

Figure 4-3. Conditions for deposition of amber. Amber is generally preserved in sediments formed in bays or estuaries, deltas, or river mouths in a continental coastal zone. Resin that exudes from the tree becomes incorporated in the soil, which gets washed into streams. Buoyant in water, resin is carried downstream with logs, eventually becoming concentrated along ocean shores. There, sediments cover the resin, which becomes amber, and the wood becomes lignite. Mangrove trees, whose stilt roots are excellent for catching material, characterize some depositional sites. Adapted from Grimaldi (1996).

Sources of Amber

Botanical Evidence

The most direct evidence for the botanical source of amber is fossilized wood or other plant organs containing resin in the tissue where it was synthesized. Most often, however, amber is dissociated from the parent tree; therefore, plant remnants contained in the amber and plant fossils in the amber-bearing strata provide circumstantial evidence for the source of the resin. Plant inclusions in amber are generally not abundant, unlike insects (e.g., figure on Part II page). In fact, knowledge of the evolutionary history of some insects has been much enhanced by their abundant and excellent preservation in amber (e.g., Larsson 1978, Grimaldi 1996, Poinar 1992a, Poinar and Poinar 1999). Plant parts found in amber include needles and small conifer cones (Plate 23), small leaves (Plate 25), isolated petals (Plate 26), catkins, stamens, and other flower parts (occasionally a whole flower) as well as mosses, liverworts, and some fungi. Surprisingly, pollen is not as common as expected. When pollen is found, it has lost its characteristic covering (exine) except in rare cases; hence, it is generally not identifiable (Langenheim and Bartlett 1971).

The relatively small number of plant parts compared to insects is unfortunate because few materials other than amber preserve in such great detail. Preservation has been so good that it has even been possible to recognize cell organelles (e.g., nuclei, ribosomes, endoplasmic reticulum, and mitochondria) by transmission electron microscopy of abdominal tissue of a 40-million-year-old fly (Poinar and Hess 1982). This amazing life-like fidelity of preservation probably occurs through rapid and thorough fixation and inert dehydration as well as other natural embalming properties of the resin that are still not understood. These properties apparently vary with the chemistry of different ambers; for example, leguminous resins seem to preserve soft tissues in much finer detail and greater consistency than coniferous ones. At least this seems to be true in some comparisons; for example, inclusions in Baltic amber are less well preserved than those from the Dominican Republic.

Using a technique for exhuming specimens, Grimaldi et al. (1994) examined in situ preservation of whole organs and tissues from inclusions in Dominican amber using scanning and transmission electron microscopy. Incredibly, they found *Hymenaea* leaf epidermal cells (Figure 4-4) and columnar palisade cells, the principal photosynthetic tissue, in their original positions in the leaf (Figure 4-5). Figure 4-6 shows the relationship of the leaf epidermis and palisade cells as well as the other major cell types that occur in a living leaf. In the insects, preserved structures include uncollapsed tracheae and various parts of the gut as well as the brain and bundles of muscle fibers in their original positions and insertions. Specialized pockets in wood-boring beetles (Platypodidae) still possessed spores of their symbiotic fungi. Even beautifully preserved bacterial symbionts, among partially digested wood in the gut of an extinct termite, have been reported from Dominican amber (Weir et al. 2002). Unlike other amber-entombed insects, including other termites, this species *(Mastotermes electrodominicus)* was found preserved with bubbles of methane and carbon dioxide emanating from the breathing holes of their bodies. The gases are a natural product of wood digestion by the bacteria in the termite's gut.

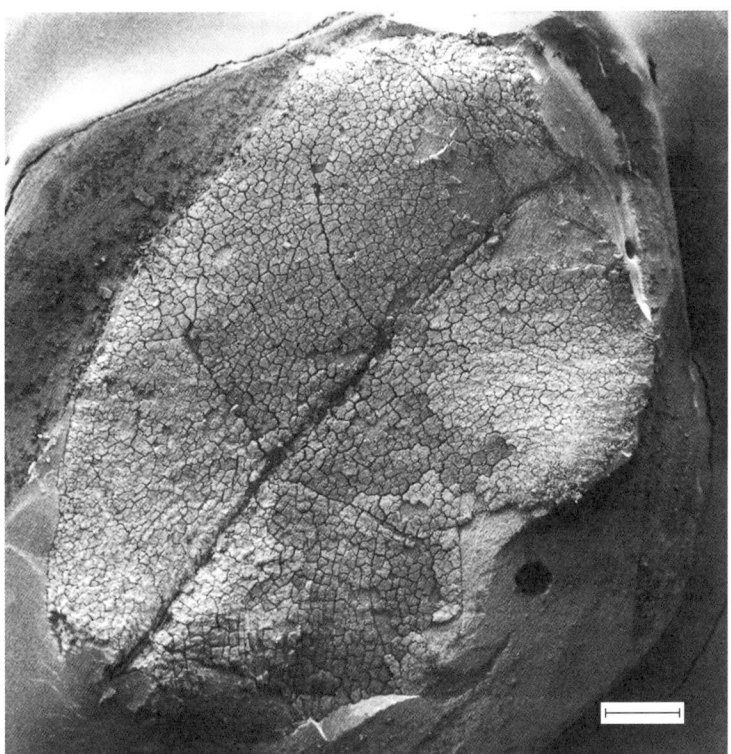

Figure 4-4. Scanning electron micrograph of exhumed Hymenaea leaf in Dominican amber, still showing the patterning of epidermal cells although the surface is dry and cracked. Scale, 1 mm. From Grimaldi et al. (1994) with permission of the American Museum of Natural History. See also Plate 25.

Figure 4-5. Scanning electron micrograph of a group of columnar palisade cells, the principal photosynthetic tissue, in their original size and position under the epidermal layer, from the leaf in Figure 4-4. Scale, 10 μm. From Grimaldi et al. (1994) with permission of the American Museum of Natural History.

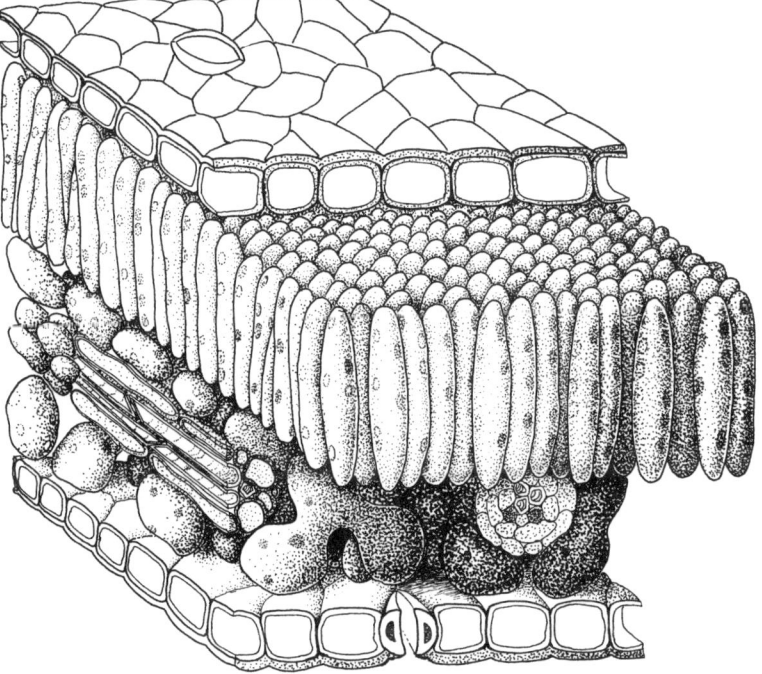

Figure 4-6. Diagram of a leaf, showing the relationship of the tissues exhumed from Dominican amber in Figures 4-4 and 4-5, with epidermal cells (top and bottom), palisade cells (extending beyond the cross section), and spongy mesophyll cells below.

The degree of cellular preservation in amber has excited amber researchers to collaborate with molecular biologists to search for fossil DNA molecules (e.g., Cano et al. 1992, DeSalle et al. 1992). The oldest report of DNA is from a weevil in amber from the early Cretaceous (125–130 Ma) of Lebanon (Cano et al. 1993). Unfortunately, attempts to replicate these results have not succeeded; the DNA from these amber organisms is fragmented, degraded, and strongly cross-linked with other cellular molecules, or it may be a modern contaminant (Austin et al. 1997, Gutiérrez and Marin 1998). Although the DNA sequences reported from a Dominican amber termite (DeSalle et al. 1992) have yet to be corroborated, the nucleotide differences between it and its closest living relative suggest the DNA is authentically ancient. Nonetheless, in lieu of any reproducibility, the prevailing view is that amber has not preserved DNA (Sykes 1997), and DNA may not even be preserved beyond 100,000 years (Wayne et al. 1999).

Chemical Evidence

Although the extraordinary preservation of organisms embedded in amber has long been recognized, there has been little awareness until more recently that fossil resins can preserve details of their own molecular structure better than any other form of sedimentary material. In fact, it has been only since the later decades of the 20th century that amber has been analyzed chemically with the goal of defining it as a plant product, a situation related perhaps in part to the development of relevant technologies. Moreover, the importance of amber as a gemstone led to its originally being characterized chemically as a mineral, and different ambers were given mineralogical names such as succinite, burmite, and glessite among many others. In fact, there has been an increase in the number of new mineralogical names in the literature, such as amekite from Nigeria and bitterfeldite from Germany, indicating an interest in describing different kinds of fossil resins (Vávra 1993). The resins are being described as organic minerals, however, using organic geochemical methods of analysis that potentially provide evidence for determining the botanical source of some of them. The use of mineralogical nomenclature is also considered by some to be better than simply naming according to geographic source. Despite some change during maturation, certain resin polymer constituents are remarkably stable, enabling establishment of chemical and botanical relationships between some modern and fossil resins. On the other hand, as evidence becomes more refined, understanding resin-producing taxa over long periods of time, including possible convergences or parallel evolution of resin polymer structures (Chapter 2), becomes ever more complex.

Because the polymerized terpenoids that constitute the mass of most ambers are largely insoluble, solid-state spectrometric techniques such as infrared spectroscopy (IR), Fourier transform IR (FTIR), nuclear magnetic resonance spectroscopy (NMR), and NMR with cross-polarization / magic-angle spinning (CP/MAS) were primarily used in initial studies of botanical origins (Langenheim and Beck 1965, 1968; Langenheim 1969; Lambert and Frye 1982; Cunningham et al. 1983; Beck 1986). Absolute identities cannot be determined by comparing spectra of modern and fossil resins using these techniques, but a characteristic fingerprint of major constituents can establish relationships at the level of plant family or genus. Identification at the level of species would not be expected. IR remains the most commonly used technique by researchers who do not have access to more expensive, sophisticated equipment.

X-ray diffraction has been used to identify the rare individual crystalline components in resin (Frondel 1967a, b). Structures of the small amounts of volatile mono- and sesquiterpenes trapped in amber can also be determined by gas chromatography–mass spectrometry (GC-MS; Thomas 1969, Grantham and Douglas 1980, Mills et al. 1984, Grimalt et al. 1988, Otto et al. 2002b). Pyrolysis GC (Py-GC) gives more informative fingerprints than FTIR, and combined with MS, Py-GC-MS has enabled even greater analysis of the structure of amber constituents (K. Anderson 1995). Another technique, direct-temperature-resolved MS (DTMS), a sensitive technique for analysis of complex mixtures of insoluble organic compounds, has been added to the arsenal of amber analysis (Grimaldi et al. 2000).

Using Py-GC-MS, K. Anderson et al. (1992) provided a chemical classification scheme for ambers (Appendix 3) that can be related to plant source and also provides some evidence for evolutionary convergence of the skeletal chemistry of the polymers. Diterpenoid labdane polymers (Class I) are particularly important because they characterize ambers occurring in major deposits. Furthermore, this basic labdane structure occurs in both conifer and angiosperm resins and is highly conservative across modern and fossil resins. Three subclasses of labdane polymers have been significant in providing information about the plant source of some ambers. Subclasses Ia and Ib characterize ambers derived from polymers or copolymers of either communic acid (Figure 1-4) or communol, as first recognized by Carmen et al. (1970). The basic polymer is the same but the diterpenoids are cross-linked with succinic acid in Ia, and not in Ib. Succinic acid thus provides an important additional characteristic for determining botanical source. Although the structure of the polymer in Class Ic is the same as in Classes Ia and Ib, it differs in the stereochemistry of the methyl group at carbon 4a and the side chain at carbon 5. Therefore, the polymer is not derived from communic acid but from ozoic acid (Figure 1-4) and/or its hydroxy derivative, zanzibaric acid, as well as biformenes and related isomers, which again can be significant botanically. The labdanoid polymer structure in ambers often contains small amounts of occluded nonpolymeric compounds such as monoterpenes. These terpenoids vary greatly in both modern and fossil resins although the relative proportions can be useful as an additional characteristic in assigning the botanical source at the level of family or genus. Their intra- and interspecific variation in modern resins is thought to be a part of the plant's defense (Chapter 5).

Class II ambers, comprising bicyclic sesquiterpenoids, especially cadinene and related isomers, also form ambers that occur in large deposits. Dimeric cadinenes may cyclize to form bicadinanes (triterpenoids). This combination of terpenoids indicates a different group of plants than those characterized by diterpenoids. Phenolic resin polymers are represented by polystyrene (Class III), which is a good indicator of plants with resin containing large amounts of cinnamic acids and esters. Classes IV and V have characteristic nonpolymer resin components that often are preserved in small amounts (Appendix 3). Class IV includes cedrane sesquiterpenoids (Figure 1-3) found in fossil leaves. Class V ambers, with primarily abietane and pimarane diterpenoid skeletons (Figure 1-4), have occurred in association with fossilized conifer cones and wood, often suggestive of botanical source, and also as small soluble granules in European brown coals (Beck 1999).

Despite increases in knowledge about the chemical properties of resin, there are several dangers inherent in using chemistry as the sole criterion to identify the botanical source of amber. First, in methods that produce a fingerprint, such as IR, different spectra may well indicate different plant sources (although not always the identity of a particular taxon), but very similar spectra do not necessarily mean the same source. Moreover, the older the amber, the more similar the spectra usually become. Second, techniques based on isolation of particular polymerizable compounds must take into account that the same compounds can occur in different resins (Chapter 1). In fact, the occurrence of the same predominant compounds in unrelated plant taxa raises interesting questions of convergent or parallel evolution (Chapter 2). This does not deny that chemical evidence alone, which may be all that is available, can be significant in suggesting botanical origin, particularly at the level of family and sometimes of genus. Concomitant evidence from included or associated plant remains as well as depositional conditions in some cases, however, is often essential for definite identification of the taxon from which the amber was derived. Lack of correlative information in addition to the complexities of evolution within some plant groups result in incompletely solved mysteries regarding the plant source of some significant amber deposits.

Geologic History of Amber-Producing Plants

Here, most deposits in which amber has been studied botanically are mentioned (Appendix 4). However, more attention is given to data that bear on perplexing problems of the plant origins of major amber deposits. The earliest evidence of resin is from arborescent cordaitaleans (early conifer relatives), which together with medullosans (seed ferns) constituted the bulk of seed plants in the widespread late Carboniferous tropical forests that formed extensive coal deposits in Europe and North America. Although seemingly resinous material occurs in parenchyma and "secretory sacs" in the cortex and secondary wood of the ancestral conifers *Cordaites* and *Mesoxylon,* it has not been substantiated as either terpenoid or phenolic resin because of difficulty in isolating the material for chemical analysis. Vertical canals (often occupied by resin rodlets) occur in Carboniferous and Permian medullosans, and IR of some of these rodlets has shown them to be neither terpenoids nor phenolics (Lyons et al. 1982). On the other hand, Lyons and coworkers identified some rodlets dispersed in shales, coal, and coal balls as terpenoid resins, and these are presumed to be from the *Medullosa* or, possibly, *Cordaites.* However, van Bergen et al. (1995) used Py-GC-MS to analyze a sample from a well-characterized medullosan petiole and another from loose rodlets probably derived from medullosans. Rodlets were dominated by phenols in the in situ sample whereas loose rodlets were dominated by (alkyl)naphthalenes, alkylbenzenes, and phenols. These chemicals are unlike those of any known resin, and van Bergen and coworkers concluded that they may constitute a diagenetically altered nonhydrolyzable tannin. They also postulated that the chemistry of these rodlets is specific for Carboniferous medullosans because these plants are unlike any living plant taxa.

Whatever chemicals were present in the secretory tissues of these late Carboniferous plants, the presence of secretory structures is ecologically and evolutionarily significant because they were probably important initially for sequestering toxic chemicals that could then be released on attack by pests (Chapters 3 and 5). Evidence of repair in stem tissue of a very early land plant, *Psilophyton,* suggests that biochemical pathways producing compounds that helped heal plant wounds existed before the Carboniferous (Taylor and Taylor 1993). Land plants may have used toxic resinous chemicals first as protection against fungi (Langenheim 1990, 1994). The pattern of decay caused by wood-decaying fungi in the late Devonian progymnosperm *Callixylon* is similar to that of modern wood rot (Stubblefield et al. 1985). Elaborate networks in Carboniferous wood suggest the activities of beetles (Labandeira 1998), although direct evidence for these insects does not occur until later, in the Triassic (Chapter 5). In fact, wood-boring and bark beetles trapped in amber become prominent much later. There is interest in looking for organismal inclusions in very early fossil resins, such as in late Carboniferous coals from Scotland, because they might provide the early history of some insect taxa (Jarzembowski 1999).

Amber from Conifers

Carboniferous Through Jurassic Conifers

The earliest remains attributable to conifers occur in the late Carboniferous (310 Ma); conifers diversified and were floristically important by the end of the Paleozoic era. Although resin has not been found in contact with these remains, there is evidence of secretory structures. Several genera of late Carboniferous and Permian conifers in Europe are characterized by resin canals (Vogellehner 1965). Of further possible evolutionary importance is that species of late Carboniferous conifers (e.g., *Europoroxylon*) had only vertical canals, like medullosan seed ferns, whereas a Triassic conifer had developed the more complex system of both horizontal and vertical canals, like modern conifers. Although the amount of exudate from endogenous canals or cysts following injury in Paleozoic plants appears small compared to later plants, these older plants apparently had considerable internal storage of resin. Thus abundant microscopic pieces (resinite) led some early coal scientists such as D. White (1914) to suggest that tropical Carboniferous coal floras may have been as "richly productive" in resins as the later (particularly Cretaceous) coal floras. An unanswered question is whether these resins would fit the definition used in *Plant Resins*. Scattered evidence suggests answers of both yes and no.

Although resin canals occur in Permian and Triassic coniferous woods, perhaps only small amounts of resin were produced during cooler and drier climates during these periods. Questionable Permian evidence for amber comes from a limestone in the western piedmont of the Ural Mountains, which contains microscopic quantities of a chemically undetermined resin. On the other hand, small pieces of positively identified Triassic amber have been found in strata in Austria (Vávra 1984). Well-preserved microorganisms, which may have formed a community in the resin-producing tree, have been described from the late Triassic (200–230 Ma) from Bavaria, Germany. Unfortunately, neither IR nor NMR proved diagnostically useful regarding the botanical origin of this amber (Poinar et al. 1993).

More information is available on late Triassic amber from the Petrified Forest National Park, Arizona, although its plant origin is unclear (Litwin and Ash 1991). Detrital amber there occurs in carbonaceous coals that may represent slack water deposits. Preliminary IR data indicated that some of the amber may have been produced by an araucarian-type tree, which traditionally has been considered to be the source of most of the large accumulation of petrified wood *(Araucarioxylon)*. Frustratingly, none of this analyzed amber occurred in direct association with *Araucarioxylon*. Although evidence of amber has generally not been reported in the four densest accumulations ("forests") of petrified wood within the park, there are fragments of wood with amber-like material on it. Turkel (1968) discussed resin secretory structures in *Araucarioxylon*, which he assumed to be araucarian. On the other hand, the wood anatomy of *A. arizonicum* suggests only equivocal relationships to Araucariaceae, clouding the issue (Ash 1987). In sum, a definite botanical source for this North American Triassic amber remains unresolved, and it is quite possible that both the resin and the resin-producing plant are not referable to any extant taxon.

There is little documented evidence of Jurassic amber although it has been reported anecdotally from Bornholm Island and elsewhere in Denmark. Zherikhin and Eskov (1999) provided evidence of amber enclosed in a conifer cone from Jurassic coals in the Republic of Georgia. Moreover, Stockey (1978) showed resin canals in silicified cone scales of *Araucaria mirabilis* from the Cerro Cuadrado Petrified Forest of Patagonia, Argentina, considered to be late Jurassic in age (Stockey 1989). It is interesting that secretory structures, and sometimes small amounts of resin, are reported in cones from Jurassic deposits, a possible early evolutionary indication that resin protected reproductive structures chemically. There has been no evidence of copious production of resin from tree trunks, which occurred later, in the Cretaceous and Cenozoic.

Cretaceous Conifers

Although some modern conifer families had made their appearance by the late Permian and Triassic, all families were present by the early Cretaceous. Thus amber from several modern families could be anticipated during this period. A major climatic warming occurred toward the end of the early Cretaceous, apparently correlated with copious resin production in some areas, as indicated by amber deposits in Austria and the Levantine belt (mountainous areas of Israel, Lebanon, and Jordan). These Middle Eastern ambers constitute the oldest extensive fossiliferous Cretaceous deposits, approximately 125–130 million years old. The numerous organic inclusions are being studied. IR suggests a probable araucarian source for this amber (Appendix 4; Vávra 1984, Nissenbaum and Horowitz 1992). Another early Cretaceous deposit rich in inclusions is in Álava, Basque Country, Spain, for which an araucarian source has also been suggested based on chemistry and pollen (Alonso et al. 2000).

Studies of the botanical origin of Lebanese amber, however, have implicated an extinct conifer, *Protopodocarpoxylon,* in the family Cheirolepidaceae (Azar 2000). This determination was based on leaves impregnated with amber, a female cone associated with the leaves, and wood containing pieces of amber. Leaves of this type occur in the majority of Lebanese amber beds (Azar 2000). Cheirolepidaceae were a large family of Mesozoic conifers, representing a great many different types of plants that inhabited a variety of ecological niches (Watson 1988). The affinities of Cheirolepidaceae remain speculative, with earlier ideas placing this diverse group within the Taxodiaceae,

Cupressaceae s.s., or Araucariaceae based principally on foliage. Taylor and Taylor (1993) indicated that until more information is obtained about the organization and diversity of the ovulate cones, Cheirolepidaceae will continue to stand as a dominant group of Mesozoic conifers whose progenitors and descendants remain equivocal. Interestingly, no cheirolepidaceous wood contains resin canals, indicating that the Lebanese amber found in the *Protopodocarpoxylon* wood must have been formed by resin cells characteristic of the Cupressaceae s.l. or was induced by injury.

By the late Cretaceous (95 Ma) there are numerous amber fossils, particularly in North America (Figure 4-2). For example, along the Atlantic Coastal Plain of the United States (from Massachusetts to Georgia) there are 24 localities at which amber has been found. Plant sources have been suggested for a number of these (Langenheim and Beck 1968; Grimaldi et al. 1989, 2000). Large deposits on Staten Island, New York, were discovered in open pits mined for clay to manufacture bricks. The amber was so plentiful that workers could pile it in barrels and burn it during winter to keep warm. Amber occurs in similar abandoned sandy clay pits in central New Jersey, where the most abundant amber in North American deposits has been found. There is an exceptionally diverse assemblage of organismal inclusions in the New Jersey amber; in fact, it is probably "one of the most significant Cretaceous deposits in the world" (Grimaldi et al. 2000). The diversity of organisms in the amber may in part be attributed to their having occurred in subtropical or warm temperate forests. The amber was deposited in a network of deltaic, slack water streams where peat accumulated in shallow anaerobic basins.

Amber occurs with lignitic remains in these Atlantic Coastal Plain deposits. Grimaldi et al. (2000) reviewed the tortuous and controversial history of the botanical source of these lignites based on anatomy of the wood and associated foliage. They reevaluated the source of the amber in the wood and the foliage as well as the chemistry of the amber. The wood, foliage, and amber remains had been considered to come from three conifer families: Araucariaceae, Cupressaceae s.l., and Pinaceae. However, previous IR and Py-GC suggested an araucarian source for much of the amber in New Jersey as well as generally along the Atlantic Coastal Plain. Grimaldi and coworkers meticulously collected amber from the commonly occurring wood of *Pityoxylon* and confirmed by Py-GC-MS that *Pityoxylon* is the source of most of the New Jersey amber. *Pityoxylon* has been considered a noncommittal coniferous genus, related at various times to the Araucariaceae, Cupressaceae s.l., and Pinaceae because of characteristics shared by these families. Stewart (1983), however, suggested that *Pityoxylon* represents the earliest wood remains of Pinaceae in Cretaceous deposits. The chemistry supported a pinaceous origin for the resin with a predominance of retene, a transformation product of abietic acid (Figures 1-4 and 4-1), which in large quantities is diagnostic of resin from Pinaceae.

Amber was also retrieved from cone scales of *Dammara,* which are juniper-like, or cupressaceous. Early workers (e.g., Hollick and Jeffrey 1909) and Penny (1947), however, maintained an araucarian affinity beyond question. Miller (1977) indicated that *Dammara* and *Protodammara* are taxodiaceous (Cupressaceae s.l.) rather than araucarian and represent a *Cunninghamia* lineage. On the other hand, the picture is not clear-cut in that the small cone scales of *Cunninghamia* may be a derived condition. In fact, Miller (1999) suggested that the taxodiaceous line that includes *Cunninghamia* is evolutionarily grouped with Araucariaceae. Thus the taxodiaceous cone scale-bract complex may still link *Cunninghamia* with Araucariaceae; alternatively, these features may have evolved in parallel. DTMS of amber within the cone scales of *Dammara* indicates a large amount of abietane diterpenoid skeleton, again suggestive of Pinaceae. Disseminated amber similar to that found in the cone scales further indicates that it originated from the same kind of amber tree, at least in the northern range of the Atlantic Coastal Plain. Thus this evidence points away from a direct and primary araucarian origin for these deposits but does not clarify the taxodiaceous–pinaceous dilemma, again demonstrating problems in sorting out coniferous identities during the Cretaceous. The study by Grimaldi et al. (2000) shows that straightforward conclusions are difficult to come by, in some measure probably because of lack of understanding of the complexities of conifer evolution during the late Cretaceous. These problems also are evident in other Cretaceous studies but they have not been evaluated as carefully as those along the Atlantic Coastal Plain.

Amber from the Hanna Basin in southern central Wyoming occurs in extraordinarily thick late Cretaceous and Paleocene sediments of unequivocal stratigraphic superposition. The 20-million-year period between the highest and lowest occurrences of amber is unprecedented in amber deposits (Grimaldi et al. 2001). The amber is preserved as small droplets in three geologic formations spanning the entire time interval. Although the amber is chemically and physically mature, probably because of age and depth of burial, only minor diagenetic changes were detected in the samples. Distinctive conifer cone scales occurred in carbonaceous to lignitic strata, representing fluvial episodes bounded by incursions of epicontinental seas. The Hanna Basin cone scales are similar to those of the Atlantic Coastal Plain for which Grimaldi et al. (2000) suggested a taxodiaceous origin. The chemistry of the amber in the scales from the Atlantic Coastal Plain indicates a pinaceous origin for the resin; the Hanna Basin resin from similar cone scales is sufficiently similar chemically to be considered from the same source. Thus, for both the Atlantic Coast and Wyoming ambers, the chemical evidence does not corroborate a taxodiaceous (Cupressaceae s.l.) origin. What is interesting, however, is the presence of similar cone scales with amber preserved in the secretory tissue, occurring across a large area of North America through 20–30 million years.

Amber of unknown Cretaceous age from Cedar Lake, Canada, was suggested to be of araucarian origin by comparison of IR and FTIR of the amber with resin from present-day *Agathis australis* as well as New Zealand amber from that source. Py-MS further supports this conclusion (Poinar and Haverkamp 1985). However, there was no plant fossil evidence to provide corroboration. Although Araucariaceae are often implicated as the source of Cretaceous amber, Lambert et al. (1996) have generalized to a greater extent based only on NMR of the molecular composition. They concluded that their spectra of amber from Alaska, Canada, France, Greenland, Kansas, Mississippi, New Jersey, Switzerland, and the Middle East "suggests a common paleobotanical source or family of sources related to *Agathis* (Araucariaceae) that had a broad geographical distribution during the Triassic, Cretaceous and Early Tertiary times." This, indeed, is a daring conclusion without supporting botanical evidence.

On the other hand, abundant pollen and associated foliage (including *Sequoia, Sequoiadendron,* and perhaps *Taxodium*) from late Cretaceous localities such as the Alaska Arctic Coastal Plain suggest a taxodiaceous source for some of the amber there (R. Langenheim et al. 1960, Smiley 1966). The paleobotanists in this study concluded that the amber was probably derived from taxodiaceous trees growing close to lakes or coastal swamps. Although IR did not clearly indicate taxodiaceous resin and tended to support an araucarian one, J. Langenheim and Beck (1968) concluded that the amber might be from more than one botanical source. There is considerable variation in IR of modern taxodiaceous resins, with *Sequoia* and *Taxodium* distinct from *Metasequoia* and *Sequoiadendron* (J. Langenheim 1969). Using Py-MS, however, Poinar and Haverkamp (1985) supported a close relationship of this Alaskan amber to *Agathis*-type resin. Poinar (1992a) also noted that *Podozamites* occurs in the amber-bearing beds along with the taxodiaceous remains. *Podozamites* is very similar to present-day *Agathis,* and Stewart (1983) indicated that it could belong to the Araucariaceae. Thus definite conclusions regarding the botanical source of the Alaskan amber remain tenuous.

Other late Cretaceous amber from Tishomingo County, Mississippi, consists of small pieces of amber embedded in fossil wood identified as Cupressaceae s.l. and Pinaceae, thus suggesting both as resin sources. Megafossil remains and IR support the taxodiaceous *Metasequoia* as the only source of mid-Eocene amber from the state of Washington and neighboring British Columbia (Mustoe 1985). Furthermore, amber has been found in situ in large secretory cysts in a *Metasequoia* log (Plate 24) in Eocene deposits from Axel Heiberg Island, Canada. Py-GC-MS shows that this amber has a Class Ib polymer (Appendix 3) similar to that of *Agathis*. These results point out the possibility of confusing taxodiaceous and araucarian sources through the occurrence of communic acid-based polymers (K. Anderson and LePage 1995). Amber from the Fruitland formation (75 Ma) in the San Juan Basin, New Mexico, again provides direct botanical evidence because amber was embedded in a tree trunk identified as taxodiaceous (probably *Metasequoia*).

Several points stand out about Cretaceous conifer amber. First is the complex evolution of modern conifers, characterized by morphological characteristics and, possibly, resin chemistry shared by members of different families. However, Graham (1999) emphasized that it is not only difficult to relate Cretaceous plants to modern taxa but also to reconstruct the fossil North American plant communities because modern analogs do not exist. Macrofossil and pollen evidence for taxa of Araucariaceae, Cupressaceae s.l., and Pinaceae indicate that the trees grew in mixed conifer forests. Graham further pointed out the importance of a diverse conifer element within Cupressaceae s.l., particularly the deciduous taxodiaceous taxa such as *Metasequoia* and *Parataxodium* (extinct) and perhaps those similar to the evergreen *Callitris* (now primarily Australian). Moreover, a *Metasequoia-Taxodium-Sequoia-Sequoiadendron* complex (Chapter 2) was abundant in swampy wetland habitats. These habitats would provide excellent opportunities for deposition and preservation of resin. In fact, these are the kinds of sites where masses of resin are collected for human use today, for example, various kinds of copals (Chapter 9).

Late Cretaceous amber generally occurs in relatively small pieces and has matured through time. Significantly, the presence of Class Ib polymers (predominance of communic acid or communol polymers) in amber characterizes either an araucarian or taxodiaceous resin and may account for some of the confusion regarding plant source when botanical evidence is lacking. Pinaceous resin has also been suggested by the presence of retene, an abietic acid derivative in some ambers. Thus the three families are likely to have produced resin preserved in some sites, but apparently with a predominance of resin from one source at other sites (Appendix 4).

The Tertiary Baltic Amber Mystery

The botanical source of the extensive deposits of early Tertiary (Eo-Oligocene) amber from the Baltic coast and other areas of Europe has long been intriguing. It is important to note that the commonly used designation, Baltic amber, may be misunderstood as implying that the amber only occurs in and around the Baltic Sea. Although the largest deposits of this amber occur in the Baltic area, it was apparently eroded from marine sediments near sea level, carried ashore during storms, and subsequently carried by water and glaciers to secondary deposits across much of northern and eastern Europe (Bachofen-Echt 1949). Thus Baltic amber constitutes the largest and most widespread deposition of amber in the world. The enormous deposits of amber have been a part of human lore since historical records began, and its origin is part of mythology (Chapter 6). It has also been the amber most intensively studied by scientists since the 19th century.

Mineralogically named succinite, 90% of Baltic amber appears to be from one plant source, although some rarer types of European amber, such as glessite, may be from another source. The botanical origin of succinite remains controversial because it has not been possible to resolve the conflict between paleobotanical evidence (from enclosed plant remains) and resin chemistry. Over the years, different coniferous sources have been suggested. Although the amber contains remains of numerous genera of Cupressaceae s.l., members of Pinaceae have classically been considered to be the source of succinite.

Preconceived notions that the amber-producing tree was a pine came from several sources. The idea, however, was formalized by Goeppert (1836) when he designated the amber tree as *Pinites succinifer,* from amber associated with wood. Goeppert used the ending *-ites* to confer an affinity, but not an identity, of the fossil wood with that of living species of *Pinus*. Although Conwentz (1890) recognized that the anatomy of the wood was unlike that of any living pine, he included several species of pine and perhaps spruce under a single name, *Pinus succinifera*. His use of *Pinus* is probably the source of the considerable subsequent confusion that modern species of *Pinus* were the source of succinite. Schubert (1961) found droplets of amber in the resin canals in woody tissue characteristic of extinct pines but unlike that of modern pines, providing the most convincing evidence for a pinaceous source. Larsson (1978) summarized by concluding that an identification as *Pinus* stretches the evidence, and *Pinites* (pine-like fossil wood) is all that can be supported paleobotanically. Nonetheless, some investigators continued to seek a chemical relationship of Baltic succinite with that of modern pine resins (e.g., Rottländer 1970, Mosini and Sampri 1985).

Several lines of chemical evidence strongly negate a relationship to species in the modern genus *Pinus,* as ably summarized by Beck (1999). First, the primary diterpenes, with abietane and pimarane skeletons, in the resin of present-day pines lack the structural prerequisite for polymerization, that is, a conjugated diene (Figure 1-4 and Appendix 3). Thus pine resins are not preserved in a quantity characteristic of polymerized amber. Under anaerobic conditions, however, abietic acid (often the predominant diterpene acid in pine) is decarboxylated and hydrogenated to retene or fichtelite. Retene has been found in amber in Cretaceous Atlantic Coastal Plain deposits. That amber is abundant but occurs in relatively small pieces, not the very large lumps in which some pieces of succinite occur. Retene also occurs in small amounts in old pine logs in brown coal. Second, the soluble and polymeric fractions of succinite are uncharacteristic of pines (Gough and Mills 1972). Carbon-13 NMR of succinite is also consonant with labdane rather than abietane structures (Lambert and Frye 1982).

On the basis of the vegetation, Takhtajan (1969) concluded that there was a subtropical climate in the Baltic region when the amber was formed, a view to which Larsson (1978) subscribed. In fact, Larsson concluded that the European amber trees grew over a large area, presumably as long as subtropical and tropical conditions prevailed. However, he emphasized the inter-mingling of plants in the amber, representing different climates. The boundary separating subtropical and temperate climates ran generally from central England, southern Scandinavia, north of the Baltic states, and through central Russia, thus occurring much farther north than at present. The majority of plants in the rich Paleocene-Eocene London Clay flora demonstrate the transition zone between subtropical and tropical climates in England. Some of the plants in this flora led Reid and Chandler (1933) to surmise that the northern shores of the warm Tethys Sea supported elements of a similar flora from Southeast Asia northwest to western Europe.

The Baltic coast Samland deposits of succinite are so large (even though restricted to an area approximately 140 km east–west and 32 km north–south) that they have been mined commercially on a large scale since the mid-19th century (Chapter 6). Although the amber has been secondarily concentrated in the famous blue earth *(blau Erde)* deposits over approximately 20 million years, the apparently inexhaustible quantities of amber and frequent occurrence of large pieces (used for making art objects) have raised questions as to how or why trees would have produced so much resin. Because the northern pines, which occur in the Baltic region today, do not exude such massive amounts of resin, and wood anatomy indicated unhealthy tissue associated with the amber (Aycke 1835), Conwentz (1890) proposed that the trees suffered from a unique pathological condition. He called this condition succinosis, relating it to succinite. He thought that succinosis resulted from injury caused by agents such as fire or insect and microbial attack. Succinosis was so pervasive, he believed, that the entire forest of amber-producing trees was affected: *"Das pathologische war die Regel, das Normale die Ausnahme,"* the pathological was the rule, the normal the exception. Conwentz presented these ideas so dramatically that they were accepted without questioning the scale of such events over space and time. Moreover, there had been so much focus on conditions in temperate zone pine forests that investigators did not give attention to the enormous amount of resin produced by certain conifers and angiosperms in subtropical and tropical habitats.

By the late 1960s, it seemed clear to numerous researchers that Baltic amber was not derived from modern pines. Furthermore, elegant chemical studies by Gough and Mills (1972) and Mills et al. (1984) demonstrated a relationship between succinite and *Agathis australis* resin. The composition of the soluble and polymeric fractions of succinite (75–85% labdane diterpenoids and only 1.5–4% abietane diterpenoids) is uncharacteristic of pines but closely matches that of *A. australis*. Approximately 80% of succinite is insoluble, and IR of the insoluble fraction of hydrolyzed amber is almost superimposable on that of polymer from *A. australis*. Py-GC-MS confirms that polymers of *A. australis* and succinite are copolymers of communic acid and communol (Appendix 3). Additionally, Mills et al. (1984) reported 19 monoterpenes in succinite that closely reflect those found in fossil *Agathis* from New Zealand. Czechowski et al. (1996) reexamined soluble fractions of amber from Poland, confirming the results of Mills and coworkers. On the other hand, significantly, *Agathis* resin does not contain succinic acid, a characteristic component of succinite. Nonetheless, from this model chemical analysis of Baltic amber, Mills and coworkers concluded that succinite is more araucarian, like *Agathis*, than pinaceous.

Other characteristics of *Agathis* led investigators to consider the possibility of an *Agathis*-like source for succinite (Langenheim 1995). For example, *Agathis* produces resin copiously under natural forest conditions in which the resin becomes incorporated into soil, especially in peat swamps. The well-documented story of *A. australis* (Figures 2-2 and 9-5) as a commercially useful resin-producing tree is particularly instructive in interpreting the massive accumulation of Baltic amber (Chapter 9). Collections of resin from *A. australis*, from surface deposits that accumulated presumably over the past several thousand years, supported a major industry important in the early economy of New Zealand (Chapter 6). The recovered resin alone, taken from the present range of *A. australis* in northern New Zealand, has been calculated as 25,000 kg/km^2 whereas Baltic amber in the great Samland deposits may have accumulated during at least 20 million years over an area as large as 1 million km^2 (Thomas 1970). Thus the vast accumulations of Baltic amber are easily explained by trees producing *Agathis*-type resin in a subtropical or tropical forest. That the northern shores of the warm Tethys Sea in Europe during Baltic amber time apparently supported a flora partly similar to that of Southeast Asia today suggests the possibility of the occurrence of an *Agathis*-type conifer in the region. However, the lack of succinic acid in *Agathis*-type resin, and clear-cut evidence of araucarian inclusions as well as the presence of amber in pinaceous-like wood, prolonged the mystery of the source of Baltic amber.

On Axel Heiberg Island, one of the northernmost in the Canadian Arctic archipelago, plant fossils containing amber were found in a series of well-preserved Eocene swamp forests, perhaps providing an alternative to a *Pinus* or *Agathis* relative as the source of succinite (K. Anderson and LePage 1995). Although the taxodiaceous *Metasequoia* (Plate 11) and *Glyptostrobus* codominated the forest, the pinaceous *Pseudolarix* (Plate 7) and *Pinus* were associated with them. Most importantly for the botanical origin of succinite, Py-GC-MS of amber from cone scales of a fossil *Pseudolarix* revealed succinic acid as a major amber component. *Pseudolarix* is the only known amber-producing conifer containing succinic acid; a relative, *Keteleeria,* also produces the acid but no amber has been recorded for it. Since *Pseudolarix* was producing resin at high latitudes at the same time some of the Baltic succinite was being formed, Anderson and LePage thought that a better understanding of resin production in *Pseudolarix* might help in determining the source of Baltic succinite.

An important question was whether *Pseudolarix,* an uncommon member of the Eocene swamp forest, produced sufficient resin to account for the large accumulations in which Baltic amber occurs. Some evidence that *Pseudolarix* was capable of producing significant quantities of resin is provided by numerous unassociated resin balls with chemistry similar to that occurring in its cone scales. Nonetheless, because *Pseudolarix* only produces resin when injured (Chapter 3), K. Anderson and LePage (1995) retreated to the concept introduced more than a century earlier by Conwentz (1890) that the trees were probably diseased or attacked by insects to produce the large quantities of amber known as succinite. Analysis by Py-GC-MS presented another problem in that that succinite has a polylabdanoid structure based on regular labdanoid diterpenes (Class Ia, Appendix 3) whereas the structure of *Pseudolarix* amber is based on enantio polylabdanes (Class Ic). Furthermore, as in the case of *Agathis,* there are no inclusions of *Pseudolarix* among the many coniferous remains discovered in the amber. Thus succinite could not be directly derived from *Pseudolarix*. However, Anderson and LePage hypothesized from the chemistry that *Pseudolarix* and the taxon that produced succinite may have been derived from a common ancestor, which probably produced a nonspecific polylabdanoid resin (including both regular labdanes and enantio labdanes) that incorporated succinic acid into the polymer.

Paleobotanical and chemical evidence—large masses of resin with inclusions of pinaceous-type wood and cones but with araucarian chemical characteristics—suggested that Baltic amber might have been derived from an ancestor common to members of both families, Araucariaceae and Pinaceae (Langenheim 1995). Miller (1977) discussed the origin of modern conifer families based on analysis of fossil conifer cones and suggested that the older Araucariaceae and younger Pinaceae are sister groups, that is, sharing a common ancestor. In 1999, however, Miller reported evidence that Pinaceae and Araucariaceae diverged from different ancestors within the clade of a Paleozoic family, Majonicaceae, thus apparently foreclosing the intriguing evolutionary explanation that succinite originated from an ancestor shared by the two families.

In sum, no definite conclusion has been reached about the botanical source of the vast deposits of Baltic succinite despite decades of study. It is clear that the amber is not derived from modern species of *Pinus*, but there are mixed signals from suggestions of either an araucarian *Agathis*-like or a pinaceous *Pseudolarix*-like resin-producing tree. Table 4-2 summarizes the pros and cons for each (Langenheim 1995, K. Anderson and LePage 1995), showing how vexing the quest is for a definite answer. Although the evidence appears to lean more toward a pinaceous source, an extinct ancestral tree is probably the only solution.

Table 4-2

Comparison of the pros (+) and cons (−) of an *Agathis*-like versus *Pseudolarix*-like source of Baltic amber

AGATHIS-LIKE (ARAUCARIAN)	*PSEUDOLARIX*-LIKE (PINACEOUS)
+ Labdane polymer same	− Labdane polymer enantiomeric
− No succinic acid	+ Succinic acid present
+ Massive accumulations of resin	− Not known to produce massive amounts of resin
− No *Agathis* remains	− No *Pseudolarix* remains
− No succinite found in araucarian wood	+ Succinite found in some pinaceous wood
+ Associated succinite flora with similarities to extant subtropical and tropical forests of *Agathis* today	+ Diversity of pinaceous components in succinite flora

Other Tertiary Ambers

Although succinite has some intriguing characteristics of *Agathis australis* resin, Oligocene and Miocene ambers from coals and lignites in New Zealand are well documented by plant remains, IR, and GC as coming from *A. australis* (Thomas 1969, 1970). IR of older (Eocene) New Zealand amber is similar to *A. australis* but shows differences in the volatile terpenes enclosed in the polymer net, suggesting a different species of *Agathis*. This kind of evidence emphasizes that the diterpenoid fraction (source of the polymer) of the resin is often more indicative of differences at the level of genus whereas variation in the volatile fraction (generally used in chemosystematic studies of extant plants) is more useful in distinguishing species or populations. Embedded leaves associated with wood in Pliocene amber from Victoria, Australia, support its origin from an Australian species of *Agathis* (Hills 1957).

Fossil remains of Cupressaceae s.s. show the wide distribution of that family in conifer forests during the Cretaceous and Tertiary. Chemical support for amber from this source is provided by cedrane sesquiterpenes (e.g., cedrene and cuperene) characteristic of the family in its restricted sense. Such sesquiterpenes occur in Oligocene amber from Devon, England, and Pliocene amber from Ione Valley, California (Grantham and Douglas 1980). Because these ambers consist of "resinous earthy material," K. Anderson et al. (1992) questioned putting them into their classification system. Ambers from early Tertiary European brown coals have been identified as either cupressaceous or taxodiaceous through association with wood specimens such as *Cupressinoxylon, Juniperoxylon,* and *Taxodioxylon,* but with no chemical support. A phenolic diterpene, ferruginol, has been useful in helping to verify other ambers as being cupressaceous (Otto et al. 2002a).

Members of the largely southern hemispheric Podocarpaceae (Chapter 2) apparently produce small quantities of trunk resin, probably explaining the sole report of podocarpaceous amber. IR and CP/MAS ^{13}C NMR demonstrate that a predominant diterpenoid acid isolated from an extinct member of the Podocarpaceae, *Dacrydiumites,* occurs in amber from mid-Eocene coals of South Island, New Zealand (Grimalt et al. 1988); several characteristic sesquiterpenes were also identified.

Figure 4-7 summarizes reported sources of amber from modern plant families, ranging from the Triassic to the Pleistocene. Although there are secretory structures in conifers in Carboniferous and Permian sediments,

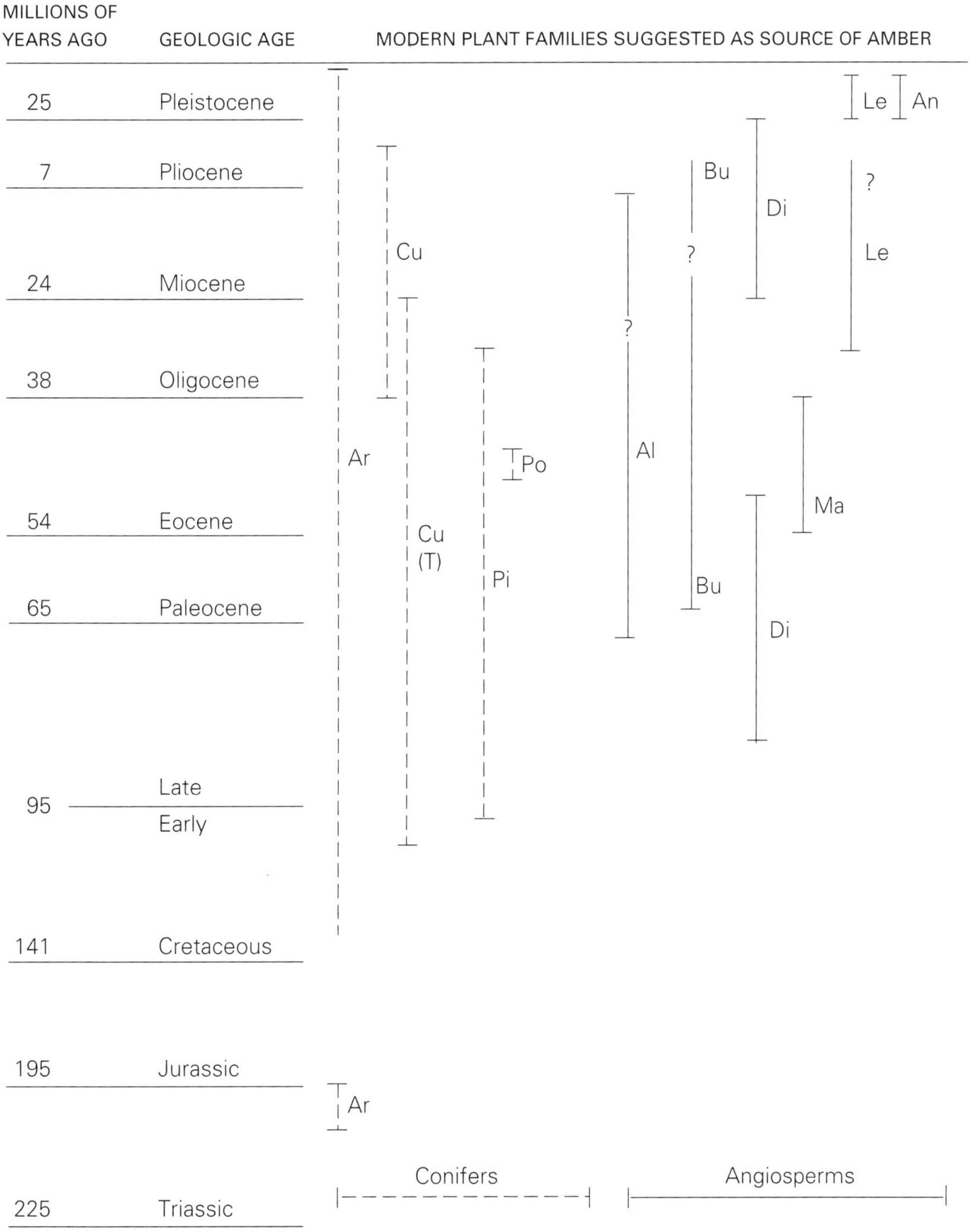

Figure 4-7. Modern plant families reported in the literature as sources of amber through geologic time: Al, Altingiaceae; An, Anacardiaceae; Ar, Araucariaceae; Bu, Burseraceae; Cu, Cupressaceae s.s.; Cu (T), Cupressaceae s.l., including Taxodiaceae; Di, Dipterocarpaceae; Le, Fabaceae (Leguminosae); Ma, Cornaceae (including Mastixiaceae); Pi, Pinaceae; Po, Podocarpaceae. Vertical line segments show the general range of time through which the family has been recorded. Specific time intervals often have not been determined by detailed stratigraphic studies, so the segments only represent approximations.

none of the associated "resins" has been assigned to a member of an extant resin-producing family. In fact, evidence for amber from any modern conifer taxon comes primarily from the early to mid-Cretaceous, around 120 Ma. Araucariaceae (essentially *Agathis*-like resins) have the longest, most extensive record as an amber source (Appendix 4). It is important to recognize, however, that many of these source determinations are based solely on chemical evidence. Despite support from various kinds of chemical analyses, there is a general lack of unequivocal substantiating botanical data for an araucarian source until the Tertiary. Amber from Cupressaceae s.l. (Taxodiaceae in particular) has an impressive record, with some botanical support from the late Cretaceous into the Miocene. This amber was most closely related to resin from *Metasequoia*, which dominated conifer forests then. Terpenoids from resin protected in plant parts, however, extend the Cupressaceae s.s. into the Pliocene. Interestingly, it is possible to confuse araucarian and cupressaceous sources chemically because certain taxa in both families have large quantities of similar diterpenoid labdane polymers; thus support from botanical evidence becomes essential. In Pinaceae, only taxa with labdane compounds that polymerize (e.g., *Pseudolarix*) are likely to produce large quantities of amber, although smaller amounts of resin from plants with a predominance of abietane and pimarane diterpenoids (e.g., *Pinus*) may be preserved under special circumstances. There are ambers considered to be pinaceous from the late Cretaceous into the Oligocene, with the huge Baltic deposits noteworthy in the Eo-Oligocene. There is only one report of amber associated with an extinct member of the Podocarpaceae in the early Tertiary. By the mid-Tertiary, the record of amber shifts to domination by tropical and subtropical angiosperm resin-producing families, accompanied by one tropical conifer, the araucarian *Agathis*.

Amber from Angiosperms

Unusual Late Cretaceous to Early Tertiary Angiosperms

Despite the modest rise of the angiosperms during the earliest Cretaceous (Crane et al. 1995), only coniferous origins for ambers have been suggested until the very late Cretaceous. Ambers representing two angiosperm families, Dipterocarpaceae and Altingiaceae, occur in sediments reported as late Cretaceous and early Tertiary (Paleocene and Eocene). Those of Dipterocarpaceae have a longer authenticated Cretaceous record than Altingiaceae, but because there are more problems associated with positive identification of the dipterocarp source, Altingiaceae are discussed first. Additionally, there is indirect evidence that a member of the Clusiaceae was producing floral resin in the late Cretaceous, related to resin-collecting bees preserved in the amber.

Liquidambar (Plate 13 and Figure 2-8; Altingiaceae, including some former hamamelids) or its immediate ancestor is considered to be the source of amber from several localities along the North American Atlantic Coastal Plain, close to the Cretaceous–Tertiary boundary but primarily in the Paleocene. The chemical evidence is based on the presence of polystyrene (Figure 4-1) and its derivatives, determined through IR (Langenheim and Beck 1968), FTIR and Py-GC (Grimaldi et al. 1989), and Py-GC-MS (K. Anderson et al. 1992). Cinnamic acid (Figure 1-8) and its esters, major constituents of *Liquidambar* trunk resin (Chapter 8), are readily decarboxylated to styrene, which can polymerize and then remain stable for millions of years. Hamamelids apparently occurred in warm temperate to subtropical mixed coniferous forests, which included taxodiaceous amber-producing trees during the late Cretaceous (Graham 1999). However, the genus *Liquidambar* has not been positively identified from plant fossils until the Paleocene. IR of the amber from the late Cretaceous Atlantic Coastal Plain is virtually indistinguishable from Montana amber (Langenheim and Beck 1968), and Py-GC shows it to be almost pure polystyrene (Shedrinsky et al. 1989–91). Authentication of amber as the source of polystyrene is essential, of course, given the occurrence of synthetic polystyrene in the environment. A *Liquidambar* source is further supported for this amber by Muller's (1981) report of *Liquidambar* pollen from Paleocene sediments in the Rocky Mountains. Thus, by inference, amber from the Cretaceous–Tertiary boundary could well be from *Liquidambar* or at least a close relative. In any event, the appearance of *Liquidambar* in North America essentially begins during the Tertiary. The Atlantic Coastal Plain amber is similar chemically to siegburgite, which had been related to *Liquidambar* by isolation of cinnamic acid and styrene. Siegburgite occurs in Eocene–Miocene Rhineland brown coals; Teichmüller (1958) considered *Liquidambar* to be a prominent component in the *Myrica-Cyrilla* marshes and *Sequoia* woods that formed these coals (Appendix 4).

More than 200 angiosperms have been recovered from clay in central New Jersey late Cretaceous deposits, which have also yielded a rich assemblage of insects preserved in amber. Flowers in the clay have been charcoalified, that is, it is assumed that the intense heat of forest fires hardened cell walls, making them impermeable to water. Furthermore, the flowers remained three-dimensional because they were charcoalified before sediments covered them. Although these flowers are not directly related to the amber, an interesting discovery was made in a fossil flower, *Paleoclusia*, a close relative of *Clusia* (Clusiaceae) that occurred in the clay (Figure 4-8). Using scanning electron microscopy, Crepet and Nixon (1998) found secretory cavities or canals in the receptacle, sepals, and petals in these 90-million-year-old flowers that are similar to those of modern Clusiaceae (Curtis and Lersten 1990). *Clusia* is dioecious, bearing male and female flowers on separate plants. In some present-day *Clusia* species, the female flowers have modified anthers that open up and release resin rather than pollen (Plate 30). Stingless meliponid bees collect pollen from male flowers and resin for nest construction from female flowers (Chapter 5). Although no amber was actually found, the *Paleoclusia* female flowers have resin-like material stretched across the openings in modified anthers. Thus *Paleoclusia* is a rare example, suggesting the presence of floral resin relatively early in angiosperm evolution. The occur-

Figure 4-8. *Paleoclusia*, Cretaceous relative of *Clusia*, which produces resin in female flowers from modified anthers, ×68. Note the petal, subtending a fascicle of long and short modified stamen filaments and the five-lobed stigma. Strands, thought to be resin, fill staminodial anthers in other specimens. Courtesy of W. L. Crepet.

rence of meliponid stingless bees in amber in nearby Cretaceous deposits, similar to those from Dominican amber (Figure 4-9), suggests the possibility of a relationship between these bees and clusiaceous flowers, mediated by resin, spanning many millions of years (Crepet 1999).

The Cretaceous and Tertiary Dipterocarp Amber Enigma

Considering the copious production of resin by genera in the tropical family Dipterocarpaceae (Chapters 7 and 8), one might expect coal deposits with massive amounts of dipterocarp amber. Large accumulations of amber in late Cretaceous coal deposits (e.g., Blind Canyon coals in the Blackhawk formation on the Wasatch Plateau, Utah) have been shown by Py-GC-MS to be composed of sesqui- and triterpenoids very similar to those of dipterocarp resins (Meuzelaar et al. 1991). These deposits have as much as 15% resin content, which is very high compared to most coal beds. An associated pollen flora dominated by *Sequoia* (Parker 1976) was deposited in a swamp with a warm temperate to subtropical climate. There is no dipterocarp pollen to support the chemistry, but *Sequoia* neither produces large amounts of resin nor has a chemistry similar to that of the amber. Py-GC-MS shows that amber

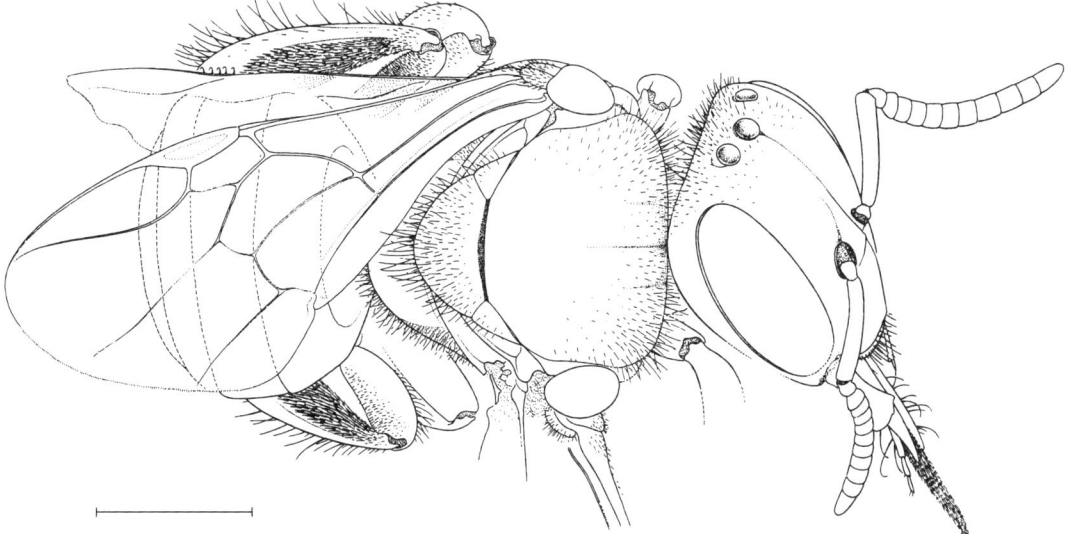

Figure 4-9. Stingless bee *(Proplebeia)* in Dominican amber. Scale, 1 mm. This meliponid bee is relatively common in *Hymenaea* amber; it becomes entangled when attempting to collect resin balls for nest construction (Chapter 5). It also belongs to the same group of bees that probably collected resin from *Paleoclusia*. From Carmago et al. (2000) with permission of the American Museum of Natural History.

in late Cretaceous beds from the Hanna Basin, Wyoming, also has chemical characteristics of Dipterocarpaceae (Grimaldi et al. 2001). The trees producing this amber are thought to have grown under environmental conditions different from most of the Hanna Basin amber, which may be of taxodiaceous origin. Additionally, IR and Py-GC-MS of some amber from the mid-Eocene Claiborne lignites in central Arkansas show that its chemistry is similar to that of resin from *Shorea* (Dipterocarpaceae). Again, dipterocarps were not represented among the primarily pinaceous pollen in these Arkansas lignites (Saunders et al. 1974).

Dipterocarp pollen is not present in the pollen that occurs in the Blind Canyon coals and Claiborne lignites, but that does not indicate that dipterocarps were not present. In fact, the combination of pollen walls that do not preserve well and a lack of distinctive sculpturing limits the value of dipterocarp pollen as fossils (Muller 1970). Moreover, the percentage of dipterocarp pollen is always low, even in present-day forests on peat completely dominated by *Shorea* (Ashton 1982).

Another hurdle for a dipterocarp source for these North American ambers is that the family today occurs predominantly in Southeast Asia, with the subfamily Dipterocarpoideae, in which resin producers occur, restricted there (Chapter 2). There is evidence, however, that the subfamily was not restricted to Southeast Asia in the past. Winged fruits in late Cretaceous sediments on the eastern coast of the United States and the early Cretaceous in England have been compared to the extinct dipterocarp *Woburnia*. Wolfe (1972) reported leaf impressions of *Parashorea* and *Anisoptera* from North American Eocene subtropical forests, genera that today grow in Asian peat swamp forests. In fact, Wolfe (1975) asserted that much of the present Southeast Asian flora can be thought of as a relict of an early Tertiary boreotropical flora that has been largely but not completely eliminated from the New World.

The discovery that the polycadinene polymer structure (Figure 4-1, and Appendix 3, Class II), thought to be distinctive for Dipterocarpaceae, also occurs in resin canals of fruits of an extinct member of the Mastixiaceae (included in Cornaceae) from Germany and England (van Aarssen et al. 1994) initially seemed to provide a tantalizing alternative source for the North American amber. Mastixiaceous plants occurred in the boreotropical flora throughout the northern hemisphere and apparently were common in mid-Tertiary European coal deposits (Jung et al. 1971). Like the resin-producing Dipterocarpaceae, however, resin-producing species of *Mastixia* today are restricted to Indomalesia, and neither mastixiaceous macro- nor microfossils are associated with the extensive North American amber deposits. Furthermore, since mastixiaceous amber is associated with fruits, it is not known if the plants produced the large amounts of resin that dipterocarps do today. Thus mastixiaceous resin clouds the issue and is not more convincing than that from Dipterocarpaceae as the source of North American amber deposits. Similarity in the polymer chemistry of the ambers suggests the possibility of convergent evolution of this polymer, as Dipterocarpaceae belong to the eurosids II whereas mastixiaceous plants belong to euasterids I (Appendix 2, Cornaceae).

In sum, the botanical source of the amber from the North American Blind Canyon, Hanna Basin, and Claiborne deposits remains something of an enigma. From the available evidence, replete with caveats, these ambers appear to have been derived from a dipterocarp source rather than any other. Certainly, they are derived from an angiosperm rather than any known conifer such as *Pinus* or *Sequoia,* as suggested by pollen found in the deposits with the amber.

In contrast to the mystery regarding the amber from the three North American localities, some Miocene coal deposits in Southeast Asia have large quantities of resin chemically similar to that of present-day dipterocarps. Lacking the biogeographic problems presented by the North American deposits, it is reasonable to assume that these deposits are from dipterocarps even though there is no supporting botanical evidence. Various dipterocarps produce enormous amounts of resin, which could easily accumulate in appropriate deposition sites (Chapter 8). In fact, the largest piece of amber in the world, which is similar to dense coal impregnated with resin, comes from Sarawak, Malaysia (Figure 4-10).

The copiously resin-producing genus *Shorea* (Plate 44 and Figure 9-1) has been suggested as the source of Miocene amber from two localities on Sumatra (Appendix 4) based on the similarity of IR of *Shorea* resin and the amber at one site (Langenheim and Beck 1965) and ^{13}C NMR of the amber, indicating characteristic sesqui- and triterpenoids, at the other (Brackman et al. 1984). Using several chemical components to characterize a taxon is more chemosystematic than depending on a single polymer, which may, of course, be important in the fossilization process. Brackman and coworkers investigated this fossil resin because it occurs in sufficiently large amounts, in finely dispersed veins and lenses, that they thought it could be produced in bulk quantity as a sideline of coal processing (Chapter 8). As in coal deposits, fossil resin of probable dipterocarp source, as evidenced by the sesqui- and triterpenes, is a common constituent of Southeast Asian petroleum source rocks (S. Stout 1995). For example, Miocene amber from the Mahakam Delta, Indonesia, is related to Miocene petroleum there and at numerous other Southeast Asian sites (Chapter 9). Although geochemists have begun to recognize the contribution that plant resin can make to crude oil, this role for resin is most prominent in the apparently dipterocarp accumulations in Southeast Asian Tertiary basins.

Figure 4-10. Excavation of the largest single piece of amber in the world, in Sarawak, Malaysia. Mid-Miocene in age, it is assumed to be derived from a member of the Dipterocarpaceae. At more than 68 kg, the piece had to be sawed into several sections to transport it to the Museum für Naturkunde, Stuttgart, where it is on display. From Grimaldi (1996) with permission of Harry N. Abrams, Inc.

Mid–Tertiary Leguminous Amber

Ambers from several mid-Tertiary deposits in the New World, and from the Eocene and Pleistocene in Africa, have been either demonstrated or suggested as having a tropical leguminous source (Figure 4-7 and Appendix 4). Concomitant chemical and botanical evidence from Mexican and Dominican amber as well as the depositional conditions of the Mexican amber definitely support *Hymenaea* as the source tree, unlike the puzzlement over the sources of Baltic amber and some ambers of possibly dipterocarp origin. First, there is compelling chemical evidence from IR (Langenheim and Beck 1965, Langenheim 1969), CP/MAS ^{13}C NMR (Cunningham et al. 1983; Lambert et al. 1985, 1989), and Py-GC-MS (K. Anderson et al. 1992) that *Hymenaea* is the source of amber in large deposits in Mexico and the Dominican Republic. Shedrinsky et al. (1989) and K. Anderson (1995), using Py-GC and Py-GC-MS, further showed that the polymers in the Mexican and Dominican ambers are essentially indistinguishable, both consisting of labdatriene skeletons with an enantiomeric configuration (Appendix 3, Class Ic). Inclusions of plant parts in these ambers also support *Hymenaea* as the source, but possibly different species in the two areas.

Major portions of Dominican amber were once thought to come from deposits of different ages: Eocene, Oligocene, Miocene, and Pleistocene (reviewed by Grimaldi 1995). Although ambers of varying age may exist (Dilcher et al. 1992), Iturralde-Vinent and MacPhee (1996) clearly demonstrated that the primary deposits were formed in a single sedimentary basin during the latter part of the early Miocene to mid-Miocene (15–20 Ma). Deposition apparently occurred near the shore, probably in coastal lagoons adjacent to low, densely forested hills (Figure 4-3) with little evidence of any extensive redeposition. Thus, according to biostratigraphic data based on invertebrate and vertebrate fossils, the major deposition of Dominican amber occurred over a 5-million-year period. This short interval, in contrast to the 20 million years hypothesized for the redeposited Baltic deposits, provides a temporal benchmark that can be used to calibrate rates of molecular evolution in amber-embedded taxa.

All floral parts of *Hymenaea* are included in Dominican amber, with petals characteristic of the African *H. verrucosa* type particularly abundant in the major deposits (Plate 26; Hueber and Langenheim 1986). Poinar (1991) described this Dominican amber species, with its African affinities, as *H. pro-*

tera. If some of the smaller deposits of Dominican amber in the western part of the country are Pleistocene in age, they were not produced by *H. protera* but could have been produced by *H. torrei.* That species has characteristics of both *H. verrucosa* and *H. courbaril* but occurs only in Cuba at present (Lee and Langenheim 1975). If the amber is even younger than Pleistocene, the resin was probably produced by *H. courbaril,* the only *Hymenaea* species that occurs in the Dominican Republic today. However, *Hymenaea* floral parts are not known from Pleistocene or younger deposits; hence, the relationships of the *Hymenaea* species producing that amber are unknown.

Although the only large deposits of amber in the Greater Antilles occur in northern areas of the Dominican Republic, there are trace and small amounts from the early to mid-Miocene in Haiti and northeastern Puerto Rico (Iturralde-Vinent and Hartstein 1998). It is assumed that these ambers are also probably derived from *Hymenaea.*

Amber from Chiapas, the southernmost state of Mexico, has been dated as late Oligocene to early Miocene based on foraminiferal zonal sequences (Frost and Langenheim 1974) and radiometry (Berggen and van Couvering 1974). The age range, 22.5–26 Ma, indicates 3–4 million years between the beginning and end of the amber accumulation period. That interval is somewhat smaller than that of the Dominican amber, with Mexican deposition starting and ending earlier. The first reported plant remains of *Hymenaea* were parts of the calyx and numerous stamens, the latter possibly resulting from bat pollination, which occurs in the taxonomic section to which *H. courbaril* belongs (Lee and Langenheim 1975). Stamens could easily have become trapped in the sticky resin exuding from the tree, as abundant stamens fall to the ground under trees following a night of pollination by bats. Leaflets and sepals of *Hymenaea* in the Mexican amber share characteristics with *H. courbaril,* now widespread, and *H. intermedia,* which presently is very restricted in Amazonia (Langenheim 1966).

Floral parts in both Dominican and Mexican amber, suggesting relationship to extant species in the primitive section *Trachylobium* of *Hymenaea,* bring up the difficult question of whether *Hymenaea* originated in Africa or the Americas. An understanding of such phytogeographic history depends on an adequate fossil record and some knowledge of the phylogeny of the taxa involved as well as geologic history (particularly plate tectonics) to assess the relative roles of oceanic dispersal and vicariance. Africa and South America were joined until the early Cretaceous but were split by the mid-Cretaceous. Unfortunately, the first unequivocal evidence for caesalpinioid legumes is from the non-resin-producing genus *Crudia* in Paleocene sediments from Nigeria. Because of the high diversity of genera as well as the high percentage of genera endemic to Africa, legume systematists (e.g., Brenan 1966, Raven and Polhill 1981, Herendeen and Dilcher 1992) have assumed that Africa was the primary site for radiation and evolution of tropical legumes, with a secondary center in Amazonia.

Although *Hymenaea* today is restricted to the eastern coast of Africa and adjacent islands, Langenheim and Lee (1974) assumed it once occurred in moist forests across central Africa but was restricted by a progressively drying climate and major uplifts in central Africa during the Tertiary and Pleistocene. This view is supported by the distributions of a close relative, *Guibourtia,* as well as *Tessmannia* and *Oxystigma,* which occur in the same East African coastal belt as *Hymenaea* but are also abundant in West Africa with other related resin-producing caesalpinioids (Chapter 2). Furthermore, there is possible Eocene evidence of *Guibourtia* in Tanzania (Herendeen and Jacobs 2001).

Langenheim and Lee (1974) suggested that an ancestor of *Hymenaea verrucosa* was dispersed to eastern Brazil via favorable ocean currents following the separation of the African and American continents. During the early Tertiary, the continents were still relatively close and rain forests were widely distributed on both, increasing the chance of successful dispersal. *Hymenaea* pods are known to float great distances in the ocean (Gunn and Dennis 1976), and seeds can germinate rapidly following transport. Herendeen and Dilcher (1992) stressed the importance of oceanic dispersal in interpreting the distribution of fossil legumes in general.

Poinar (1991), on the other hand, proposed vicariance (plants carried to present positions on a landmass) to explain the geographic origin of the genus. Poinar thought that *Hymenaea* originated in the Americas during the Cretaceous when South America and Africa formed a common landmass. Poinar (1992a) further elaborated the view that "*Hymenaea* expanded and speciated during the late Mesozoic from a distributional center located along the equator. Diversification resulted in the genus extending from Mexico across South America and into the adjoining African continent." This alternative view is not supported by either the fossil record or relationship to other resin-producing legumes. Although evidence from petals places both the

Dominican and Mexican *Hymenaea* amber species in the primitive section *Trachylobium,* understanding the evolution of the Mexican species appears to be more complex than that of the Dominican one. All specimens of the Dominican *H. protera* indicate a close relationship to *H. verrucosa;* however, evidence from sepals and leaflets in the Mexican amber suggest similarities to both *H. intermedia* and *H. courbaril* (both in the more advanced section *Hymenaea*), with *H. intermedia* having characteristics intermediate between *H. oblongifolia* and *H. courbaril.*

The complexity of understanding the evolution of Mexican *Hymenaea* increases: *H. verrucosa* is thought to be ancestral to *H. oblongifolia,* and *H. intermedia* is thought to represent an intermediate stage in the evolution of *H. courbaril.* Although Langenheim and Lee (1974) recognized two morphologically based sections of *Hymenaea*—one primitive and the other more advanced—they considered them as "trends with within a basically homogeneous genus." Thus the primitive *H. verrucosa–H. oblongifolia* stock may have been evolving during the Oligo-Miocene in Mesoamerica, perhaps tending toward *H. courbaril,* which is the only species in Mexico today.

Interestingly, with the primary Dominican amber deposits now thought to be Miocene in age (Grimaldi 1995), the Mexican amber would be slightly older than the Dominican amber. Therefore, did a *Hymenaea* amber species evolve first in Mesoamerica from a basic *H. verrucosa–H. oblongifolia* ancestor and then disperse to the Dominican area? Or did the Mexican and Dominican species evolve in separate areas? Separate evolution is suggested by the close resemblance of *H. protera* to *H. verrucosa* and the fact that *H. torrei* shares characteristics of *H. verrucosa* and *H. courbaril.* As in Mesoamerica, *H. courbaril* is the only species currently found throughout the Caribbean, except for *H. torrei* in Cuba. From this information, one can speculate that evolution toward *H. courbaril* was possibly occurring from an ancestor of *H. verrucosa* in the Caribbean.

Deciphering the relationship between these *Hymenaea* amber species hinges partially on plate tectonics during the early Tertiary in the Caribbean and Gulf of Mexico. Although 13 mobilist models (concerned with movement of the plates) have been presented (Pindell and Barrett 1990), Graham (2003) pointed out that consensus among paleobotanists is developing around a more recent model of the history of the flora in this region. The primary difference between this model and the older ones is that land forming the proto-

Antilles originated as islands and was never a continuous or nearly continuous land bridge connecting or nearly connecting North and South America. This model suggests a complex pattern of collision and separation, and submergence and emergence, providing opportunities for both vicariance and oceanic dispersal in the migration, evolution, and speciation in the flora.

Only additional evidence from Cretaceous or early Tertiary fossils can eliminate speculation regarding an African versus New World origin for *Hymenaea*, and it will require additional well-dated and more complete floral remains to answer questions about the evolutionary relationship between the Dominican and Mexican *Hymenaea* amber species. It now appears that they evolved separately from an ancestor similar to both in some respects. That *Hymenaea* pods (Figure 5-6) float well and germinate easily following oceanic transport also suggests the possibility of a complex history of land and sea migration during the Tertiary.

Although evidence from both the chemistry of the amber and enclosed floral remains is conclusive regarding *Hymenaea* as the source of the Mexican amber, there is additional information from the depositional conditions of the amber. The amber occurs primarily in lignites and associated marine sandstones. Pollen from these sediments indicate that the amber was deposited in a fluctuating shallow sea fringed by a complex mangrove vegetation, or in brackish estuaries or back swamps dominated by mangroves (Langenheim et al. 1967). *Hymenaea courbaril* trees today occur either on slopes above or lining mangrove-dominated estuaries and swamps along the Mexican Pacific coast. There, the resin can be washed from the soil, where it often collects, directly into the mangroves or indirectly via streams that enter the mangroves (Figure 4-3). The stilt roots of mangroves are excellent at trapping resins, thereby forming a good deposition site. Thus some species of *Hymenaea* occur today near deposition sites that duplicate those in which the amber was discovered (Langenheim et al. 1967).

Resin from *Hymenaea verrucosa,* considered by some to be amber, is known from Zanzibar and Dar es Salaam, Tanzania (Schlüter and von Gnielinski 1988). Although the ages of the deposits are unverified, the resin apparently occurs in coastal or reef limestones, thought to be Pleistocene, as well as amid sandy clay sediments considered by some as Tertiary (Pliocene or perhaps Miocene). Schlüter and von Gnielinski suggested that this resin was deposited in a mangrove estuarine situation similar to that proposed for the Mexican amber (Langenheim et al. 1967). *Hymenaea verrucosa* resin from Madagascar has also been called amber, but ^{14}C analysis of some pieces has shown it to be as young as 50 years (Alex Brown and G. O. Poinar, pers. comm.).

In addition to the large amber deposits in Mexico and the Dominican Republic, smaller ones thought to be Miocene in Pará, Brazil, and Medellín, Colombia, have been identified as derived from *Hymenaea* by IR (Langenheim and Beck 1968); no included remains have been found at these localities. Large amounts of *Hymenaea* "amber" (identified by IR and Py-GC-MS) from the Santander region of Colombia have become available from amber traders. However, these are now considered to be of recent origin (180 ± 50 years old from ^{14}C dating from one site; K. Anderson 1995). Although some of the material may be older, Poinar (1992a) indicated that physical characteristics such as softness and melting point also support the view that at least large amounts of this resin are not fossilized. The resin contains numerous insects, including an abundance of bark beetles that may have initiated the resin flow (Figure 5-1).

The occurrence of so much amber from *Hymenaea* results from the trees' production of large quantities of resin and the presence of labdatriene diterpenoids (Figure 1-5), which polymerize and persist under depositional conditions. The large deposits of mid-Tertiary Neotropical leguminous amber raise questions regarding the conditions under which large amounts of resin accumulate in particular areas. Were there relatively local outbreaks of herbivores, such as wood- and bark-boring beetles, that led to injuries resulting in massive exudation of resin? There is ample evidence relating beetle activity and large accumulations of resin in *Hymenaea*. A section of a trunk of *H. verrucosa* from Zanzibar shows almost pure resin between the bark and heartwood (Plate 27). Beetles in the families Platypodidae (ambrosia beetles) and Scolytidae (bark beetles), which bore into wood or bark, are well represented in *Hymenaea* amber. Ambrosia beetles excavate tunnels and galleries in the sapwood or in the phloem just under the outer bark. Sawdust produced from these excavations gets compressed into plugs, which are preserved in the amber along with the beetles. A fungus occurs in specialized pockets in the body of the beetle; pockets in ambrosia beetles exhumed from Dominican amber are still filled with spores and filaments of this symbiotic fungus (Grimaldi et al. 1994). Although boring by the ambrosia beetle does not kill the tree, the fungus carried by the beetle is inoculated into the wood, where it develops and carpets the galleries. These beetles feed on the fungus (their ambrosia), which in turn kills the tree. Although scolytid bark beetles live and feed within the bark, like ambrosia beetles they carry fungi that can kill the tree (Figure 5-1). Today, resin-producing trees are known to produce large amounts of both constitutive and induced resin, defending against both bark beetles and their fungi (Chapter 5).

Other Tertiary and Pleistocene Resin Sources

Copaifera, a tropical legume noted for its production of abundant resin in Africa, similar to that of *Hymenaea* (Chapter 9), is the source of resin from northeastern Angola dating from the Pleistocene to about 8000 B.C. (Langenheim 1972). These dates place this resin in the subfossil to fossil range. IR of the amber and modern resins from the African tropical genera *Canarium, Copaifera, Colophospermum, Daniellia,* and *Guibourtia* suggests that *Copaifera mildbraedii* or a close relative was the source. Although *C. mildbraedii* today occurs about 200 km north of the site at which the amber was collected, it could well have been a component of the gallery forests (rain forest trees that occur along rivers outside the rain forest proper), even extending along the interfluves until around 880 B.C. The botanical source of this amber supports other evidence for restriction of forests in this part of Africa, suggesting that the disappearance of this gallery forest tree probably was the result of a drying and cooling climate, aided by various kinds of human activity during the last millennium.

Although few amber deposits are known in Africa, the discovery of additional deposits is to be expected. Resin from numerous leguminous genera has collected in large surface accumulations in West African rain forests (Chapters 2 and 8) in much the way that *Agathis* resin collected in New Zealand (Chapters 7 and 8). These deposits are of the type that could accumulate over long periods of time and become amber. There is Eocene amber from Umuahia, Nigeria (Poinar 1992a, Grimaldi 1996), most likely derived from one of the numerous West African leguminous genera such as *Daniellia, Guibourtia,* and others analyzed in the Angolan study.

A burseraceous source for Tertiary amber from three localities has been suggested by X-ray diffraction that demonstrates the presence of a crystalline triterpenoid alcohol, α-amyrin (Figure 1-5), characteristic of elemi resins from species of *Bursera, Canarium,* and *Protium* (Frondel 1967b). The presence of this alcohol in some specimens of amber called Highgate copalite from the

London Clay flora along the southeastern coast of England, with its associated burseraceous seeds, has suggested either tropical *Canarium* or *Protium* as the source. The affinities of the Paleocene-Eocene London Clay flora are predominantly subtropical to tropical, with many elements of the flora allied to present-day tropical rain forests in Southeast Asia; *Canarium* and *Protium* produce large quantities of elemi-type resin in these forests today (Chapter 8).

A burseraceous source for Highgate copalite was questioned by K. Anderson and Botto (1993), using GC-MS. However, IR indicates that specimens called Highgate copalite from a number of collections may be derived from several botanical sources (Beck and Shennan 1991). Therefore, Anderson and Botto's results may not invalidate Frondel's (1967b) conclusions from X-ray diffraction since the samples may have been from a different source.

α-Amyrin occurs in glessite, a rare variety of Eo-Oligocene amber associated with the common succinite in European deposits, suggesting a burseraceous origin for this amber, too (Frondel 1967b). Glessite, however, has been shown to be very similar to succinite by several analyses, including NMR, FTIR, and Py-GC-MS (Langenheim 1995). On the other hand, Vávra (1990, 1993), using GC, supported the presence of both α- and β-amyrin in an authenticated source of early Miocene glessite from Bitterfield, Germany, and in Eocene amber from Corinthia, Austria. Amyrins (Figure 1-5) occur in other plants, but Burseraceae are the only ones known that produce them in resin. A burseraceous source for these ambers further supports Southeast Asian floral affinities during this period of Tethyan connection of European and Asian floras. Additionally, deposits of quayaquilite of unknown age (probably Tertiary) in Ecuador contain α-amyrin, and a burseraceous source is supported by similarity of its IR to that of resin from *Protium* (Langenheim and Beck 1968), species of which produce large quantities of resin in Ecuador today (Chapter 8).

Pistacia (Anacardiaceae) was reported as the source of Pleistocene amber from Israel (Paclt 1953). This amber is representative of plant families that generally produce either fluid resin or resin as a component of latex, which probably do not preserve well (Chapter 2). However, *Pistacia* is exceptional in Anacardiaceae in producing triterpenoid-dominated resin, called mastic (Chapter 8), rather than a more fluid phenolic resin or latex (Chapter 10). Furthermore, the monoterpene myrcene in *Pistacia* has been shown to form a polymer, which could help explain the preservation of this resin (van den Berg et al. 1998; Chapter 8). With revival of scientific interest in amber, families such as Anacardiaceae may become better represented through analysis of ambers from smaller deposits.

Figure 4-7 shows the ages of ambers reported from six angiosperm modern families, from the late Cretaceous to Pleistocene. It also depicts the general shift from conifers to tropical angiosperm ambers in the mid-Tertiary. Large North American deposits of late Cretaceous and early Eocene ambers have been identified as dipterocarp solely by polycadinene chemistry (Appendix 3). Although the chemistry is similar to that of small amounts of European amber found in the fruits of a mastixiaceous genus (Cornaceae), the North American ambers are more likely to be derived from Dipterocarpaceae because dipterocarps produce copious amounts of trunk resin. Furthermore, dipterocarp genera occurred in North American floras during the late Cretaceous and Eocene, although they have not been associated with the amber. Chemical determination of Miocene and Pliocene ambers (as *Shorea*) in Southeast Asia is more conclusively dipterocarpaceous, however, because that family dominates the forests of the region.

North American amber from the Cretaceous–Tertiary boundary has been assigned to *Liquidambar* (Altingiaceae), based on a distinctive polystyrene chemistry. It primarily occurs from the Paleocene to the Miocene, however, from which there is also both chemical and botanical evidence. The occurrence of burseraceous amber from the Paleocene to early Miocene in Europe is based mainly on crystalline triterpenoid chemistry, with some associated botanical support. Conclusive chemical (diterpenoid labdane) and botanical evidence indicates a leguminous source *(Hymenaea)* of Oligo-Miocene amber in Mexico and Miocene amber in the Dominican Republic. The large angiosperm-derived Dominican amber deposit is considered analogous to the large conifer-derived Baltic amber deposit in terms of size and diversity of enclosed organisms. Another leguminous source *(Copaifera)* has been suggested for amber from the Pleistocene or younger beds in Angola. Amber derived from *Pistacia* (Anacardiaceae) has been reported from the Pleistocene in Israel.

Amber of Unknown Botanical Source

Microscopic particles of resin, or resinites, are ubiquitous in most American coals below the medium-volatile bituminous range. Although generally minor components, they can constitute as much as 15% of such soft coals, as discussed for late Cretaceous deposits in the Wasatch Plateau, Utah (Crelling 1995). In most Appalachian and Midwestern U.S. coal seams of Carboniferous age, these resinites occur as ovoid bodies at the time of deposition, whereas in western coals the resin has been mobilized, flowing into fractures during coalification. Resinites from both geographic areas are assumed to be derived from the plants making up the coal, but their source has not been determined through correlated chemical and paleobotanical analyses.

There are significant amounts of amber from other deposits in which the plant source has not been studied; these are included in Figure 4-2. Some deposits are listed here but are not discussed in detail because of lack of information regarding the resin-producing tree. A few of these ambers, however, are discussed in Chapters 6 and 10 because of their use by humans. Smaller amounts of botanically undetermined amber are pointed out, however, in regard to other important aspects such as the occurrence of resin at a particular geologic time.

One of the largest deposits of mid- to late Cretaceous amber occurs in northern Siberia (Taymyr Peninsula). Although the late Cretaceous ambers primarily occur in Siberia, Zherikhin and Eskov (1999) reported that amber is also found in Transcaucasia (Azerbaijan and Armenia). The amber contains many included organisms, particularly insects, but the botanical source remains unknown. However, this amber is based on copolymers of communol and biformene (K. Anderson 1994). Because this structural characteristic is not known for any other fossil resin, a new botanical source might be expected.

Two Cretaceous deposits occur in Japan: Chōshi amber is early Cretaceous whereas Kuji amber is later Cretaceous and characterized by very large pieces (16–18 kg) prized for carving into art objects (Chapter 6). Early Cretaceous amber (100 Ma) from northwestern France contains insects that resemble those in the younger Cretaceous amber (90–94 Ma) from certain deposits in New Jersey (Grimaldi et al. 1989). Significant late Cretaceous deposits also occur near Goldsboro, North Carolina, as well as at Grassy Lake, Alberta, Canada. Very large pieces of amber from northern Myanmar (Burma) are distinctively colored, ruby red, and have been extensively used for carving fine art objects especially prized by Chinese mandarins. Although generally reported as Eocene or even Miocene, evidence now supports a Cretaceous age for them (Zherikhin and Ross 2000). Although the amber is excavated from mid-Eocene sediments, many of the insects in it are quite primitive, and some belong to groups that occur exclusively in the Cretaceous (Rasnitsyn and Ross 2000). David Grimaldi (pers. comm.) reports that the amber formed under distinctly more tropical conditions than any other fossiliferous Cretaceous amber. Botanically unidentified Paleocene amber occurs on Sakhalin Island. Eocene amber from Fushun, Liaoning province, China, occurs in small pieces in coal beds. Amber from Sicily (simetite) is probably younger (Oligocene to mid-Miocene), and its deposits are much smaller than those of Baltic amber. Cretaceous and Tertiary deposits of amber occur in the Carpathian Mountains of Romania. Most of this amber (rumanite) is similar to simetite and differs from succinite in not having succinic acid.

The Floras of Amber Forests

Because deposition of amber depends on water transport (Figure 4-3), in most cases little can be said about the definite characteristics of the forest in which the resin was produced based on where it is found. There are exceptions, however, where the source trees may have occurred at least partially in or near a swamp or bay, as in the cases of the mid-Tertiary Mexican and late Cretaceous central New Jersey ambers. In most deposits, however, resin was probably transported some distance. Repeated transport may result in the presence of amber at a site younger than the material itself. In some large deposits, it is not known if the resin-producing trees dominated the forest or if less common trees produced copious resin that was either transported over long distances or accumulated over long periods of time. On the other hand, some idea of the flora of the amber forest may be provided by organisms (or parts) entombed in the resin. Evidence for the flora, however, is biased toward small materials and fortuitously proper conditions for their entrapment, for example, pollen (Langenheim and Bartlett 1971). Inferences about the flora can also be made from insects whose existence is restricted to certain plants (Grimaldi 1996, Poinar and Poinar 1999). Embedded organisms may have affinities with modern floras and faunas, giving an indication of the climatic conditions of the forests in which they lived.

Baltic Amber Forests

The flora of the Baltic amber has been the most extensively studied, partly because of the huge quantities of amber mined from the rich deposits (Chapter

6). During the peak of mining activity, German scientists developed the paleontological study of amber fossils. Grimaldi (1996) suggested that these studies were accelerated by the German perfection of the microscope, allowing analysis of very small organisms. Between 1830 and 1937, 750 species of spore- and seed-producing plants were described. Most of these botanical studies were brought together in five major monographs (Goeppert and Berendt 1845; Goeppert and Menge 1883; Conwentz 1886, 1890; Caspary and Klebs 1907). Plates from the quarto volumes by Goeppert and Berendt, and Conwentz, display the magnificent detail recorded by these researchers (Figure 4-11). Of the 750 plant species, however, Czeczott (1961) reported that only 216 are valid, probably because many descriptions were based on isolated plant parts. Czeczott reported 63 cryptogams (molds, parasitic fungi, mosses, etc.), 52 gymnosperms, and 101 angiosperms.

The large number of extinct species in the Baltic amber flora provides an important evolutionary insight. The affinities of most of the conifers appear to be with modern taxa in North America, China, Japan, and Africa, with a greater diversity of taxa in Cupressaceae s.l. than the Pinaceae. Coniferous remains are more abundant than those of angiosperms, perhaps because of the ease of incorporation of small pieces of conifer remains (particularly cupressaceous ones) or the dominance of conifers in the forests at the time. On the other hand, there is a greater diversity of angiosperm taxa. The angiosperms also have affinities to modern taxa from subtropical (including dry Mediterranean) and tropical climates. In fact, Czeczott (1961) concluded that 23% of the families she analyzed were tropical (e.g., Dilleniaceae, Lauraceae, and Myrsinaceae). Examples of now extinct plants with closest relatives in tropical or subtropical Asia, Australia, or the Americas include *Trianthera*, in amber, related to *Eusideroxylon* (Lauraceae) today in Borneo and Sumatra, and *Drimysophyllum*, in amber, most closely resembling *Drimys* (Winteraceae) in Malaysia, New Caledonia, New Zealand, and tropical America. Of the Baltic amber plants, 12% have a temperate zone distribution today, and the remainder are either cosmopolitan or have an anomalous or discontinuous distribution. Stellate hairs are particularly abundant and often attributed to oaks (*Quercus*, Fagaceae) because flowers of oaks (or their relatives) are abundant in Baltic amber. Subsequent analyses indicate that there are five types of these hairs, all of which occur today in representatives of the Fagaceae, which has a cosmopolitan distribution. Czeczott pointed out that

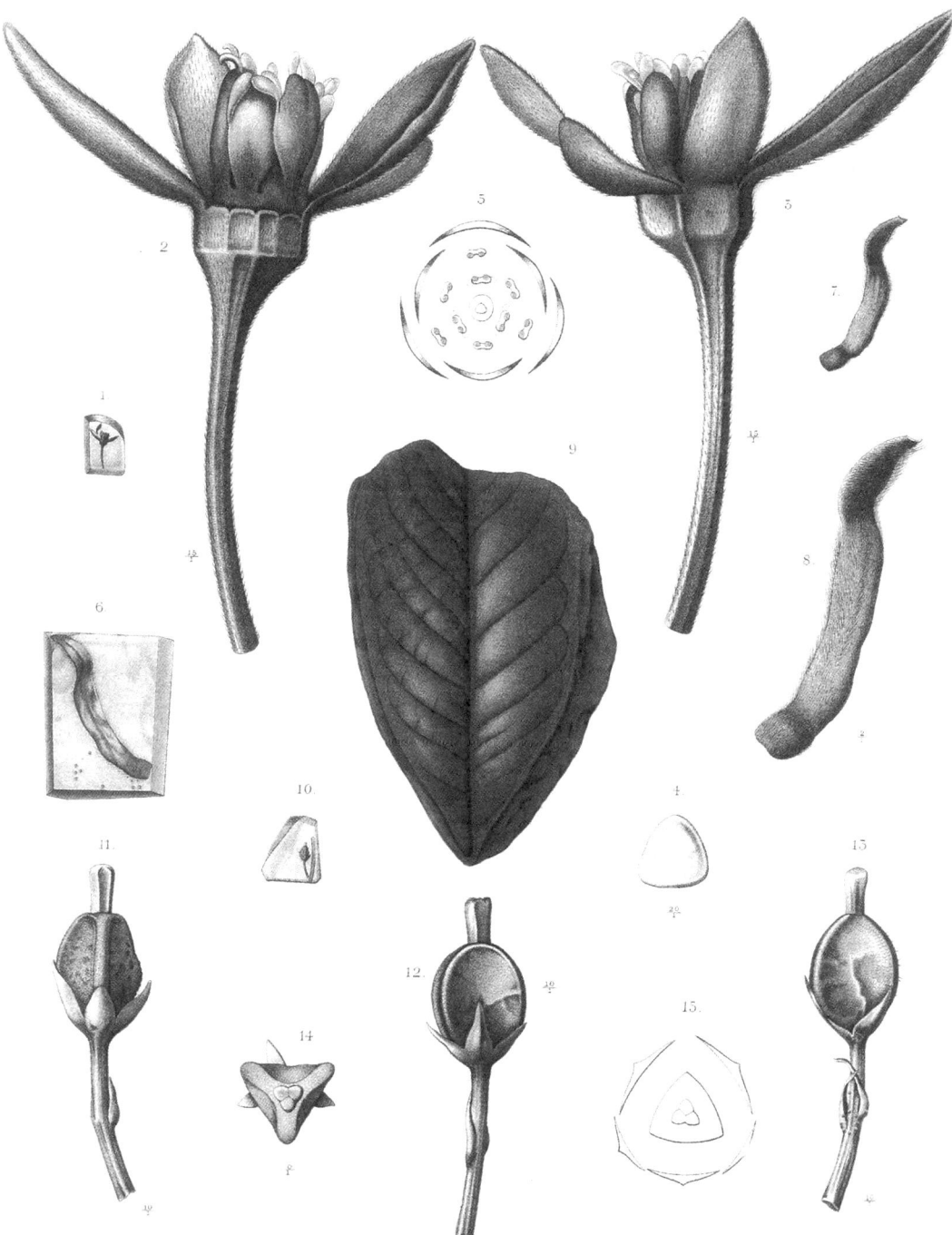

Figure 4-11. A plate of drawings from Conwentz's monograph (1886) of dicotyledonous plant parts found in Baltic amber. *Cinnamomum* (Lauraceae) flower at top; leaf in center and stipules to the right from an extinct member of the Magnoliaceae; fruits at bottom from an extinct member of the Cistaceae. The plant inclusions are represented at their actual size (reduced here) in the pieces of amber at the left.

this "intermingling" of plants, representing different climatic regimes, is "a peculiar feature of this flora" in Baltic amber.

Several hypotheses have been presented to explain this mixture of plants. Schubert (1953) and earlier workers thought that the amber forests could have been similar to those in southern Florida, where islands (hammocks) of tropical plants occur amid forests constituted of subtropical and warm temperate species. Ander (1941) thought, on the other hand, that the amber forest was denser and the climate moister than at present in Florida; hence, he envisioned the area as mountainous, with more temperate species occurring at higher elevations and tropical taxa on south-facing slopes. As noted, Takhtajan's (1969) view, that "both systematic and ecological characteristics indicate that the above [Baltic amber] flora has been derived from a subtropical flora of the Assam-Burma-Yunnan type" (Larsson 1978), would include a diversity of habitats, which could explain the intermingling of conifers and angiosperms. Therefore, just as the identity of the amber-producing tree(s) is a mystery, there is puzzlement over the nature of the plants constituting the Baltic amber forest(s).

Dominican and Mexican Amber Forests

Compared to the rich flora found in Baltic amber, relatively few plants are known from other amber deposits. Although ambers from Mexico and the Dominican Republic are both from species of *Hymenaea*, more floral remains have been described from Dominican than Mexican amber. This larger amber flora, like that of Baltic amber, may be partly the result of the greater availability of Dominican amber for study. In contrast to the many extinct taxa in Baltic amber, many genera in the Mexican and Dominican ambers are closely related to modern ones, and affinities with extant species can even be established for some. Legumes are well represented, as they are today in Neotropical forests. In addition to *H. protera*, other caesalpinioids reported by Poinar and Poinar (1999) include flowers of *Prioria* and *Peltogyne* as well as flowers of the mimosoid *Acacia*. Poinar and Poinar also listed *Trichilia* and *Swietenia* in the mahogany family (Meliaceae), and *Peritassa* (Celastraceae). Moreover, Grimaldi (1996) had identified flowers in the nutmeg family (Myristicaceae), balsa family (Malvaceae), grass family (Poaceae, or Gramineae), Euphorbiaceae, and Thymelaeaceae. Leafy liverworts and a moss have also been found (Hueber and Langenheim 1986). These latter, very small plants grow on tree trunks and are to be expected as they could easily be incorporated into the resin as it flowed along the trunk.

Some plants are inferred to have occurred in these forests from the presence of embedded insects that depend on specific types of flowers, fruits, or leaves that themselves do not occur in the amber. For example, figs (*Ficus*, Moraceae) are known to have been present because of agonid wasps in the amber. These wasps have formed a symbiotic association with fig flowers; the female wasp pollinates the flowers, the inflorescence in turn providing a secure place to rear her young. Similarly, although no evidence for palms (Arecaceae) per se has been found, they are assumed to have occurred because of the presence of the thaumastocorid palm bug, whose flattened body allows it to live securely between tightly closed leaves of the young fronds. Epiphytic bromeliads are indirectly indicated as members of the forest by certain kinds of damselflies and beetles whose larvae live in the bromeliads' tanks of water. Orchids are suggested by orchid bees in the amber. A few grass spikelets are corroborated by lygaeid bugs.

Poinar and Poinar (1999) proposed a structure for this Dominican amber forest. They thought that *Swietenia* trees emerged above the canopy. *Hymenaea* was prominent in the canopy along with *Peltogyne* and the palm *Roystonea* whereas *Ficus* occurred in the subcanopy and understory, *Acacia* and various palms occupied the shrub layer, and a tropical grass grew on the forest floor. They thus assumed that the moist tropical forest was similar to that occurring today in Panama. This is a quizzical assumption because the New World relatives of *H. verrucosa* (the presumed ancestor of *H. protera*) presently occur as emergent trees in Brazilian coastal and Amazonian rain forests (Langenheim et al. 1973). Only *H. courbaril*, whose flower is unlike that of the Dominican amber *H. protera*, occurs in Panamanian forests today. Furthermore, there would be no reason to assume *Ficus* in the subcanopy and *Acacia* in the shrub layer in New World rain forests. Additional botanical evidence is needed to reconstruct the Dominican amber forest adequately.

Miranda (1963) identified a leguminous leaf in the Mexican amber as a species of *Acacia,* and a flower as a species of *Tapirira* (Anacardiaceae), *T. durhamii,* which appears to share characteristics with the widespread South American *T. guianensis* and the Mexican *T. mexicana. Acacia,* a pantropical and subtropical genus, has been found in Dominican as well as Mexican amber. Thus both *Tapirira* and *Hymenaea* amber species display affinities with modern South American species as close as those to current Mexican species. A leaf and stellate hairs, perhaps, of oak have also been found, and the first fruit pod of an orchid. A liverwort (Jungermanniales) is of the type that occurs epiphytically on trees in moist tropical forests, and the oaks and acacias can grow in either moist or dry tropical forest. If the forest was somewhat dry tropical or in a transitional mountain zone, then the resin was transported by streams (Figure 4-3) to a lower region of tropical mangrove estuary (Sousa and Delgado 1993). The great diversity of insects characterizing Dominican and Mexican amber forests has been admirably summarized by Grimaldi (1996) and Poinar and Poinar (1999).

Renewed Interest in Amber Research

The study of amber is developing rapidly, based on a long history of research that has been revitalized by more recent discoveries and newer analytic techniques. The renewed interest spans several disciplines. The unique properties of fossil resins, such as containing compounds that are structurally distinct, make them noteworthy as potential markers of paleoenvironments, sedimentary conditions, and geochemical processes (K. Anderson 1995). Compounds derived from plant resins are major components of some petroleums formed in oceanic sediments derived from weathering of terrestrial rock (Murray et al. 1998). There is also considerable interest in the chemistry of Paleozoic and Mesozoic fossil resins that would help explain the early evolution of resin biosynthesis and the defensive roles these complex materials play. These research opportunities arise from continually developing techniques of chemical analysis, allowing more detailed molecular characterization of fossil resins than before. In another arena, some of these chemical techniques are used to detect amber forgeries (Grimaldi et al. 1991, Shedrinsky et al. 1993). Such detection is particularly important for jewelers and art collectors. Furthermore, when organisms have been fraudulently embedded in the resin, scientists could be led to seriously wrong conclusions regarding their identity and evolution.

Since amber preserves insects and plant materials better than any other medium, questions arise regarding the special properties of resin that allow it to preserve delicate tissues through millions of years and why this occurs better in some ambers than others. Despite some pessimism regarding the possibility of finding intact DNA in organisms older than 100,000 years, there will likely be a search for improved techniques of analysis of DNA from organisms (primarily insects and some plants) entombed in the amber. The recovery of live bacterial spores from amber (Cano and Borucki 1995), if reproducible, would have significant implications for molecular paleontology.

Discoveries of amber with enclosed organisms continue to provide evidence for phylogeny of insect lineages, in particular, as well as for different kinds of ecological interaction in ecosystems. Furthermore, the fossils in amber, which seem to have come predominantly from tropical or subtropical forests, can be used to describe the relationships of these kinds of forests to those that exist today. High rates of decomposition, thus low potential for fossilization, result in relatively few macrofossils from the lowland moist tropics, a problem exacerbated by meager rock exposures. There is considerable evidence available for reconstructing Neotropical vegetation, particularly from pollen (Burnham and Graham 1999). Some major plant elements of rain forests may be poorly represented by pollen, however, because they are animal pollinated and thus do not leave a pollen record. Amber, on the other hand, with its large associated insect record, adds another dimension, helping create a more interactive ecosystem perspective in the reconstruction of some tropical and subtropical forests.

From a botanical perspective, it is evident from Figure 4-2 and Appendix 4 that the botanical source of many ambers has not been determined, nor has information been obtained regarding the plant communities in which the plants grew. The early history of the plants that produced what has been called resin (Carboniferous–Jurassic) and the secretory structures in which these substances were stored need to be analyzed from correlated paleobotanical and chemical perspectives. The botanical origin of Baltic amber, one of the most extensive and famous deposits, is a continuing mystery. Although it has been intensively studied, particularly in regard to embedded insects, much remains to be learned about the Baltic flora and the forest in which the resin-producing trees lived. More detailed analyses could probably be made of floras of other amber deposits, as plant inclusions have often been overlooked in favor of the more obvious insect remains.

CHAPTER 5

Ecological Roles of Resins

Why did trees begin several hundred million years ago to produce resin, a mixture of compounds that are not physiologically important? Although a few chemists still view secondary compounds, including resins, as plant waste products, chemical ecologists early hypothesized that resins provide trees with a versatile set of defenses against an ever-increasing number of organisms attacking them (Langenheim 1969). This hypothesis, however, does not necessarily imply that resins were initially synthesized for defense. The geologic record provides evidence of the essentially simultaneous evolution of secretory structures in which resins are stored. Storage of these potentially toxic compounds in an intercellular structure may have been an adaptation to avoid autotoxicity (Chapters 1 and 3), with the resin providing the secondary benefit of defense against pests. It is difficult to determine the evolutionary origin of a defensive trait, and in most cases the best we can do is to try to understand something of the evolutionary factors that maintain a defensive role (e.g., Rausher 1992, Berenbaum and Zangerl 1995). Here I review the ecological and evolutionary studies that demonstrate the roles resins play in particular interactions and explore hypotheses of how resins may have evolved to serve different functions.

Ecologically Important Properties of Resin

In thinking about ecological roles of resins, it is important to understand the special properties of resins that enable them to mediate interactions with different organisms. This also involves the different kinds of secretory structures in which resin is stored (Chapter 3). Variation in the mixture of chemical compounds plays a central role in ecological interactions of resins. Therefore, a general discussion of this variation precedes a presentation of exemplary studies of resin ecology in conifers of the temperate zones, and angiospermous trees of the tropics, and shrubs of arid regions.

Complex mixtures of volatile and nonvolatile components in terpenoid and/or phenolic resins (Chapter 1) provide ecological versatility to the plants producing them. Mixtures of toxic resinous compounds may confound a wide range of herbivores or pathogens to evolving resistance to the effects of all the individual compounds or groups of compounds, thereby slowing the breakdown of plant defense (Langenheim 1994, Cates 1996). The volatile mono- and sesquiterpenes or phenolic compounds also may attract benefactors such as parasitoids or predators of the herbivores that attack the plant, and this is considered an indirect defense.

The ratio of volatile to nonvolatile components affects the physical and defensive properties of rcsin. The volatile and more fluid components of resin enable movement of the more viscous nonvolatile components. The nonvolatile di- and triterpenoid or phenolic components increase the viscosity and rate of crystallization (hardening) of the resin. These properties affect the rate of resin flow and the ability of the resin to trap and immobilize enemies or coat wounds in tree trunks, which has been considered a first line of defense (Figure 5-1). The nonvolatile compounds also often coat the surfaces of young leaves and stems in arid-zone shrubs, providing additional protection against desiccation, ultraviolet radiation, and high temperatures.

Despite the defensive advantages of a mixture of chemicals, are there are trade-offs to these benefits? The metabolic cost of synthesizing, transporting, and storing the compounds may be disadvantageous. First, there is the cost of any extra biosynthetic machinery. However, additional enzymes (synthases) are not always needed to generate a mixture of terpenoid compounds. Some common volatile monoterpene are produced by a single synthase, reducing the cost of synthesis. For example, camphene, α- and β-pinene, limonene, and myrcene, which occur frequently in preformed resins of numerous taxa, are all produced by the multiproduct pinene cyclase (Chapter 1). Also, several single synthases produce large numbers of sesquiterpenes in an induced response to wounding. Moreover, some compounds are only induced in response to herbivore or pathogen attack, and the enzymes involved occur in plants for only restricted periods of time. Multiple roles such as direct plant defense against herbivores and pathogens, indirect defense through attraction of herbivore parasitoids, as well as allelopathy and pollination, further reduce any initial synthetic cost of mixtures for the plants.

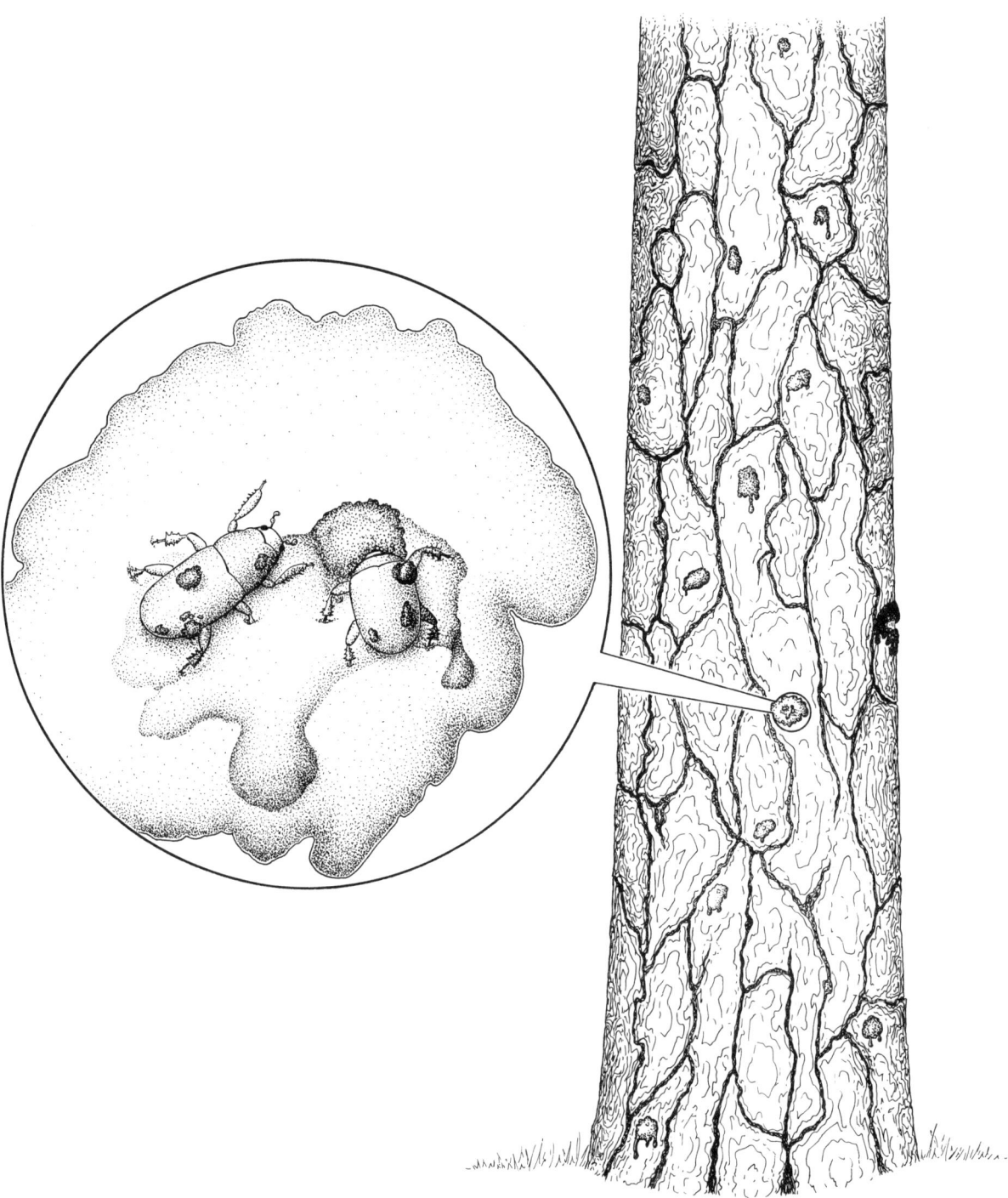

Figure 5-1. *Dendroctonus* bark beetles attack trees, for example, *Pinus ponderosa,* ponderosa pine, and may initially be repelled by the tree's response with a flow of resin, as shown here by numerous small masses on the trunk. An enlargement of one of the masses shows beetles mired in the resin.

Variation in Resin Composition

A mixture of compounds in resins provides the possibility of varying the chemical composition (relative proportions of the constituents). This variation is usually expressed as percentages of the constituents in characterizing the compositional pattern but is expressed as concentrations of the constituents per unit weight or volume in analyzing the dosage effects on herbivores and pathogens. The volatile composition of terpenoid resin has been analyzed in considerably more detail than the nonvolatile, probably because of the ease of identification of monoterpenes using mass spectroscopy and quantification of the composition using gas chromatography. Furthermore, genetic studies of monoterpenes provide information crucial for both ecological and evolutionary interpretation.

Constitutive and Induced Resins

Differences in composition between constitutive (preformed) resins and those induced in reaction to wounding can be used to characterize different kinds of chemical defense (Tollrian and Harvell 1999). Karban and Baldwin (1997), however, claimed that constitutive and induced plant defenses represent a continuum rather than discrete categories. They argued that all chemical defenses require a sequence of steps for production; where defense falls along the continuum depends on how far the plant has proceeded in the sequence of chemical events before herbivory occurs, activating the defense. Plants rely less on transcriptional control over synthesis and more on preformed precursors in constitutive defense, whereas the opposite is true in induced defense. On the other hand, different synthases are involved in the production of components in constitutive and induced conifer resins, as shown in grand fir *(Abies grandis),* thus the distinction is meaningful biochemically (Chapter 1). Furthermore, constitutive resins are stored in secretory structures whereas induced resins may occur either in special traumatically produced structures or are synthesized at the site of injury without a secretory structure (Chapter 3). The presence or absence of a secretory structure and the different suite of compounds characterizing induced resin are part of the multilayered defensive properties usefully distinguished by the terms constitutive and induced. Therefore, these terms are used for resins by biochemists (e.g., Lewinsohn and Croteau 1991, Phillips and Croteau 1999, Trapp and Croteau 2001) and chemical ecologists working with resin-producing plants.

Chemosystematic Studies and Ecology

Chemosystematic studies of temperate zone conifers (e.g., von Rudloff 1975) and tropical angiosperms (e.g., Martin et al. 1976) pointed out the importance of compositional variation as well as the repeatability of constitutive resin compositional patterns to ecologists. Chemical ecologists then began to emphasize the ecological significance of variation in chemical mixtures (Berenbaum 1981), with resins receiving particular attention (Langenheim 1994, Cates 1996).

Chemosystematic analyses of widespread, economically important North American conifers such as *Pseudotsuga menziesii* (e.g., Zavarin and Snajberk 1975, von Rudloff and Rehfelt 1980) and *Pinus ponderosa* (e.g., Smith 1964, Zavarin and Cobb 1970) provided data on interplant and intraspecific compositional variation over large areas of western North America in monoterpenes in xylem (wood) and phloem (bark), which ecologists subsequently interpreted in continuing experiments. Similarly, chemosystematic analyses of leaf sesquiterpenes of species of the tropical legume *Hymenaea* showed that variation is greater intraspecifically than interspecifically in many cases (Martin et al. 1976). The study by Martin et al. (1974) exemplifies the high degree of both inter- and intrapopulational variation in a single angiosperm species, *H. courbaril,* which ranges throughout the entire New World distribution of the genus. These patterns of compositional variation in various *Hymenaea* species provided valuable information for ecological studies.

The first studies of compositional variation of resins, from both systematic and ecological perspectives, were based primarily on field observation and collection of constitutive resins stored in secretory structures. Studies of induced resins later focused on the plant's chemical response to wounding. As ecological studies progressed, more emphasis has been placed on experimental approaches to interpret the relative roles of constitutive resins and those induced by diverse enemies that attack plants (Raffa 1991, Cates 1996).

Genetic Control of Composition

Knowledge regarding the genetic control of resin composition is important for systematic, ecological, evolutionary, and commercial research. Over the years, compositional variation of some common individual monoterpenes such as β-pinene, myrcene, limonene, terpinolene, and Δ-3-carene has been shown to be under single-gene control (Hanover 1966a, b, 1971, 1992; Squillace 1971; Yzadeni et al. 1982; Bernard-Dagan 1988). Monogenic control has also been proposed for sesquiterpenes such as β-caryophyllene and longifolene (Marpeau et al. 1975). Although there has been controversy about the statistical bases for some of these conclusions (Birks and Kanowski 1988), the results are generally considered valid. Additionally, closely related genes that encode terpene synthases have been isolated using molecular techniques. Some of these enzymes, which have the unusual ability to synthesize multiple terpenes, maximize the chemical diversity using only minimal genetic machinery (Chapter 1). Moreover, some monoterpenes are produced in fixed ratios, contributing to the control of their relative amounts in the composition.

Despite the increase in data on various aspects of genetic control of composition, researchers caution that other factors may confound the interpretation of variation. These factors include change at different stages of development of an organ, and environmental variation. In natural populations, the underlying genetic basis for variation in resin composition provides the material on which selection can act. Understanding the relative aspects of genetic and developmental control as well as the environmental influences on resin composition is important both systematically and ecologically.

Mosaic of Intra- and Interplant Variation

Because plants are perceived differently by pests and attacked on many different spatial and temporal scales, they are best defended by a mosaic of chemical variation. A different suite of terpenes may predominate in the resin compositional pattern in different tissues or organs of a plant, for example, sesquiterpenes in the leaves and diterpenes in the pods and trunk of *Hymenaea*. Within an organ, however, only the relative amounts or percentages of the individual components characterizing the compositional pattern vary. Thus there is both qualitative and quantitative intraplant spatial chemical variability. The relative concentrations of components may change with the development of the organ but stabilize with maturity; for example, caryophyllene may be low in young *Hymenaea* leaves but high in mature leaves, even characterizing the compositional pattern. Compositional patterns also vary between trees in a population as well as between populations. Hence, the intra- and interplant chemical variation is spatially and temporally heterogeneous, providing a defensive mosaic effect. As Whitham (1983) so aptly commented, "Variation as a defense may place plant pests in an evolutionary 'squeeze play' that even rapidly evolving pests cannot easily surmount."

Resin can operate defensively through a variety of mechanisms, producing toxic, deterrent, or inhibitory effects. Defensive dosages may involve individual components, combined components (acting additively or synergistically), or total quantity of the resin. Within the compositional pattern, individual constituents potentially have multiple ecological effects (Langenheim 1994). Although some individual compounds apparently have broad-spectrum effects, most are probably selected for particular effects on different organisms in different ecosystems. Compositional patterns that deviate qualitatively or quantitatively from a population average confer the greatest deterrence or resistance properties on some trees (Cates 1996), probably because pests do not have enough encounters to develop adaptations to them.

Role of Natural Selection

Variation in inherited traits such as resin composition is subject to natural selection, which favors genotypes with favorable traits. Selection responds by favoring genotypes for traits better adapted to new conditions. The degree of adaptation, however, is limited by the variability in traits of the individuals that constitute the population.

Since variation in some resin compounds is under genetic control, and since variation in composition may be effective against herbivores and microbes, selection pressure by these pests may lead to different kinds of selective pressures. When pests feed preferentially on a particular portion of the plant, defensively successful resin components can result in directional selection for these compounds. Furthermore, some evolutionists support the concept that selection plays a major role in maintaining variation, such as in resin compositional patterns (e.g., Endler 1986, Linhart and Grant 1996). Attack of phenotypically different host trees of a single species by a variety of dependent species such as insects, vertebrates, and microbes can result in multidirectional, diversifying selection of resin characteristics within plant populations. In some cases, diversifying selection can occur when a single pest species attacks populations of several host plant taxa in an area with similar but varying resin patterns.

Variation in Resin Quantity

The complex ways in which mixtures of compounds can mediate interactions is central to the defensive role of resin, but the capacity to produce copious quantities of resin is also important. First, the ability to sustain resin flow over a long period of time is particularly significant as a first line of defense and as a defense against repeated attack (Figure 5-1). Trees with greater resin flow, including its rate, tend to be more resistant than trees with similar resin composition that have lesser flow (Smith 1966). This initial response tends to occur in trees that have resin in an interconnected canal system (Chapter 3). Second, because the defensive success of individual compounds or groups of compounds is dosage dependent, the dose increases with greater quantity of resin. The total increase does not affect the *relative* amounts of the composition, only the concentration of the individual compounds in the resin. As total resin quantity increases, the individual component's concentration increases in proportion to its percentage of the composition. For example, although α-pinene may constitute 50% of leaf resin composition, with an increase in total quantity of resin from 2 to 5 mg (usually expressed per unit area or volume), the concentration of α-pinene increases from 1 to 2.5 mg. α-Pinene may not be effective at the lower concentration, but may be at the higher.

Although a complex set of genes is probably involved in the original allocation of carbon to the total pool of resin (Lerdau and Gershenzon 1997), abiotic factors such as light intensity and moisture availability influence the total quantity of terpenoid resin (Langenheim 1994). This phenotypic plasticity adds flexibility to the success of constitutive resins in that the genetically controlled composition may be constrained in coping with the numerous enemies without increase in the concentration of effective compounds through increase in the total quantity. Moisture availability and temperature can also influence exudation pressure (Lorio and Hodges 1968). Phenotypic plasticity of total resin production and exudation pressure (an aspect of the flow rate) adds another layer of complexity to the defensive properties of resin.

Resin Defense of Conifers

The potentially defensive role of genetically determined variation in constitutive resin composition is evident from examples of intratree, intrapopulation (tree-to-tree), and interpopulation differences in long-lived plants, which face onslaught from multiple enemies over time. In some conifers, localized seed dispersal generates a genetically patchy distribution, presenting pests with a mosaic of trees that vary in their attractiveness and resulting in a patchy distribution of pest damage. Trees left alone by one pest are often susceptible to another, and differences in susceptibility to various pests may generate local diversifying selection for monoterpene variation (Linhart et al. 1989, Linhart 1991). Tree-resin-mediated relationships between organisms have been documented in the greatest detail in economically important conifers such as ponderosa pine *(Pinus ponderosa)*, slash pine *(P. elliottii,* syn. *P. caribaea)*, and Douglas fir *(Pseudotsuga menziesii)*. Stem resins have been studied most because of interest in resin protection of valuable timber and because these resins are the ones most commonly used industrially (Chapter 7). Needle resins have received some attention because severe defoliation also has economic consequences. Although chemical defense characteristics have been little used in managed plantations, biotechnology looms on the horizon as a way to manipulate these defensive properties (Chapter 11).

Ponderosa Pine as a Model System

Ponderosa pine *(Pinus ponderosa)* is an excellent model system for investigating the evolutionary dynamics of interactions in which resin characteristics are subject to natural selection by different kinds of herbivores and parasites. More than 200 different insect species as well as various birds, mammals, fungi, and parasitic plants attack ponderosa pine in some part of its range (Linhart 1989). Long-term study plots of ponderosa pine are available for studies of herbivory, with data on tree fitness and genetics as well. That monoterpene composition and resin flow rate are under genetic control is known, and variation in monoterpene composition has been compared in six tissues of ponderosa pine in one of the study plots (Latta et al. 2000). The monoterpene composition of resin in wood, roots, and trunk exudate was similar within trees. Phloem and cone monoterpenes were similar, but needles were different. Exuded resin showed the least variability, and needles the most. Because of variation between tissues, different herbivores attack different tissues. To study defense mechanisms, resin variation in the attacked tissue has been and should be emphasized. For broad studies of the relation of natural selection and monoterpene variation in ponderosa pine, however, Latta and coworkers suggested that it would be useful to focus on resin exuded from the trunk because these monoterpenes are most characteristic of the tree as a whole.

Bark Beetle Interactions

Bark beetles (Scolytidae) are the single most destructive pests of conifers worldwide (Berryman 1972). Scolytid beetles of the genera *Dendroctonus*,

Ips, and *Scolytis* are the most common causes of tree mortality in North American conifer forests (S. Wood 1982). Species of these genera attack living trees exclusively, and some must kill their hosts to complete their life cycles. That, in addition to the beetles' having a highly specialized adaptive response to the host's defense, puts heavy selection pressure on conifers in developing their defense (Berryman et al. 1989). How trees actually defend themselves against attack by bark beetles is not completely understood, but resins are usually invoked to various degrees. The first line of defense is exudation of resin at the site of attack, with success dependent on flow rate and viscosity, which determine whether beetles are either pushed out or trapped (Figure 5-1). Following this initial defensive action, variation in resin composition begins to play a prominent role in the bark beetle–tree interaction.

Although evidence for resins goes back to Carboniferous time (Chapter 4), there is no evidence for scolytid bark beetles until the late Cretaceous. The first definitive record of the important genus *Dendroctonus* occurs even later, in the middle Eocene (Labandeira et al. 2001). *Dendroctonus* currently consists of three pinaceous host-specific lineages. Molecular studies of the phylogenies of the subfamilies of Pinaceae and species of *Dendroctonus* indicate phylogenetically congruent colonization of the plants by these bark beetles (Kelley and Farrell 1998). Because of their high level of host specificity and well-described life habits, *Dendroctonus* provides an almost ideal system for detecting coevolutionary associations. Labandeira and coworkers suggested that although historical data are tentative, "the interplay of modern-aspect host conifer and bark beetle terpenoid compounds (Pinaceae and *Dendroctonus*) has been in existence during the past 45 million years at the minimum."

Tree defense and beetle adaptation have been demonstrated by study of variation in monoterpene composition of constitutive xylem resin in California ponderosa pine (Sturgeon and Mitton 1982), attacked by western pine beetle (WPB; *Dendroctonus brevicomis*), and in Colorado ponderosa pine, attacked by mountain pine beetle (MPB; *D. ponderosae*). In California, where *Pinus ponderosa* is the sole host for WPB, Sturgeon (1979) reported that trees with rare monoterpene patterns apparently survived infestations at her study sites better than trees with common patterns. Rare trees were characterized by low levels of α-pinene, a pheromone precursor for WPB, and high levels of limonene, a toxin known to repel WPB. The presence of limonene alone, however, was not sufficient to deter WPB. Trees with high levels of limonene were tolerated by the beetle as long as they had sufficiently high levels of the pheromone precursor, α-pinene. The successful combination for the tree was high concentration of limonene and low concentration of α-pinene; thus WPB appears to simultaneously exert frequency-dependent selection pressure (for the rare monoterpene phenotype) and directional selection pressure (for the defensively successful combination of high limonene, low α-pinene). By selecting and rejecting trees on the basis of monoterpene composition, WPB can act as a powerful selective agent for California ponderosa pines. In the part of the range of the pine where WPB was most destructive, the resin was also characterized by high myrcene, required to synergize a beetle pheromone, and high β-pinene, which is least toxic. Sturgeon and Mitton (1986) concluded that, in California, the key to the multiple roles of the monoterpenes (pheromone precursors and synergists for the beetles, and antiherbivore defense for the tree) is great between-tree compositional diversity, which is maintained by the feeding habits of WPB.

Pinus ponderosa var. *scopulorum,* attacked by MPB in a Colorado study area, presents a scenario different from that in California (Sturgeon and Mitton 1982, 1986). Low compositional variation apparently did not affect resistance or susceptibility to attack by MPB in the relatively pure stands of these ponderosa pines. In contrast to WPB, MPB in Colorado has three major host species: ponderosa pine, lodgepole pine *(P. contorta* var. *latifolia),* and limber pine *(P. flexilis).* Sturgeon and Mitton suggested that MPB might not discriminate between chemical variants of ponderosa pine in Colorado because monoterpene compositional variation is less than one-thousandth that in California. Therefore, MPB may be indifferent to ponderosa pine variants in the biochemically uniform part of its range, and ponderosa pine, lodgepole pine, and limber pine may appear no more different to MPB in Colorado than three ponderosa pine chemical variants appear to WPB in California. This prediction was supported by finding no significant difference in the ability of MPB and WPB to digest pine terpenes, by analyzing their polysubstrate monooxygenase activities (Sturgeon and Robertson 1985). If MPB discriminates between resins of the three hosts rather than between variants of a single host, diversifying selection would lead to greater chemical differences between hosts and decreased chemical differences between variants of a single host. Hence, Sturgeon and Mitton (1982) suggested that feeding habits of these beetles help maintain patterns of monoterpene compositional variation in the resins of the three host pines in Colorado.

Dwarf Mistletoe Interactions

Dwarf mistletoe *(Arceuthobium vaginatum* subsp. *cryptopodum)* uses ponderosa pine *(Pinus ponderosa* var. *scopulorum)* as its primary host, with trees parasitized to different degrees. Mistletoe is a parasitic angiosperm that produces haustoria, similar to fungal mycelia, that penetrate the host tree and ramify into slender branches. Dwarf mistletoe stimulates the host to produce clumps of twigs and branches called witches'-brooms and can exert significant selection pressure on ponderosa pine populations. It is a highly damaging parasite on these trees in the southwestern United States, with large trees often supporting dozens of witches'-brooms. It commonly reduces growth and reproduction by 50% or more compared to uninfected trees and often kills the host over time. Snyder et al. (1996) found significant differences in xylem monoterpene composition (α-pinene and camphene) as well as resin flow rate between parasitized and nearby nonparasitized ponderosa pines. These differences point to the possibility of selection pressure by the dwarf mistletoe on resin composition, adding to the complexity of selection pressures on ponderosa pine resins exerted by bark beetles.

Abert's Squirrel Interactions

The relationship of the feeding pattern of Abert's squirrel *(Sciurus aberti)* to constitutive monoterpene variation is particularly interesting because the squirrel is a highly specialized and selective herbivore, restricted to stands of ponderosa pine *(Pinus ponderosa* var. *scopulorum)* in the southwestern United States and the Sierra Madre in Mexico. Few mammals are so dependent on a single plant. From fall to spring, the squirrel's diet is composed solely or primarily of inner bark (mostly phloem but including cambium and recent xylem tissue) stripped from twigs. The high degree of selectivity of these squirrels for particular trees (called feed or target trees) while adjacent, apparently similar trees are left alone (Figure 5-2) has been well documented at several locations.

Snyder's (1992) detailed study of the biochemical basis for feed tree selection at several Colorado sites showed that xylem resin flow rate and concentrations of β-pinene and β-phellandrene (Figure 1-3) are significantly lower in

Figure 5-2. Abert's squirrel feeding on young shoots in the upper part of a Colorado ponderosa pine tree *(Pinus ponderosa)*. The upper portion of two trees in the background have been heavily defoliated by the squirrels but another has been left undamaged. Such undamaged trees have been shown to be partially protected by resin characteristics of the inner bark (active phloem).

feed than nonfeed trees. Feeding packets laced with different concentrations of the monoterpenes, simulating the compounds as they would be exuded from the xylem into the phloem, provided direct evidence of the dosage-dependent deterrent effect for both β-pinene and β-phellandrene. Snyder (1993) further demonstrated that ponderosa pine trees, defoliated as a result of inner bark feeding by squirrels, suffered as much as 90% reduction in fitness components such as incremental growth, male and female cone production, and seed quality. Since resin flow rate and xylem monoterpene composition are under strong genetic control, Snyder concluded that Abert's squirrels may be important selective agents for monoterpene variation in the pine stands. In fact, Abert's squirrel and ponderosa pine provide a unique opportunity for assessing the importance of mammals in generating directional selection in host populations and a offer a context for comparison with the "more diffuse selection pressures" imposed by generalist mammalian herbivores (Snyder 1998).

Zhang and States (1991) suggested that differences in monoterpene composition of feed versus nonfeed trees at the Colorado sites and at their study site in Arizona might indicate that Abert's squirrels are adapted to monoterpenes characteristic of ponderosa pines in different geographic regions (Smith 1977).

Abert's squirrel is differentiated into six geographically disjunct subspecies. Snyder and Linhart (1997) used the opportunity to compare host selection by two squirrel subspecies, each closely associated with stands of ponderosa pine trees representing two geographically differentiated chemical races. Furthermore, there had been two major bark beetle epidemics and repeated dwarf mistletoe infestations at the Colorado site in the previous century. In contrast, beetle and mistletoe activity had been much less at the Arizona site. Thus the squirrels, varying themselves, were faced with a different suite of conditions, which could produce squirrel–tree interactions with different characteristics and consequences. Different chemically mediated feeding patterns, distinct for the two squirrel subspecies, suggested geographically differentiated herbivore–host interactions, as predicted in general by Zhang and States (1991). Snyder and Linhart concluded that geographic differentiation is potentially significant evolutionarily in chemically mediated plant–vertebrate interactions, which is generally not considered for such specialized herbivores.

Aphids, Deer, and Porcupines

The diversity of metabolic and physiological challenges that plants face from different organisms, representing a broad spectrum of behavior, has the potential of generating selection for different kinds of defense evolutionarily. With evidence from other studies for selection of ponderosa pine monoterpene characteristics by bark beetles, dwarf mistletoe, and Abert's squirrel, as discussed, Latta and Linhart (1997) analyzed the strength and mechanisms of selection for xylem resin flow and monoterpene composition through herbivory by several other organisms: aphids, deer, and porcupines. Although the patterns of resin characteristics and fitness components in the ponderosa pine population they studied provided some evidence for natural selection of resin characteristics, most patterns were mediated by some factor independent of these particular herbivores. However, Latta and Linhart suggested that past attack by herbivores no longer active in the population may have generated the observed association between resin characteristics and fitness.

Other Conifer Resin Interactions

The suite of studies of interactions between ponderosa pine resin and herbivores or parasites demonstrates the difficulty in generalizing about specific selective agents across a number of populations in even a single resin-producing tree species. In the ponderosa pine system, studies have emphasized various kinds of biotic selection pressures on constitutive resin characteristics. Here, the role of flow rate in southern pine–bark beetle interactions is discussed. Moreover, there is focus on the differences between constitutive and induced resins in conifers with particular reference to bark beetle–fungus interactions in trunk resin. Interactions with organisms attacking needles and the response in resin mixtures also receive attention.

Bark Beetles and Resin Flow Rate

Although resin flow was considered in some of the ponderosa pine studies, the emphasis was on the role of compositional variation in constitutive resin. Researchers in the southern United States have struggled for many years to understand the role of resins in interactions of the southern pine beetle *(Dendroctonus frontalis)* and southern pines such as *Pinus elliottii, P. palustris,* and *P. taeda.* These pines are the primary species used by the naval stores industry in the United States (Chapters 6 and 7). Lorio and Hodges (1985) focused on physiology, using the growth–differentiation balance hypothesis to explain the role of resin in the seasonal onset of beetle outbreaks. This hypothesis predicts that plants tend to grow when abiotic conditions are optimal for growth, and they differentiate (which includes production of secondary chemicals such as resins) when conditions for growth are suboptimal. The balance determines when during the growing season and in what tissues resin canals are formed and constitutive resin is produced, as well as factors affecting resin synthesis and flow rate from wounds. Thus seasonal activity of cambial meristems and seasonal production of resin form a basis for understanding the tree's initial resistance to beetle attack, and thus perhaps in preventing outbreaks.

According to the growth–differentiation balance hypothesis, trees should produce relatively little new resin in spring because conditions are favorable for growth. As water deficits develop, growth slows, photosynthate accumulates, and differentiation processes dominate, permitting production of resin (Lorio and Sommers 1986). Warm temperatures during summer should also facilitate resin flow from wounds through lower viscosity of the resin. Studies of the three southern pine species show that maximum resin yield occurs in mid- to late summer, with increased flow rate also associated with high temperatures (Lorio and Hodges 1985), as found by Smith (1977) for ponderosa pine. These results suggest that resistance to beetle attack should be low in spring and high in summer because of resin flow, a pattern substantiated by Lorio and Hodges. Thus these studies (which influenced thinking on and testing of the growth–differentiation theory of plant chemical defense; Herms and Mattson 1992) emphasized resin flow rate rather than compositional variation in defense. It seems likely, however, that compositional variation would prove to be an additional factor in southern pine chemical defense against bark beetles.

Resin, Bark Beetles, and Fungi

The interactions of scolytid bark beetles and fungi with resin provide excellent examples of the complexity of interactions relating to the mixture of terpenoid resins (with some compounds repelling and others attracting herbivores) and the importance of the plant's ability to defend against both insects and the fungi they carry. It is interesting to consider the history of the resin–fungus, bark beetle–fungus interaction in the geologic record. Resinous chemicals may initially have provided land plants some protection against fungi (Langenheim 1990, 1994). This is suggested by reports of wood-decaying fungi in silicified specimens of the late Devonian progymnosperm *Callixylon*, which show a pattern of decay similar to modern-day wood rot (Stubblefield et al. 1985). "Resinous-appearing material" in the wood is thought to be a response to the fungi. Evidence of resin appears in arborescent plants later, during the Carboniferous (Chapter 4). Several groups of Carboniferous swamp plants display bored wood (Scott and Taylor 1983), but there is no direct evidence of bored wood attributed to beetles until the Triassic, and none for scolytid bark beetles until the Cretaceous. Thus resin could have been a response first to fungal attack of the wood rather than to herbivore damage.

Scolytid bark beetles are not only the best-studied group that attack woody tissue, they also have a history of relationships with fungi and the trees that produce resins. Berryman (1989) suggested that primitive scolytids exhibited a saprophagus habit in wood that brought them into intimate contact with fungi and bacteria, then the bark beetles formed an association with plant pathogenic fungi. Survival of trees attacked by bark beetles depends not only on readily available constitutive resin, with its specific physical and chemical properties (as discussed), but also on the general physiological vigor of the tree, the number of attacking beetles, and the fungal spores they carry on their bodies. These fungi, which are introduced into the tree during beetle attack, are often pathogenic to conifers (Strobel and Sugawara 1986, Rane and Tattan 1987). In fact, Phillips and Croteau (1999) indicated that they are apparently the "principal and immediate agents of tree death." Death may be caused by fungal growth in conductive cells and production of toxins. Thus fungal infection aids the beetles by providing an aggressive mechanism for overcoming the tree's resinous defense, weakening the tree and thereby hastening its death. Monoterpene components of conifer resin have been shown to function in defense against both beetles and their associated fungi (Cobb et al. 1968, Hintikka 1970, Schuck 1982, Raffa et al. 1985, Bridges 1987). *Ceratocystis* (syn. *Ophistoma*) and *Trichosporum* are among the most common of the pathogenic fungi introduced into conifers by bark beetles. They stain the wood either blue or black and are generally referred to as blue-stain fungi.

Pine species store large quantities of constitutive resin in an interconnected canal system (Chapter 3) and there is generally only a small induced response to bark beetle attack and fungal infection (Cates 1996). On the other hand, conifers such as *Abies* that store resin in blisters (Chapters 3 and 8) lack the massive amounts of constitutive resin present in the extensive canal system of many pines. Thus *Abies* and some other conifer genera rely on rapid de novo synthesis of resin near the site of the infection. Induced resin synthesis begins immediately upon attack, and its monoterpene composition often differs from that of constitutive resin (Raffa and Berryman 1982, 1983; Raffa 1991; Raffa and Klepzig 1992). In grand fir *(A. grandis)*, for example, there can be significant qualitative and quantitative changes in monoterpene composition between constitutive phloem resin and resin induced by fungal attack (Figure 5-3). Responses to mechanical wounds in *A. grandis* resulted in less

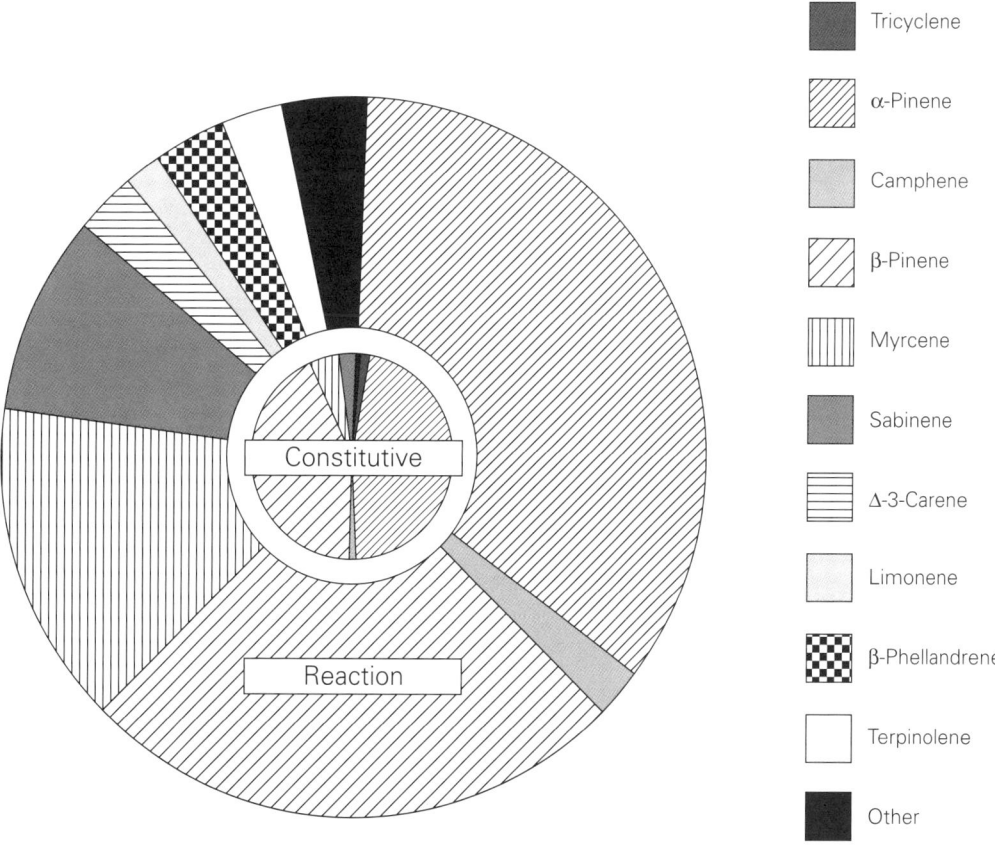

Figure 5-3. Comparison of constitutive and reaction (induced) monoterpene composition of *Abies grandis*, grand fir, phloem. Reaction resin is induced within 7 days after inoculation with *Trichosporium* fungus. The proportions represent the percentage of each compound determining the monoterpene composition. Defensive responses (full circle, both constitutive and reaction) involve both total and relative changes in composition. Adapted from Raffa and Klepzig (1992) with permission of Springer-Verlag.

resin accumulation and no proportionate chemical changes (Raffa and Berryman 1982), indicating a specific chemical response to presence of the fungus. In fact, numerous studies have shown that mechanical wounding does not produce the complete response that occurs with fungal infection (Cates 1996). Gijzen et al. (1992) also showed that the enzymatic machinery for making these induced compounds is not active in undamaged fir trees. Generally, terpenes in constitutive resins that occur in the largest quantities are induced the least, whereas the rarer monoterpenes are induced in greater quantity.

Time-course biochemical studies, using grand fir as a model system, have demonstrated that monoterpenes, sesquiterpenes, and diterpene acids are induced sequentially (Steele et al. 1998b). Monoterpenes appear first (within 24 hours of wounding) and continue to accumulate, thus serving a dual role in providing immediate toxicity and, later, acting as a solvent for the diterpenes. Maximum production occurs in approximately 10 days. Although delay in producing diterpene acids might appear maladaptive for entrapment of bark beetles, the delay fits the time course of their attack. The slower induction of sesquiterpenes could be significant in providing a secondary defense system, that is, allowing conversion of some compounds into those that are fungicidal or that act as juvenile hormones, disrupting insect development.

As induction proceeds, a lesion surrounds the invading beetle and fungus, the phloem tissue becomes soaked with resin, nutrients decrease, and digestibility of the tissue is greatly reduced. As this process continues, phloem and epithelial cells die, and eventually in vigorous trees, the necrotic tissue containing the bark beetle and fungus is sealed off from the nonlesion area by a periderm. Beetles die in the resin-impregnated lesions, and fungal growth is arrested. The complexity of this timed, multicomponent, multitargeted defense strategy could represent a mechanism for responding to the many generations of insects and fungi that conifers encounter during their long lives.

Pines such as lodgepole pine (*Pinus contorta* var. *latifolia*) that have little constitutive monoterpene variation and a generally low flow of resin in response to *Dendroctonus* attack are vulnerable to infestation by these bark beetles. In fact, heavy infestation often leads to more than 90% mortality where lodgepole pine is a pioneer or early-successional, fire-adapted species (Chapter 2). Nebeker et al. (1995), however, showed that total flow of resin from lodgepole pine is influenced by the interaction of wound size, tree diameter, and whether the tree is diseased. Generally, resin flow was greater from larger wounds and larger trees as well as from trees diseased with mistletoe, blister rust, and root rot. This increased flow of constitutive resin was important in resistance to bark beetles. However, Raffa and Berryman (1987) hypothesized that beetle epidemics actually favor reestablishment of this pine. Epidemics produce many dead trees, which are then susceptible to large-scale fires. Lodgepole pine has a serotinous cone, that is, resin holds the seeds in the cone until the resin melts during the fire. Following fire, seeds are so abundant that they often result in even-aged, pure stands of this pine.

In contrast to lodgepole pine, spruces such as Norway spruce *(Picea abies)* produce a complex, anastomosing traumatic canal system following wounding (Chapter 3). Nagy et al. (2000) showed experimentally that this system develops within 18 days of massive inoculation with a pathogenic fungus *(Ceratocystis polonica)* carried by a bark bectle *(Ips typographus)*. The time it takes for development of the complex system of ducts is too slow to protect against massive attacks of bark beetles, which may involve hundreds or thousands of attacks on a tree within a few days. If attacks are slow, however, Nagy and coworkers thought that this traumatic response system could reduce both beetle and fungal propagation (Figure 5-4). Their conclusion is consistent with observations that traumatic canals are formed both at wound sites and at a considerable distance from them.

Tomlin (1998) further showed that weevil-resistant white spruce *(Picea glauca)* produces multiple rings of traumatic resin canals. These studies of the development of the resin secretory system also point out the differences in kind of response in different kinds of trees even within the same family (Pinaceae) and the ecological importance of the kind of secretory tissues in a particular taxon. The Norway spruce study also suggests that defense by terpenoid resins is augmented by polyphenols, at least in the induced system.

Needle Resin Induction by Herbivores

Litvak and Monson (1998) reported the first evidence for induction of monoterpenes in conifer needles. It had been assumed that herbivore induction of these terpenoids did not occur in leaves (Cates 1996). Litvak and Monson found that monocyclase activity was induced 4–8 days after simulated herbivory on needles of ponderosa pine *(Pinus ponderosa)*, lodgepole pine *(P. contorta* var. *latifolia)*, white fir *(Abies concolor)*, and Engelmann spruce *(Picea engelmannii)*. In ponderosa pine, herbivory by tiger moth larvae *(Halisdota*

Figure 5-4. Constitutive resin exuded from *Picea* ×*lutzii,* Lutz spruce, a natural hybrid of *P. glauca* × *P. sitchensis,* abundant in forests of the Kenai Peninsula, Alaska. Moderate levels of attack by bark beetles *(Dendroctonus rufipennis)* in this case have allowed the tree to respond with abundant defensive resin flow. Courtesy of R. Wheeler.

ingens) induced a significantly larger response (4.5-fold increase in mono-cyclase activity) than simulated herbivory (2.5-fold increase). Despite the increase in monoterpene synthesis, there was not a significant increase in the total monoterpene pool size in wounded needles compared to controls. Litvak and Monson also observed large increases in monoterpene volatilization in response to wounding and concluded that volatile losses caused by tissue damage compensated for herbivore-induced synthesis, resulting in no change in pool size. This result may account for previous studies that reported no induction in conifer needles because researchers did not analyze cyclase activity but only measured pool size.

Foliar Spruce Budworm Effects on Phloem Resin

Wallin and Raffa (1999) showed that foliar feeding by the budworm *(Choristoneura pinus)* altered constitutive and induced monoterpene content and composition in jack pine *(Pinus banksiana)* stem phloem. Total monoterpene concentrations were always higher in induced resin than in constitutive resin, but the extent and rate of monoterpene accumulation and proportionate changes in monoterpenes during induction varied with defoliation intensity. Arrival rates of bark beetles *(Ips grandicollis)* were more strongly associated with the chemical composition of constitutive resin than with that of induced phloem resin. Chemical changes during the induced host responses were also associated with *Ips* behavior. The ratio of α- to β-pinene during the early days of induction was consistently associated with *C. pinus* defoliation and suscep-tibility to *I. grandicollis*. These results indicate that direct injury to one tissue, foliage, can also cause changes in another, stem phloem. In fact, such changes can be greater in the tissue not directly affected. The influence of defoliation intensity on monoterpene accumulation suggested a link between production and allocation of carbon within *P. banksiana,* but duration of the stress (defo-liation) also appeared to alter the relationship between stress and terpene profiles. These results further supported the need for integration of growth–differentiation balance and plant stress theories (Herms and Raffa 1995).

Foliar Spruce Budworm and Sawfly Interactions

Major defoliators of conifers such as spruce budworms and sawflies are influ-enced by qualitative and quantitative variation in resin components and mix-tures, and differences in these between branches, canopy levels, individuals in a population, populations, and species (Cates et al. 1991, Cates 1996). Resistance patterns observed in both laboratory and field studies have sometimes shown that one or two terpenoid compounds significantly reduce either spruce budworm *(Choristoneura)* or sawfly *(Rhyacionia* or *Neodiprion)* fitness. More commonly, however, toxic or inhibitory effects on different life stages of these defoliating insects are the result of a mixture of several compounds. For example, Cates et al. (1983) showed that bornyl acetate increased western spruce budworm *(C. occidentalis)* mortality 88% on Douglas fir (*Pseudotsuga menziesii*, Plate 4); however, bornyl acetate is often accompanied by camphene, tricyclene, terpinolene, myrcene, α-pinene, and other monoterpenes (Figure 1-3) in a mixture that is an effective inhibitor of *C. occidentalis*. Mattson et al. (1983) found that a suite of monoterpenes (e.g., camphene, β-pinene, β-phellandrene, and terpinolene) from balsam fir (*Abies balsamea*, Plate 2) was effective against *C. fumifirana*. Ikeda et al. (1977) reported that a single diterpene acid was highly effective against two sawfly *(Neodiprion)* species on white spruce *(Picea glauca)* whereas several of these acids were necessary for successful defense against the sawfly *Pristiphora erichsonii* on American tamarack *(Larix laricina)*. Generally, higher concentrations of diterpene acids in young needles were correlated with avoidance by many sawfly larvae (Cates 1996).

Gambliel and Cates (1995) studied terpenoid changes during needle maturation throughout the canopy of Douglas fir in relation to western spruce budworm. Changes in the relative composition of major compounds (tricyclene and camphene, and bornyl acetate) occurred during needle development. Leaf maturation varies with canopy level, that is, beginning in the lower level and proceeding upward. Differences between canopy levels in the timing of onset of greater concentrations of tricyclene and camphene, and bornyl acetate, are significant because a high concentration of these compounds is known to affect western spruce budworm adversely. Furthermore, changes in terpenoid concentration during needle maturation often correlate with budworm feeding patterns (Cates and Redak 1988). Although budworm densities were usually highest in upper and middle canopy levels, bornyl acetate accumulated at lower canopy levels and more slowly during the first 2 weeks following bud break, coinciding with the time the budworms were feeding heavily on young foliage. Such variation results in an ever-changing chemical mosaic confronting western spruce budworm and raising the possibility of selection for some of the resin variation patterns.

Foliar Endophytic Fungus Interactions

Fungi that have endophytic portions in their life cycles, colonizing aerial tissues, are ubiquitous in conifers. Although some of these fungal symbionts may be lethal, many do not harm plants during their endophytic phase. However, some otherwise harmless endophytes may cause disease when plants are under stress. Most leaf endophytes in conifers may remain inactive after colonizing the leaf, their active growth and reproduction occurring when the leaf senesces or is damaged (Carroll 1988). Espinosa-García et al. (1996) investigated the possibility that the volatile fraction of resin has a role in maintaining this inactivity in needles of coastal redwood *(Sequoia sempervirens)*. Four redwood monoterpene compositional patterns were uniformly inhibitory to some of the most common endophytic species, and differently so to others, with susceptibility varying widely within and between fungus species (Espinosa-García and Langenheim 1991a). This diversity of response suggested that redwood terpenoids may have differential intraspecific significance in regulating pathogenic and nonpathogenic activity in foliage. Sabinene and γ-terpinene, prominent compounds in the compositional patterns that were the most inhibitory to endophytes, acted additively on each of six species of endophytes when assayed both singly and together (Espinosa-García and Langenheim 1991b). Furthermore, the dose–response curve to the volatile terpenoids for the nonpathogenic, host-specific, and suspected redwood mutualist *Pleuroplacanema* sp. was finely tuned to changes in terpenoid concentrations occurring in mature and in senescent needles (Espinosa-García et al. 1993). Although terpenoids inhibited the activity of a pathogenic conifer specialist, *Pestalotiopsis funerea,* they did so less than for the possible mutualist *Pleuroplacanema* sp. Thus the effects of the terpenoids were dosage dependent in relation to the degree of specialization of the fungal endophyte (Espinosa-García et al. 1996).

Ecological Roles in Tropical Angiosperms

Copious Resin Production in Tropical Trees

A question that has intrigued chemical ecologists is why the greatest diversity of tree taxa producing copious amounts of endogenous resin occurs in the moist tropics or subtropics (Chapter 2). Langenheim (1975, 1984, 1990) hypothesized that conditions favoring active photosynthesis occur through-out the year in the moist tropics, providing abundant carbon for growth. Fur-thermore, low utilization of carbon in slow-growing trees in mature forests (characteristic of most copious tropical angiosperm resin producers) on typ-ically low-nutrient tropical soils would make ample carbon available for copi-ous resin production (Chapter 1). A significant addendum was provided by Lincoln and Couvet (1989), who demonstrated that carbon availability alone cannot account for increased terpenoid synthesis; hence, they suggested that copious production may also be influenced by defensive pressures. There-fore, a further hypothesis supported the initial one, that resin provides trees with defensive properties, which would be more necessary against the great diversity of phytophagous insects, mammals, and pathogenic fungi that flour-ish in the favorable conditions of the humid tropics. In a corollary to this defense hypothesis, Langenheim (2001) posited that some defensive proper-ties, such as compositional variation, could also evolve under biotic selective pressures, basing this corollary on the fact that plants have a variety of depen-dent organisms that could potentially produce either directional or diversify-ing selection if the trait were under genetic control and related to the fitness of the tree. Copious resin production and compositional variation could be important in the coevolutionary struggle of these organisms, as shown in the temperate zone *Pinus ponderosa*.

Despite the diversity of copiously resin-producing tropical angiosperms, less ecological information has been available about them than temperate zone conifers. Progress is being made regarding defense against insects and fungi in three important tropical plant families: Fabaceae (or Leguminosae), Burseraceae, and Dipterocarpaceae. The bee pollinator–resin reward system also has been investigated in Clusiaceae and Euphorbiaceae. In some of the research, an evolutionary perspective has been added by using more recently developed molecular and phylogenetic techniques.

Hymenaea and *Copaifera* as Model Systems

Langenheim and coworkers selected *Hymenaea* (Plate 16 and Figure 2-9) and *Copaifera* (Figures 5-5 and 7-9) as model systems for long-term studies of resin production in tropical angiosperms because of their prominence as members of one of the most important tropical resin-producing families (Fabaceae; Chapter 2) and their history in producing amber since at least the

Oligo-Miocene (Chapter 4). They also were selected because of their widespread Neotropical distribution and economic importance (Chapters 7 and 9). Although both genera occur in Africa (*Hymenaea* in East Africa and *Copaifera* in West Africa), they are primarily distributed in the New World. Amazonia is the center of their distribution, from where *Hymenaea* radiated to the limits of the tropics, into Mexico to the north and throughout Brazil to the south, with *Copaifera* extending almost as far (Langenheim 1973). Throughout southern Central America and South America, co-occurring species across a wide range of tropical lowland ecosystems (rain forest to desert thorn forest) provide opportunities for interesting ecological comparisons.

Figure 5-5. *Copaifera pubiflora,* which commonly co-occurs with *Hymenaea courbaril* in Venezuelan dry forests. Inset shows the common microlepidopteran *(Stenoma)* and the damage that this major defoliator does to *Copaifera* and some *Hymenaea* species; the larvae tie leaves together to seek protection from predators and feed on the surface of the leaves within the mass.

Similarities and Differences in Resin Chemistry

Although the nonvolatile fraction of conifers and caesalpinioid legumes consists of diterpenes, monoterpenes predominate in the volatile fraction of conifers, and sesquiterpenes in legumes (Figure 1-3). *Hymenaea* and *Copaifera* have essentially the same suite of predominant sesquiterpenes but different diterpenes and different ratios of sesquiterpenes to diterpenes in some organs (Langenheim et al. 1978). This intraplant variation suggests interactions with different herbivores and pathogens known to be associated with particular organs. In the leaves and primary stem tissues, sesquiterpenes predominate over diterpenes in both *Hymenaea* and *Copaifera* (Langenheim 1981, 1995). In secondary stem (trunk) tissue, however, diterpenes that control resin viscosity occur predominantly in *Hymenaea* but only in African *Copaifera*. Labdane diterpenes, which readily polymerize, result in an excellent wound sealant as well as permitting the formation of amber (Chapters 4 and 9). In the New World, fluid sesquiterpenes predominate in most species of *Copaifera*. The difference in composition of *Copaifera* trunk resins in Africa and the New World is reflected in their different uses by humans (Chapters 7 and 9).

In both *Hymenaea* and *Copaifera*, diterpenes are primary constituents of pod resins, but in *Copaifera* the pod is dehiscent and thin-walled. The indehiscent pod of *Hymenaea* has generated ecological interest, beginning with the difference between New World and African pods (Figure 5-6). In Africa the pods have an exaggerated verrucose surface loaded with resin and a structural group of diterpenes different from that of pods in the New World. In all species, various amounts of resin have been observed to exude following wounding, sometimes to the extent of encasing the pod. In the New World, *H. courbaril* pods resist bruchid beetle attack but can be heavily attacked by rhinochinid weevils. The possibility that differences in the structure of the surface, amount of resin, and chemistry of African and New World *Hymenaea* pods are at least partially the result of selection pressures from different predators is supported by Janzen's (1975) study of pod resins and rhinochinid weevil attack. In Puerto Rico, where there are no pod predators, the pods have very little resin. On the other hand, in Costa Rica, where rhinochinids are abundant, pods have correspondingly high levels of resin.

Resin synthesis occurs at different times in different organs during the early stages of ontogeny of *Hymenaea,* raising the possibility of conflict in the metabolic costs of defense versus growth. The hypocotyl (stem axis, bearing the root primordium, below the first leaves in the embryo) and young roots of *H. courbaril* grow rapidly, the roots attaining depth quickly, but there is evidence neither for terpenoids nor their secretory structures until secondary development of the root (Langenheim 1967). On the other hand, the epicotyl (stem primordia above the first leaves in the embryo) and aboveground young stem and leaves grow slowly, and the secretory structures in them are rapidly

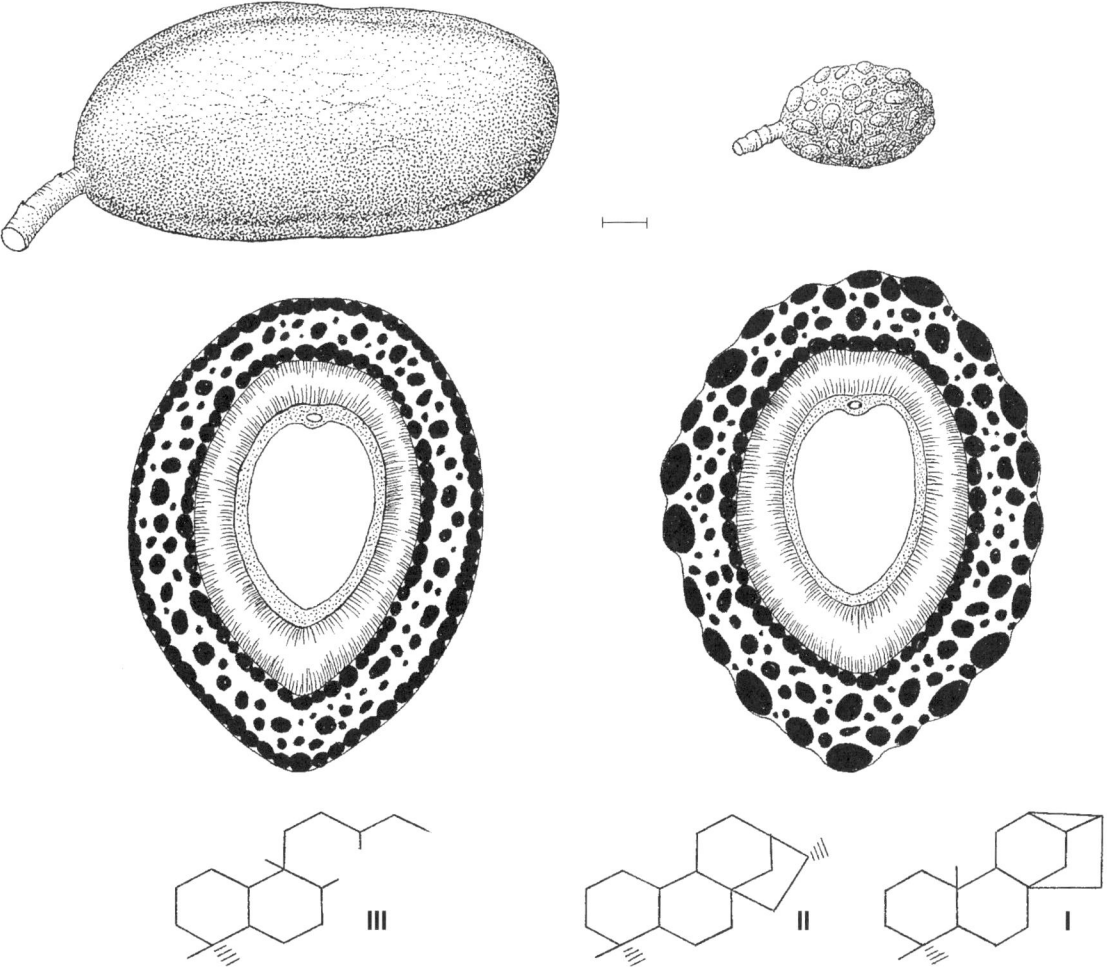

Figure 5-6. Comparison of pods of *Hymenaea courbaril,* representative of the New World species (left), and the African *H. verrucosa* (right). Scale, 2 cm. Basic morphology of the pods is shown in cross section, revealing the several layers of resin pockets (black) in the pod wall, which provide several defensive barriers against insects attempting to reach the seed. The labdane diterpene structures that predominate in the pod resins are contrasted: New World (**III**), Africa (**I, II**)

formed and filled with resin. Thus rapid root growth in *H. courbaril*, which often occurs in relatively dry habitats, is apparently more important to survival than defensive chemistry. Above ground, however, the immediate investment in terpenoids and their secretory structures may be a response to potential defensive demand. Resin production in the root occurs with secondary tissue development, and there is a shift from primarily sesquiterpene to diterpene compounds. Stem resin is not abundantly produced in saplings but increases with maturity of the tree; there is also a shift to predominantly diterpene compounds from sapling to mature tree, as in the root.

Genetic Control of Sesquiterpene Compositional Patterns in Leaves

Repeated, discrete, quantitative compositional patterns (called compositional types or chemotypes) of predominant foliar sesquiterpenes occur in different populations of the widespread *Hymenaea courbaril* and other *Hymenaea* species (Langenheim et al. 1978). Langenheim and Stubblebine (1983) noted greater compositional diversity in leaves of rain forest populations with progressively less diversity in populations in drier forests. Although some phenotypic variation was found, there was no significant variation in relative amounts (i.e., the percentages) of the compounds defining the type. In subsequent studies, *Hymenaea* and *Copaifera* were found to have a number of compositional types in common, leading to consideration of whether similar chemistry in co-occurring species of the two genera defends against similar herbivores.

In contrast to the known genetic control of many monoterpenes in conifers, the genetic background was unknown for most sesquiterpenes in *Hymenaea* and *Copaifera* resins. Since it was unlikely that information about genetic control of these sesquiterpenes would become available in the near future, Langenheim and coworkers designed experiments to assess the phenotypic effects of abiotic conditions on resin composition as a prelude to sorting out possible biotic selective effects. Sesquiterpene compositional types in mature leaves of individual *Hymenaea* and *Copaifera* saplings did not change with experimental alteration of light duration and intensity, temperature, moisture stress, total nutrients, and nitrogen plus light intensity (Langenheim 1995, 2001). In contrast, total quantity of resin in mature leaves increased significantly with increased light intensity (Langenheim et al. 1981), an effect that has been shown for monoterpenes in other plants (Gershenzon and Croteau 1991). However, there was no significant intracanopy and seasonal variation in sesquiterpene composition of mature leaves on adult trees (Langenheim et al. 1977, Macedo and Langenheim 1989b). On the other hand, leaves at different developmental stages, with their differences in composition (Crankshaw and Langenheim 1981, Langenheim et al. 1986b), provide variation through the canopy as they do in Douglas fir (Gambliel and Cates 1995).

The essential constancy in concentrations of the major constituents under a diversity of abiotic conditions does not deny that some variation may occur in minor components not involved in defining the compositional types. Minor compounds (often defined as less than 1% of the composition) might have additive or synergistic effects but are difficult to evaluate because they may fall within both analytic and sampling errors.

Sesquiterpene Variation and Interactions with Insects and Fungi

Lack of significant phenotypic plasticity (caused by abiotic factors) in the leaf resin compositional types of individual *Hymenaea* and *Copaifera* trees suggested a possible role for biotic selection in maintaining variation between types in different trees attacked by different pests. Lepidopteran insects and leaf pathogenic fungi were used to test the defensive properties of leaf resin in *Hymenaea* and *Copaifera*, to see if variation in compositional type could be a response to biotic selection pressures. Stubblebine and Langenheim (1977) demonstrated that differences in total quantity of leaf resin in *H. courbaril* had differential dosage effects on mortality of larvae of the generalist lepidopteran *Spodoptera exigua* in laboratory experiments. In a subsequent experiment, Langenheim et al. (1980) showed significant differences on larval, pupal, and adult mortality as well as time to pupation of *S. exigua* by the compositional types of *H. courbaril* that occur most commonly in Central America and much of South America. Varying dosages of the dominant compounds defining the compositional pattern (caryophyllene in one type and α- and β-selinene in another; Figure 1-3) in the mixture of about 15 sesquiterpenes had different effects on the insect's life stages.

The leaf-spotting fungal pathogen *Pestalotiopsis* (syn. *Pestalotia*), which occurs on *Hymenaea* and *Copaifera* species throughout their New World distribution (Arrhenius and Langenheim 1986), was used to analyze possible selective effects of fungi on sesquiterpene composition. Arrhenius and Langenheim (1983) demonstrated in cultural experiments that caryophyllene oxide (rather than caryophyllene) is the resin compound most inhibitory to various *Pestalotiopsis* species and is also an effective fungistatic agent against a wide range of fungi. Thus caryophyllene oxide is a resin compound that may function in general resistance, that is, affecting a broad spectrum of fungi on plants at specific dosages for different fungi. Hubbell et al. (1983) reported that caryophyllene oxide defended *H. courbaril* leaves against fungus-growing, leaf-cutting ants *(Atta)*, which can extensively damage undefended trees in a short time. The authors thought that the ants avoided cutting and collecting *Hymenaea* leaves because caryophyllene oxide probably killed or inhibited the growth of the fungi, which they farm for food. The culture medium in which the fungi grow consists of a fermenting mass of chewed-up leaves prepared by these ants. Howard et al. (1988) subsequently showed that caryophyllene from *Hymenaea* and other tropical trees had no detectable effect on survival of adult *A. cephalotes* ants, although it decreased growth of its mutualistic fungus by 50%. On the other hand, caryophyllene oxide produced strong deterrent effects on the ant and completely inhibited fungal growth.

These studies of caryophyllene and caryophyllene oxide demonstrate that components of a resin mixture can play different roles in interactions with several organisms, as shown for other chemicals in plants (Langenheim 1994). Caryophyllene and its oxide are interesting in that caryophyllene is a monogenic compound common in plants. Caryophyllene may seldom occur in sufficiently high concentration in most plants to be effective as a defense, however, which may result from the fact that the dominant allele codes for low concentration of the compound. Even in *Hymenaea* and *Copaifera,* when caryophyllene dominates the compositional type in terms of percentage, its actual concentration may be lower than dominant compounds in other compositional types.

Despite the presence of essentially the same leaf sesquiterpenes in co-occurring populations of *Hymenaea* and *Copaifera* in Brazilian cerrado (woodland and savanna), compositional differences between the genera (along with other parameters such as leaf toughness) partitioned the lepidopteran herbivore community (Langenheim et al. 1986a). In the savanna, the leaf compositional type of *H. stigonocarpa* var. *pubescens* differed from that of *C. langsdorfii*. During leaf development, *H. stigonocarpa* attained the highest concentration of terpenoids before the leaf was fully expanded (Crankshaw and Langenheim 1981) whereas *C. langsdorfii* attained the highest levels when the leaf was fully mature (Langenheim et al. 1986b); *Hymenaea* leaves also attained twice the degree of toughness of those of *Copaifera*. *Hymenaea stigonocarpa* had very low levels of herbivory during the study period, primarily by an oecophorid leaf-tying moth with a wide distribution on *Hymenaea* species, and other geometrid lepidopterans. In contrast, *C. langsdorfii* was heavily attacked, primarily by a specialist oecophorid restricted to *Copaifera,* and a generalist gelechiid lepidopteran.

Significant differences in foliovore activity on *Copaifera langsdorfii* occurred over a 5-year period at two woodland sites only 75 km apart and with similar climate and vegetation (Macedo and Langenheim 1989a). At one site, an outbreak of the specialist oecophorid leaf tier *Stenoma* aff. *assignata* (Figure 5-5) resulted in consumption of 48% of the mature leaf area of the trees, followed by only sporadic occurrence of these insects in later years. Heavy attacks by particular herbivores one year, followed by essentially an absence of them the next, have occurred in other tropical forests (Wolda 1978). Similarly, in another year at the same site, a generalist gelechiid defoliated 52% of the trees with immature leaves. However, this attack occurred sufficiently early in the rainy season to allow reflushing of leaves. In contrast to the site where *C. langsdorfii* had been heavily damaged by the two different kinds of infestations, the trees at the other site sustained for 5 years what Coley and Aide (1991) considered the "normal range" of herbivory for tropical trees. Total amount of sesquiterpenes and concentration of caryophyllene were significantly higher in trees with less damage from oecophorids at both sites. Much higher total amounts of sesquiterpenes and more variable concentrations of three components tree to tree may have been important in restricting herbivores such as oecophorids and gelechiids on *C. langsdorfii.* As in ponderosa pine and Douglas fir, this kind of tree-to-tree variation makes the adaptation of insects more difficult. Furthermore, differences between these sites may reflect historically different intensities and types of herbivory. These *Copaifera* studies reinforce the importance of recognizing the variability of both resin characteristics and herbivory in apparently similar nearby sites, a situation of interest for both ecological understanding and utilization of the trees for resin.

Macedo and Langenheim (1989a) expected that the fitness of the trees at the heavily defoliated site might have been affected compared to the site that sustained moderate herbivore damage. Although the canopy sizes of the trees at the two sites were similar, crown vigor and diameter of the trees were significantly lower at the heavily damaged site. Growth-ring data, which could have assisted in evaluating herbivore effects on diameter growth, are generally unreliable in most tropical habitats unless carefully analyzed over a long period of time, posing a problem in obtaining immediate long-term fitness data.

Ontogenetic Differences in Sesquiterpene Leaf Defense

Higher levels of chemical defense for the protection of vulnerable juvenile plants might be expected, but this relationship has received relatively little attention except in studies of interactions with mammals in the subarctic (Bryant et al. 1991). Juvenile plants of Amazonian species of *Hymenaea* and *Copaifera* stand out as being highly successful under the parent tree, and seedling success is known in other rain forest resin-producing trees, for example, *Eperua* (Forget 1992) and *Gossweilerodendron* (Hart et al. 1989) as well as those producing essential oils, for example, *Nectandra* (Sánchez-Hidalgo et al. 1999). Higher rates of mortality under conspecific trees have been recorded for a range of species, however. In fact, Janzen (1970) and Connell (1971) independently proposed that seed and seedling predation by specialized herbivores and pathogens greatly reduce the chances of plant establishment under the canopies of conspecific trees. They further concluded that this could explain the high species diversity and large distances between conspecific adult trees characteristic of rain forests.

Numerous seedlings and saplings of several *Hymenaea* species *(H. courbaril, H. intermedia,* and *H. parvifolia)* surviving under conspecific trees in Brazilian Amazonia primarily had leaf sesquiterpene compositional types different from that of the parent tree (Langenheim and Stubblebine 1983). In fact, after several years at one Amazonian site, seedlings with the compositional type of the parent tree had been culled. Langenheim and Stubblebine hypothesized that the culling occurred because genetically controlled sesquiterpene compositional types different from that of the parent tree conferred a selective advantage against specialized insect herbivores. In this scenario, specialist lepidopteran larvae, which were adapted to the foliar chemical type of the parent tree, migrated to the ground and preferentially fed on plants with the chemical profile to which they were adapted. This could result from either drastic selection of early instar larvae or developmental habituation. Previously, Langenheim et al. (1980) demonstrated experimentally that lepidopterans can be differentially deterred by different compositional types.

Using chemical similarity analyses, Sánchez-Hidalgo et al. (1999) also found that leaf terpenoid profiles in most seedlings of *Nectandra ambigens* (Lauraceae) were significantly different from that of the nearest adult in a Mexican rain forest. They suggested that the cause of the chemical differences is probably differential mortality of seedlings, resulting from biological agents similar to that proposed by Langenheim and Stubblebine (1983).

Amazonian rain forest study sites presented difficulties for continued detailed study of the high level of sesquiterpene variation there. Since seedling success under parent trees was also prominent in woodland sites, Macedo and Langenheim (1989b) studied the relationship of *Copaifera langsdorfii* parent trees and progeny seedlings to their primary oecophorid foliovore. This field experiment demonstrated that sesquiterpenes protect leaves of seedlings better than leaves of the adult tree. Mean total sesquiterpene concentration of leaves was three times greater in seedlings than in adults and resulted in 48% mortality of oecophorid larvae or pupae of the specialist leaf-tying moth *(Stenoma)* reared on seedlings compared to no mortality on the adult. In this experiment, however, lack of mortality on the adult did not suggest that leaves on the tree could be devoured at will because in some leaves the amount of total sesquiterpenes exceeded the threshold of tolerance observed over the range of variation in seedlings.

This long-term, pioneering study of two tropical angiosperm genera lays a foundation for comparing the defensive properties of such tree resins with those that have been investigated in more detail in temperate zone conifers. Since *Hymenaea* and *Copaifera* are commercially important, further studies of the defensive role of terpenoid variability may be useful in the management of natural forest and silvicultural systems of various resin-producing trees (Cates and Redak 1988, Langenheim 1990, Barnola et al. 1994, Cates 1996). Although these kinds of studies in the tropics have received relatively little attention by chemical ecologists, ecologically and evolutionarily related studies of tropical angiosperms have increased since the 1980s.

Other Angiosperm Resin Interactions

Bursera and Beetle Coevolution

Bursera is a predominantly New World genus whose major foliovore is a chrysomelid beetle *(Blepharida)* with numerous species that are feeding specialists. When leaves are damaged, some but not all *Bursera* species release abundant terpenoid resin. Stored under pressure in canals, it is released either as a squirt (Plate 28) or a stream that bathes both surfaces of the leaf. Evans et al. (2000) and Becerra et al. (2001) compared the resin chemistry of *Bursera* species that squirt with those that are nonsquirting, and the behavior of *Blepharida* species that are specialized on them. The resin composition of squirting species is more volatile, dominated by one or two monoterpenes, whereas the composition of the nonsquirting species is more complex, comprising a less volatile mixture of mono-, sesqui-, and diterpenes. The defensive significance of complex chemical mixtures has been discussed previously for various taxa, but the adaptive value of a simple chemistry has generally not been described. Becerra and coworkers thought that the explanation in the case of *Bursera* may arise from a few volatile and more liquid monoterpenes that can be squirted, producing a physical as much as a chemical effect. Evans and coworkers found that larvae of *Blepharida* species cut the canals of resin-squirting *Bursera* species to avoid the squirt; in contrast, larvae on nonsquirting species do not sever the canals. Furthermore, *Blepharida* species feeding on *Bursera* with simple chemistry are monophagous (feeding restricted to one species) whereas *Blepharida* species feeding on *Bursera* species with a more complex chemistry are polyphagous (feeding more generally on several species).

Becerra and Venable (1990) and Becerra (1994) studied the response of an undescribed *Blepharida* species to resin release from *Bursera schlechtendalii,* a succulent shrub in arid areas of central Mexico and Guatemala. Becerra and Venable have called the release of the great quantity of resin that coats the leaf surface when a canal is broken "the rapid bath response." Becerra (1994) referred to terpenes that were squirted in a syringe-like stream 5–150 cm (Plate 28) as a squirt-gun defense. There is variation between plants, and within a plant, in the proportion of leaves releasing resin in the bath or syringe manner following damage. The strength of the plant's response depends on water potential (algebraic sum of the solute potential and the pressure potential or wall pressure; the potential energy of water) and, hence, varies with moisture availability.

In their counterdefense, *Blepharida* larvae may damage plants substantially because they can avoid the squirt and bath responses by either mining the leaves between the resin canals or by cutting the canals and letting them drain before feeding on the leaves. Nevertheless, Becerra (1994) showed that resins still protect *Bursera* against this beetle. Both naturally occurring and experimentally placed larvae had a higher mortality on highly responsive plants. Furthermore, on highly responsive plants, larvae spent more time either cutting the resin canals or trying to clean themselves (if they got a non-lethal dose) than feeding, and grew more slowly. Therefore, although *Bursera* resin responses do not provide a completely effective defense, Becerra thought that it resulted in a "handling-time cost" on *Blepharida* larvae, reducing their growth rate and further increasing their risk of predation. Thus the variability in the plant's response and the insect's success constitute a coevolutionary struggle between the two.

Becerra (1997) also used the *Bursera-Belpharida* system to test the evolutionary importance of chemistry on herbivore–plant host shifts. Factors that direct the evolution of host shifts by phytophagous insects have been central to the study of plant–insect interactions since the seminal paper by Ehrlich and Raven (1964). Ehrlich and Raven postulated that shifts from old to new plant hosts are mediated by chemical similarity between hosts, and thus that host plant chemistry should leave its trace in phylogenetic patterns of host shifts. However, detailed quantitative studies to test their hypotheses had to await the development of modern molecular and phylogenetic techniques to reconstruct accurate host and herbivore evolutionary trees. To test the ancient and diverse *Bursera-Blepharida* system, Becerra reconstructed their molecular phylogenies and obtained the monoterpenoid profiles of the host plant resins. Statistical analysis showed that historical patterns of host shifts strongly correspond to host chemical similarity. Becerra concluded that resin chemistry in this system played a significant role in host shifts by the phytophagous insects. Becerra and Venable (1999a) further quantified the host range coincidence of *Bursera* relative to its chemistry and its phylogenetic relatedness. Overall, *Bursera* chemistry best explained *Blepharida*'s macroevolutionary patterns of host use. These studies of *Bursera* provide examples of the combination of ecological and evolutionary perspectives, enhancing the understanding of the mediating role of resin monoterpenes in tropical angiosperms.

Dipterocarp Resins, Termites, and Fungi

Members of the Southeast Asian plant family Dipterocarpaceae produce copious amounts of resin, which is used commercially (Chapters 7 and 9), and their timbers are well known for resisting various kinds of biological attack. Untreated dipterocarp timbers are reported to be very resistant to fungal invasion, and volatile terpenes from *Vateria indica* inhibit bacterial growth (Bhargava and Chauan 1968). *Shorea* is highly resistant to termites (Richardson et al. 1989). Richardson and coworkers showed that small quantities of four sesquiterpenoids closely related to the major constituent, α-gurjunene, are responsible for the termiticidal and antifungal activity of the resin of *Dipterocarpus kerrii* (Plate 40). The paucity of ecological studies of resin–insect and resin–pathogen interactions in this very large family leaves an open door for ecological analyses.

Bee Nest Construction with Wound and Flower Resins

Many insects collect plant resin for some phase of their nest construction. Among the most abundant and diverse are members of the superfamily Apoidae (bees), particularly the families Apidae (including the Euglossini and Meliponini) and Megachilidae. Resins from numerous genera of tropical angiosperms are collected by megachilid, euglossine, and meliponid bees. They may collect resin either from wounds on trees or from flowers of vines or other kinds of plants, an interaction that has led to a variety of ecological studies.

Megachilid bees are known to use wound resin from various genera of Dipterocarpaceae. Euglossine bees, on the other hand, collect resin from wounds of plants in several families: *Bursera, Protium,* and *Trattinnickia* (Burseraceae), and *Hymenaea* and *Prioria* (Fabaceae). Meliponid bees use resin from *Bursera* and *Hymenaea* but also from *Calophyllum, Clusia,* and *Garcinia* (Clusiaceae), from wounds that the bees sometimes produce (Armbruster 1984). All bees observed collecting plant resin use it as a primary nest construction material, building cells and closing off the nest. Euglossine bees collect resin to seal their hives and reinforce the combs. The collection of this resin demands accurate timing by the bee: resin must be malleable but obtained before the bee gets caught in the hardening mass. That many bees get trapped is evident by their abundance in amber (Figure 4-9).

Skutch (1971) suggested that lining the entrance to some nests with resin also plays a role in nest defense. In fact, one hypothesis for the relative scarcity of bees in the humid tropics is that fungal attack on pollen stores and immature bees is a major mortality factor (Michener 1974, Roubik 1983). Thus Messer (1985) analyzed the effects of volatile sesquiterpene components of fresh dipterocarp *(Hopea papuana)* resin, used by bees to build nests, on fungi that occur in pollen. Messer's experiments demonstrated the anti-fungal properties of these sesquiterpenes, which could protect the bee nests. Newly constructed nest cells of the megachilid *Chalicodoma pluto* (Plate 29) give off a pungent, resinous odor. Although these volatile components are lost as the resin hardens, the communally nesting bee continuously relines cells with a fresh layer of aromatic dipterocarp resin (Messer 1984). Resin also may reduce other causes of mortality such as parasitism. In Jamaica, the resin-using *C. rufipennis* completely resisted parasitism by the wasp *Melittobia* whereas two leaf-using megachilids had a frequency of 25–30% parasitism (Jayasingh and Freeman 1980).

Many species in the genera *Clusia* (Clusiaceae) and *Dalechampia* (Euphorbiaceae) produce resin in flowers as a reward for pollinators (Armbruster 1984). In *Clusia*, which are dioecious plants, male and female flowers are borne on different individuals. The resin-secreting structures of flowers of both the male and female plants in most *Clusia* species are sterile stamens, called staminodia. Euglossine, meliponid, and megachilid bees collect these floral resins. Megachilids are able to collect pollen and resin in one visit whereas meliponid and euglossine bees have to make separate foraging trips. These latter bees collect resin from the surface of the staminodia and roll it into balls that are passed on to the corbiculae (pollen baskets) for transport (Plate 30). Nectar is not secreted by either male or female plants. Therefore, meliponid and euglossine bees collect the resin as the sole nonnutritive nest-construction reward for pollination (Bittrich and Amaral 1996, 1997).

As discussed, dipterocarp wound resin serves not only as nest-building material; it has important antimicrobial properties that may reduce risk of pathogenesis in the nest. Since insect pollinators are known to exert selective force on numerous floral characteristics, Lokvam and Braddock (1999) questioned not only whether *Clusia* resin is microbicidal, but if so, whether floral resins from male and female plants show differences that could have resulted from their evolution under different selective environments. They found that resins, containing polyisoprenylated benzophenones and other unidentified compounds (de Oliveira et al. 1996), from male and female plants of *C. grandiflora* (Plate 30) exhibit a marked chemical dimorphism. Both male and female resins had pronounced antimicrobial activity, but female resin had zones of inhibition of gram-positive honeybee pathogenic bacteria more than twice the size of those produced by male resin. Lokvam and Braddock concluded that this divergence in *C. grandiflora* reward resins might be in response to different selective regimes mediated by the pollinating insects.

Euphorb Resins, Pollinators, and Herbivores

Since a resin reward system for pollinator bees has been observed in only a few tropical genera, Armbruster (1984) embarked on a long-term study of the vine *Dalechampia* (Euphorbiaceae), a large, nearly pantropical genus (Plate 31), to answer the question of how this unusual mutualism arose. *Dalechampia* flowers are unisexual, but the male and female flowers are united into a functionally bisexual blossom, subtended by usually large showy bracts. Floral resins in *Dalechampia* are mixtures of oxygenated triterpenes, generally secreted by blossom glands (Figure 5-7) associated with the male flowers (Armbruster et al. 1997). They include dammadienol, dammadienone, β-amyrin (Figure 1-5), β-amyrone, and 24-methylene cycloartenone and/or β-cycloartenone. Most triterpenes are solid at room temperature, but those in *Dalechampia* floral resin remain liquid several weeks after secretion. Because these floral resins are waterproof, slow to harden, and produced regularly and predictably, they are particularly valuable resources for the female bees that use them for nest construction. A few species of *Dalechampia* also produce volatile monoterpenes as pollinator rewards; these fragrant monoterpenes attract and reward male euglossine bees, who probably collect them as precursors for sex pheromones (Armbruster 1993).

The widespread defensive role of resin in other plants suggested to Armbruster (1984, 1993) that floral resin secretion in *Dalechampia* may have originated as a defense system and then, secondarily, assumed a reward function after resin-collecting bees began to visit flowers incidentally to obtain resin for nest building. Armbruster assumed that the initial functional shift from resin defense to resin reward probably required little or no genetic change. Some of this reasoning was based on the fact that chemically mediated interactions may evolve rapidly because (1) organisms have highly specific responses to chemicals in that minor changes may lead to changed ecological activity, and (2) a single plant compound may affect more than one interacting species, as discussed previously.

To evaluate possible evolutionary links between plant–herbivore and plant–pollinator, Armbruster (1997) mapped defense and reward characteristics as well as pollination ecology onto a morphologically estimated phylogeny of 42 species of *Dalechampia*. For a clear description of this general process, see Armbruster (1992). This procedure generated historical hypotheses regarding the evolution of defense and pollination systems. In this manner, he found that multiple lines of defense seem to have evolved in *Dalechampia*. He hypothesized that the first defense system was use of the triterpenoid resin to defend male flowers from florivorous insects, constituting a preadaptation (or exaptation) that allowed evolution of the resin-based pollinator reward system. This suggestion is supported by the most primitive *Dalechampia* species, in which resiniferous bractlets are dispersed throughout the staminate

Figure 5-7. Blossom of *Dalechampia tiliifolia* (see also Plate 31) visited by *Eulaema cingulata,* a euglossine bee known to be a pollinator but that also collects resin from the flowers for nest construction, ×2. The resin-secreting gland is above the bee's head. Male flowers are obscured by the bee. The bee's abdomen is touching a stigma of a female flower, hence pollination is likely occurring. From Armbruster (1984) with permission of the *American Journal of Botany.*

inflorescence, covering the male floral buds. Only in advanced *Dalechampia* species, pollinated by resin-collecting bees, are the bractlets aggregated into a gland (Figure 5-7), facilitating resin collection. Armbruster et al. (1997) assumed that this reward system represented a novel mutualistic relationship with resin-collecting bees and led to a radical shift in the nature of pollination in *Dalechampia,* apparently affecting the course of evolution of the genus. Hence, pollination by resin-collecting bees is a new function that replaced an old one (transfer exaptation in evolutionary language). This hypothesis is supported by antiherbivore functions of floral resins and by chemical similarity of floral defense and reward resins. After resin defense of the flowers was lost by conversion into a reward system, a sequence of defensive innovations followed, some of which did not involve the resin.

Armbruster further concluded that additional lines of resin defense evolved in *Dalechampia,* such as defense of developing ovaries and seeds as well as leaves and growing shoot tips (addition exaptations, i.e., new functions added to the old). Support for this scenario came from the chemical similarities of sepal, foliar, and floral resins as well as the antiherbivore properties of foliar resins. Experimental studies demonstrated that floral and foliar resins greatly deter significant feeding or leaf cutting by a broad sample of Neotropical insects, including two generalist herbivores (*Orophus* grasshoppers and *Atta* leaf-cutting ants) and four *Dalechampia* specialists (the chrysomelid beetle *Syphraea,* and the nymphalid lepidopterans *Ectima* and two species of *Hamadryas*). Armbruster et al. (1997) concluded from these experiments that chemical (resin) exaptations in *Dalechampia* played a major role in the evolution of plant–insect interactions. Adaptations reducing herbivory influenced the evolution of plant–pollinator relationships, and adaptations for pollination affected the evolution of plant–herbivore relationships. Armbruster pioneered the link between the role of resins in plant defense and pollinator interaction; his research is an outstanding example of the integration of modern phyletic methods with ecological data to provide a long-term evolutionary perspective.

Lokvam et al. (2000) explored chemical and biological similarities between the resin and latex components that occur in the laticifers extending through all tissues in *Clusia;* theirs is the only ecological analysis that has attempted to study the resinous component in a latex. They showed that resin polyisoprenylated benzophenones and novel latex benzophenones are the agents responsible for bioactivity of the *C. grandiflora* pollinator reward, thus enabling some comparisons with the *Dalechampia* system.

Roles of Surface-Coating Resins

In contrast to the internally produced resins that occur in coniferous and tropical trees, resins exuded to coat the surfaces of stems, leaves, and sometimes reproductive organs, occur commonly in shrubs, occasionally herbs, in arid habitats. They are also present in northern hemispheric boreal and subarctic deciduous angiosperm trees. Although both terpenoid and phenolic resins are produced, there is a greater representation of predominantly phenolic resins (often accompanied by some terpenoid volatiles) than in either conifers or tropical angiosperm trees. The prominence of resins in desert or semidesert perennial evergreen shrubs in Australia led Dell and McComb (1987) to suggest that such resins should be included among the xeromorphic features of plants. The question then arises as to why surface resins also characterize young tissues of common northern hemispheric deciduous trees such as birches, poplars, and alders. Ecological roles have been studied in these two kinds of surface-resin-producing plants, with antiherbivory taking some precedence over ultraviolet screening and protection against water loss and high temperatures.

Shrubs and Herbs in Xeric Communities

Ecological studies of surface leaf resins in xeric shrubs in North America have focused on five species: *Larrea tridentata* (Zygophyllaceae), *Mimulus aurantiacus* (syn. *Diplacus aurantiacus*, Scrophulariaceae), and *Eriodictyon californicum* (Boraginaceae) produce abundant leaf resin characterized by flavonoid aglycones. On the other hand, *Flourensia cernua* and *Grindelia camporum* (Asteraceae) produce resins constituted predominantly of terpenoids accompanied by some flavonoids. Several xeric annual asteraceous herbs (e.g., *Holocarpha*) also produce a mixture of terpenoid and phenolic resins that coat leaves, stems, and floral bracts.

Desert Shrubs

Creosote bush *(Larrea tridentata),* the dominant shrub over vast areas in the hot deserts of the southwestern United States and northern Mexico (Plate 14), is regarded as the most drought-tolerant higher plant in North America.

Although the resin makes the plant smell vaguely like creosote, 80% of *Larrea* (both *L. tridentata,* and *L. cuneifolia* from deserts of Argentina) resin is composed of phenolic aglycones. The major component is nordihydroguaiaretic acid (NDGA), a lignan (Figure 1-8), which is accompanied by a complex mixture of partially *O*-methylated flavones and flavonols as well as some common mono- and sesquiterpenes (Rhoades 1977a, Mabry et al. 1977). Resin is secreted from trichomes in the young leaf (Plate 32), but as the leaf matures, resin secretion ceases. Although resin secretion apparently ceases during leaf maturation in a number of shrub species (Chapter 3), this cessation was clearly recorded during senescence of the resin-synthesizing trichomes in *L. tridentata* (Figure 5-8).

This sequence of events may partially account for avoidance of young leaves and preference for mature leaves by some insects. Rhoades (1977b) showed a general deterrent effect of resin on grazing by the generalist feeding grasshopper *(Cibolacris parviceps)* at all concentration levels experimentally. However, for a specialist feeding on creosote bush, the geometrid moth *Semiothisa colorada,* resin had an attractive effect at low concentrations but a deterrent effect at high concentrations. Rhoades concluded that the moth may have adapted to use resins as feeding cues while still being deterred by them at high concentrations. In Argentina, the most abundant insect was a branch-mimicking grasshopper *(Astroma quadrilobatum),* which reaches high, damaging densities on *Larrea cuneifolia* (Plate 32) but prefers to cut leaves of low resin content (Rhoades 1977a). Rhoades (1977b) further thought that the mechanism of action of *Larrea* resin is reduction of proteolytic enzyme activity. Because the resin has strong affinity for binding protein in vitro, he hypothesized that the resin has a tannin-like effect of reducing nitrogen digestibility when *Larrea* foliage is consumed, thus forcing herbivores to feed in a manner to minimize resin intake.

In subsequent herbivory research on *Larrea tridentata* in New Mexico, Schultz and Floyd (1999) studied the effect of resin chemistry and predation on insect abundance and diversity of the insect community. They found, overall, that plant chemistry and predation contributed equally to these community parameters. As to the effects of the resin, an unidentified creosote bush caterpillar was more abundant on bushes with low resin content (as Rhoades had found), a grasshopper could be found on shrubs with high resin content, and resin levels did not matter to a katydid (Tettigoniidae). A bee scraped resin from stems, which it used to build a nest, just as tropical bees do, and as Wollenweber and Buchmann (1997) found in *Ambrosia*, *Baccharis*, and *Gutierrezia* in the Sonoran Desert. All the creosote bush insects are well camouflaged, protecting them from predators.

Despite cessation of active resin synthesis in the mature leaf of the creosote bush, the resin secreted over the young leaf produces a shiny surface, which Rhoades (1977b) analyzed experimentally to assess its antidesiccant action and ultraviolet screening properties. Since cuticular waxes are known

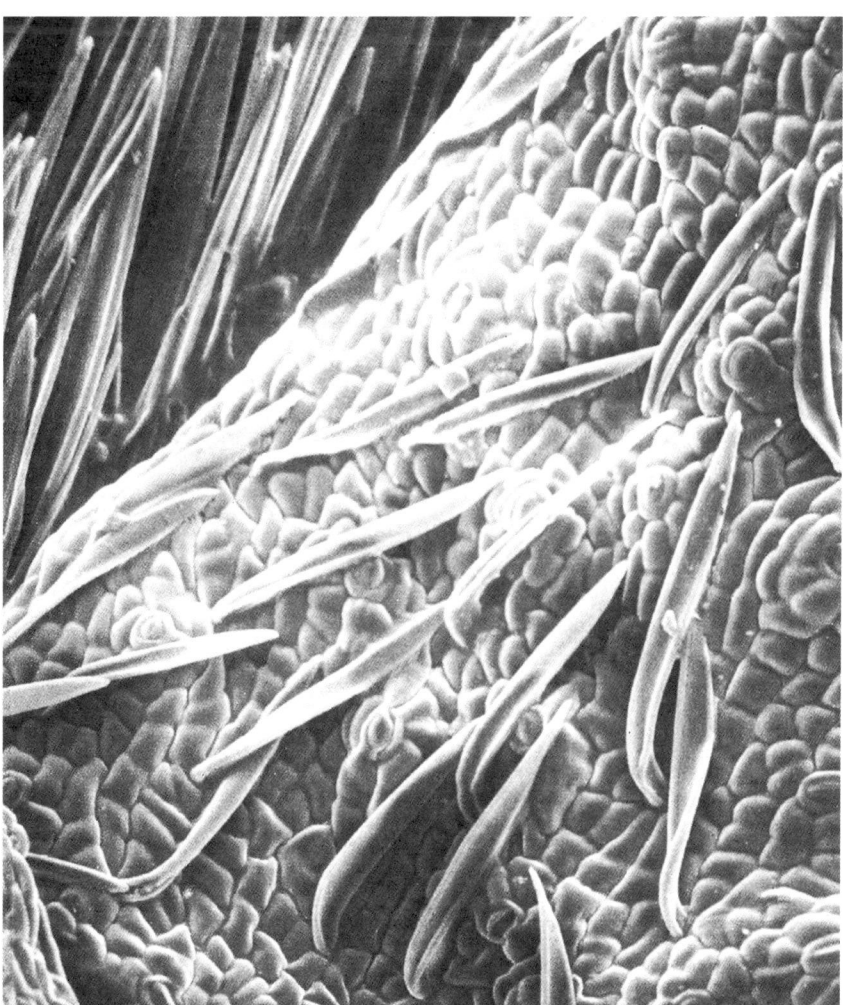

Figure 5-8. Scanning electron micrograph of flattened, collapsed trichomes on a mature leaf of *Larrea tridentata*, creosote bush, indicating that resin is only produced by young leaves and, hence, restricting the defensive role of resin in mature leaves; ×390. From Thomson et al. (1979) with permission of the University of Chicago Press.

to significantly reduce transpirational loss from leaves and other organs (Martin and Juniper 1970), Rhoades isolated the wax (5% of the resin) so as not to confuse the action of the resin and the wax (Figure 1-9). He found that when this resin acted as an antidesiccant, it seemed to require both lipoidal and nonlipoidal components, but the reason for this was unknown. In subsequent, more intensive studies of the antidesiccant effects of the leaf resin, Meinzer et al. (1990) showed that the resin behaves as an ideal antitranspirant because it clearly diminished transpiration rate more than assimilation rate. When resin was removed, the net effect on total gas exchange was to reduce water use efficiency by altering the coupling between carbon assimilation and carbon dioxide conductance. Rhoades had also demonstrated that creosote bush resin not only strongly absorbs ultraviolet radiation (expected for flavonoid-based substances), it is relatively transparent in the visible and photosynthetically active region of the spectrum and, hence, does not inhibit photosynthesis. Rhoades's integrated study of the resin defense system in *Larrea* provided the initial basis for the optimal defense theories presented by him and coworkers (e.g., Rhoades and Cates 1976, Rhoades 1979), which have had a sustained influence on research in chemical ecology.

Ernest (1994) investigated the relative influences of constitutive and browsing-induced resistance on patterns of mammalian herbivory in *Larrea tridentata*. Only two species of mammals commonly eat the vegetative portions of creosote bush in North American deserts. Black-tailed jackrabbits *(Lepus californicus)* consume *Larrea* in the Chihuahuan, Sonoran, and Mojave Deserts, and the desert wood rat *(Neotoma lepida)* consumes it in the Mojave Desert. Individual creosote bushes vary considerably in resistance to jackrabbit browsing. Shrubs with a history of heavy browsing (low constitutive resistance) are more likely to be browsed again than individuals with a history of light browsing (high constitutive resistance). Ernest confirmed experimentally the importance of both constitutive and induced resistance to jackrabbit browsing. The rabbits browsed more heavily on plants with low than high constitutive defense but less heavily on artificially clipped shrubs than on controls. Ernest interpreted the enhanced resistance of clipped shrubs as an induced response to herbivory and suggested that this may be responsible for the great variation in browsing levels on individual shrubs.

M. Meyer and Karazov (1989) indicated that the antiherbivore mechanism of creosote bush resin against desert wood rats in the Mojave Desert is probably the result of toxicity rather than reduction in digestibility, as suggested by Rhoades's early insect studies. Although *Larrea tridentata* constituted more than 75% of the rats' diet, preference trials demonstrated that they consistently selected plant parts with low resin content. When resin was added to laboratory chow, it caused the food to be avoided to the point of starvation, but there was no evidence of reduction in protein digestibility.

Tar bush *(Flourensia cernua)* is a Chihuahuan Desert shrub with a resinous leaf surface containing volatile terpenes that influence feeding of livestock. Although tar bush leaves contain more nutrients than most desert shrubs, depending on season and availability of other shrubs, it is consumed only in limited amounts by free-ranging cattle, sheep, and goats. Field experiments indicate that the leaf surface chemistry of individual tar bushes is related to the degree of defoliation by livestock (Estell et al. 1998).

Of the great diversity of shrubs with surface resins in the Western Australian deserts (Chapter 2), most have not been studied for antiherbivore properties. On the other hand, resin appears to protect leaf buds against two desert environmental extremes: low levels of water availability at times of high transpiration, and high leaf temperatures (Dell and McComb 1978). The resin layer apparently decreases transpiration, particularly in the young bud. Moreover, leaf temperatures were lowered by the high reflectance produced by the resin. Resinous surfaces in these shrubs are often shiny, and removal of resin from *Eremophila fraseri* (Myoporaceae; Plate 21) resulted in marked reduction of total reflectance (Dell 1977). The general appearance of the leaf is matte-like in *Beyeria viscosa* (Euphorbiaceae), but on exposure to high temperatures the adaxial resin melts and rapidly coalesces, resulting in a leaf with a shiny, mirror-like appearance. The consequent reflectance may lower temperatures.

California Chaparral Shrubs

Sticky monkey flower *(Mimulus aurantiacus,* syn. *Diplacus aurantiacus),* a major component in California coastal sage and chaparral (Plate 22) that endures summer drought, has many attributes similar to those of *Larrea tridentata* (Mooney et al. 1980). In both shrubs the number of leaves depends on moisture availability; they drop all but terminal leaves during drought. Furthermore, the large quantities of resin (to 25–30% dry weight) on leaf surfaces are greatest on the young terminal leaves that, amazingly, are present during drought. Sticky monkey flower resins are composed primarily of five prenylated flavonoids, also called monoterpenoid-flavonoids. Lincoln (1980) proposed the structures of the orthodihydroxylated diplacol and diplacone (Figure 1-8), the most abundant compounds; later, Lincoln and Walla (1986) isolated additional constituents that are methoxylated derivatives of the two parent compounds as well as a third new flavonoid. Although flavonoids with C_5 or C_{10} side chains are known, none had previously been identified from other members of Scrophulariaceae.

The mechanism of this phenolic resin's reduction on larval feeding by the checkerspot butterfly *(Euphydras chalcedona)* has been unclear. Inconclusive feeding experiments led K. Williams et al. (1983) to question whether the avoidance of high resin leaves by this major herbivore might result from mechanical difficulties in feeding caused by the sticky resin coating the leaves. In later field experiments, however, Lincoln (1985) found that larval growth of *Euphydras* was enhanced by high nitrogen content but inhibited by high resin content dominated by the hydroxylated diplacone and diplacol in *Mimulus aurantiacus* leaves. Larvae fed less on leaves near the branch tip, which contained a higher resin content. These results were concordant with prior laboratory studies (Lincoln et al. 1982) indicating that dietary content of nitrogen and leaf resin are major determinants of the butterfly's larval growth. Lincoln concluded that the flavonoid leaf resin of *M. aurantiacus* may both deter and inhibit the herbivory of *Euphydras*. Lincoln and Walla (1986) also suggested that resin on the upper leaf surface may prevent cuticular transpiration and retard water loss, which may be related to the methoxylated components of the resin. In further experiments with sticky monkey flower, Han and Lincoln (1994) indicated that allocation of carbon to these predominantly flavonoid resins may respond to natural selection and that the phenotypic cost of resin production may have a genetic basis in this plant.

Hare (2002) studied southern Californian populations of *Mimulus aurantiacus*, grown from clones in a common garden, that spanned a range of water availability and difference in attack by the checkerspot butterfly. The study was designed to test the suggestions from the research of Lincoln and co-workers on a northern population of *M. aurantiacus* that different components in the resin might contribute differentially in plant protection. The patterns of genetic variation in resin composition found in the northern population were not supported in the southern populations, that is, differences in chemical structures (hydroxylated versus methoxylated flavonoids) did not explain either the role of herbivore deterrent or antidesiccant. Hare concluded that the differences between his results and those of Lincoln and coworkers might be due to differences in genotypes between the northern and southern populations, reflecting differences in adaptation to local environments, or alternatively, to the differences in test environments used by the researchers. Hare is continuing experiments to try to understand how the components of resin in *M. aurantiacus* contribute to protect the plant from different environmental stresses.

Yerba santa *(Eriodictyon californicum)* is a shrub that occurs with *Mimulus aurantiacus* in California coastal chaparral (Figure 2-17). *Eriodictyon* resin also includes flavonoid aglycones (Chapter 1) but in a mixture of seven major compounds (N. Johnson 1983). The flavones luteolin and chrysoeriol, and five methylated derivatives of the flavanone eriodictyol, constitute approximately 80% of the mass of the resin. The flavanones were divided more or less equally between those with 7-OCH$_3$ and 7-OH groups. Johnson thought that the optimal leaf resin for a particular environment would be a mixture of compounds, but requirements for secretion to allow leaf growth may restrict possible combinations of the flavonoids. Change in composition of flavonoids during leaf development might reflect different roles for the predominant components (N. Johnson et al. 1984). One flavonoid predominated during peak herbivore activity (chrysomelid beetles *Hemiglyptus* and *Trirhabda*) whereas another increased when it could serve as a ultraviolet filter and antidesiccant.

Eriodictyon shares several features of its habitat and physiology with *Mimulus aurantiacus*, and the major herbivores of each, although different, are present in the winter–spring growing season. Therefore, N. Johnson et al. (1984) concluded that such shared features may have led to convergence in leaf surface flavonoid resins. They also suggested that this convergence implies similar functions for resins in other chaparral shrubs. This hypothesis seems plausible, especially considering the number of shrubs (in many genera and several families; Chapter 2) that produce terpenoid and/or flavonoid resins in Western Australian deserts (Dell and McComb 1978).

Summer Annual Herbs in Special Dry Habitats

Some herbaceous annuals in the Asteraceae are characterized by a glandular, viscid, heavily scented herbage, leading them to be called tarweeds (Chapter

2). One of these, *Holocarpha* (Figure 5-9), has four species that occur in geographically separated populations in valley grassland and foothill woodland in California. A distinctive characteristic of these tarweeds is their late summer flowering; thus they grow and reproduce out of phase with their grassland associates (Boersig and Norris 1988). In doing so, the tarweed is exposed to herbivores throughout its long reproductive cycle in an environment essentially devoid of other green herbs. Furthermore, it must tolerate more xeric conditions and more ultraviolet radiation than most Californian annual plants.

Although the ecological role of the resins of *Holocarpha* has not been studied in detail, resin appears to be important to the plant's survival. The resin, comprising two terpenoids and about 10 flavonoids, is secreted in a sessile gland on the leaf tips as well as in stalked trichomes on leaves, stems, and bracts (Boersig and Norris 1988). The quantity of resin on *Holocarpha* varies through the year, depending on the developmental stage of the plant and its response to the environment. Inland species produce 70–80% more resin than coastal species, suggesting that copious production may be advantageous in hot, dry habitats. Experimentally, resin may lessen water loss by reducing the amount of moisture in the air that passes through open stomata (Boersig 1981). The resin also contains flavonoids that absorb ultraviolet light, levels of which would be expected to be high during the growing and reproducing season of this plant. Boersig and Norris suggested that resin in *Holocarpha* may deter vertebrate feeding because the plant is palatable during its rosette stage, but grazing declines as the plants develop, with mature plants left untouched throughout the summer. No phytophagous insects have been found thriving on the plant. Although final conclusions must await more definitive ecological studies, the resin has all the indications of playing a significant role in the survival of this plant.

Subarctic and Boreal Trees

Woody plants such as species of birch *(Betula)*, poplar *(Populus)*, and alder *(Alnus)* are important food sources for wild mammalian herbivores in northern hemispheric boreal and subarctic forests. During winter, snowshoe hares *(Lepus americanus)* are highly selective in their feeding habits; winter-dormant twigs of a juvenile tree are generally less palatable than twigs of the mature plant. In every case studied, the low palatability of young trees was

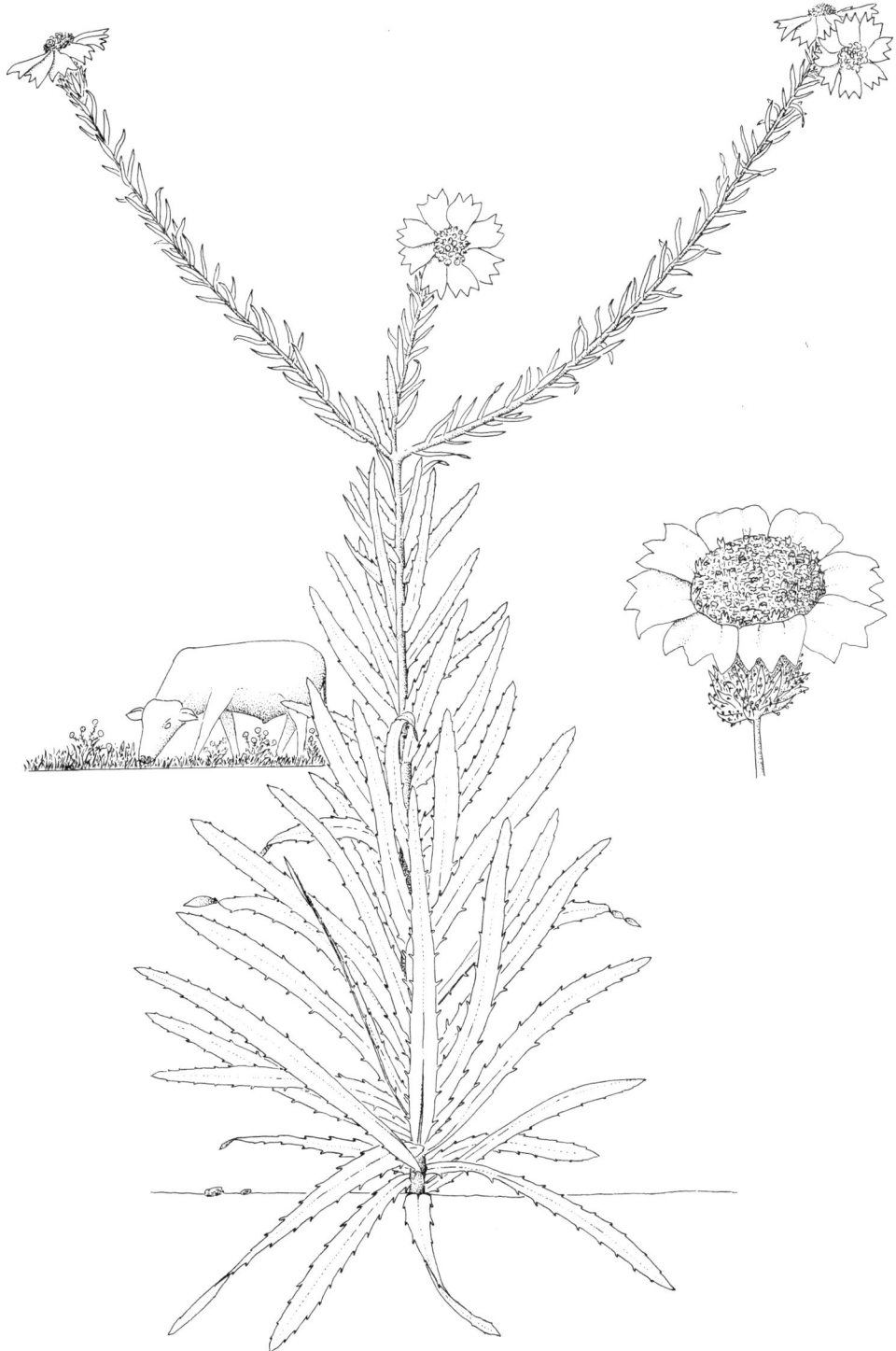

Figure 5-9. *Holocarpha macradenia,* a tarweed in California grasslands, a rare example of a resin-producing annual. Close-up of flower head shows the resinous trichomes of the bracts. Grazing cattle are thought to be deterred from eating *Holocarpha* by the sticky resin.

related to increased concentrations of resinous antifeedants in internodes of the branches (Bryant et al. 1992). The importance of understanding the development of resin secretory structures, changes in chemistry, and ontogeny of young organs of birch to answer perplexing questions of hare herbivory is discussed in Chapter 3.

Many of these hare antifeedants are terpenoid components of surface resins. For example, Reichardt et al. (1984) demonstrated that juvenile Alaska paper birch (*Betula resinifera*, Plate 33) is defended against winter browsing through triterpenes such as papyriferic acid. Furthermore, papyriferic acid accumulates on the surface of juvenile twigs only in winter, when the plant is vulnerable to snowshoe hare attack. Resin is synthesized late in the growing season and thus is available for defense of the overwintering juvenile twigs. Resin concentrations drop 25-fold in the mature internodes when herbivores no longer damage the tree. Feeding experiments demonstrate that papyriferic acid is highly distasteful to this major herbivore, with concentrations in the tree more than sufficient to explain the absence of herbivory during juvenile growth stages. In older trees, concentrations may decrease because the tissue is out of reach from the hares. Papyriferic acid can act as a deterrent to moose and certain rodents as well as snowshoe hares. Furthermore, Reichardt et al. (1990) demonstrated that buds of balsam poplar *(Populus balsamifera)* in Alaska are defended from hares by the terpenoids cineole and α-bisabolol. The phenolic resins, including salicaldehyde and 6-hydroxycyclohexane, deter hare feeding on winter-dormant internodes of juvenile balsam poplar, with the current year's growth of internodes more heavily defended chemically than older internodes or bark (as is the case in *B. resinifera*).

Alaskan green alder *(Alnus crispa)* produces a resin in buds, catkins, and internodes in which the primary constituents are pinosylvin and pinosylvin methyl ether (Clausen et al. 1986). Concentrations of pinosylvin methyl ether by itself on older buds and catkins are sufficient to account for their rejection by snowshoe hares (Bryant et al. 1983). In winter, however, internodes of both saplings and mature plants are defended against hare browsing by both pinosylvin and pinosylvin methyl ether, with the current year's growth of the saplings containing three times more of both compounds than that of adults. These birch, poplar, and alder studies emphasize the importance of surface resin in young tissue, providing defense against mammals, as they do in xeric shrubs against insects.

Bryant et al. (1994) analyzed biogeographic patterns of chemical defense against snowshoe hares in birches circumboreally. This followed the discovery that chemical defenses of juvenile woody plants against browsing by mammals in winter vary markedly across the vast subarctic forests of Eurasia and Alaska (Bryant et al. 1989). Bryant and coworkers then questioned whether similar large-scale spatial variation occurs in subarctic and temperate North America. Feeding trials showed that Alaska paper birch *(Betula resinifera)* is the least palatable to snowshoe hares, followed by eastern boreal paper birch *(B. papyrifera)* and then the temperate forest yellow birch *(B. alleghaniensis)* and black birch *(B. lenta)*. The most palatable birches did not contain the triterpenes known to deter snowshoe hares. In fact, twigs of the unpalatable Alaska paper birch contained at least three orders of magnitude more triterpene resin defense than twigs of the more palatable North American birches.

Bryant et al. (1994) postulated the following for this pattern of resin defense: Climatic variation that developed across subarctic North America after the Ice Age resulted in a geographic pattern in North American wildfires such that the arid boreal interior of Alaska burned more than the moist boreal forest of eastern Maine, and the temperate hardwood forests of Connecticut burned even less. Greater fire frequency resulted in more successional species than mature forest species, which in turn influenced the intensity of selective browsing by hares. All this, they suggested, resulted in biogeographic variation in chemical defense of juvenile woody plants against browsing by mammals in winter. Thus Bryant and coworkers in essence suggested that chemical defense has been selected by the activity of hares, as exhibited by differences in chemical expression of this defense over a broad geographic area. These extensive studies of the role of resin in defense of boreal and subarctic trees against vertebrates (Bryant et al. 1992) not only provide support for the carbon–nutrient balance and resource availability theories of plant chemical defense (e.g., Bryant et al. 1983, 1985; Coley et al. 1985), they direct attention to testing the theories in other interactions.

Ecosystem Interactions of Resins

Resin can be useful to organisms other than the plants producing it, for example, the bees that collect resin from plants to construct nests, using it to protect them against fungi and termites. Use of resin by one organism, such as bees, may directly or indirectly influence the life of other organisms, constituting a type of mutualism, which is particularly noteworthy in the tropics. Armbruster (1984) referred to the process by which some bees collect floral resins for nest construction and thereby pollinate the flower as a "diffuse mutualism" because some plants may thus be indirectly dependent on floral resin for their reproduction. Plowden et al. (2002) pointed out that in burseraceous plants (e.g., *Protium* in Amazonia) there are additional complex invertebrate mutualisms associated with the exuded wound resin used by the bees. For example, assassin bugs (Reduviidae) dab their forelegs with resin to help capture bees as they harvest resin. Curculionid weevils burrow into soft resin lumps, where they shape a protective chamber for their development (Chapter 8). Syrphid fly larvae also develop in these resin lumps. Furthermore, other arthropods such as ants and spiders, among others, nest or forage in hardened lump chambers vacated by the weevils. Pharmaceutical use of resins by mammals and as insect pheromones are other ways in which resin is valuable to various members of ecosystems. Other resin interactions influence nutrient cycling and tropospheric chemistry.

Resin Use by Bees in the Temperate Zones

Although some tropical bees collect resin from plant wounds, bees in the temperate zones collect resin from buds and young leaves, primarily of deciduous trees. Flavonoid resins are collected from buds and unfurling leaves of *Aesculus* (Figure 2-14), *Alnus*, *Betula* (Plate 33 and Figure 2-10), *Populus* (Figure 10-8), and *Salix* by *Apis* honeybees to seal their hives and reinforce the combs (Michener 1974, 2000). This resin, used by honeybees, is called propolis, from the Greek *pro*, for, and *polis*, the city, that is, defense of the city, the hive. Propolis generally refers to resin used by *Apis*, not the tropical megachilid, euglossine, and meliponid bees, but Bankova et al. (2000) referred to resin used by tropical bees as propolis. Some tropical bees mix the resin and beeswax with soil, a product that Tomás-Barberán et al. (1997) and Bankova et al. (1998) call geopropolis. Although there is not as much evidence for resin mutualisms involving temperate zone invertebrates as there is in the tropics, humans have a long history of sharing the honeybees' use of propolis (Chapters 10 and 11).

Poplar bud resins have always been considered the most important source of honeybee propolis. Even within the genus *Populus*, bud exudates are gathered from a limited number of species (Greenaway et al. 1990); all compounds collected were typical of *Populus* section *Aigeiros*, which includes *P. nigra* (black poplar), *P. deltoides* (eastern cottonwood), and poplars related to *P. deltoides (P. fremontii, P. sargentii,* and *P. wislizeni).* Exudates from other widespread poplars such as *P. balsamifera* (balsam poplar) and its subspecies *trichocarpa* (black cottonwood, Figure 10-8), principal members of section *Tacamahaca,* have major components (terpenoids and dihydrochalcones) not common in section *Aigeiros* and seldom seen in propolis. Even in New Zealand, introduced poplar species are the source of propolis.

There are reports suggesting that in the future, honeybees will be shown as collecting resins from a greater variety of temperate zone plants than known at present. For example, some species of *Cistus* have been found to be the source of propolis in Tunisia (Martos et al. 1997), as has *Ambrosia deltoidea* in the Sonoran Desert (Wollenweber and Buchmann 1997). Resins from *Araucaria, Baccharis,* and *Xanthorrhoea* have been suggested but not authenticated as collected by bees.

The factors that trigger propolis collection by honeybees are not completely understood. Ghisalberti (1979) noted that some researchers think that seasonal factors are involved whereas others think that inherent behavioral changes are responsible or related to the use of resin. The bees apply the resin in a thin layer to the internal wall of their hive or other cavity they inhabit. It is used to block holes, repair combs, and make the entrance to the hive weathertight and easier to defend. It is also used to embalm the carcasses of hive invaders that bees kill but cannot transport out of the hive. Like resins collected by tropical bees, propolis is thought to have antimicrobial properties, which have been demonstrated in its use by humans (Chapter 10).

Pharmaceutical Use of Resin by Coatis

Attention has focused on the possibility that mammals may use chemicals from specific plants for their potential pharmaceutical, or medical, value. For example, Gomppers and Hoylman (1993) observed white-nosed coatis *(Nasua narica)* grooming themselves and other group members with resin from *Trattinnickia aspera* (Burseraceae) in mature Panamanian forests, where the tree is frequent. Bands of coatis approach the tree, and individuals scrape the resin that has exuded from open wounds with their paws and vigorously groom the resin into their fur. Gomppers and Hoylman hypothesized that this may result from the resin's deterring the wide range of ectoparasites carried by coatis, antifungal benefits, treatment of external irritations, facilitation of social interactions, or the resin may simply smell good to the animals.

Resins as Beetle Pheromones

Although emphasis has been placed on the role of resin in defending conifers against bark beetles, the use of resin chemicals by these and other beetles should not be ignored. The volatile monoterpenes of conifer resin are involved in intraspecific host-finding pheromonal communication, with aggregation pheromones (signaling mass attack) frequently being allylic oxidation products of host monoterpenes (D. Wood 1982). Although the compounds in the tree's mixture of constitutive monoterpenes may directly deter beetles, some may also signal that the tree is unfavorable as a host. For example, the pine shoot beetle *(Tomiscus pinperda)*, which feeds on European *Pinus sylvestris* (Plate 1), is able to recognize a tree that is unsuitable for colonization during flight because of the release of verbenone from a colonized tree. Verbenone is released in increasing quantities as beetle attack progresses, eventually completely inhibiting the attractiveness of the host's monoterpenes (Byers et al. 1989).

Chararas et al. (1982) indicated the importance of the mixture of terpenoids by comparing mono- and sesquiterpene compositional patterns in 10 conifer species to the establishment of six genera of Scolytidae. They concluded that bark beetle attraction resulted from synergism of different terpenes, which they thought was necessary for attraction to prevail over the generally repellent properties of some of these terpenes (when they occurred at sufficiently high concentrations). Ratios of individual monoterpenes have also been shown to be crucial for attractiveness of pine needles to ovipositing insects. For example, pine beauty moths *(Panolis glammea)* use the ratio of β- to α-pinene as a rough guide for selecting the correct host plant, then use other monoterpenes as indicators for suitability for egg laying (Leather 1987).

The complexity of interactions involving stereoisomers of the monoterpenes is also intriguing, for example, the interaction of the bark beetle *Ips typographus* with Norway spruce *(Picea abies)* in which (+)-*cis*- and (+)-*trans*-verbenol are produced from (+)- and (−)-α-pinene, respectively (Phillips and Croteau 1999). Since only (+)-*cis*-verbenol serves as an aggregation pheromone for this species of *Ips*, the synthesis of both stereoisomers of α-pinene by the tree undermines the effectiveness of pheromonal activity because both isomers are transformed by the beetle. The verbenols can be further oxidized to the ketone verbenones, which act as dispersal pheromones. This pheromone-based dispersal regulates bark beetle attack density, as discussed for pine shoot beetles. Therefore, when a host tree has been colonized by mass attack, other beetles are warned to attack other trees, an interesting beetle society phenomenon. Myrcene can be converted to either (+)- or (−)-ipsdinenol, both apparently important in communication for all species of *Ips*. However, simple oxygenated monoterpenes that mediate host location and bark beetle aggregation also mediate long-range attraction and ovipositing of *Rhizophagus* predator beetles (Gregoire et al. 1992). In such complicated series of ecosystem interactions, the selection pressure on the host for altering resin chemistry is multidirectional, involving attraction, repulsion, toxicity, and pheromonal communication as well as interactions with predators and parasites of the beetles.

It seems paradoxical that bark beetles are attracted to and exploit terpenes that can be toxic and may also attract predators and parasites, but the benefits to the beetle must outweigh the disadvantages. Although the chemical, physiological, and behavioral components of pheromonal communication have been described for numerous bark beetles species, Raffa (2001) pointed out that there is still limited understanding of how these signals function under natural conditions. He reviewed published data from 13 scolytid-conifer systems studied since the 1960s and concluded that evidence increasingly refutes the traditional view emphasizing "the fidelity and constancy of these signals." Rather, there are numerous factors influencing infraspecific variation in bark beetle pheromones, "in which complex mixtures vary within temporal scales of host colonization, seasonal phenology, and predator feedback, within the spatial scales of between forest stands, and throughout geographic ranges, and where a wide array of predators, parasites, competitors and abiotic influences affect their functioning." Thus these conflicting selection pressures can accentuate variation in the bark beetle communication system. Understanding this variation remains a challenge.

Role of Resin in Ecosystem Nutrient Cycling

Monoterpene components of resins have been shown to be significant in nitrogen and carbon cycles in coniferous forests (C. White 1994, Ward et al. 1997). In the litter and soil beneath trees, monoterpenes can act as a substrate for a portion of the heterotrophic microbial community, which contributes to net carbon mineralization but promotes nitrogen immobilization in the microbial biomass. Because conifer trees typically dominate forests over large areas (Chapter 2), they can provide an extensive terpene-rich fallen needle litter. In these coniferous ecosystems, which have high carbon-to-nitrogen ratios in the litter, microbial reduction of available nitrogen reduces net decomposition rates and net nitrogen mineralization (conversion to ammonium). By inhibiting specific monooxygenases, monoterpenes may inhibit the nitrification process (conversion to nitrites and nitrates) and even methane oxidation. By promoting fire, monoterpenes further interact with the nitrogen and carbon cycles directly through combustion and indirectly through changes in rates of nutrient cycling following fire (Plate 34). Thus White suggested that components of conifer resins may significantly affect ecosystem-level properties despite occurring in concentrations of only parts per million in the forest soil.

Herbivore-Induced Terpene Emissions and Tropospheric Chemistry

Although the monoterpenes in conifer leaf resins are stored in internal canals or cysts, their high vapor pressures result in some volatilization into the atmosphere. Herbivory can result in an immediate increase in the rate of monoterpene emission. Litvak et al. (1999) measured monoterpene emission rates from undamaged and damaged needles of ponderosa pine *(Pinus ponderosa)* and Douglas fir *(Pseudotsuga menziesii)* trees. Fluxes from ponderosa pine forests with 10% and 25% damaged foliage were potentially 2- and 3.6-fold-higher, respectively, than from forests with no damaged foliage. Model simulations suggested that fluxes resulting from even low-level damage are sufficient to increase local tropospheric production of ozone and organic nitrates and to suppress hydroxy radical concentration. In both forests, the predicted magnitude of perturbation increased linearly with extent of the foliar damage and was dependent on local mixing ratios of nitrogen oxides. The presence of isoprene in conifer forest air diminished the role of herbivory in generating local ozone. Litvak and coworkers thus concluded that their results were conservative in estimating the potential effect of defoliation in conifer forests on tropospheric chemistry. They further suggested that conifer defoliation should be considered an important potential control over local oxidative tropospheric chemistry and thus in the perturbation of local ozone dynamics in many rural conifer forests.

Future Ecological Research on Resins

Chemical ecology of resin-producing plants is a fast-moving field as a result of rapid progress in chemical analytical technology, accrued background data, and interest in integrating basic and commercially oriented research. Chemical technology facilitates rapid determination of the total composition of the resin, and computer programs enable quantification and statistical analysis not possible as recently as 1990. These technologies make possible an even more realistic assessment of complex interactions in the future. Increased background data allow analysis of the multiple roles that the complex mixtures constituting resins can play. Molecular genetics will increase the understanding of biosynthesis, the sites of synthesis, and the temporal relationships in the production of resin components as they function defensively for the plant (Berenbaum 2002). The complexity of interactions with prominent and economically significant herbivores, such as bark beetles, are being unraveled, but with a future major challenge to understand the variation that occurs in chemical communication at different temporal and spatial scales.

Increasing attention will be given to integration of modern phyletic methods with ecological data on resin-producing plants to answer long-term coevolutionary questions. For example, paradigms of coevolution assume an increase in plant chemical diversity and complexity as an underlying mechanism for insect feeding adaptation, and plant and insect diversity. Some resin-producing plants provide ideal model systems for analysis of this mediation, where the resin chemistry is known and time-calibrated plant phylogeny could quantify chemical trends.

Ecological research on tropical resin producers will probably receive more attention in the future, especially those with resins of economic value to local people. Emphasis will probably center on the best means of their utilization in the context of maintenance of tropical biodiversity (Chapter 11). Additional research can be anticipated on the arid-area shrubs and herbs in which the ecological importance of leaf- and stem-coating resins has been recognized more recently than that of internally produced resins of trees. Again, economic interest in some of these resins may lead the way for basic ecological research.

PART III

The Ethnobotany of Resins

Overleaf: harvesting frankincense from a large *Boswellia* tree in southern Arabia.

CHAPTER 6

Historical and Cultural Importance of Amber and Resins

Resins have played important roles in cultures around the world for a diversity of reasons: utilitarian, aesthetic, and economic. Resin-producing plants occur in many of the diverse environments in which humans have lived, and they have been put to a wide variety of uses. In fact, anthropologists have claimed that no other material was so versatile in the preindustrial world as resin. Moreover, resins preserve and travel well; hence, they have been traded between societies around the world. Figure 6-1 is a time line of the significant events in the history of amber and resin from the Stone Age to the present. For example, an extensive and lucrative incense trade by Arabians was in progress by at least 1000 B.C., and incense was probably being used by the New World Maya by 600 B.C. and by the Scythians at about the same time. European amber trade has been documented back to the Stone Age whereas there is evidence of the Chinese obtaining amber from Myanmar (Burma) by 200 B.C., and of Mayan use probably by the first century A.D. Use of *Cannabis* resin (hashish) can be traced back to the Bronze Age, and its trade influenced cultures on almost every continent. Naval stores resins were significant to the commercial and military success of seafaring nations back to at least the Bronze Age. Britain needed naval stores for wooden ships to connect a vast empire, and she looked to the forests of the American colonies, beginning in the 1600s, to supply these resins. Later, the naval stores industry had a noteworthy effect on both the economy and culture of the southern United States. Copals also influenced the development of various countries, such as New Zealand, where the "gum" (copal) industry led to a major immigration of people in the mid-1800s. These are a few examples of how resin and amber have played a variety of significant roles in shaping the history of different countries.

257

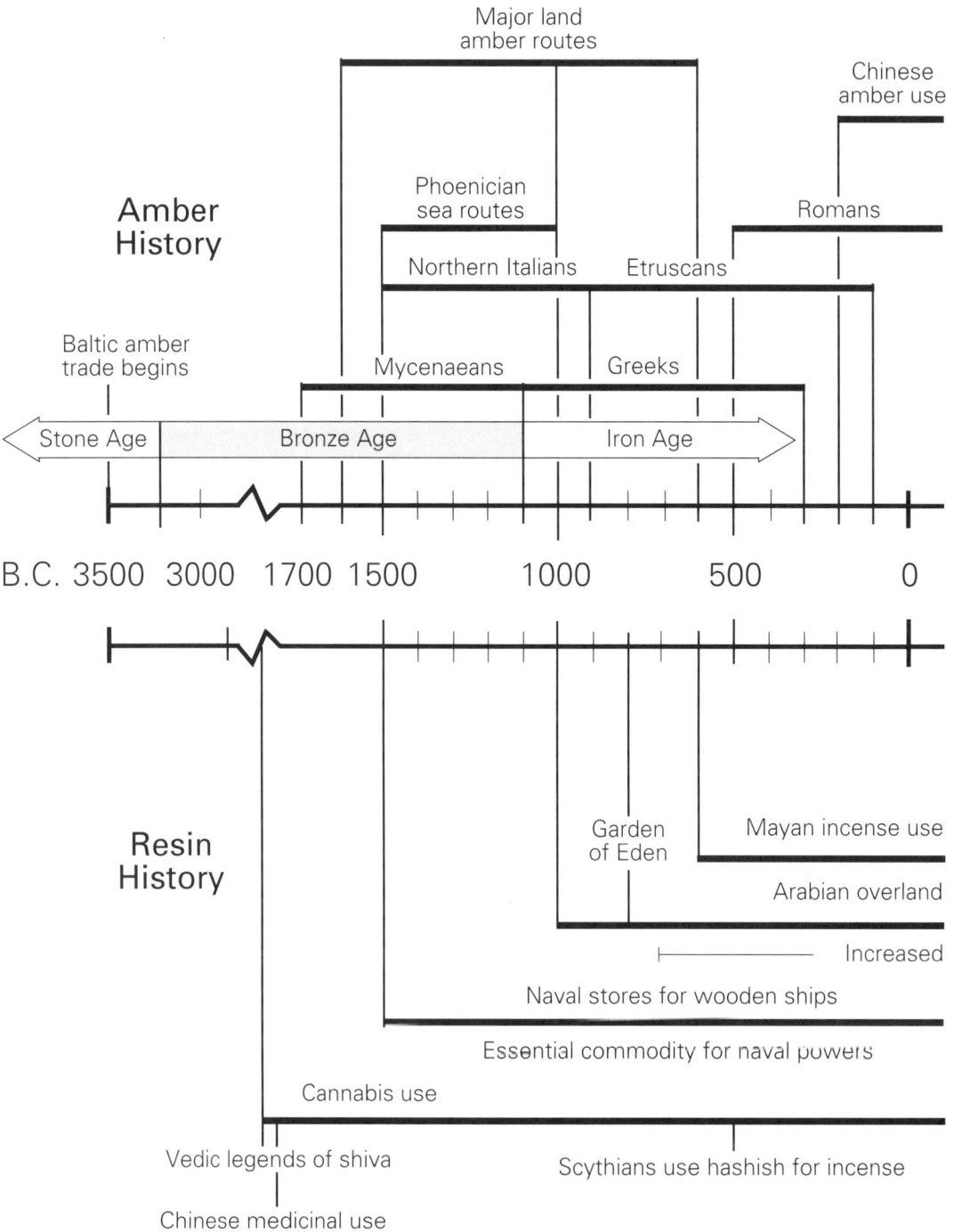

Figure 6-1. Time line of important events in the history of human use of amber (above) and resin (below). Although some evidence for use of amber extends back to the Old Stone Age, the time line begins with the period when trade routes were well established and when there is a good record for the use of nonfossilized resin.

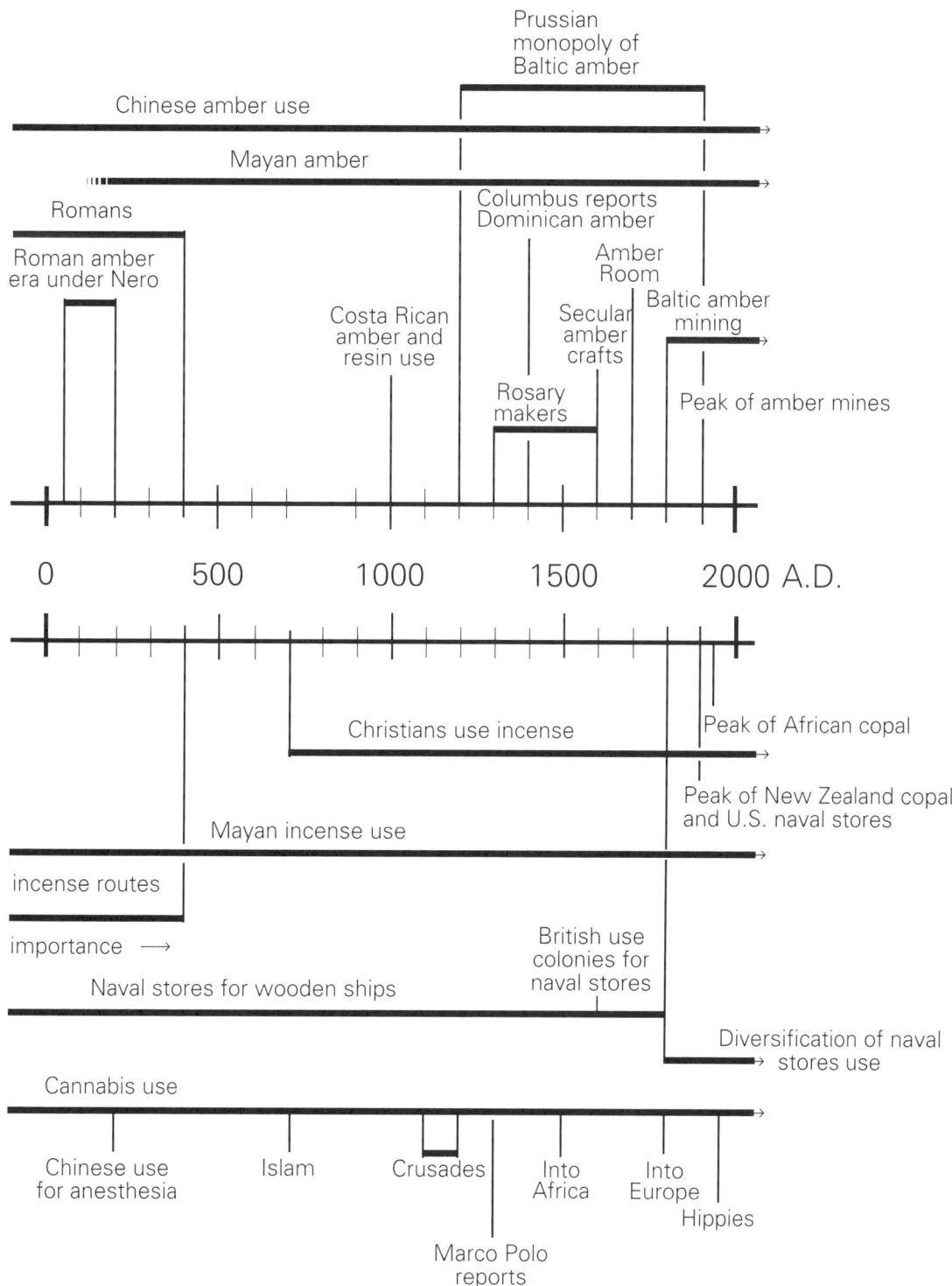

Prussian
monopoly of
Baltic amber

Chinese amber use

Mayan amber

Columbus reports
Dominican amber

Romans

Amber
Room

Roman amber
era under Nero

Baltic amber
mining

Secular
amber
crafts

Costa Rican
amber and
resin use

Rosary
makers

Peak of amber mines

| | | | | | | | | | | |

0   500   1000   1500   2000 A.D.

Peak of African copal

Christians use incense

Peak of New Zealand copal
and U.S. naval stores

Mayan incense use

incense routes

British use
colonies for
naval stores

importance ⟶

Naval stores for wooden ships

Diversification of naval
stores use

Cannabis use

Chinese use
for anesthesia

Islam

Crusades

Into
Africa

Into
Europe

Hippies

Marco Polo
reports

In this chapter, the historical and cultural roles that resins have played are discussed, providing a framework for Chapters 7–10 in which detailed information on the biology and chemistry as well as methods for obtaining and processing specific resins are presented. Although synthetic chemicals have replaced some natural resins for industrial use, the likelihood that resins will continue to be important to humans in so many ways in the future is analyzed in Chapter 11.

Amber Trade from the Stone Age to the Classical Age

The geologic history of amber, its chemistry, and the kinds of trees that produced the resin as well as the forests in which the trees lived are discussed in Chapter 4. Here, the focus is on the cultural importance and historical influence of amber. Although amber has been used wherever it has been found in abundance, its long European history has been so pervasive that it has been intensively studied.

Old Stone Age to the Iron Age

For at least 10 millennia, essentially since the time of the hunter-gatherers, European people have adorned themselves with amber beads and amulets from the extensive amber deposits around the Baltic Sea (Chapter 4). In this northern region of long and cold winters, a substance with sunlit colors and a warm feel when handled or worn has been especially coveted. A belief that amber possessed supernatural powers may have been related to sun worship, which was widespread if not universal in ancient times (Rice 1987). This belief is illustrated by archeological finds of amulets of sun wheels (Figure 6-2). Stone Age humans were superstitious, and the attractive force of the negative electrical charge that accumulates on amber when it is rubbed, as well as the remains of life entrapped in it, may have contributed to its mysterious reputation. These unique characteristics, in addition to amber's imperishability, made it such an important commodity for barter that it became a major factor in the development of Baltic civilization. As Grabowska (1983) reiterated, the "simply phenomenal" demand for amber made it inseparable from the cultural history of the Baltic region.

Amber from the Baltic region was traded throughout most of Europe. However, because amber occurs naturally in smaller deposits in other areas of

Europe, including Romania, Sicily, and France as well as other places (Figure 4-2), archeologists needed to find a way to distinguish amber artifacts from different locales. Chemical analysis has been critical in distinguishing ambers (Chapter 4), infrared spectroscopy in particular because it requires only small samples from valuable artifacts (Beck 1970). That all known succinite, the predominant amber from the Baltic area, contains more than 3% succinic acid has turned out to be the safest guide to the authenticity of this amber. Therefore, the absence of this acid from an archeological artifact or low amounts of it definitely demonstrate that it is not made of Baltic amber. Beck et al. (1966, 1967) provided an annotated bibliography of the provenance of Baltic amber archeological artifacts that extends from the eighth century B.C.

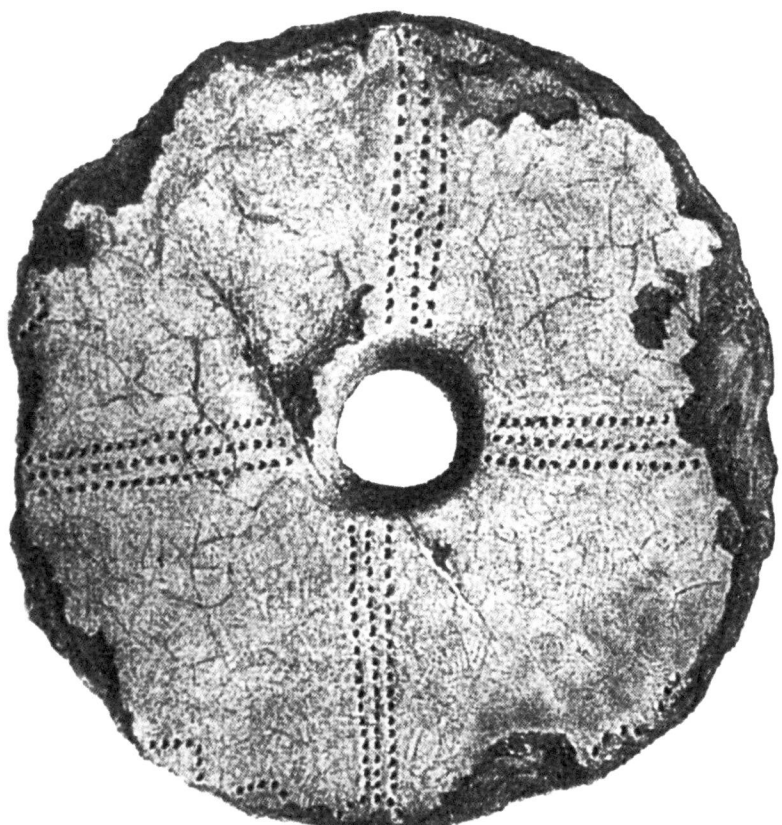

Figure 6-2. Late Stone Age amber disk from Samland Promontory, actual size. The cross-and-ray design indicates the cult of the sun wheel. The small holes in the design were filled with resin, thereby producing an ornamental contrast with the smoothly polished surface. Disks such as this are the oldest known symbols of sun worship from the ancient Balt culture. From Klebs (1880).

to assist archeologists and chemists in determining trade routes from the northern source of the amber to southern parts of Europe.

The oldest amber artifacts, considered Old Stone Age (11,000–9000 B.C.), are from caves in England, where the amber had probably washed up on the shore from a Baltic origin farther east. Actually, this was still at a time, at the end of the Ice Age, when the British Isles were partially connected to the European mainland. Amulets and distinctive double-headed-ax beads are known from southern Scandinavia during the middle periods of the Stone Age that extended to about 4000 B.C. However, large-scale production of amber artifacts only became evident around 3400–3100 B.C. during the Late Stone Age. Archeologists have found artifacts from grave sites and peat swamps in areas of present-day northern Poland, Lithuania, and Latvia. In fact, approximately 100 Late Stone Age burial sites are known (Rice 1987). Stone Age dates vary for different cultures, and the dates for Baltic cultures are later than those of other European cultures (Gimbutas 1963). In fact, Rice indicated that Baltic cultures continued into the Late Stone Age until 1800 B.C., whereas most of Europe was well into the Bronze Age by that time. During this period, polished stone implements were manufactured, and flint and bone tools used to cut lumps of amber into amulets and items of adornment (Figure 6-2). Juodkrantė (Schwarzort), in present-day Lithuania, is one of the oldest and most celebrated Late Stone Age archeological sites (Figure 6-3); it included stylized forms and numerous animal figurines (Plate 38) as well as beads, buttons, small disks, and pendants. Another rich Late Stone Age site near Lake Lubāns, Latvia, contained buttons and amulets similar to those from Juodkrantė. The Baltic tribes also traded raw amber and artistic works. For example, amber ornaments have been found at Woldenberg (Dobiegniew, Poland), east of Berlin, and in remains from lake dwellers of Switzerland and France. In fact, Late Stone Age amber artifacts have been found in almost every country of Europe (Rice 1987). Some amber even reached the shores of Africa, as Baltic amber ornaments in Egyptian tombs date to the Sixth Dynasty (3200 B.C.).

Baltic tribes established trade routes in the Bronze Age (about 3200–1200 B.C.) over which amber was traded for metals. The archeologist W. Arnold Buffum (1897) stated, "The amber trade route is the original trade route along which luxuries of life went out in search of the necessities." Rice (1987) thought that the Baltic Bronze Age commenced with transcontinental amber trade for tin and copper. Although the coastal amber traders were thus able

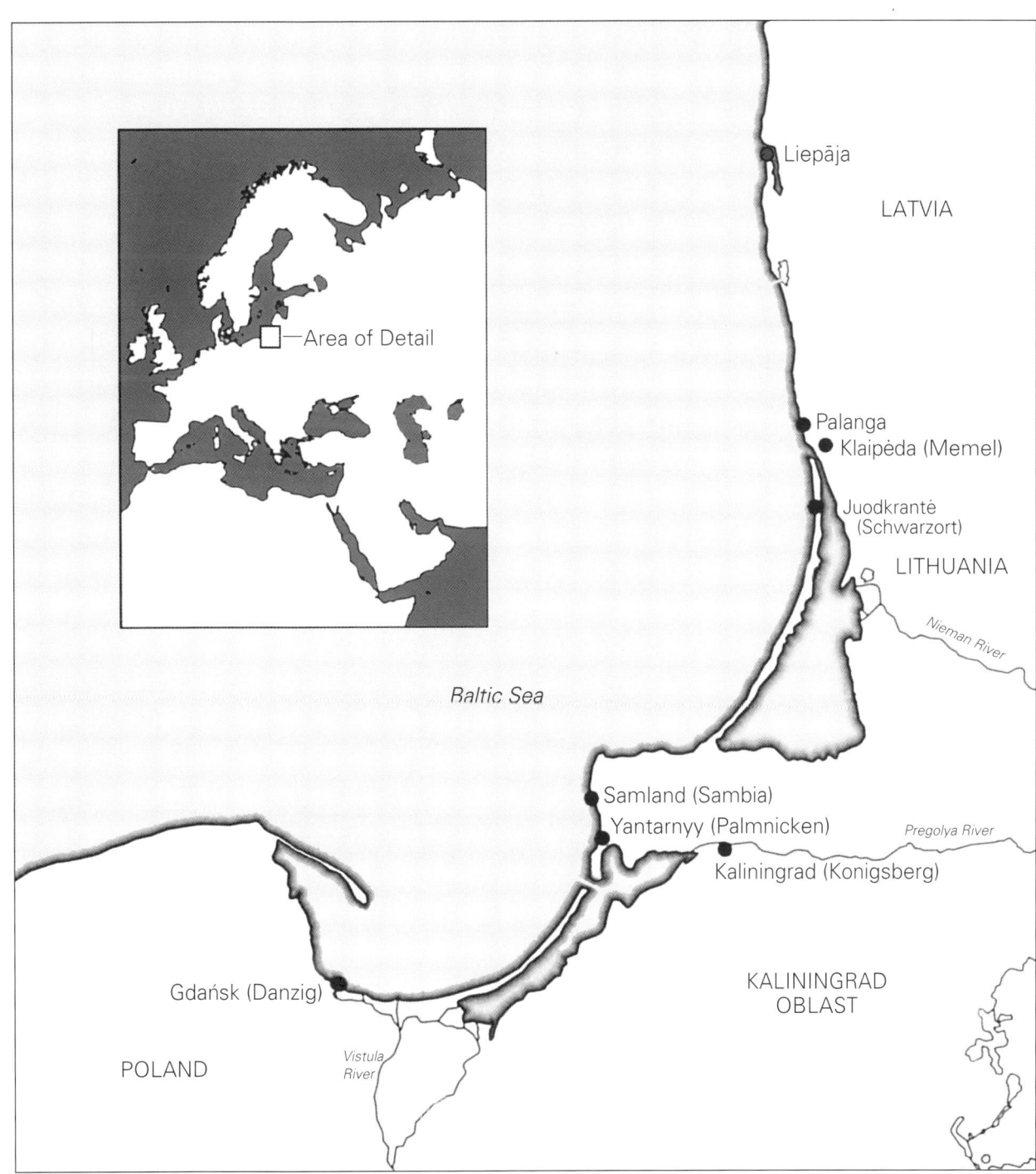

Figure 6-3. Major amber-collecting areas along the Baltic coast and Samland Peninsula from the Stone Age to the present. Place names are current, with the German names common in older literature in parentheses.

to import metals for tools, the inland Balts were still using stone tools as late as early Iron Age (Gimbutas 1963). This active amber trade by the coastal Balts resulted in cultural development faster than that of the inland people. The distribution of Baltic amber artifacts indicates a widely established trade for this highly valued material; various routes were established, mainly along European rivers from the Baltic to southern regions (Spekke 1957). During the middle Bronze Age (1800–1200 B.C.), two western branches, one of which traversed central Europe, began in Jutland, Denmark, and passed along the Elbe River southward with the main branch following the Saale River, then continued along some small rivers to the Danube, crossed the Alps, where it joined other rivers into Italy, with centers around the Po River (Figure 6-4). Smaller branches followed the Elbe into the Czech Republic and the Danube into Austria. A major western branch diverged from the more central one on the Saale, passing westward to the Rhine River, then southward along rivers to Massilia (now Marseilles).

The amber traded into central Europe during the Bronze Age was then traded to the Mycenaeans (about 1700–1200 B.C.) and early Greeks (1200– 900 B.C.; Figure 6-1). In early Greek cultures, amber was an important luxury of the educated and ruling classes, and amber beads, pendants, amulets, and other articles were important burial items of the Mycenaean kings of Crete. The Mycenaeans, around 1300 B.C., fostered enterprise, forging links deep into Europe for trade of amber, bronze, and gold—some of which they reexported to islands in the Aegean Sea and eastern Mediterranean.

A possible link between the Mycenaeans and the early Bronze Age Wessex culture in the British Isles is suggested by special amber beads and elaborate necklaces found in both areas. Amber was particularly cherished in the Wessex culture; two amber cups from this society are considered to be among the greatest archeological amber finds (Beck 1974, Beck and Shennan 1981). Amber from the early Bronze Age and extending back to the Late Stone Age was also found in mounds near Stonehenge. The people who built Stonehenge were sun worshippers and thought of amber as a substance of the sun.

The Phoenicians probably brought sun worship with them from their homeland, present-day Syria and Lebanon, when they ventured into the Mediterranean (Buffum 1897) and therefore were excited by amber as the most sunlit of any gem. The Phoenicians, the dominant trading power in the Mediterranean from 1500 to 1000 B.C., may have been the first sailors to trade amber during the late Bronze Age (Figure 6-1). They pioneered sea routes to northern Europe to obtain amber in exchange for bronze, then sold amber throughout the Mediterranean for the price of gold; thus the name, gold of the North, became common in referring to the highest-quality Baltic

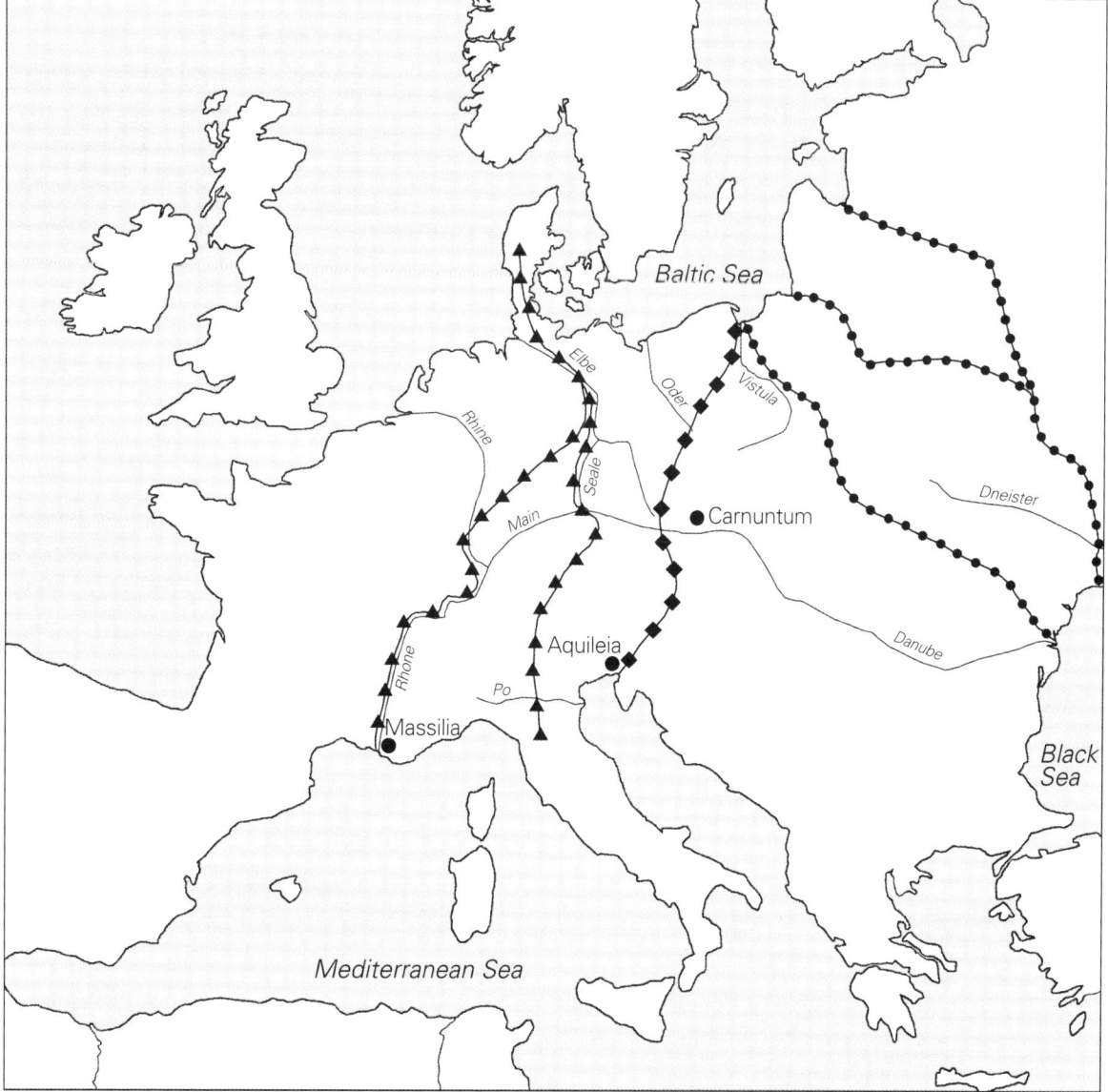

Figure 6-4. Approximation of some important amber trade routes from Baltic areas in the Bronze and Iron Ages, which followed rivers across Europe to the Mediterranean and Black Seas: ●, eastern branches; ■, central, later Roman, branch; ▲, western branches. Information from Spekke (1957), Gudynas and Pinkus (1967), and Rice (1987).

amber. They may also have obtained some of their amber at Massilia from the Ligurians at the mouth of the Rhone River, a Mediterranean terminus for one western branch direct from the Baltic during the late Bronze Age (Figure 6-4). Amber in this region was at one time called "ligurian." The Phoenicians concealed the sources of their amber, often telling tales of great danger encountered at sea as a way to protect their lucrative trade.

The seafaring Etruscans, from Etruria, a league of city-states in Italy centered around Tuscany, were avid traders who exported iron and imported amber from the Baltic. Many of the early Italic ambers are Etruscan because amber preserved well in the extensive necropolises around Tuscany (Rice 1987). People in ancient northern Italy who later became Etruscans and who were predecessors of the Romans began trading amber as early as 1500 B.C. and developed extensive commercial relations with the early Greeks and Phoenicians throughout the Mediterranean. The civilization of Etruria flourished between approximately 900 and 100 B.C.; the height of popularity of their artistic creation, including a combination of amber with gold, was reached around 500 B.C. during the Iron Age. There was also a focus on carved animals and human figurines, with some pendants considered symbolic of fertility.

It was during the Iron Age, which began about 1200 B.C., that the most famous and extensive amber trade routes were established. These routes created a major north–south axis in amber trade that linked the Baltic with the northern Mediterranean (Spekke 1957). During the early part of the Iron Age, a route was established beginning at Gdańsk (Danzig) in the heart of the amber region (Figures 6-3 and 6-4). This route followed the Vistula River, thence to the Oder River, over the Carpathian Mountains, with a main branch that finally terminated at the Adriatic Sea. The western routes from the Jutland coast (established during the Bronze Age) or the central route from the Samland Peninsula (established during the Iron Age) formed the principal amber trade routes in central Europe to Italy and areas around the Mediterranean (Spekke 1957, Rice 1987).

However, several eastern routes from the Baltic deposits, thought to have been established during the Iron Age (Rice 1987), extended to the Black Sea (Figure 6-4). Trade occurred in the central Caucasus from 700 to 400 B.C. The amber was called "slav," or "byzantine" from Byzantium (now Istanbul), which was sometimes an intermediary in trade. Although chemical analysis has authenticated amber in Assyria as originating from the Baltic region, the exact route by which this amber reached there remains a matter of speculation. The first references to Arab sources are from 900 B.C., and archeological finds along the Tigris River support Middle Eastern trade occurring by that time. Arab chroniclers of 700–400 B.C. indicated that amber was one of the most prized materials brought from distant lands (Grabowska 1983).

Greeks and Romans

Amber was important in Greek trade from the time of the early cultures at the beginning of the Iron Age to when Greek civilization was flourishing (900–300 B.C.; Figure 6-1). However, very little amber has been documented from Greece after 600 B.C. The Greeks wanted amber mainly for its decorative qualities; Homer mentioned amber several times in his *Odyssey* (1000 B.C.), referring to "amber beads that glowed as if with sunshine." The ability of amber to attract small bits of lint, straw, or other light objects after it is rubbed was noted by the Greek philosopher Thales about 600 B.C., which led to the incorporation of familiar words into the English language. The Greek word for amber is *elektron*, derived from *elektor*, sun's glare; it is from this early discovery of amber's attractive properties that *electron* and *electricity* come.

The Greeks were also responsible for some of the most famous legends that have been preserved in Western literature regarding the source of amber. One of the best-known Greek myths, the Tears of Heliades, which attributes a divine origin to amber, was recorded by the Roman poet Ovid. The legend recounts the adventures of Phaëton, who asked his father, the sun god Helios, to allow him to drive the chariot of the sun across the heavens. The inexperienced boy lost control of the fiery horses and they came too close to Earth, causing a terrible drought. To save Earth, Zeus struck Phaëton dead with a thunderbolt. Phaëton's sisters, the Heliades, wept so heavily that they became rooted and were changed into poplar trees, with their hardened tears becoming amber.

There were numerous other Greek legends about the origin of amber (Wheeler 1954). For example, Nicius thought that amber was a liquid produced by the sun's rays striking the surface of the earth, which then solidified. Sophocles believed that amber was produced in countries beyond India from tears shed for Meleager by his sisters, who were transformed into birds called Meleagrides who traveled afar. Also, Plutarch thought that amber formed from the urine of lynx.

Although the Greek philosopher Aristotle suggested the botanical origin of amber as "petrified poplar gum" about 300 B.C., it is the Roman, Pliny the Elder, much later in the first century A.D., who is usually acknowledged not only for discrediting Greek myths but for proclaiming that amber is a plant product, "produced from a marrow discharged from trees . . . like a gum from a cherry or resin from an ordinary pine."

The city-states of Etruria declined around 100 B.C. as the Romans assumed dominance in Italy. For several centuries before and after the birth of Jesus, however, Baltic traders did not transport amber along the routes to the south. Rather, they traded with neighboring Germanic tribes east of the Vistula River despite the tolls levied by those tribes. Meanwhile, the Romans had established the seat of a garrison at Carnuntum in central Europe near the Danube, and it became a natural center for trade and commerce with the central Germans and tribes beyond (Figure 6-1). The Germans and Romans had different names for amber: the German *Bernstein* referring to stone that burned, the Latin *suc(c)inum* referring to its being an exudate.

During the height of the Roman Empire, the Romans regarded amber so highly that Pliny the Elder stated (about A.D. 77) in his *Natural History,* "the price of a figurine in amber, however small, exceeded that of a healthy slave." Amber pendants with an enclosed insect were considered protection against evil; the belief was so widespread during Pliny's time that children wore amber around their necks to protect them against evil powers. In the early years of Nero's reign (A.D. 54–68) the demand for amber was so great that he sent a Roman officer to the north to find its source. It is not known, but it is assumed that the officer at least partly followed the Iron Age route to the Gulf of Danzig (Gdan´sk). Nero was particularly interested in providing amber amulets for his gladiators to wear to assure victory in their exhibitions. About 6000 kg of amber were brought back by the officer as a gift to Nero from the Balt king. Some historians think this is one of the most significant events of the Roman amber era because it opened direct trade with the Baltic, bypassing the German middlemen (Rice 1987). Thus, during Roman time, most amber came from Samland, along the Vistula, then via other rivers to the Adriatic— essentially the central route established during the Iron Age (Figure 6-4).

By the end of the first century A.D., the Romans controlled an industry of amber carving with its principal center at Aquileia near the head of the Adriatic Sea. Archeological finds, mostly from graves of Roman women, include toilet articles, rings, pendants, knife and mirror handles, figurines, and small containers and vases. Roman women often carried small lumps or stylized carvings of polished amber to enjoy its touch. By the third century A.D., however, the amber trade to Italy was reduced as a result of the decline of the Roman Empire and the presence of warring Goths. With the fall of the Roman Empire later in the third century, the trade of Baltic amber to the Mediterranean ceased.

Baltic Amber from the Middle Ages to the Present

Whereas the history of amber use has been preserved in the literature of Mediterranean cultures, the story of amber in the region where it was produced has only become evident through relatively recent studies of archeological sites in Latvia and Lithuania, dated A.D. 100–1200. These excavations attest to the importance of amber in the lives of the Balts (Rice 1987). The number of Roman artifacts suggest an active trade with the Romans and that the coastal inhabitants, where there was an abundance of amber, enjoyed luxuries resulting from its trade. In the period A.D. 100–400 the coastal Baltic culture was superstitious, as indicated by objects made to dispel evil spirits. From 400 to 800 there was a changing culture, with class systems developing. During this period the amuletical (protective) power of amber kept it from being usurped by other gemstones when the wealthy developed a taste for more elaborate forms of jewelry. Later, 900–1200, amber is rarer in grave sites, probably because the coastal Balts began cremating their deceased or because the amber was used primarily in exchange for imported bronze and iron tools rather than for making ornaments locally.

Only in the 13th century, around the middle of the medieval period, is there evidence of significant use of Baltic amber again throughout Europe. From then through the 20th century, monopolies were created to control the supply of this amber. In fact, Williamson (1932) thought that all changes in production and sale of Baltic amber for the past 800 years were dominated by monopolies. This amber has been the greatest source of material for art objects, everything from necklaces, rosaries, and worry beads to cabinets, chandeliers, and paneling.

The importance of amber in the Medieval Period (5th–15th centuries) is even evident in the word. It is an early medieval loan from the old Arabic *andar* or *ambar*, which entered the English and Romance languages as the

Spanish *ambar* or *ambeur,* French *ambre,* and Italian *ambra* (Ley 1951). The Arabic word refers to ambergris, a waxy substance (actually a mixture of cholesterol and steroids) regurgitated by sperm whales that has long been used as a fixative in perfumery. Although ambergris is a substance completely different from amber, confusion of the two persists. Some scholars suggest that the Arabic *andar* was originally used for perfume and incense, and because amber was also used as incense, the word came to be applied to both amber and ambergris. The French distinguish the two as *ambre jaune* (yellow amber) and *ambre gris* (gray amber).

Medieval and Renaissance Periods

The arrival in East Prussia of the order of Teutonic Knights *(deutschen Ritter)* in the 13th century ushered in an amber era important in European history. In early times, Baltic amber was the property of the finder. Later, the dukes of Pomerania claimed all amber as theirs as far east as the coast of Danzig (Gdańsk), including the vast deposit of the historic Samland (Figure 6-3). Control of the amber was the reward given to the Teutonic Knights for their victory over the Baltic Old Prussians in 1283. The monopoly over amber was extended by the Teutonic Order, later called the Amber Lords, along the entire Prussian coast (Williamson 1932). Interestingly, a special permit was even needed to walk the beach (Ley 1951). Fishermen and others living near the seashore had to swear to the Amber Oath every third year. There was even a special Amber Oath for priests and a special Amber Court to administer the oath and punish offenders. Collecting amber from the beaches or fishing for it was strictly forbidden except under supervision. Punishment was harsh and sometimes included hanging, as shown by the gallows in the background in a 17th century etching (Figure 6-5).

Paternoster beads, or prayer beads, were a mainstay of the amber economy by the mid-13th century and have a rich history in themselves (Grimaldi 1996). Early Christian tradition included amber as a sign of the presence of God; amber rosary beads, threaded into symbolic sets of numbers, were a part of everyday life by the mid-13th century. The sets had evolved into the Catholic rosary (five decades, or sets, of beads, one large bead and 10 small ones in a set) by the second half of the 15th century. Thus amber had again become a lucrative trading commodity with southern areas, and as early as 1302, shipments of amber were sent to the first guild of paternostermachers (makers of Lord's prayer beads) in Bruges. In 1310 another amber guild was established in Lübeck, and by 1312 the Teutonic Knights had assumed a monopoly on trade in Baltic amber. During this period, the Bruges and Lübeck guilds produced rosaries for the entire Christian church (Figure 6-6). The paternostermacher guilds were so important that they had their own patron saint, St. Adalbert, the first apostle of the amber region (Williamson 1932).

During the latter half of the 15th century, the influence of the Teutonic Knights weakened, and by 1466 the major Prussian cities (now Germany and

Figure 6-5. Etching from the first book on amber, *Succini Prussici Physica et Civilis Historia* (Hartmann 1677), illustrating licensed amber fisherman, with gallows in the background used for those who were caught collecting amber illegally.

the Baltic States) were allied with Poland. In 1480 the king of Poland granted Gdańsk (Danzig) the right to its own paternostermacher guild, which became the largest guild of amber craftsmen. Nonetheless, the Knights provided competition by establishing amber turners in Königsberg (now Kaliningrad). Albert of Brandenburg, the Hohenzollern ruler and grand master of the Teutonic Knights, secularized the order under the duchy of Prussia and became a Protestant in 1525. Subsequently, the guild in Königsberg began to promote secular items such as game boards, goblets, and tankards as well as carvings of religious figures and crucifixes to offset the loss of the rosary business.

Figure 6-6. Etching of a paternostermacher workshop, from Weigel (1698).

Additionally, an active amber trade to the Middle East was founded to supply Mohammedan prayer beads *(tasbih)*, 99 beads, one for each martyr of the Muslim faith, and especially valued when made of amber. In further attempts to make amber trade profitable again, Duke Albert asked his court physicians to investigate the medicinal uses of amber. Their investigations produced the earliest scientific treatise designed to promote amber as a remedy for various ailments (Chapter 10). They recommended both internal and external uses, especially with white amber as an ingredient. Having been used as a talisman for centuries, amber gained a more scientific credibility from their studies. Nonetheless, widespread decline in the use of rosaries after the Reformation ultimately led to the collapse of the amber guilds.

Seventeenth Through Nineteenth Centuries

In the 17th through 19th centuries, northern Europe (Prussia in particular) supported many amber workshops and artisans. The shops were headed by great master craftsmen whose sophisticated skills have never been matched by later amber workers. Large decorative, religious and secular objects such as chests (Plate 35), inlaid cabinets, alters, lidded tankards, and chandeliers were created. Most of these objects were made by fitting pieces of amber, varying in color and transparency, together in a mosaic for an inlaid façade. Ivory or bony amber was also inlaid with transparent amber to create beautiful contrasts. For example, faces of figures were carved of a single transparent amber, with hands carved of bony amber. These objects could never have been made without large supplies of amber because pieces were selected for color, transparency, size, and shape. Therefore, the great workshops developed where amber was abundantly available, such as in the towns of Danzig (Gdańsk) and, especially, Königsberg (Kaliningrad), along the southern coast of the Baltic Sea (Figure 6-3). Moreover, this grand era of amber work could not have developed without the patronage of an aristocracy, including the czars of Russia (Peter the Great and Catherine the Great), Prussian rulers (the Fredericks), and Louis XIV of France. As Grimaldi (1996) stated, "The splendor and opulence of the age were captured in amber."

The pinnacle of amber creation came in the 18th century with the Amber Room, which required many skills of the amber craftsmen. Inspiration for the paneling of a room probably came from the chests and house altars of inlaid multicolored amber that were common in Königsberg and Danzig in the 17th and 18th centuries (Plate 35). It was commissioned as a banquet room for a Berlin palace in 1701 by Frederick I, king of Prussia. Panels with a base of oak were made of 100,000 pieces of amber laid into mosaics of floral designs, royal heralds, and profiles. Delicate carvings included scenes from the life of fishermen as well as full-figure sculptures. Soon after its completion in 1712, the room was admired by Peter the Great. Frederick William I, son of Frederick I, presented the Amber Room to Peter in 1716 to commemorate the signing of the Russo-Prussian Alliance. The room was dismantled and the pieces crated and taken by sleigh to St. Petersburg. The first home for the Amber Room was as a study in the Old Winter Palace, but the panels were moved in 1755 to an especially designed room in the Ekaterininsky Palace (Plate 37). The differences in the warm amber colors, from dark to light, apparently created an inimitable charm to the whole room. Numerous other amber art objects, including chests, candlesticks, and crucifixes, lent to this charm.

Although amber had been harvested all along the beaches of the Baltic coast, the greatest amount came from the Samland Peninsula (Figure 6-3). This peninsula, only about 1000 km², has yielded 90% of European amber. During one day alone, 1815 kg were collected from the beaches of Palmnicken (now Yantarnyy, Kaliningrad) in the mid-1800s. Thus it is not coincidental that the most productive amber mine ever known was later established at Palmnicken to supply the great demands of artisans (Prockat 1932).

The history of amber mining is impressive (Bellerman 1913; Andrée 1937, 1951). Stantien, a Memel innkeeper, began successful dredging operations for amber in the lagoon near Schwarzort in 1855 (Figure 6-3). He joined with Becker, a merchant, to form the mining firm of Stantien and Becker. By 1865 they were operating 22 steam barges to recover amber by dredging and in 1868 recovered an extraordinary amount of amber: 83,900 kg (Grimaldi 1996). In 1850, however, a geologist, George Zaddach, began a geological survey of the Samland amber-bearing beds and in 1867 described how amber is concentrated in the *blau Erde* (blue earth, which is actually greenish, formed of glauconite) in the Palmnicken area. He reported that the amber washed up on shores had been cast off from the blue earth beds just 5 m below sea level but that the amber-bearing beds extended 40–47 m below the topsoil. Thus a rich underwater source of amber remained to be exploited. By 1870, Stantien and Becker began open-pit mining and procured about 4500 kg of amber yearly. The famous Palmnicken mine opened in 1875 and produced 204,120 kg of amber that year. There, the whole amber-bearing layer could be removed via shafts, tunnels, and galleries. Its yield increased continuously until 1895 when, incredibly, 544,320 kg were extracted.

In 1899, Stantien and Becker sold the mine back to the Prussian government and it became the Royal Amber Works at Königsberg. In addition to mining, the state operated a factory to produce jewelry and other items as well as chemical plants to produce amber oil and amber varnish from subquality pieces. Although ambroid (formed through fusion of small, unusable pieces of amber under heat and pressure) had been labeled as imitation by the Royal Amber Works, the works subsequently purchased the rights from the inventors at a high price. Except for flow lines that develop with age, ambroid is difficult to distinguish from unworked amber; moreover, it is harder than amber and thus is particularly suitable for smaller articles such as buttons and smoker's accessories. Therefore, ambroid was a highly profitable investment for the works at the turn of the century, since half of all amber production was for the manufacture of smoking items. Ambroid was especially popular for smoker's mouthpieces because it was considered healthier than horn, bone, or ivory.

Twentieth and Twenty-first Centuries

As underground mining costs began to increase and decreasing amounts of amber were found, open-pit mining was adopted at Palmnicken. After World War I, the Samland amber-producing mines were under the control of the German government. Demand increased during the 1920s, with amber second only to diamonds in U.S. gem imports.

By 1930, amber extraction at Palmnicken was largely mechanized; buckets of blue earth were carried by train over grates and spilled into a spray house where the material was blasted with high-pressure hoses (Prockat 1932). Nearly 90% of the hundreds of thousands of kilograms extracted was of poor quality and suitable only for chemical processing; the remainder was pressed into ambroid, used for carvings and jewelry, or that containing fossilized plant or insect remains was saved for scientific study. For chemical processing, the amber was dry-distilled in huge iron retorts, which yielded 60–65% amber varnish, 15–20% amber oil (used in medicines, coatings, and highest-grade varnishes), and 2% distilled acids (used for medicines and dyes).

With industrialization of amber crafts, the cult of artistic hand production greatly decreased. There was demand for lathe-turned beads produced on a mass scale. Nevertheless, during the period that hand amber working fell into decline, the traditions of some Baltic cultures such as those of the Kurp region of Poland were preserved in folk art (Grabowska 1983). The works of Kurp folk artists showed that amber-working techniques, yielding jewelry such as the necklace in Figure 6-7, had remained unchanged for centuries and were still used. Not only had the age-old amber working by the Kurps survived, but also the use of amber for barter payment instead of money. Above all, inhabitants of the Kurp region believed in the magical and healing properties of amber. Amber was burned and the resinous aroma inhaled in the belief that it was of use in treating all kinds of maladies, such as eye complaints and headaches. Wearing amber also was considered to be a certain remedy for infertility. When an old Kurp died, people said he had gone to the other world to dig for amber.

Amber production dropped greatly with the political unrest when Hitler came to power in Germany. The National Socialist Party did attempt to help the amber industry by requiring that all sporting societies use amber prizes and decorations for sports awards. The amber industry was nearly destroyed during World War II. Moreover, when the Nazis invaded Leningrad in 1941, the Russians tried to hide the panels of the Amber Room in the Ekaterininsky Palace by papering them over. According to Rohde (1942), however, two Nazi officers, art historians in civilian life, realized the value of the room as one of the world's great art treasures and had the panels installed in the main castle in Kaliningrad (old Königsberg). In 1945, when Kaliningrad was bombed by the Allies, the panels were presumably taken down again and hidden.

The survival and whereabouts of the Amber Room panels remain a mystery. Imaginative explanations for their disappearance abound and it is difficult to separate fact from fiction. With a few exceptions, the panels have not been found, and private and governmental detective work continues, trying to locate this great art treasure, assuming that at least parts of it still exist. The search was abetted by the finding of part of a panel in 1997. The Russians consider the Amber Room so much a part of their cultural history that Russian artisans began the incredible task of exact replication of the panels in 1979. The replicas are based on a single color photograph (Plate 37) and several black-and-white photographs taken before the Nazi invasion, fragments of paneling, and a few drawings. Work has been impeded by the cost of the large amounts of high-quality amber required, which still comes from the Samland mines, as well as the unimaginable skill and labor. One of the last persons to have seen the Amber Room intact described it as like stepping into a fairy tale.

After World War II, the amber mines and pits of Samland were incorporated into the Lithuanian amber industry. The mining was entirely open pit, spread over 1200 m to a depth of 36–40 m, with the amber-bearing blue earth 5–7 m thick. Amber-processing shops opened in towns such as Palanga,

Figure 6-7. Folk art necklace from Poland, probably made during the 18th century. Handwork such as this continues today in the Kurp region. Courtesy of H. Czeczott.

Klaipėda (Memel), and Vilnius; amber is considered the gem of Lithuania (Rice 1987). When Lithuania and Latvia were incorporated into the Soviet Union, much of the processed amber was shipped to the official Russian wholesale outlet that controlled the sale of amber, especially for shops catering to tourists. An instructional center for amber processing, however, was established in Liepāja, Latvia, and graduates there and in Lithuania processed the amber for domestic sale and export. The processing of amber has clearly been important culturally and economically over the long history of the Baltic.

The current problems of the amber industry in the Samland area are similar to those at the end of the 19th century (Kostiashowa 1999). During most of the 19th century, Prussia maintained a monopoly over most amber extraction and processing in Samland, but by the end of that century most amber from the region was in the hands of foreign manufacturers. Today the amber industry in the Kaliningrad Oblast of the Russian Federation is just the source of raw material for foreign companies. Kostiashowa thought that the main reason for the present crisis was a repetition of the old problem: creation of a monopoly with the government's establishing a giant enterprise, the Kaliningrad Amber Factory, that could only operate under Soviet planning. When the factory went bankrupt after the breakup of the Soviet Union, private companies and joint ventures that involved foreign capital concentrated on export of raw amber for creation of amber products outside the area. In 1997, however, a new program was implemented to put amber processing, craftsmanship, and restoration of the famous Königsberg amber collection on a firm financial basis.

Amber in Other Areas

Although other parts of the world did not experience the equivalent of the extensive cultural and economic importance of Baltic amber in Europe, amber was of cultural significance in China and a number of Latin American countries.

Burmese Amber into China

Amber is used for jewelry or other items by local people wherever it is found in sufficient amounts, but China has the richest history of artistic use of amber outside of Europe (Figure 6-1). Burma (now Myanmar) was the origin of most of the amber used in Chinese art objects, at least prior to the 19th century.

After that, Chinese trade with Europe made Baltic amber more available than Burmese. Amber was collected from shallow mines in the Nangtoimow Hills in northern Burma (Chhibber 1934) and the major portion was sent to trade centers such as Mandalay and Mogaung. The amber was then probably brought by traders to Yunnan province in China where it was known to Chinese craftsmen from as early as the first Han dynasty (206 B.C. to A.D. 8). Although China had established complex trade links with Europe by the middle of the first century B.C., just prior to the Roman amber era initiated by Nero, there is no evidence that Baltic amber reached China during that period.

Some of the Burmese amber has always been particularly prized for its deep red color. One of the favorite figures carved by Chinese craftsmen was the fish, representing long life and good fortune (Plate 36). Amber was also known to the Chinese as *hu-p'o*, soul of the tiger. Thus the Chinese regarded amber as possessing many strong qualities of the tiger and symbolizing courage (Rice 1987). Special care had to be taken to fashion the traditional necklaces worn by the mandarins; all beads were required to be round and uniform in size, and they had to number exactly 108 because each bead had its own religious significance.

An opaque, dark brown Burmese variety is called root amber. The various mixtures of brown shading result from penetration of calcite into openings and cracks in the amber, giving it a mottled appearance. In China, root amber is used for ornaments with carved designs utilizing the natural swirls and color variation to create special effects. This type of carving is called *ch'iao-tiao*, clever or ingenious carving. Interestingly, this amber was so cherished in China that the carvers followed the natural contour of the piece as they considered it wasteful to lose any in the creation of an object.

Because of the high value placed on amber in China, it has often been counterfeited. As early as the sixth century A.D., Táo Hungching warned against false amber, using the electrostatic ability of amber to attract straw as a means of distinguishing amber from imitations. By around 1910, however, Bakelite plastic imitations reproduced these electrostatic properties. Furthermore, the red color of Bakelite is similar to Burmese amber, making the distinction even more difficult. Large carvings made from transparent red synthetic resin have been claimed to be "reconstructed Burmese amber." Although the majority of these large objects are made of synthetic resin such as Bakelite, pressed amber blocks had been used for carving in China for many years (Rice 1987).

Pre-Columbian Amber Trade in Mesoamerica

Amber from Chiapas, southern Mexico (Chapter 4), was traded by the Mayan Zinacanteca Indians northward to the Aztecs in pre-Columbian time (Figure 6-1; Blom 1959). Zinacantán was the most important trade center from which the Zinacantecas monopolized trade with the Aztecs and probably other Indian nations. In fact, Blom stated that the Zinacantecas "certainly used a gangster method to protect this trade by murdering the Aztec intruders if they caught them."

The rich deposits in central Chiapas were the source of the amber so widely used by the pre-Columbian Indians of Mexico. The amber was collected most commonly from landslides, following rains. It was formed into lip plugs for warriors to show that they did not fear death and were greatly skilled in the art of war. Amber also was worked into the beads, necklaces, and amulets such as the ones shown in Figure 6-8, found in a tomb at Monte Albán, Oaxaca. Blom (1954) thought that these objects had been traded by the Zinacantecas to the neighboring Zapotecas in the Valley of Oaxaca, an area under Aztec governmental administration. Amber is still carved into the forms of hands, teardrops, and other shapes for amulets to protect babies against the evil eye. It continues to be sold in town markets, but the Zinacantecas no longer hold a monopoly on this commerce.

Resin Figurines from Costa Rican Burial Sites

Various objects carved or molded from resin, or possibly amber, occur in aboriginal burial grounds in Costa Rica. The Chorotega people, who spoke a Chiapanec language, were among those people generally migrating to northwestern Costa Rica from farther north, A.D. 700–1400 (Figure 6-1). Langenheim and Balser (1975) suggested that these Chorotega people could have brought amber objects into the area from Chiapas, Mexico.

Resin figurines occur in numerous burial sites along the slopes of both the Pacific and Atlantic coasts of Costa Rica. Infrared spectroscopy suggests that most of these figurines were carved from *Hymenaea* resin, probably *H. courbaril,* which only grows on the Pacific side of Costa Rica. Some, however, could have come from *Hymenaea*-derived Chiapas amber. Interestingly, *H. courbaril* is not only restricted to Pacific slopes but is the only species that produces the quantities of resin needed for carving the figurines of animals in the southern Diquis region on the Atlantic side of Costa Rica (Figure 6-9).

Therefore, the resin had to be traded, probably from a southern, Pacific area to several Diquis sites. Metalworking in the Diquis region has been dated about A.D. 1000, and resin objects associated with it in burials could have appeared about the same time. The Diquis region borders the Panamanian cultural area of Chiriquí, where resin was also used for artistic objects as well as for industrial and medicinal purposes.

Figure 6-8. Amber bead necklace and amulets found in a tomb at Monte Albán, Oaxaca, Mexico, from perhaps as early as A.D. 100. Note the similarities to the necklace made in 18th century Poland (Figure 6-7). Redrawn from Blom (1959).

Figurines were of sufficient interest as grave offerings that the aboriginal peoples of Costa Rica traded to obtain either them or the resin to make the objects themselves. Balser (1960) suggested that these figurines could have been intended for burning as incense after death. Animal figures were carved from amber during the Stone and Bronze Ages in the Baltic region (Plate 38), and carved animal objects in eastern Asia were believed to increase virility in men and fecundity in women (Plate 36). Such convergence in symbolism in objects created from resinous material in different parts of the world is intriguing.

Dominican Amber

Columbus, on his second voyage to the New World (1496), described an amber-mining region on Hispaniola and received gifts of amber from the Taino people when he arrived on the northern shores of that island. This was the earliest report of amber from the New World. There is no evidence that amber was traded with other people in the region, however, nor that it was exported commercially until much later. Dominican amber was also not known scientifically until the mid-1940s. The Dominican Republic has become the most plentiful source of amber, and its amber is the scientifically best studied outside the Baltic area (Chapter 4).

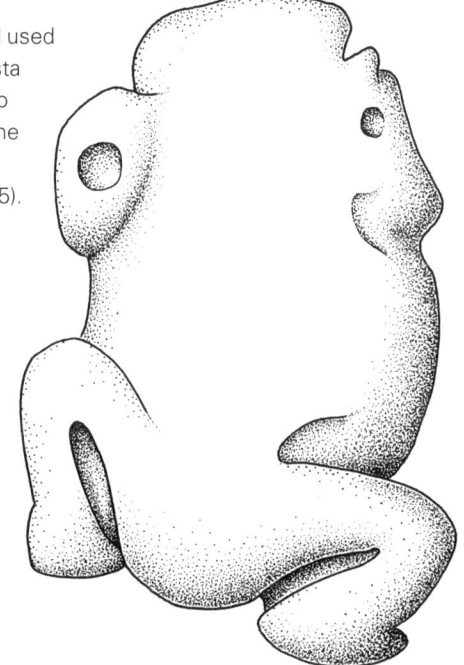

Figure 6-9. Resin figurine of a toad used as grave offering by aboriginal Costa Rican people. Note the similarity to Stone Age animal figurines from the Baltic regions (Plate 38). Redrawn from Langenheim and Balser (1975).

Dominican amber is collected from numerous small mines, usually named after a nearby village. When the Dominican government first began amber mining, a great deal of raw material was sold to the German amber industry. In the 1950s, however, the Cooperative of Industrial Artists was organized to provide a government-supported trade school to teach amber-working techniques to local workers. The government then restricted the exportation of unworked rough amber to foster the development of a cottage industry using one of the region's few natural resources. As a result, there are now numerous home-based artisans who have become expert in working amber. Jewelry designed by Dominican artists tends to be distinctive in reflecting the past culture of the Taino Indians (Rice 1987). Pendants also are carved to resemble the Haitian tiki god.

Incense Trade Routes

For more than a millennium, trading caravans supplied incense resins and other luxury goods from the southern coast of Arabia to insatiable markets in the Mediterranean region and Mesopotamia (Figure 6-1). During the height of imperial Rome, trade in incense assumed importance along with trade in amber. Roman trade outside the empire was founded on supply of five commodities that were a part of the Imperial way of life: amber and incense along with ivory, pepper, and silk (Spekke 1956). Some historians, such as Hoots (1993), suggested that the material wealth of the Arabian Peninsula in the 20th century would "never rival the relative prosperity" there at the height of the incense trade 2000 years before. Arabia was known at that time as Arabia Felix, happy Arabia or Arabia the blessed. As Pliny the Elder claimed, "They are the richest races in the world."

The incense was produced by two genera of Burseraceae—frankincense from *Boswellia* and myrrh from *Commiphora*—that occur primarily in southern Arabia and neighboring areas (Chapters 2 and 8). Dates for the earliest incense trade, however, remain an enigma. Hoots (1993) thought that the original spark to the trade apparently came from pharaonic Egypt in 3000 B.C. because oil from myrrh was used to embalm the dead. Since myrrh-producing plants do not occur in Egypt, he assumed that the product arrived by trade. An Egyptian inscription about 2500 B.C. records the purchase of "80,000 measures" of myrrh (Majno 1975). "Whatever the measure," Majno contended, "this is a lot of myrrh," implying the necessity for trade. The Egyptians were not the only early users of incense. The Phoenicians so esteemed incense that the name of the principal deity, Baah Hamman, denoted "Lord of the perfumed altar." The Babylonians, Sumerians, and Assyrians also valued incense highly, but Groom (1981) questioned whether these people were using frankincense and myrrh for incense or, rather, other resins from local plants and aromatic woods. He conceded, however, that small amounts of frankincense and myrrh could have been brought to these areas, having been traded hand to hand northward from southern Arabia but not by any organized form of transport.

Groom (1981) further argued that the earliest references to the transport of frankincense and myrrh were much later, in a 1500 B.C. inscription in Queen Hatshepsut's temple near Thebes. This inscription records an expedition sent down the Red Sea to collect incense from the land of Punt, an ill-defined region considered by some to have been what is now the Somali coast and a portion of the Arabian coast opposite. In what has been described as the world's first plant collecting expedition, 31 incense trees were brought back (Figure 6-10) and planted at the temple of Karnak on the banks of the upper Nile in Egypt. The carving on one of the walls of the temple shows how the plants were loaded onto a ship and transported in wicker baskets. Although it is not clear whether these were frankincense or myrrh trees, Groom suggested that they were probably myrrh, of the species common in Somalia today.

Domestication of the camel provided the means for an organized overland incense trade from southern Arabia to the Mediterranean; that donkeys may have been used as pack animals before camels has been generally unacceptable to historians (Groom 1981). Although camels were domesticated in southern Arabia by 2000 B.C., it is improbable that the camel was put into significant use to transport incense northward before 1000 B.C. (Figure 6-1). Camel transport allowed larger quantities of material to be carried more quickly across the Arabian deserts, thereby accelerating trade to meet the demands of the Greeks and Romans who used the incense to win the favor of the gods. It was the smoke from incense that carried aloft the fragrance of a person's gift, a gift regarded as actual food for the gods who would starve without it (Majno 1975). Atchley and Cuthbert (1909) pointed out that incense was also used in other important ways, that is, as a sacrifice to deceased humans, to drive away evil spirits, as a symbol of honor to a living person, and as an accompaniment to festivities and processions. Among some peoples, such as Egyptians, Persians, and others, the right to offer incense was only the prerogative of the priesthood (kings were included). On the other hand, Greeks and Romans regarded use of incense as a duty incumbent on all people, but it could only be presented to a deity or deified man. To offer incense was to acknowledge the deity of that person. Incense resins were further sought for use in medicines and perfumes; the word perfume comes from *per fumum*, by smoke.

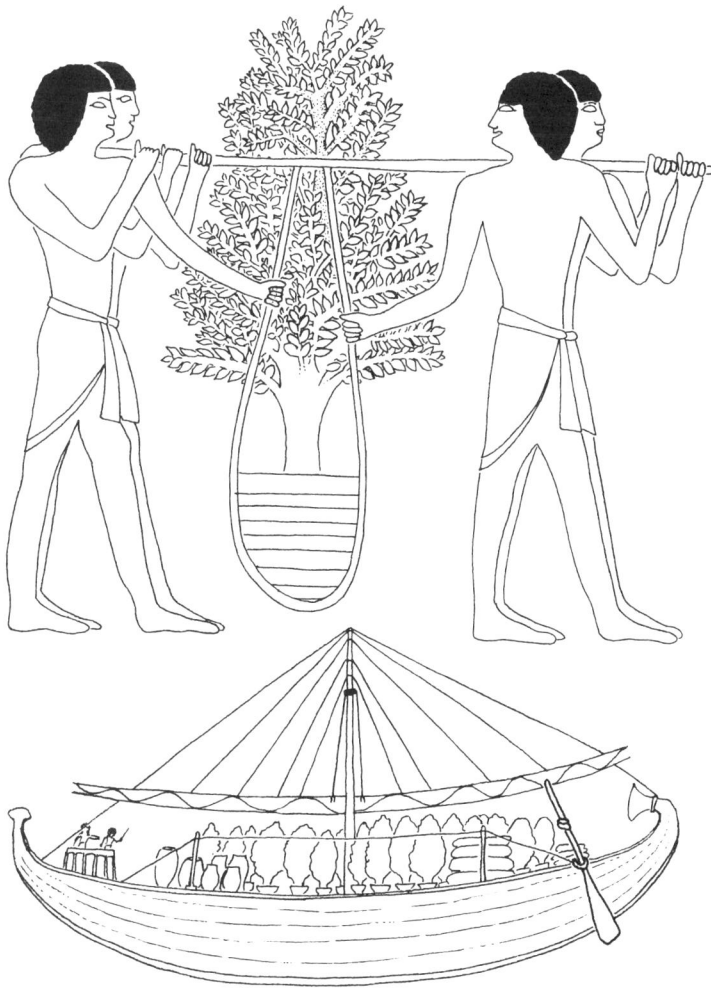

Figure 6-10. Inscription in a temple near Thebes of Queen Hatshepsut's expedition to land of Punt to obtain incense. Myrrh trees are carried to the ship (shown below in smaller scale and typical of the times, with trees on board) for planting in Egypt. Redrawn, in part, from Groom (1981).

Arabian commerce with the Mediterranean region became more organized. Herodotus, the Greek father of history, is the first classical scholar to provide evidence for the use of incense from Arabia, demonstrating that a substantial trade to Greece existed by 500 B.C. At the time when Athens was at its peak, he mentioned Arabia's aromatic resins: "The whole country is scented with them—and exhales an odor marvelously sweet." Theophrastus, the Greek father of botany, around 300 B.C. wrote about the first eyewitness account of incense trees and the harvesting of incense from the reports of reconnaissance ships sent out by Alexander the Great. It was the Roman, Pliny the Elder, near the end of the first century A.D., however, who has provided most of the information regarding transport of incense from its source in southern Arabia and its processing in the north. Thus the literature reflects this Roman influence.

Bowen (1958) pointed out that references, including maps, to *the* incense route should not be taken literally because there were numerous alternate pathways for reaching various destinations. There was clearly a long main route to the Mediterranean, however, with a major side route to Gerrha and smaller side routes to the Red Sea and around Palestine. For this reason, some researchers (e.g., Groom 1981) prefer to distinguish the major route as the Incense Road. Caravans suffered losses from bandits en route, which also dictated local changes in the actual routes.

As security was established for the caravans, more of other kinds of goods were carried with the incense, including spices, ebony, silks from India, and rare woods, skins, and gold from the nearby African coasts. Caravans returned to southern Arabia with various goods from Greece and Etruria, then from the Roman Empire, leading to the emergence of highly centralized wealthy states. The first and best known in southern Arabia was Saba, or Sheba. Saba used its newfound wealth to increase an already rich agriculture with further development of an irrigation system. A great dam built in the eighth century B.C. by the rulers of Saba held water for the wonderful orchards and gardens around Mar'ib, thought by some to be the Garden of Eden mentioned in the Bible and Koran. As the capital of Saba, Mar'ib remained an important stopping point on the Incense Road (Figure 6-11), even after the fifth century B.C. when several small states began to exert their independence from Saba.

The Incense Road continued to grow in importance in the centuries before the Christian era (Groom 1981). Caravans of 2000–3000 camels were not uncommon. Resin from the Dhofar region (now southern Oman), Soma-

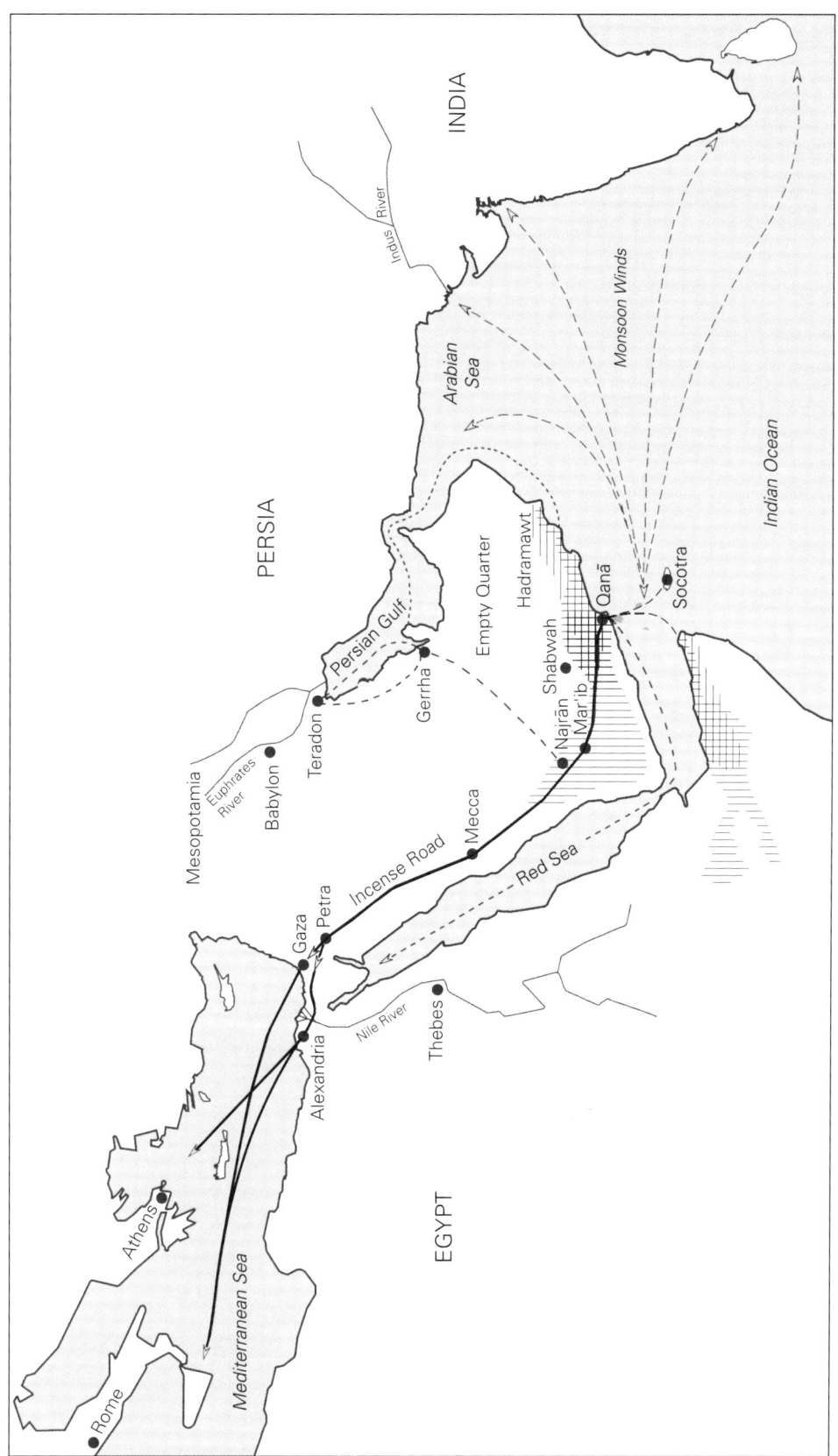

Figure 6-11. The Incense Road from the region where the myrrh (vertical hatching) and frankincense (horizontal hatching) plants grew—today, southern Oman, Somalia, and the island of Socotra—to the Mediterranean Sea, thence to Rome and Athens (with a branch to Persia and Mesopotamia). A few of the side routes, overland and sea, are indicated with dashes.

lia, and the island of Socotra, was transported via sea to towns such as Qanā in the Ḥaḍramawt region of eastern Yemen (Figure 6-11). By the time the resin reached the port town, it had been transported more than 800 km by sea. From the seaport it was 260 km across the mountains to the emporium at Shabwah, where the resin was taxed, then more kilometers to the frontier of the incense lands at Najrān. From there, merchants went overland with frank-incense to a transshipment site thought to be Gerrha. From here, the incense was transported either by land or sea to reach Persia and Mesopotamia, where it was used lavishly. The main route, estimated to take an average of 78 days (Groom 1981), went almost 2000 km to Gaza on the Mediterranean coast. In Palestine the resin was highly valued, as reflected by its mention in the Bible 22 times (Moldenke and Moldenke 1952). Some incense was probably shipped to Rome from Gaza but most of it was sorted and processed in Alexandria, which was the industrial center of the Roman Empire. In the Roman processing factories in Alexandria, according to Pliny, the incense was so precious that workers were stripped and searched to keep them from stealing it. At the time of Jesus, "frankincense was not only more costly than gold—it may have been the most precious substance on earth" (D. Roberts 1998). Rough calculations based on figures provided by Pliny indicate that in his time, 2.5–3 million kilograms of myrrh were produced (Groom 1981). Myrrh commanded three times the price of frankincense, but the demand for frankincense was five times as great (Abercrombie 1983). From Alexandria it was another 2100 km to Rome by sea; thus the incense had traveled about 6500 km to reach Rome from its source in southern Arabia.

The Arabs also were exporting frankincense and myrrh along with other resins such as storax and Socotran dragon's blood to India and China (Chapters 8 and 10). Since most of the civilized world demanded incense, there were two harvests tied to the two monsoonal seasons, which also determined some of the shipping schedules in the Indian Ocean.

Geographic knowledge was so poor that most purchasers of incense in the cities of the Mediterranean were not sure where it came from. It was only a matter of time, however, before they attempted to find the source and seize control of the lucrative trade. In 24 B.C., Aelius Gallus, the Roman governor of Egypt, led a force of 10,000 men through the Arabian deserts in search of the location of incense trees. The Romans were forced to retreat because of lack of water, malaria, and the fierce tribes, who also spread rumors that the incense-producing areas were guarded by winged serpents who spared no intruders. Although this Roman expedition was unsuccessful, it was a sign of the increasing knowledge of geography, which would soon contribute to the decline of the Incense Road and the Arabian kingdoms that depended on it.

Many explanations have been offered for the decline and collapse of the incense trade (Groom 1981, D. Roberts 1998). A significant event was the discovery of the Indian Ocean monsoons by Roman sailors (Figure 6-11), allowing ships from the Mediterranean to sail down the Red Sea to ports in India, Africa, and Ḥaḍramawt on the Arabian coast. By the first or second century A.D., Roman traders had at least partially succeeded in bypassing the expensive middlemen and taxes along the overland road, thereby breaking Arabian monopoly of the incense trade.

Another factor in the decline was neglect of agriculture in the source areas because of the enormous wealth coming from the incense trade. The agricultural decline was abetted by the growing number of livestock (incense trees made excellent fodder for camels and goats), demand for wood (resulting in cutting of the trees), and a decrease in rainfall around the turn of the Christian era. These factors led to the disappearance of frankincense trees (e.g., *Boswellia sacra*; Chapter 8) over much of southern Arabia. D. Roberts (1998) stated, "Despite their importance to human history, frankincense groves are so elusive that even today no one knows their full range."

As wealth diminished, the southern Arabian kingdoms became a less desirable place to live, leading to a steady migration to the highlands of Yemen and the formation of another centralized state. Himyar commanded the developing ports of the Red Sea and was able to defeat the kingdoms formerly dependent on the overland incense trade.

The decisive blow to the incense trade, however, was the spread of Christianity in the Mediterranean region, climaxed by its becoming the official religion of the Roman Empire. At the end of the fourth century, the Roman emperor Theodosius I forbade the pagan practice of making offerings, including incense, to household gods. Incense was still used in funerals because those were civil ceremonies, but internal crises and severe inflation were paralyzing a Roman economy that previously allowed the purchase of such luxuries. Groom (1981) indicated that it is doubtful that the Incense Road survived the fourth century, and sea trade continued on a very much smaller scale. Once paganism had been vanquished, the Christian church slowly reintroduced the use of incense. During the eighth century, it began to be used in the Roman Catholic rite during lauds and vespers, and as a mark of respect for the altar, the priests, and the faithful (New Catholic Encyclopedia 1967).

With the arrival of Islam in the seventh century, the incense trade from southern Arabia had been completely transformed. Despite a slight revival in the fifth and sixth centuries, when wars between the Byzantine Empire and the Sussanians in Persia made sea routes too risky, the incense trade was a mere shadow of what it had previously been. During this time, Constantinople became the main center of perfumery. A sixth century description of the cathedral of Saint Sophia refers to hundreds of perfumed lamps. Continued export of frankincense and myrrh into the early Middle Ages is revealed in the commercial laws established by Emperor Leo VI in 895, which included them as perfume ingredients.

Cannabis and Trade in Its Resin

Cannabis resin has been traded across cultures spanning almost every continent. According to a Late Stone Age Chinese legend, the gods gave *ma (Cannabis)* to humans to fulfill all needs. As discussed in Chapter 9, this remarkable plant not only provides psychoactive resins but also durable fibers and nutritious seeds. The female flowers and floral leaves contain the resin that makes up marijuana (Figure 10-3), and the resin extracted from the trichomes is hashish (Chapter 3). The resin rivals alcohol, caffeine, and nicotine as the most widely used nonmedical drug (although it has been and is used medicinally; Chapter 10).

Old World Hashish Cultures

There are several hypotheses regarding the geographic origin of *Cannabis*. Although prehistoric dispersal has probably completely obscured its exact geographic origin, *Cannabis* most likely originated in central Asia or possibly in India along the foothills of the Himalaya (Clarke 1998). If it did originate in central Asia, it was dispersed from there to China, Persia, and India during the conquests of the Aryan and Scythian peoples. The Chinese apparently were the first to cultivate it (1500 B.C.) but both China and India were primary sites for its cultivation (Clarke 1998). Although the Chinese knew about the psychoactive properties of the resin, they were mainly interested in using *Can-*

nabis medicinally. The early use of *Cannabis* in medicine probably developed from the Chinese use of the seed for food (Li 1974). *Cannabis* is mentioned in the herbal *Pên-ts'ao ching,* attributed to Emperor Shên-Nung about 2000 B.C. (Figure 6-1). Li thought that the *Pên-ts'ao ching* was apparently based on early traditions that had been passed along even from prehistoric times. However, its later compilation in the first century A.D. in the late Han dynasty is often erroneously referred to as the first medicinal record by the Chinese.

The psychoactive properties of *Cannabis* resins were first exploited in India. *Cannabis,* "sacred grass," is mentioned in the Hindu Atharvaveda (2000–1400 B.C.) along with legends of Shiva, the "Lord of *bhang*" (Figure 6-12). The vedas describe Shiva as bringing *Cannabis* from the Himalaya for the enjoyment of the Indian people (Abel 1980). The plant was called *indracanna,* food of the gods. Indians maximized production of *Cannabis* for resins by growing the plants widely spaced, removing male plants to prevent fertilization, and prolonging the flowering of female plants, on which the resin-filled trichomes occur. *Ganja* is the Indian word for marijuana; *charas* is used for hashish. *Cannabis* resin in India is either eaten or drunk in a milk-based beverage, both of which are called *bhang.* This widely used drink is concocted with ground *Cannabis* leaves, along with sugar and various spices.

Figure 6-12. The Indian god Shiva was called the "Lord of *bhang*" because the Vedas credited him with giving humans *Cannabis. Bhang* is the name of the drink that includes the resin.

One of the earliest uses of hashish may have been as incense. In addition to frankincense and myrrh, *Cannabis,* amber, and gum arabic were added to incense burners (Clarke 1998). Burning hashish produces large quantities of thin, fragrant smoke, and there are many early references to breathing this smoke from an open vessel or censor. For example, Herodotus (about 500 B.C.) described how Scythians and other central Asian tribes breathed the combustion vapors. Clarke noted that the intent was to obtain both the mystical benefits from inhaling the fumes and to appreciate the aroma of the burning hashish. Simple approaches to obtaining the resin predated the more complicated process of harvesting and drying the plant, followed by sieving out the resin. The simplest involved hand-rubbing the floral bracts of the female plant. Colorful stories abound regarding other ways to obtain the resin. For example, naked men or those clothed in leather were reported to have run through flowering *Cannabis* fields in Nepal, then scraping off the resin that collected on their bodies or their clothes. In Persia and Afghanistan, plants were beaten on carpets or rolled in a carpet and danced upon; the carpets were then washed to free the resin glands.

Persia (now Iran) is often cited as the original home of hashish even though the ancient Zoroastrian texts refer only to *bhang* and not to hashish. Mechoulam (1986b) thought that *Cannabis* intoxication for shamanic use probably originated in the Zoroastrian traditions of ancient Persia, spread through central Asia, and influenced Islamic mysticism. The tale of the Old Man of the Mountain (Sheikh Hasan ibn al-Sabbah) and the Hashishin, hashish users, or Assassins, is probably one of the most widespread legends concerning early use of hashish. Although there is no evidence to support the story, it became a Western classic after Marco Polo brought it back to Europe in the late 13th century (Kimmens 1977). There are many versions of the legend, but reference to verdant gardens is common to all of them (Clarke 1998). These gardens were so magnificent that they fulfilled the prospects of paradise in Mohammed's teaching. The Old Man of the Mountain drugged young candidates with hashish to join his band of followers and brought them to the mountain garden. There, all pleasures were made available to them. Initiates were promised that as long as they remained completely loyal to the Old Man and their fellow Hashishin, they could return to the garden both in this life and after death. For this, the dedicated Hashishin would dispose of whomever the Old Man condemned. There was a succession of eight leaders of the Hashishin, beginning with Hasan ibn al-Sabbah in A.D. 1090 and ending with Khwurshah in 1256 (Burman 1987).

When European contact with the Arabs was reestablished during the Crusades (11th–13th centuries), marijuana use was common throughout Africa and Asia. Hashish was introduced to many people in Europe through the tales of the *Arabian Nights* (written sometime between 1000 and 1700). In the 15th and 16th centuries, Arab traders introduced marijuana to Africa, where its use spread quickly. The Kafirs called the plant *dagga,* the name later applied to a culture that made widespread use of *Cannabis.* Africans originally either chewed the dried plant parts or mixed them into beverages.

Introduction of tobacco smoking into Europe and Asia during the early 16th century changed the pattern of use of both marijuana and hashish. Until tobacco was introduced from the New World, marijuana and hashish were consumed by mixing with food and drink. By the 18th century, techniques of harvesting, drying, and sieving plants to make hashish became widespread because mass production was necessary to meet the demands of a rapidly increasing Eurasian trade. Once people began to smoke rather than eat hashish, many more began to use it. Smoking hashish became popular among affluent and well-traveled members of Western society during the late 19th and early 20th centuries.

When Napoleon's army was stationed in Egypt toward the end of the 18th century, soldiers experimented with *Cannabis* resins and returned home with samples and reports of its wondrous effects. A French psychiatrist, Jacques Moreau de Tours, had read of the medicinal uses of hashish. He thought he might be able to use the drug to reproduce symptoms of psychosis and, hence, learn about the causes of mental illness. As a part of his studies he asked an artist friend, Théophile Gautier, to experiment with the effects of hashish on his creative senses. Eventually, a group of artists met monthly and consumed hashish while being observed by Dr. Moreau. This group called themselves Le Club des Hachichins, following the legend of the Old Man of the Mountain. Eventually, it evolved into meetings of the well-known artists and writers of the day, including Victor Hugo, Alexandre Dumas, and Eugène Delacroix. Thus, despite Moreau's ultimately concluding that hashish could have deleterious effects on overall mental health, the resin continued to attract users. The interest in hashish traveled across The Channel, where intellectuals such as Oscar Wilde and W. B. Yeats experimented with it to see if it enhanced creativity.

The introduction of *Cannabis* to North America centered on its use as as fiber (hemp). Before 1800, the British had ordered colonists in North America to cultivate *Cannabis* for hemp to meet the military need for rope-making fibers. Later use of the resin for its psychedelic effects followed a pattern similar to that in France and England. First, curiosity by intellectuals and artists, followed by drug-inspired art, then by more general use. In the early 20th century, jazz musicians in the United States used marijuana, which they called moota, to enhance their music without the stupefying effects of alcohol. Their appreciation for marijuana was expressed with songs such as Benny Goodman's "Sweet Marijuana Brown" and Cab Calloway's "That Funny Rope Man" among others. Marijuana became more prevalent than hashish in North America for several reasons. There were no traditional centers of hashish production in the Americas, and *Cannabis* could be easily grown for marijuana in favorable climates in large, available, sparsely populated areas. In fact, marijuana during the late 20th century was the largest cash crop in the United States, with maize a distant second (Clarke 1998). Mass marketing of marijuana was primarily an American occurrence, beginning in the 1960s and flourishing by the end of the 1970s.

Beginning in the late 19th century and early 20th, the patterns of hashish supply in Eurasia changed. Large-scale production in central Asia shifted from Russian Turkestan to Chinese Turkestan, eventually into Kashmir, and thence into Afghanistan. Greece, Syria, Nepal, Lebanon, and Turkey also became major exporters. By the late 1960s, however, Morocco was the last major country producing hashish in sufficient quantities for export.

Clarke (1998) pointed out that when the Western world's counterculture learned about hashish manufacture and trade, their travels significantly and permanently affected the world's hashish cultures (southern Asian, middle Asian, and Middle Eastern). Clarke called their destinations a Hippie Hashish Trail, which was essentially a collection of pilgrimage routes (trod before the hippies by Sufis) that were traveled to seek adventure and hashish. The first Westerners came in the early 1960s, and by 1969 the hippies had traversed the trail heavily. The Sufis, following the principles of a special form of Islamic mysticism, traditionally used hashish for religious insight. The itinerant Sufi fakirs told the hippies about the long cultural history of use while they shared their hashish. The hippie travelers so popularized hashish consumption in the West that great demand for it was created. This influx of

Western influences permanently changed the methods of hashish production across Eurasia.

Prohibition

South Africa had the first anti-*Cannabis* law in 1870, and in 1925 the League of Nations agreed to an international statute against its use. In the late 1920s and early 1930s, several states (e.g., Texas, Illinois, and Louisiana) banned the use of marijuana, and in 1937 its use came under federal jurisdiction. The legend of the Old Man of the Mountain was revived and twisted in such a way that hashish was considered to cause irrational behavior that could lead to criminal activities. Nonetheless, all attempts failed to halt the spread of the use of marijuana. Most medical studies failed to find detrimental effects when marijuana was used moderately. Sanctions against smoking marijuana only seemed to increase its popularity, as evidenced by the worldwide spread of its use by young people in the 1960s. In 1970 the U.S. Congress passed the Controlled Substance Act, which assigned psychoactive drugs to five schedules. *Cannabis* was placed in the most restrictive Schedule 1, that is, of no medical use and high potential for abuse, and unlawful for use even under the supervision of a physician.

In 1972, however, a U.S. federal study recommended decriminalizing marijuana because there was no evidence of damage from medicinal use of the drug. In 1981, another study by the U.S. National Academy of Sciences concluded that there was no evidence of permanent deleterious effects with light or moderate use of marijuana. Although the panel was concerned that heavy use was associated with some psychological and physiological problems and could lead to "hard drug" usage, they concluded that, overall, marijuana was safer than the legal, more widespread use of alcohol and tobacco. They suggested that legal sanctions against possession and trade be removed. Despite these studies, marijuana and hashish remain illegal not only in the United States but most of the Western world. Use for medical purposes, however, is receiving considerable attention (Grinspoon and Bakaler 1997) with eight U.S. states having passed laws legalizing medical marijuana. *Cannabis* is even grown for this purpose in the Netherlands. Recognized medical benefits as well as accelerated research to understand these further are discussed in Chapters 10 and 11.

Resins in Indigenous Cultures

Resins have been important in most of the cultures in which resin-producing plants grow naturally. Only relatively recently, however, have most been studied from a cultural perspective by anthropologists and other scientists.

Mesoamerica and the Maya

The Maya have a long history in Mesoamerica; it is now generally agreed that Mayan civilization dates as least to 600 B.C. (Henderson 1997). There is evidence for the use of amber from Chiapas for ornament by the Maya perhaps as early as A.D. 100, as discussed. The roles of resin in Mayan cultures are various, extending from natural resource use and management to religious and social organization.

The use of resin as incense has been particularly important in Mayan culture. In fact, it has been used throughout Mayan history, as supported by evidence from censers dating to around 600 B.C. (Grove 1984). Maya from different geographic areas use different resins, concentrating on the ones that are locally available. At higher elevations it may be *Pinus, Liquidambar,* or *Bursera* (Breedlove and McLaughlin 1993). There, *copal,* from the Nahuatl term *copalli,* may be used generally for resin, or *pom* for resin particularly from Burseraceae. *Pom,* unlike *copalli,* is a Mayan term meaning that which is to be burned, referring to its use as incense. In lowland areas, *pom* tends to refer specifically to resin from the burseraceous *Protium copal* (Tripplett 1999). Resin used as incense may also be called *copal pom.*

To the Maya, the significance of *copal pom* involves a web of relationships from the time the resin is collected through its ritual use (Tripplett 1999). The use of *copal pom* as part of pan-Mayan belief in the transformative processes that occur during ceremonial activities has been substantiated by data from ethnography, linguistics, and folklore. Tripplett thought that these processes are critical to understanding the use of copal resins in both ancient and modern Mayan societies.

The act of burning copal, accompanied by what Gossen (1974) called "language for rendering holy," brings about two aspects of the Mayan world view. First, *copal pom* smoke generates interactions with deities and ancestors, thus perpetuating and maintaining relationships with them. Second, the incense initiates a series of transformative processes that characterize the

Mayan religious and cosmological beliefs. These processes are initiated because *copal pom* is perceived as a symbolic analog of blood (plant exudate being the blood of the tree) as well as a sacred substance. Moreover, by extension, the resin (copal) is considered an effective medicine for many illnesses. Much of the efficacy of a resin remedy in Mayan medical tradition is considered to result from association with a sacred person or object.

There is an important connection of *copal pom* as "food for the gods" in religious ceremonies of the Maya (Stross 1994–96). The gods do not eat as mortals do; rather, they imbibe the products of human ritual, primarily the smoke of incense, which is similar to the ritual concept of incense by the pagan Greeks and Romans (Atchley and Cuthbert 1909). Nonetheless, there apparently has also been a symbolic connection between maize, the primary food of the Maya, and copal resin. The evidence comes from numerous sources, including the modeling of maize ears from copal, lake offerings of maize-shaped lumps of resin, and miniature tortilla-shaped incense disks wrapped like tamales. Small disks (called pesos) the size and shape of a coin are treated as ceremonial money (Wisdom 1941). Burning this sacrificial money offered it in return for a request, such as a good hunt.

Tripplett (1999) wondered how Mayan beliefs about incense escaped the notice of Spanish priests who campaigned against all native idolatries as they proselytized for conversion to Catholicism. By not eradicating the use of incense (as the early church in Rome had), the Spanish inadvertently permitted the continuation of traditional Mayan concepts and practice. Tripplett further questioned whether the Spanish may have viewed native use of incense as similar to their own use of incense in Roman Catholic ceremony, not realizing the multiple layers of meaning of incense to the Maya. Therefore, the *use* of incense overlapped in two religious ceremonies from quite different cultures, but the *meaning* of its use remained distinct. The similarity in incense use brings up an interesting question as to whether it helped the Maya adapt to doctrinal changes imposed by the Catholic church by being able to maintain some of their fundamental convictions.

Southeast Asia and the Semelai

Until the 19th century, large tracts of Southeast Asian primary rain forest yielded many forest products, including resins (Dunn 1975). Population densities away from the coast and major waterways were sufficiently low that people who lived in these areas had either hunting-and-gathering or swidden agriculture economies. Therefore, these inhabitants, such as the Malays, had at their disposal considerable reserves of resinous products. The resins were not only valuable for local people but for others living some distance away. The Chinese, in particular, were willing to pay high prices for several kinds of resin. In fact, historical and anthropological studies, such as those by Dunn and by Wolters (1967), suggest that collection and trade of certain types of resin may have contributed to the development of complex societies in Malaya.

Although logging, increased population density, and introduction of exotic plantation cash crops such as rubber and oil palm have replaced much of the lowland rain forest that yielded resins, there are areas in Malaya (peninsular or West Malaysia today) where resins are still used by local people (Gianno 1990). For example, the Semelai in the lake area of Tasek Bera traditionally hunt, fish, use swidden agriculture, and trade forest products, employing many of their old ways while adjusting to the changing conditions of modern Malaysia. Gianno focused on the Semelai in her studies as a representative culture that could provide clues for interpretation of resin technology in prehistoric, protohistoric, and historic Malaya. She did not think that Malays in the past had a pattern of resin use and trade identical or even similar to that of the Semelai, but that her research could generate useful hypotheses for future research. Use of *Dipterocarpus kerrii* for keruing oil (Plate 40) by the Semelai is discussed in Chapter 7, and use of several species of *Aquilaria* for gharu wood in Chapter 10.

Resin in the Economies of the United States, New Zealand, and Africa

The great economic value of resin had a significant effect on the early development of some countries where resin-producing trees were native and thus readily available for exploitation. This was especially true for oleoresins from pines in the United States, and hard copals from *Agathis* in New Zealand and from various legumes in Africa.

Naval Stores in the United States

The availability of resins directly affected the fortunes of any country dependent on wooden ships in trade and war, since resins were needed to caulk the seams of vessels, to waterproof rope and tarpaulins, and for numerous other uses. Although the Greek botanist Theophrastus only described the Macedonian method for making tar or pitch from resins for ships around 300 B.C. (Chapter 7), it is assumed that these products were widely used well before this time. In fact, Meiggs (1982) reported tar making by the Minoans, Bronze Age inhabitants of Crete, around 3000 B.C. Tar probably was an essential commodity for ancient centers of power in Greece, Macedonia, Asia Minor, and Egypt because much of the trade was based on long-range sea transport (Figure 6-1).

Later, it was necessary for seafaring countries to have a dependable supply of resin (called naval stores), keeping afloat the wooden naval ships that in the case of Britain connected a vast empire. Because supplies were not available at home, Britain depended on countries around the Baltic Sea for naval stores. This supply, however, was precarious because the entrance to the Baltic Sea was sometimes cut off by war, and home consumption had increased in the Baltic countries, which often levied duties on customers. Thus Britain looked to naval stores from pine forests in the Plantations, as the American colonies were called.

In 1584, Sir Walter Raleigh sent two explorers to assess the naval stores potential along the coast of present-day North Carolina. When Jamestown was colonized in 1607, Raleigh's third most important goal was naval stores, which the English needed to decrease their dependence on Scandinavian sources. In fact, colonial settlers, particularly in Carolina and Virginia, were charged naval stores products as payments for their charters. In 1608, several experts in the recovery of pine resins were sent to Virginia, and later that year the first shipment of naval stores was sent to England from the colonies. This resin was probably obtained from *Pinus taeda* and *P. virginiana* (L. Williams 1989). The importance of colonial naval stores to the English is indicated by Captain John Smith's having been given the first set of government guidelines for conservation measures in obtaining resin from living trees in 1610 (Drew 1989). The fledgling industry, however, did not prosper in Virginia because of other, more lucrative endeavors such as growing tobacco.

The industry fared better in New England, where the Pilgrims made a profit for themselves while obtaining political favors by assisting England. Naval stores products were such valuable trade items that in 1670 the General Court of Massachusetts established the first recorded monopoly in the colonies, initially over the purchase of the resins, and subsequently over their production. Additionally, the first actual conservation legislation in America was passed in Massachusetts in 1715, requiring permits to cut pines or debark them for resin collection. During the industry's short duration in New England, its significant contribution to the early economy is illustrated by the pine tree's being engraved on shillings.

Despite conservation legislation, as sources of resin for naval stores disappeared in New England, the industry moved southward, following the first permanent settlement in Carolina in 1665. Southern pines yielded more resin than those in New England, and the Carolina pioneers found resin collection to be a good source of money. The industry declined during the Revolutionary War, but the industry rapidly followed the new southern and western frontiers when Britain turned back to Scandinavian supplies. Nonetheless, pine trees in North Carolina were the main source of naval stores in the United States for almost 150 years, until long after the Civil War. The crude form of manufacturing pitch and tar (Chapter 7) did not require expertise and capital, and could be carried out by farmers dispersed through the forests. It was time-consuming and tedious but made use of the African slave labor force (Figure 6-13). When other seasonal tasks lagged, slaves could be assigned to this profitable sideline. At its peak in 1840, North Carolina had more than 1500 naval stores operations, and it is said that three-quarters of the population was engaged in the tar business. Around the end of the 19th century, the forests of *Pinus palustris* and *P. elliottii* were being depleted along the North and South Carolina coasts, and second-growth pines became available following abandonment of farming and timbering in southern Georgia and northern Florida. At that time, the center of the naval stores industry moved to those areas, which have since dominated the business in the United States.

By 1900, the entire world had become dependent on the southern United States for most of its naval stores; hence, rosin and turpentine were important in the foreign commerce of the United States. These products had become essential to diverse industries in every developed nation. The growth of the naval stores industry in Georgia and Florida was rapid in the early 1900s as many experienced naval stores workers from North Carolina migrated southward with their families. This migration resulted partially from use of the box technique to tap the resin (Figure 7-2), which was a significant factor in the destruction of the North Carolina long-leaf pine *(Pinus palustris)* forests.

Figure 6-13. Cheap labor (at that time provided by African slaves) was essential in the labor-intensive collection of resin by tapping southern pines in the United States. Here, workers are renewing the cut faces as well as collecting resin from containers on the tree, putting it in barrels for transport to the stills to process it into turpentine and rosin (Chapter 7).

The Georgia and Florida forests did not suffer the same fate because more conservative collection techniques were used beginning in 1903.

The influence of the naval stores industry on these southern states was significant. For example, there were about 39,300 wage earners and more than 120 still sites for processing the resin in just three northwestern Florida counties by 1909 (Butler 1998). Most of the early naval stores operators in Georgia and Florida were full-time turpentiners and not involved in farming. These operators owned or leased several thousand acres of pine forest and established a camp, which constituted a small, isolated community devoted to collecting the oleoresin and processing it into turpentine and rosin. As many as 2000 fire stills were scattered throughout the turpentining region, but these were replaced by central stills in the late 1930s and 1940s. After the development of central stills, many farmers with timber became turpentine farmers, supplementing their farming income.

Despite improved methods of collecting resin, there has been a continued decline in the naval stores industry, based on tapping trees, in the United States (Chapter 7). In large measure this is the result of lack of cheap labor, which African Americans once provided (Figure 6-13). The United States has changed to refining wood pulp chemical by-products, accompanied by advanced chemical manufacturing processes. China and South America provide most of the global supply of resin from tapping pine trees. Thus the methods of collecting pine resin for the naval stores industry in the early history of the United States is essentially repeating itself in some developing countries, where there are still pine forests to be tapped by cheap labor.

Kauri Resin in New Zealand

Although the Maori of New Zealand used resin (misnamed as gum) from *Agathis australis* in various ways, it was not considered a gold of a different sort until recognition of its special varnish qualities and its extraordinary abundance, in trees and preserved in peat swamps. Large quantities of the very hard resin, called copal commercially (Chapter 9), had been exported to the United States by 1845, and there were predictions of a great future for world trade. The two biggest industries in New Zealand during 70–80 years of foreign rule were collection of this resin and timbering of the magnificent kauri tree (Figure 2-3). These industries, however, produced an environmental disaster, stripping vegetation and denuding the soil from Northland, with the kauri tree now reduced to less than 0.5% of its former extent.

Nonetheless, extraction of the resin also resulted in a culture made richer by the influx of European workers, from the work of which some of New Zealand's major industries would arise (McNeill 1991). By the 1880s, kauri gum diggers flooded in from southeastern Europe, England, France, Germany, Malaya, and China. British diggers and runaway soldiers from the New Zealand land wars worked alongside Bosnian Muslims (mistaken for Turks) and Finns (mistaken for Russians). There were thousands of gum diggers (20,000 at the peak of production in the 1890s to the turn of the century) who for the most part were unsettled workers, similar to an average colony of gold diggers, who arrived in New Zealand in the 1860s (Reed 1946). Gum digging was a form of fortune hunting with a minimum of equipment needed to pursue it.

The most obvious group of gum diggers were those known as Austrians, later as Dalmatians; they were actually a population of ethnically distinct peoples from Dalmatia, Macedonia, Bosnia-Hercegovina, Croatia, Slovenia, Serbia, and Montenegro. Most had emigrated from Dalmatia to escape problems at home, caused in part by foreign domination. The total number of these emigrants working in the gum fields at any one time was around 3000 (Trlin 1979). The first of these arrivals were often single men who lived under harsh conditions to earn as much money as possible and return home (Figure 6-14). Like the Chinese in the New Zealand gold fields, the Dalmatians were persecuted. Despite concessions made to British, Maori, and naturalized New Zealander gum diggers, racial prejudices persisted, and suspicions about the Dalmatians as enemy aliens erupted at the beginning of World War I.

Nonetheless, increasing numbers of law-abiding Dalmatians settled permanently in New Zealand and made contributions to the development of the country. Often settling on small landholdings, they relied on gum digging as a source of capital to finance property development. Their vineyards are examples of the conversion of gum lands into highly productive areas (Trlin 1979). In fact, as indicated by the title of Trlin's book, *Now Respected, Once Despised*, this group of immigrants, who were resented during their participation in the booming kauri gum industry, stayed on to influence the development of New Zealand culturally and economically.

Copal in Africa

East African traders exchanged copal from leguminous trees *(Hymenaea,* syn. *Trachylobium)* for cotton cloth with Indian traders in "early times" (Rakoto-Ratsimamanga et al. 1968). The Portuguese then developed this trade in the later 18th century. The very hard resin, like that from *Agathis,*

Figure 6-14. Northland New Zealand gum digger fitted out with a spear for prodding the soil for kauri resin, a bucket for sluicing, a billy for midday brew, and a knapsack, or pikau, for carrying the day's haul. The small flour bag in the digger's hand was kept handy for storing small pieces of resin he found while digging for larger chunks. With permission of the Alexander Turnbull Library, National Library of New Zealand, Te Puna Mātauranga o Aotearoa (Northwood Collection, negative G 9781 1/1).

was particularly esteemed for varnish and exported to China for that purpose (Chapter 9). The resin trade from Madagascar was sufficiently important in the later 19th century that the government established a monopoly over transactions. At one time, most of the East African copal was shipped from Zanzibar; hence, it was called Zanzibar copal although later the resin acquired names from other East African ports.

The hardness of this copal produced keen demand, leading to high prices for it before the copal industry in the Belgian Congo, based on various leguminous species (e.g., *Copaifera* and *Guibourtia*), was highly developed (Chapter 9). The systematic exploitation of Congolese deposits by the Belgians reduced the price of copal and increased the output to such an extent that the Congo had an almost complete monopoly of the African copal trade early in the 20th century, especially during the 1920s and 1930s. Of the African copal exports at that time, 97% originated from the Congo (Mantell et al. 1942). Howes (1949) thought that the supplies of the resin in this region "may be inexhaustible." Nationalist sentiments forced the Belgians to grant the colony independence in 1960. Subsequent civil unrest in the region (renamed Zaire in 1965 but changed to Congo again in 1997) has not encouraged the utilization of these resources.

CHAPTER 7

Oleoresins

Oleoresins, and when very fluid sometimes referred to as oily resins, are comparatively fluid terpenoid resins. They have a higher ratio of volatile to nonvolatile terpenes than some other resins such as balsams and copals. There is sufficient variation in the amount of volatile compounds, however, even between species of the same genus, such that some are referred to as oleoresins or oils whereas others are called balsams (Chapter 8). The volatile fraction consists of mono- and/or sesquiterpenes, which are sometimes referred to as essential oils (Chapter 1). Diterpenoids are the primary constituents of the nonvolatile fraction in conifers and most angiosperms, but triterpenoids are primary in the Dipterocarpaceae.

Oleoresins from Pinaceae are the basis of a long-important forest products industry called naval stores. That name was used initially applied to all the stores on ships but later referred to resins, which were essential for waterproofing wooden ships and nautical gear such as sails and ropes (Chapter 6). The name became so well established that even though such ships have become obsolete, "naval stores" refers to a diversified industry based primarily on pine resins. The volatile mono- and sesquiterpenes as well as nonvolatile diterpenes are used. Around the world, the naval stores industry differs regionally in the pine species and processing methods used as well as consumption patterns and industry structure. The differences are particularly evident between developed and developing countries.

A smaller, cedarwood oil industry is based primarily on sesquiterpenes from Cupressaceae although those from the true cedars (*Cedrus*, Pinaceae) have been used as well (Chapter 2). Species in certain genera of the tropical families Dipterocarpaceae and Fabaceae produce fluid resins, often referred to as oils, that consist predominantly of sesquiterpenes. Although these oleoresins have never been utilized to the same degree as pine resins, some have had export value, and they continue to be used in the tropical areas where the trees occur.

Naval Stores from Conifers

Various terms have been used for the crude oleoresin from conifers, mostly pines, and its naval stores products. The word pitch often denotes crude resin, but it has also been used, more restrictively, for resin that has been cooked and thereby darkened. Pitch has a long history of use, from early shipbuilders and seafarers such as the Minoans and Phoenicians, who employed pitch to seal their ships (Meiggs 1982). Pine tar is a darkened product of destructive distillation (described later), a practice used by the ancient Macedonians; it is a by-product of charcoal made from conifers. Another term, colophony, derived from Colophon (ancient Ionian city and by extension referring to the coastal region of western Asia Minor and adjacent islands), has been widely used for centuries to denote either crude pine exudate or tar. The term naval stores, referring to resins, first appeared in 17th century English records when large quantities of pitch and tar were used to keep wooden sailing vessels seaworthy. English sailors were known as Jack Tars because they had to use tar frequently to waterproof rigging and caulk ships. Tar was used to caulk ships as late as the 19th century.

Although the word pitch is still sometimes used colloquially to refer to crude resin, turpentine and rosin are the most important naval stores products. Turpentine refers to the distilled, volatile mono- and sesquiterpene fraction of the resin, and rosin (tar in past use) to the nonvolatile diterpene fraction. Stills used in the Scottish liquor industry that were imported in the 1830s for turpentine production led to the term, spirits of turpentine.

All pines produce resin (Chapters 2 and 3) but only about 30 of the 90 or so species have been tapped for their oleoresin in various parts of the world. In fact, the bulk of the world's supplies of naval stores is produced from 8 species distributed in the warmer range of *Pinus*, particularly in the subtropics and tropics (McReynolds et al. 1989).

United States

The importance of the naval stores industry in the early history of the United States and its later influence on the economy of the southern states is discussed in Chapter 6. Here, the industry's products are emphasized. For more than a century, the United States was the world's largest producer of turpentine and rosin from pine oleoresin. During this time, however, the methods for obtaining the resin and the uses for it changed considerably.

Changing Methods of Production

The method of making tar by fire, as chronicled by Theophrastus about 300 B.C. in his *Enquiry into Plants,* was commonly used in the United States from the 17th century into the 20th (Zinkel 1975). Tar production in North America was documented by the expedition of the Spaniard Pánfilo de Narváez to present-day Florida in the 16th century, thus the activity has been considered one of the oldest U.S. industries. Later, much of the naval stores in North Carolina, one of the most productive areas for many years, was obtained by controlled, air-deprived burning of piles of pine wood that were built into the shape of a hive and covered with earth. Because the pine tar adhered to the shoes of workers as they collected it at the base of the hive, the citizens of North Carolina became known as Tarheels. An outgrowth of this process of tar burning was a more refined destructive distillation in which resinous wood was heated in large retorts, producing charcoal in addition to turpentine and tar (or rosin).

With the abandonment of wooden ships by navies and the merchant marine, the use of resins specifically as naval stores declined, but industrialization created other uses. New demands required both increased supply and a higher-quality product. Tar burning and destructive distillation were supplanted by a process for recovering resin from virgin pine stumps from which the sapwood had rotted away. Resistant heartwood, which contains about 25% resin extractives, is chipped, shredded, and extracted through steam distillation to produce wood turpentine and wood rosin (Figure 7-1). The rosin is subsequently purified from the remaining resin following distillation of the turpentine.

Beginning in the 1800s, cutting (tapping) the tree trunks and collecting the resin, called turpentining, became another important method of obtaining resin for processing into turpentine and rosin (Figure 7-1). The exudate from the tapping is referred to in the industry as gum resin, but it is different from true gum resins, in which cell wall carbohydrates get incorporated into the resin (Chapters 1, 8, and 9). The most commonly used terms, however, refer to the distilled products, gum turpentine and gum rosin, thus distinguishing them from wood turpentine and wood rosin, in which the resin has been obtained from tree stumps (Figure 7-1). In the 1800s, turpentining was done primarily using pines in the Carolinas. As these forests became exhausted as

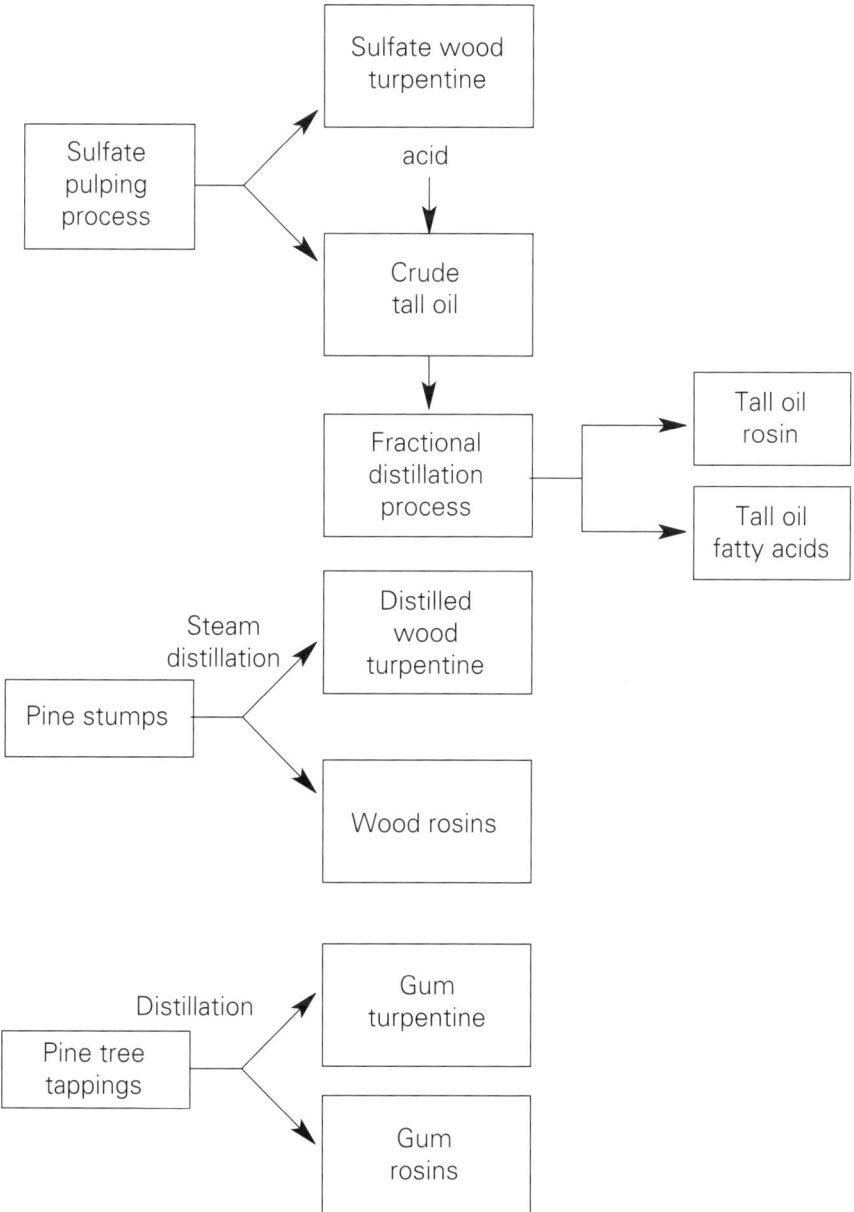

Figure 7-1. Current procedures for obtaining naval stores, with specialized terminology for the products indicated.

a result of the destructive collecting methods such as boxing (Figure 7-2), forests in Florida and Georgia became more important (Chapter 6). Georgia is the center of the industry today.

Since tapping is a controlled wounding of the tree, the type of wounding is important for sustained yields and the health of the tree. It involves concentrating the wound only in the area where the secretory tissue is located—in wood or bark (Chapter 3). It also is necessary to choose the most effective manner to refresh the wound and stimulate further flow of resin. Extensive tree mortality results not only as a direct consequence of the tapping procedure but by the indirect effects of insects, fire, and wind (blowing down trees

Figure 7-2. Stumps illustrating different tapping methods used in gum turpentining in the United States. Left, the destructive boxing method, with deep cuts or boxes cut into the wood to collect the resin. Axes and the hogal (hanging on the stump) were used for chipping. Center, Herty cup method, with much less damage to the wood both in chipping and collecting the resin. In the foreground are mauls and cupping axes. Right, the even less destructive ball cup method in which cutting is only done near the area of the cambium. Sulfuric acid is in the bottle on top of the stump and is used to maintain the flow of resin. Progressive change to less damaging methods resulted in longer life for the producing trees.

weakened by resin collection). McReynolds et al. (1989) summarized the many problems regarding methodology in collecting marketable resins from southern pines. Ecological studies of factors controlling the sometimes devastating seasonal attacks by bark beetles on these pines have led to important hypotheses as to how trees balance the carbon expended for growth versus synthesis of resin to defend themselves against enemies (Chapter 5).

Long-leaf pine *(Pinus palustris)* and slash pine (*P. elliottii*; Figure 7-3) were the two species most used for turpentine and rosin in the United States. These pines covered large areas in the South, growing on sandy soils too infertile for crops. The two species have different advantages and disadvantages for resin

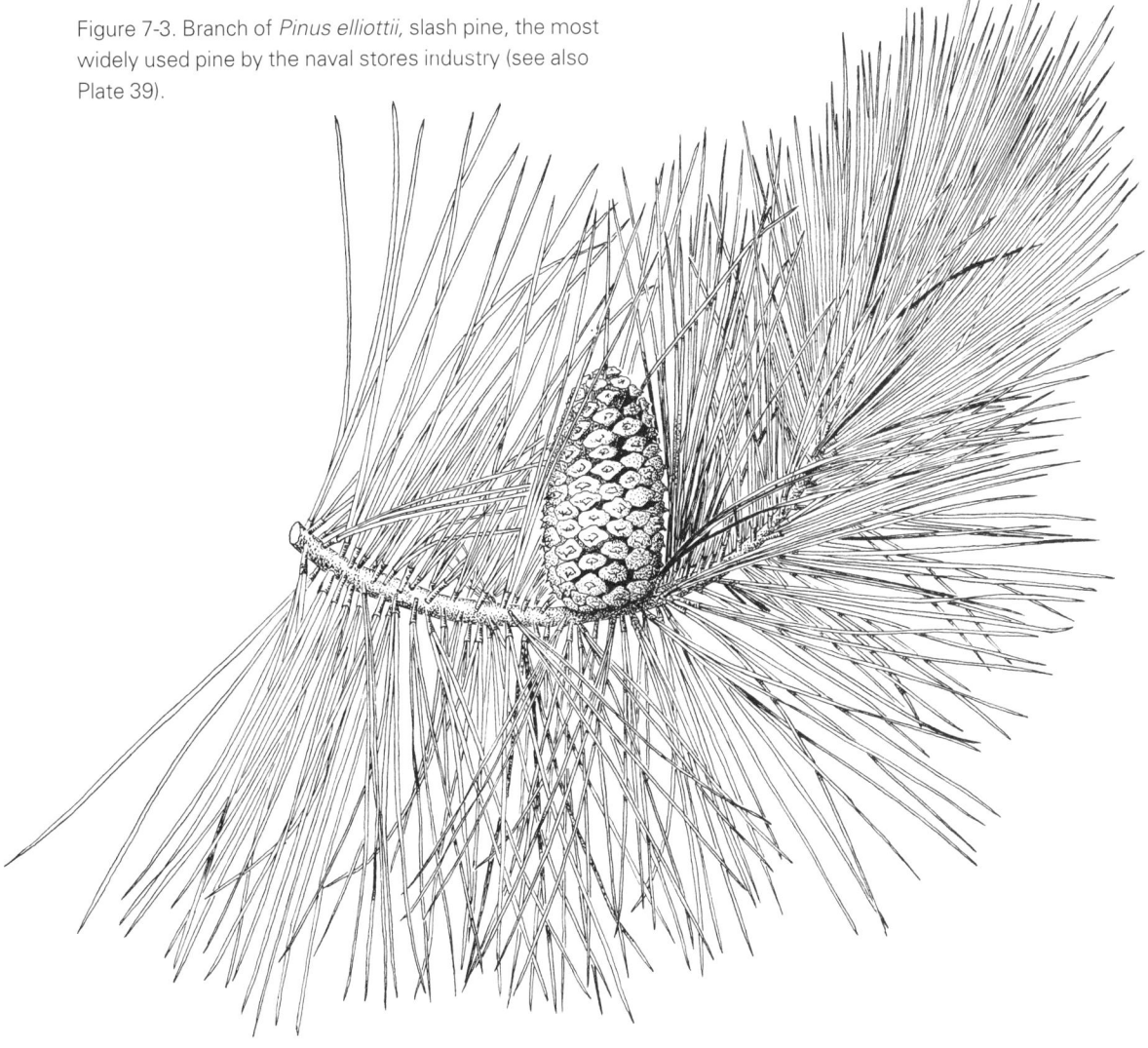

Figure 7-3. Branch of *Pinus elliottii,* slash pine, the most widely used pine by the naval stores industry (see also Plate 39).

use. Long-leaf pine tends to grow in pure, open, savanna-type stands, but it develops more slowly than slash pine in its early years. Although slash pine is one of the most rapidly growing forest trees of North America, it grows in mixed stands and is sensitive to fire in its early years. On the other hand, its relatively high resin yields, which are more liquid and thus produce little oxidized resin on the cut face, has led to its being favored for tapping. Furthermore, it can be grown in plantations (Plate 39) in which it has been used as a multipurpose tree: Resin is tapped for the first 20 years, then the boles are harvested for timber followed by resin extraction from the stumps. *Pinus elliottii* has been extensively planted throughout the world for resin production, particularly in developing tropical countries, as discussed later.

Other pines that have been tapped for turpentine and rosin either experimentally or on a small scale in the United States include loblolly pine *(Pinus taeda)*, short-leaf pine *(P. echinata)*, pond or pitch pine *(P. rigida)*, ponderosa pine *(P. ponderosa)*, pinyon pine *(P. edulis)*, sugar pine *(P. lambertiana)*, and lodgepole pine *(P. contorta)*.

Until about 1935, all pine resin produced by tapping (Figure 7-1) was converted into turpentine in a fire still. The still was usually made of copper and heated directly by a wood fire. A continuous stream of water was added to the still during distillation to provide the steam, which acted as a carrier gas to remove the turpentine.

In the United States, naval stores from the tapping of trees accounted for more than 80% of global rosin and turpentine production in 1930 but declined rapidly to about 4% by the late 1970s (Zinkel 1981). Escalating labor costs, which accompanied cultural change in the South, along with depletion of high-quality trees have been the primary factors in decreased production from turpentining. Neither improved tapping practices and chemical stimulation (Figure 7-2) nor development of high-yielding progeny (Squillace 1971) slowed the decline in U.S. turpentining. Genetic studies of high-yielding slash pines, however, provided valuable information for researchers studying the ecological and evolutionary roles of resins (Chapter 5).

Major decreases in production also resulted from the shortage of virgin pine stumps for distilled wood turpentine and rosin (Figure 7-1). As the tree grows in girth, dead tissue (heartwood) forms in the center of the trunk (Figure 3-1). As the heartwood develops, resin stored in the disintegrating canal structure saturates the dead tissue. Thomas (1970) suggested that heartwood resin is analogous to wound exudate, but secreted into the dead space of the heartwood rather than to the outside of the tree. Second-growth stumps are not satisfactory because the heartwood lacks enough resin for economical processing. The finite supply and nonrenewability of virgin pine stumps fostered interest in artificially induced resinous wood in the U.S. naval stores industry. In the 1970s, researchers found that treating pines with paraquat or diquat (dipyridyl herbicides) stimulated synthesis and subsequent diffusion of oleoresin into the wood. An extractive content as high as 40% was found in some wood samples (D. R. Roberts et al. 1973). Such resin-soaked wood is known as lightwood because it kindles and burns readily and at one time served as a source of light (candlewood). Lightwood occurs not only in the immediate area of treatment; the zone of resin-soaked wood formation extends to the pith and many meters along the trunk above the point of treatment. It is interesting that only pines give the lightwood response. Although research on lightwood was extensive in the late 1970s, Hays and Cottle (1989) suggested that much more understanding was needed to develop and fully exploit its potential for increased production.

Sulfate naval stores (Figure 7-1) are produced as by-products from the kraft pulping process. As pine chips are cooked to render pulp for paper, the volatilized gases are condensed to yield sulfate turpentine. Since 1950, sulfate rosin and turpentine have replaced gum and wood naval stores (Figure 7-1) in the United States (Zinkel 1981). Pulping by-products in 1980 accounted for more than 80% of the turpentine and 60% of the rosin produced in the United States as well as yielding some fatty acids. The supply of naval stores in the United States is now driven by overall output of the pulp and paper industry, especially by the kraft process, as well as by processors that determine the intermediate products (Hodges 1997). During the pulping process, the alkaline pulping liquor saponifies the fats and converts fatty and resin acids to sodium salts. The soaps can be skimmed from the surface of the liquor at various points and acetified to yield crude tall oil (saponified fats that are subsequently acetified to obtain fatty acids; Figure 7-1; Drew and Propst 1981). The term tall oil is derived from the Swedish *tallolja*, pine oil. A literal translation would have caused confusion with the turpentine-derived pine oil, which was already a commercial product. Thus *tallolja* was simply transliterated to tall oil (Zinkel 1975).

During the early years of the recovery process, most of the crude tall oil was burned for fuel. However, most is now fractionated into tall oil rosin (40–60%) and fatty acids (40–55%). Among the fatty acids, the unsaturated C_{18} oleic and linoleic fatty acids predominate (Duncan 1989). In the 1980s, fractionating production for tall oil in the United States was nearly three times that of the rest of the world. The large market for tall oil fatty acids as protective coatings and intermediate chemicals, in soaps and detergents and as flotation agents, has been an increasingly important part of the naval stores industry. Additionally, sterols derived from tall oil contribute to a rapidly growing market for cholesterol-reducing food additives (Hodges 1997); a margarine product that incorporates sterol esters made from tall oil, Benecol, initially sold in Finland, has been very successful in the United States (Chapter 10).

Changing Uses of Turpentine and Rosin

Turpentine was the most important naval stores product in the early to mid-1800s, but the price of rosin was so low that it was often discarded in streams, lakes, and holes in the ground. However, one of the soldiers on Sherman's march through North Carolina in 1865 recognized the potential of the buried rosin and returned after the Civil War to begin a rosin-mining venture. By the 1900s, demand for rosin had increased enough that streams and lakes were mined for the discarded material (Zinkel 1981), and it remains one of the most important naval stores commodities.

Rosin from pine wood consists primarily of diterpene resin acids of the abietane and pimarane skeletal types (Figure 1-4). Differences in the properties and uses of commercial rosins depend both on the kinds and quantities of neutral components (e.g., anhydrides, phenolics, and sitosterol) and on the relative composition of the resin acids (Zinkel 1975). High-quality rosins throughout the world tend to be similar in general composition, however, with only minor variation. Furthermore, for many purposes, rosins and rosin derivatives are interchangeable regardless of source because they are used mostly in some chemically modified form.

Although 38% of rosin was used for yellow-bar laundry soap in 1938, this use has become negligible. Rosin is now used in many unrelated industries, especially for intermediate chemical products such as tackifiers, emulsifiers in synthetic rubber, and paper sizing to control water absorptivity. Rosins also are used in adhesives, surface coatings, printing inks, and chewing gums. Although rosin solidifies, it becomes sticky when warmed, which makes it useful for minor purposes. For example, musicians use it on bows of stringed instruments. The stickiness increases friction between the bow and strings, the increased contact enhancing the musical notes. Also, baseball pitchers often use a rosin bag, which when handled results in a stickiness that enables them to grip the ball better, possibly increasing the accuracy of the pitch. The United States contributed about 37% of the world's rosin production in 1975; in 1996 it produced only 29% (from tapping, pulping, and stump processing). A summary of rosin use is shown in Figure 7-4.

The major use for turpentine was once as a solvent in paints, but this market has declined to comparative insignificance with the advent of less-expensive white spirits derived from petroleum. Turpentine has become more important to chemical industries around the world as a basic feedstock for the manufacture of a wide range of derivatives. Because fractionated components have particular uses, the composition of turpentine is particularly important. In addition to the initial variation of the mono- and sesquiterpene composition in different pine species, the product composition varies according to the technique of extraction (Figure 7-1). For example, composition of

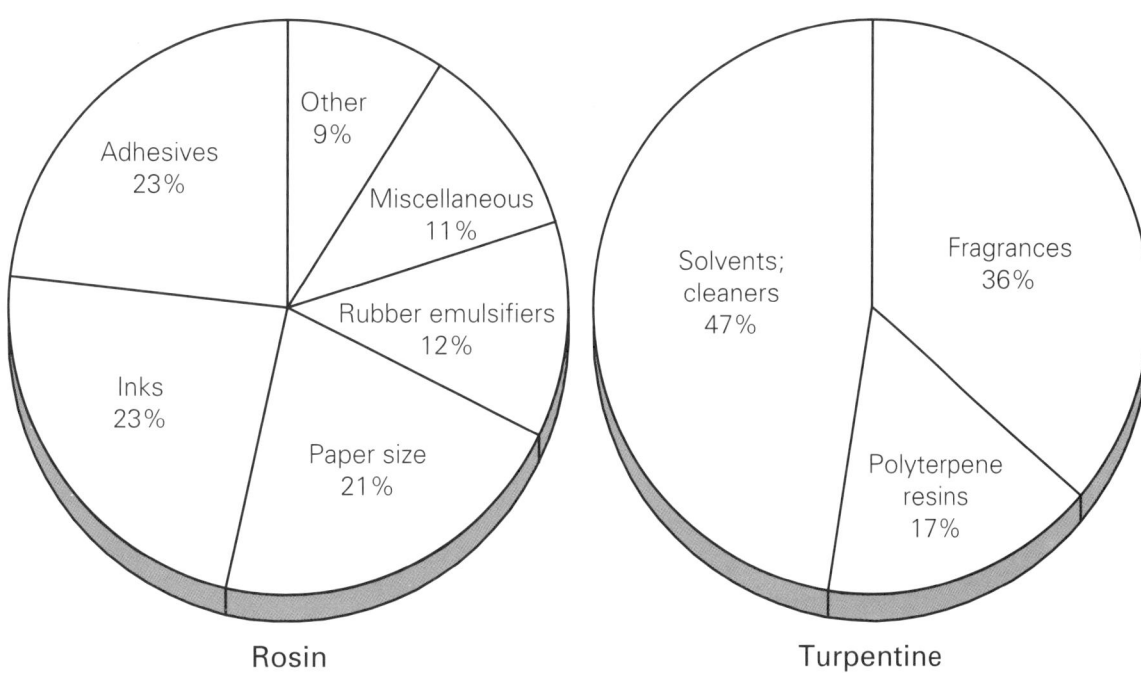

Figure 7-4. Relative commercial use globally of rosin and turpentine in 1996. Data from Hodges (1997).

sulfate wood (pulped) turpentine varies because of the number of conifer species mixed in the pulping and the extensive geographic range from which many are taken (Zinkel 1981). In contrast, the composition of distilled wood turpentine is comparatively more consistent because the raw stump material is primarily from *Pinus palustris*. Sulfate processing adds sulfur to the product, which can be a problem in some chemical transformations.

Major uses of turpentine tend to focus on the large quantities of α- and β-pinene (Figure 1-3), often the major volatile components in pine resins. Some pines have a higher proportion of either α- or β-pinene and are thus used as different feedstocks. One important use of turpentine involves converting α-pinene with aqueous mineral acids to synthetic pine oil, in which the primary constituent is then α-terpineol. More than 80% of the pine oil produced is synthetic; the remainder is natural pine oil, obtained from fractionation of stump wood extractives. Variations in synthetic conditions, fractionation, and blending provide pine oils that have different properties (Zinkel 1975, 1981). These pine oils are used in mineral flotation, textile processing, and as solvents, deodorants, and bactericides. The composition of pine oil in the United States is different from Swedish pine oil, which is obtained from needles of various conifers.

Another large and growing use of turpentine is for production of polyterpene resins (Figure 7-4). Various low-molecular-weight (600–1500) polymers are produced from β-pinene, mixtures of α- and β-pinene, and pyrolyzed α-pinene, which are used particularly in the compounding of pressure-sensitive adhesives such as transparent tapes (Zinkel and Russell 1989). Terpene insecticides based on camphene, obtained directly from wood turpentine or by isomerization of α-pinene, were once used. Insecticides are manufactured by chlorinating camphene or mixtures of camphene and α-pinene. Dry cleaning and paper sizing are some other uses.

Perhaps one of the most interesting and increasing areas of turpentine use is as feedstock for the synthesis of flavor chemicals, for example, those of lemon, lime, peppermint, spearmint, and nutmeg (Derfer 1978). In fact, turpentine from pine resins worldwide yields 10-fold greater amounts of volatile terpenes than any other source (Plocek 1998). For fragrance chemicals, species that have large relative amounts of β-pinene in their resin, such as *Pinus elliottii*, are preferred because several synthetic routes involve it as a starting compound (Zinkel 1975, 1981). For example, tonnage quantities of *l*-menthol, used in flavoring for its cooling effects and in cosmetics, are synthesized from ℓ-β-pinene through a sequence of reactions involving α-citronellol.

Figure 7-5 shows an example of a major synthetic route that leads to a number of aromatic compounds. In this sequence, pyrolysis of β-pinene forms the addition product myrcene, to which hydrochloric acid is added and then reacted with sodium acetate to produce the acetates of linalool, geraniol, and

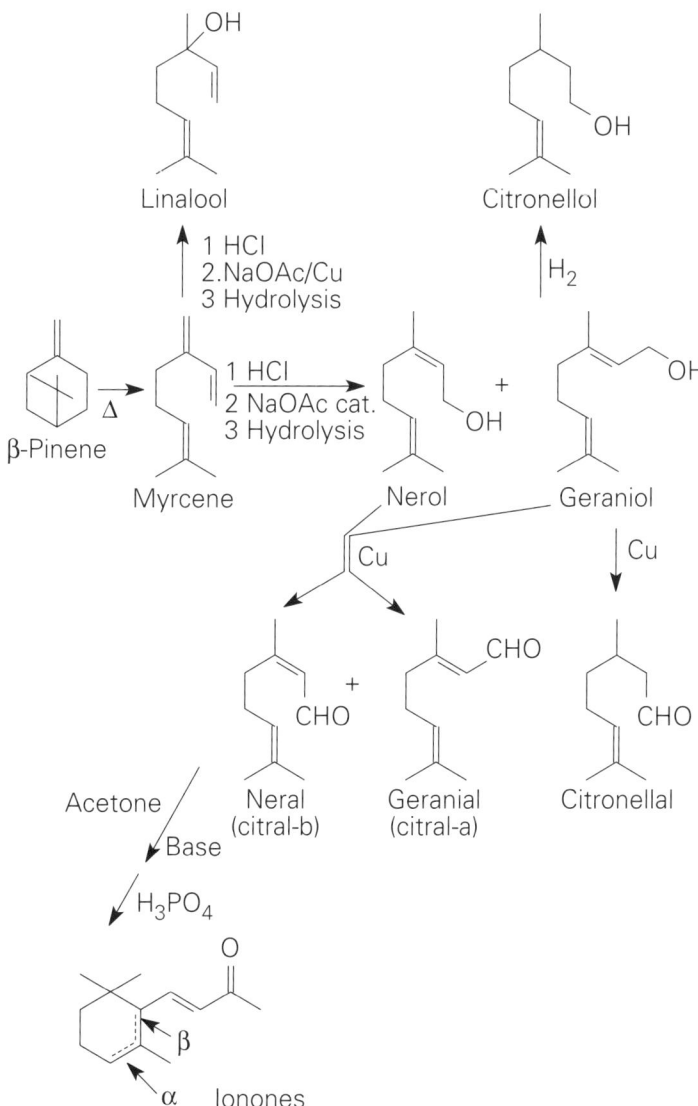

Figure 7-5. Synthesis of some flavor and fragrance chemicals from β-pinene. From Zinkel (1981).

nerol. The acetates can then be hydrolyzed to the corresponding alcohols. The reaction conditions can be controlled to produce either linalool or geraniol-nerol products. Linalool and its esters have the fragrance of lilac whereas the esters of geraniol and nerol, as well as their hydrogenation product citronellol, are similar to rose. Rearrangement of geraniol-nerol over a copper catalyst produces either citronellal or a mixture of citrals, depending on conditions. Citronellal is used as an odorant, and citral for its strong lemon characteristics in preparation of citrus flavors and fragrances. A major use for the β isomer of citral (neral) is as an intermediate in the synthesis of vitamins A and E because this synthesis requires tonnage quantities of materials. Condensation of citral with acetone, followed by isomerization, produces a mixture of violet fragrances known as ionones. Hodges (1997) reported that 30% of such aroma compounds is used in household products, for example, dishwashing liquids, bar soaps, fabric softeners, laundry aids, and air fresheners. Another 23% is used in cosmetics, and 47% for miscellaneous uses.

In 1996, the United States produced 46% of the world's turpentine, primarily from the sulfate pulping process. However, turpentine from this process may decline because of the use of thermomechanical rather than chemical pulping methods used in new paper manufacturing plants (Plocek 1998). Although 41% of turpentine is consumed by the flavor and fragrances industries in the United States, worldwide this use is only 36%. Aroma and polyterpene resin use accounted for most of the turpentine consumed in industrial countries whereas use of pine oil as a cleaner and solvent predominates in developing countries. In 1996, worldwide, in addition to use for fragrance, 17% of turpentine was used for polyterpene resins, and 47% for solvents and cleaners (Figure 7-4).

An insufficient supply remains the long-term problem for turpentine and rosin use in the United States (Zinkel 1975). In the 1970s, the industry was in the midst of transition, from traditional patterns of obtaining the raw material and from older kinds of processing to supplying chemically more sophisticated products. Zinkel and Russell (1989) summarized the status of naval stores production and utilization. By the turn of the century, naval stores competed directly with other industries that supply functionally equivalent or substitutable chemical commodities, including petroleum-based synthetic hydrocarbon resins, citrus oil terpenes, and vegetable fatty acids (Hodges 1997). The petroleum hydrocarbon and vegetable oil industries are significantly larger than the naval stores industry. Thus the naval stores industry has lost significant market share to these competitors in a number of segments, but it has retained significant niche markets for specialized uses.

Europe

The Minoans, Phoenicians, Greeks, and Romans used resin to seal and preserve wooden ships (Chapter 6). Resin was an important commodity for the establishment of extensive empires, trade within Europe, and trade between Europe and the Middle East during the Greco-Roman period (Thirgood 1981). Since most of this trade was seaborne, there was a large demand for resin to keep the wooden vessels seaworthy. Both the Greeks and Romans probably obtained their resin from Aleppo pine *(Pinus halepensis),* from local sources (Meiggs 1982).

An era of exploration and colonization by European nations during the 15th–17th centuries was driven by a desire to establish control over the sources of traded goods. The establishment of colonies and expansion of trade necessitated large fleets of both naval and merchant vessels, resulting in an increase in destruction of European pine forests to supply resin for naval stores. Later, the Napoleonic Wars created a sustained demand for both timber and resin for large naval fleets. Finally, there was the belated recognition that European forest resources had become severely depleted.

Tar making based on *Pinus sylvestris* (common or Scots pine, Plate 1) has long been an industry in Norway, Sweden, and Finland (Gamble 1921). Although resin production existed in Finland before the 16th century, the primitive and very destructive tapping method of stripping trees produced tar sufficient only for local consumption (Tsoumis 1992). In northern Sweden, on the other hand, careful processing of the resin led to worldwide recognition of a superior product. When wooden ships were replaced with metal ones, the naval stores industry in Scandinavia declined. Demand for large quantities of resin for a diversity of products could not be met by tapping the low-yielding *P. sylvestris.* Today, however, Sweden, Finland, and Norway, as major paper manufacturers, are again the European leaders in naval stores production; the resin is obtained as a by-product of pulping (Hodges 1997).

The resin industry in France is apparently very old, with evidence predating 100 B.C. of trees wounded for oleoresin (Drew 1989). Resin was exported during Celtic times. The ancient Gauls produced and traded resin in southern France until the invading Vandals destroyed their industry in A.D. 407. The later history of the French industry serves as a model for a sustainable approach, that is, for what can be done to combine utilization with conservation of renewable resources. In the early 1800s, Nicolas Brémontier proposed to restore a previously forested area in the Landes region of southwestern France that was so badly depleted by timbering and grazing that farms and even towns were uninhabitable, and waterways were unusable because of drifting sand. Planting pines and draining the area during the mid-1800s converted the once barren area into the center of an important industry. More than a million hectares were devoted to pine cultivation, with the chief business centers of the resin industry in Dax and Bordeaux. Thus a perpetual forest was started, some of which still exists.

The maritime pine *(Pinus pinaster,* syn. *P. maritima)* has been the species mainly cultivated for resin in France, although the Aleppo pine *(P. halepensis),* the most widely distributed pine along the Mediterranean coast, was also used. *Pinus pinaster* grows well near the sea and is one of the best pines for sandy soils. Its resin product was sometimes referred to as Bordeaux turpentine. The success of the French industry for many years resulted primarily from efficient management, procuring sustained resin yield while maintaining tree vigor (Howes 1949). A three-stage management was used. The first stage consisted of reseeding and growing seedlings in areas where old exhausted trees were removed. At 20 years, the second stage began. Undesirable trees were turpentined severely *(gemmage à mort),* cut out, and sold for timber. In the last stage the trees were tapped conservatively for the remainder of their 60–70 years of useful life. To achieve this *(gemmage à vie),* only a few cuts were made per tree, thereby maintaining the health and consequent growth of the tree as much as possible. The managed forests thus not only provided resin but wood for various purposes and made the forests of this southwestern region second in importance only to the southern United States at one time. The management method established a standard for sustainable tapping or turpentining that was modified for use in numerous other countries. Today, France lags behind Scandinavia and Portugal in oleoresin production, apparently because of problems similar to those that decreased gum turpentining (tapping trees) in the United States (Hodges 1997).

For many years, Spain followed France in European production of turpentine and rosin. The industry developed about the mid-19th century, centering around Madrid. In the early days, destructive exploitation caused much damage to the trees, but later, improved French methods were adopted. As in France, *Pinus pinaster* has been the main tree used, although *P. halepensis* and Corsican pine *(P. nigra* subsp. *laricio)* were also tapped. By 1996, however, Spain's production did not meet its consumption needs (Hodges 1997).

Oleoresin production developed during the 20th century in Portugal, primarily from *Pinus pinaster* and stone pine *(P. pinea),* which occur widely in the central and western portions of the country. In 1988, Portugal was the leading European producer of gum rosin (Figure 7-1), most of which is exported, but an increasing amount is used internally. Although its production has declined, Portugal continued to lead Europe in 1995, particularly because of the availability of cheap labor for tapping trees. However, Finland and Sweden produced more rosin than Portugal as a by-product of the pulping process (Hodges 1997).

The production of resin from *Pinus halepensis* in Greece goes back to ancient times; in fact, some of the earliest records of naval stores are chronicled in Greek history (Chapter 6). In these former times, *P. brutia* and *P. nigra* were also tapped. The ancient Greeks also used resin for lighting and in religious ceremonies. In the seventh century B.C., resin was used in war as liquid fire, an early form of napalm (Drew 1989). In the first century A.D., Dioscorides described the use of retsina, wine steeped with pinecones and wine that had been mixed with resin, for lung and stomach complaints and for coughs. Resin is still used in the wine industry, where it is added to provide a characteristic flavor and to aid in preserving the quality of Greek retsina. The resin may have been used originally as a sealant of Greek urns, giving the wine a flavor that the Greeks began to savor. Decline in resin production is mainly the result of the unavailability of labor and the occurrence of forest fires (Tsoumis 1992). Production of resin in Greece is subsidized. Rosin also continues to be produced, and in 1996, Greek production was similar to levels in France (Hodges 1997).

Common European or Norway spruce *(Picea abies)* is the most widespread mature forest tree in mountainous parts of western Europe, but only limited amounts of resin have been obtained from it. The Germans produced what was called Burgundy pitch and Jura turpentine from the trees in the 18th century, in an industry located in the region of the Black Forest; at the beginning of the 19th century, competition from the U.S. naval stores industry closed it down. During times of war, Germany tried experimentally to stimulate the flow of resin from the more abundant but low-yielding *Pinus sylvestris*. The business has not prospered since, although the Germans have produced some pine needle oil (Stephan 1990).

In Russia, production of turpentine and rosin has been primarily from pulping or stump wood rather than from tapping living trees. *Pinus sylvestris* covers vast areas of Russia and has formed a basis for a substantial naval stores industry. The main areas for production are in the Irkutsk and Sverdlovsk regions, the central parts of the Krasnoyarsk territory, and the central and northern portions of European Russia (Coppen and Hone 1995). The highest levels of productivity from pines, however, has been obtained in warmer, southern areas such as Crimea in Ukraine. In 1992, Russia was a distant second to China in production of rosin and fourth in turpentine, and apparently consumed rather than exported most of its production (Hodges 1997).

The common European larch *(Larix decidua)* was the source of oleoresin from which Venice turpentine was distilled. Venice turpentine used to be very important and was collected primarily in the Tyrol of Austria (Howes 1949). Unlike most conifers, the more fluid larch resin was obtained by boring holes into the trunk rather than cutting into the cambial area. This was done in spring. The hole was then closed until fall, when the resin was collected. Venice turpentine has a very characteristic composition, matched only by that of the northeastern Asian *L. gmelinii*. There is a large proportion of neutral labdane compounds: epimanool, larixol, and larixyl acetate. The latter two compounds distinguish larch resin, not having been observed in any other genus (Mills and White 1994).

The production of pine needle oil, distilled from needles and young terminal branches, has long been an industry in Europe, particularly Sweden, Austria, Germany, Hungary, Switzerland, and Russia (Drew 1989). Interestingly, the needle oils from American pine species are too similar to turpentine to be of value (Zinkel 1981). Despite the many kinds of conifers used (cedar, fir, spruce, hemlock, juniper, etc.), the product is nonetheless called pine needle oil commercially. Production of true pine needle oil, called Swedish pine oil, began in Scandinavia from *Pinus sylvestris*, but present production from pines is most commonly from Swiss mountain pine *(P. mugo)* and black or Austrian pine *(P. nigra)* in Austria, Bulgaria, Albania, and the republics of the former Yugoslavia. Oxygen-containing monoterpene alcohols and esters, which provide highly desired fragrance characteristics, tend to be less in pine needles, however. Therefore, Siberian fir *(Abies sibirica),* whose needles have about 30–35% bornyl acetate, is a particularly attractive source of needle oil for the fragrance industry. These terpenes are used in scenting soaps and bath preparations, room sprays, deodorants, and similar preparations (Kelly and Rohl 1989).

Conifer foliage offers a possible additional benefit: chemicals not found in xylem resin, such as unusual resin acids. Considerable research on use of conifer foliage has been conducted in Russia, where work has focused on obtaining classes of desired chemicals rather than particular individual compounds. The Russians have also synthesized vitamins A and E from conifer needle feedstocks.

Asia

There has been a long history of tapping pine trees for oleoresins in China (Su 1984). Pine resin was described as a medicine for improving one's health in Emperor Shên-Nung's herbal, *Code of Medical Herbs,* compiled in the first century A.D. but based on traditional use of plants as medicines going back to 2000 B.C. (Chapter 6). Tapping of pine resin also was illustrated in *Tian Gong Kai Wu* by Song Ying-xing of the Ming dynasty. Chemical analysis of rosin more than 800 years old from a tomb indicated that the primary constituents were similar to those of present-day resin from *Pinus massoniana,* still the most commonly used resin-producing tree in China.

For many years, gum rosin (Figure 7-1) was handled in a primitive way in southern China, resulting in a product of inferior quality. After World War II, efforts were made to improve the naval stores industry, and by the late 1950s China had become the world's second largest producer (Zhan-Qian 1989). China is the world's leading producer of gum rosin, taking up the slack from a decline in resin production from tapping pines in the more developed countries. In 1996, China produced about 57% of the world's supply of rosin; in fact, Hodges (1997) indicated that China is key to the global supply. With a rapidly growing economy, China not only exports rosin but is the world's largest consumer of it.

Pinus massoniana is the source of 95% of China's oleoresin, but there are other high-yielding species (Su 1984). All the pines used occur in either subtropical or tropical regions. *Pinus massoniana* has a wide distribution through central, southern, and southwestern China; *P. yunnanensis* occurs in Yunnan,

Sichuan, and Guizhou provinces and the Guangxi autonomous region, and *P. kesiya* grows in Yunnan province. The yields from these species vary; production can be two times greater in the high-yielding than the low-yielding pines. As part of intensive reforestation, the American *P. elliottii* (Plate 39 and Figure 7-3) has been planted, and tapping of it is increasing in Fujian, Yunnan, and Hunan provinces and Guangxi (Hodges 1997). The work is carried out by an estimated 200,000 Chinese, who often migrate, settling temporarily near the forests being tapped during the production season.

High yields and the relative amounts of α- and β-pinene are both important for turpentine production. For the aroma industry, β-pinene-rich turpentine is usually priced 40–50% higher than α-pinene-rich turpentine (Plocek 1998). Turpentine from *Pinus massoniana* is rich in α-pinene whereas *P. kesiya*, like *P. elliottii*, is rich in β-pinene. Therefore, increased production of the latter pines in southern China may lead to greater use of Chinese turpentine by the aroma industry (Chapter 11).

For ages in subcontinental Asia, Himalayan people have tapped the extensive pine forests for resins. Chir *(Pinus roxburghii)* is the most important pine in northwestern India. As the best-yielding Indian pine, it has been the basis of the Indian naval stores industry (Gibson and Mason 1927). Systematic tapping was initiated by the Indian Forest Department as far back as 1888 and then built upon by foresters trained in French methods for sustainability (Tsoumis 1992). The inaccessibility of the areas of turpentined trees was a major difficulty in developing the Indian industry for a long time. With some transport problems solved, resins became important in the nation's economy from 1960 into the 1970s (Stauffer 1989). However, crude resin production peaked in 1976 and fell steadily thereafter. Apparently, the decline resulted from heavy mortality of trees through poor traditional tapping methods and frequent fire in the chir forests (Chaudhari et al. 1992). The loss of substantial indigenous production of crude resin, and the demands of Indian industry for naval stores products, have led to India's being a net importer of both turpentine and rosin. The greatest single use of turpentine in India is for the production of synthetic camphor. In 1996, India was a distant third in production of oleoresin entirely from gum turpentining whereas second-place Russia produced it from gum, wood, and pulping (Hodges 1997).

To try to meet India's increasing domestic demand for resin, to discourage importation of resin as a raw material for industry, and to earn foreign exchange through export, efforts were made to increase production (Chaudhari et al. 1996). First, a new method of tapping to reduce damage to trees and increase yield was instituted. Tapping procedures on *Pinus roxburghii* were also being assessed in neighboring Pakistani forests (Sheikh 1981, 1984). Second, pines other than *P. roxburghii* were chosen to be tapped in India. Himalayan blue pine *(P. wallichiana)* and Khasi pine *(P. kesiya)* grow in the foothills of the Himalaya and show potential because of their high α- and β-pinene content. However, difficulties result from the geographic location of the trees and problems in securing proper labor. Planting of fast-growing exotic pines (e.g., *P. caribaea, P. elliottii,* and *P. patula)* is also being considered. These pines have some advantages over *P. roxburghii* because of their rapid growth and the more commercially important resin constituents.

Resin production from Tenasserim or Mindoro pine *(Pinus merkusii)* has received attention in Asia. *Pinus merkusii* is the most tropical of all pines and often occurs in dipterocarp forests in Myanmar (Burma), Thailand, Cambodia, Vietnam, and the Philippines. It was planted extensively by the Dutch on Sumatra and other Indonesian islands; Mirov (1967) reported that these plantations were often mistaken as native stands. In fact, government experiments on Sumatra demonstrated that young second-growth trees are the most satisfactory for tapping, and the plantations were sustainable when French methods were used. This situation was different from other pine species in which mature trees are needed to produce sufficient quantities of resin for commerce. In 1996, Indonesia was second only to China in production entirely from gum turpentining (Hodges 1997).

In the Philippines, both *Pinus merkusii* and Benguet pine *(P. kesiya,* syn. *P. insularis),* have been used. *Pinus kesiya* forms extensive forests in the mountainous regions of northern Luzon, where the yields are high because trees can be tapped throughout the year. Both resin composition and yield vary considerably from tree to tree, as commonly found for other resin-producing trees in systematic and ecological studies (Chapter 5). In attempts to improve tapping methods, bark chipping produced higher biweekly yields than the standard sapwood chipping, and trees in the 40- to 50-cm diameter class were the best size (Orallo and Veracion 1984). These results are similar to those from other studies, showing the importance of tapping methods that do not damage the cambium, and that medium-sized trees are the most productive. *Pinus merkusii* resin is generally higher in α- than β-pinene, however, which may contribute to production being used primarily for internal consumption for solvents rather than for export to the expanding aroma industry.

Latin America

Pines occur throughout the mountainous regions of Central America (parts of Guatemala, Honduras, Nicaragua, El Salvador, northern Costa Rica, and Belize) as well as at higher elevations on the larger West Indian islands (Cuba and the Dominican Republic; Chapter 2), but Mexico is the most important resin-producing country. At least two dozen species of pine occur in Mexico, many of them exceedingly resinous (Mirov 1967, J. Perry 1991). *Pinus arizonica, P. ayacahuite, P. caribaea, P. chihuahuana, P. montezumae* (Figure 7-6), *P. oocarpa, P. ponderosa,* and *P. teocote* are among the species considered suitable for resin production (Martínez 1959). Early destructive tapping techniques and wasteful exploitation led to laws to prevent these practices. In 1996, Mexico trailed only Brazil in gum rosin and gum turpentine (Figure 7-1) production, slightly exceeding that of Argentina (Hodges 1997).

The Oxford Forestry Institute in England has worked closely with national forestry programs in Central America to further research on *Pinus caribaea, P. oocarpa,* and other species since the 1950s (Barnes 1988). The institute has organized teams to explore natural populations of these pines, analyzing natural variation and bringing seeds to Oxford for testing. Furthermore, a Central America and Mexico Coniferous Resources cooperative program has provided networking assistance since 1980.

Although numerous pines have been tapped in several Central American countries for resin at some time, Honduras remains the major producer (Coppen and Hone 1995). Most resin is obtained from *Pinus oocarpa;* a smaller amount comes from *P. caribaea* var. *hondurensis*. Production of crude resin peaked in the early 1980s, declined, but subsequently stabilized. Stanley (1991) reported on the potential of resin production from native pines by local farmers, some of whom are members of a national federation of tappers. He pointed out that experimental community agroforestry has promise, but its future depends on governmental and donor support to help manage and market the production. Most of the rosin is exported to Europe, with Germany the largest importer.

Pine oleoresins have been harvested only relatively recently in South America. Although there are no native conifers that are copious resin producers, the introduction of pines for plantation culture has led to development of sizable industries in several countries. Brazil is the largest South American producer of naval stores with large areas planted with different pine species (Sanz 1989). *Pinus caribaea* and *P. oocarpa* are grown in the tropical northern regions, and *P. elliottii* and *P. taeda* in more subtropical southern ones.

Figure 7-6. *Pinus montezumae,* heavily tapped for resin in mountainous areas of southern Mexico. Here, resin flow increased as a result of forest fire.

Most production has been from *P. elliottii,* however, with large-scale tapping beginning in the late 1970s and increasing through the 1980s until the early 1990s. *Pinus caribaea* var. *bahamensis* yielded more resin than *P. caribaea* var. *hondurensis, P. elliottii,* or *P. oocarpa* (Zobel et al. 1987). Although most of the processed products are consumed domestically, significant quantities are exported. Coppen and Hone (1995) reported that replanting of *P. elliottii* was not keeping pace with the loss of trees at the end of their tapping life. Therefore, they thought that *P. caribaea* planted in the tropical areas would be increasingly targeted as a source of resin.

In Argentina, plantations of *Pinus elliottii* are tapped in the northeastern provinces. Substantial amounts of both rosin and turpentine are converted into derivatives for both domestic consumption and export. Venezuela also produces a relatively substantial amount of crude resin from *P. caribaea.* Plantations of *P. radiata* have begun experimentally in Chile; it is more suited to the environmental conditions there than *P. elliottii.* However, yields from *P. radiata* have not been high enough to encourage growth of the industry.

Africa

Africa is a relatively recent producer of gum naval stores (Figure 7-1), although large areas of pines have been planted for timber or pulp for some time. Zimbabwe pioneered African production; tapping of *Pinus elliottii* in the eastern highlands of Zimbabwe began in 1976 (Coppen et al. 1984). Because the pine resources are under increasing pressure for use as timber, however, resin output may fall. Small amounts of rosin have been exported intermittently to South Africa, but most has been consumed domestically by the paper industry.

South Africa and Kenya both began production in 1986 (Coppen and Hone 1995). Resin is produced from three pines in Kenya: *Pinus elliottii* in the south provides most of the resin, *P. caribaea* var. *hondurensis* is tapped in southern coastal areas, and *P. radiata* is tapped at higher elevations. Total resin production shows an upward trend; rosin is converted to a modified form and sold locally to paper mills. South African tapping is centered on extensive plantings of *P. elliottii* and *P. caribaea* var. *caribaea* in northern KwaZulu-Natal, and its resin production is the highest of the three African countries. Most South African rosin and turpentine is consumed domestically.

Several other African countries have extensive areas of underexploited pines that have potential for gum naval stores production (Coppen and Hone

1995). For example, Malawi has large areas of pine, including *Pinus elliottii*, and resin production was planned. Commercial tapping of *P. caribaea* began on a small scale in Uganda in 1994.

Cedarwood Oil

The botanical source of cedarwood oil is somewhat confusing in that *cedar* is used for taxa in both the Pinaceae and Cupressaceae (Chapter 2). The confusion may have begun with the ancient Greeks, who used *kedros* to designate any resinous tree. Species of *Cedrus* (Pinaceae) are considered the true cedars. Although several species of *Cedrus* produce cedarwood oil, the greatest quantities are derived from various species of *Juniperus* and *Cupressus*, also commonly called cedars, in the Cupressaceae.

Cedarwood oil does not contain true oils but consists principally of volatile heartwood sesquiterpenes. The use of the word oil probably derives from essential oils, referring to mono- and sesquiterpenes (Chapter 1). Production levels vary from the highest in *Juniperus ashei* and *J. virginiana* in the United States and *Cupressus funebris* in China; *Cedrus atlantica* in the Indian Himalaya and *C. deodara* in the Atlas Mountains of Morocco produce lower amounts (Lawrence 1985). Heartwood in members of the Cupressaceae has a similar qualitative composition of sesquiterpenes across the family, so similarity in compounds from different genera is not surprising (Adams 1991). However, quantitative compositional patterns, that is, the relative amounts of different compounds, vary (Chapter 5) and are responsible for differences in oils from different localities.

Cedarwood oil from *Juniperus* is an important source of sesquiterpenes such as α- and β-cedrene, thujopsene, cuparene, cedrol, and widdrol, which are used directly in fragrance compounding for scenting soaps, room sprays, disinfectants, and so on or as a source of raw components in the production of other fragrances (Adams 1987, 1991). Juniper heartwood sawdust also has extremely high termiticidal activity (Adams et al. 1988). The heartwood, bark, sapwood, and leaves of various juniper species are reported to have antimicrobial properties (Clark et al. 1990). Antiseptic and spermaticidal properties were recognized in classical times when "cedar oil" was used as a contraceptive; other medicinal uses for *Juniperus* are discussed in Chapter 10.

In the United States, *Juniperus ashei* in Texas and *J. virginiana* in the East provide most of the cedarwood oil. Although extensive mature forests of *J. virginiana* do not occur, their seedlings are pioneers in abandoned fields (Figure 7-7). These young trees subsequently form forests that provide an adequate biomass for commercial production. *Juniperus ashei* also invades abandoned fields and overgrazed rangeland, forming almost continuous stands for hundreds of kilometers in some places. Junipers have invaded an estimated 8.5 million hectares of grassland in Texas alone (Adams 1987), an extent that opens possibilities for use of *J. ashei* as well as some other invasive juniper species for cedarwood oil.

In a survey of 11 *Juniperus* species, those from arid lands generally yielded less cedarwood oil than those such as *J. ashei* and *J. virginiana* from mesic areas (Adams 1987). *Juniperus erythrocarpa*, however, from extremely arid

Figure 7-7. *Juniperus virginiana,* red cedar, invading an abandoned field in Texas. Later, the trees may provide adequate biomass for cedarwood oil production. Courtesy of R. P. Adams.

areas, had the highest yield, and the desert *J. scopulorum* also had a high yield. These yields point out the lack of knowledge regarding environmental factors that control sesquiterpene yield. In regard to age, the highest sesquiterpene yields in *J. virginiana* have been obtained from older trees with a greater ratio of heartwood to sapwood (Runeberg 1960). Adams confirmed these results, showing that yields were less than 0.5% from sapwood but 3.8% dry weight from heartwood.

Although yields of cedarwood oil from *Juniperus erythrocarpa* and *J. scopulorum* are comparable to those from *J. ashei* and *J. virginiana*, these junipers are not competitive in either the most desirable sesquiterpene composition or biomass for large-scale production. Nonetheless, Adams (1987) concluded that arid-area junipers such as *J. erythrocarpa* and *J. scopulorum* (Plate 5) could support small, local distillation facilities. Cedarwood oil is extracted in various ways, from home-built stills to more sophisticated laboratories. Distillation time, the size of wood chips extracted, and temperature and pressure of the steam can make large differences in cedarwood oil composition. Thus distilling practice as well as the species used determine the marketable product (Adams 1991).

Oily Resins from Tropical Angiosperms

Some species of the tropical families Dipterocarpaceae and Fabaceae produce resins with a high percentage of volatile sesquiterpenes. As with cedarwood oil, they do not contain true oils (Chapter 1) but are often referred to as oils. This is in contrast to the majority of resins from these families, which are used for their nonvolatile fraction as dammars or copals (Chapter 9).

Dipterocarps

A fluid resin, often called wood oil, is obtained from various species of *Dipterocarpus* and a few of *Anisoptera*, members of the family Dipterocarpaceae in Southeast Asia. The fluidity is maintained by sesquiterpene hydrocarbons that do not form polycadinene polymers (Chapters 1 and 4). This is in contrast to other dipterocarp resins (dammars; Chapter 9) in which these sesquiterpenes only serve to dissolve the predominant triterpenoids.

Because Europeans were primarily interested in dammar resins for varnishes during the 19th and 20th centuries, they had relatively little interest in dipterocarp wood oil, except in small amounts as an adulterant of the leguminous copal resins *(Copaifera* or *Guibourtia)* from Africa for use as hard varnishes (Chapter 9). On the other hand, local people considered dipterocarp wood oil one of the most important resins. It was used as an illuminant until relatively recently and was "among the first products to be exploited in the forests for sale to settled populations" (Burkill 1966). It was only after the advent of kerosene and other such products as well as electricity that wood oil was no longer used as an illuminant and traded among the local people.

The process of tapping, called boxing, and then firing *Dipterocarpus* trees to obtain oleoresin is still used in many forested areas of Southeast Asia and India. In some areas where the resin is obtained in this manner it is known locally as keruing oil or minyak oil (Ibrahim 1988). Despite being an oleoresin, however, the resin from *Dipterocarpus* has been called *gurjun* balsam in parts of India and Myanmar (Burma), possibly because the resin is less fluid there than in other areas. Tapping by the Semelai of the Malay Peninsula (peninsular Malaysia) was described in detail by Ibrahim et al. (1987) and Gianno (1990). A hole is chopped to about one-third the thickness of the trunk of a large tree, such as *D. kerrii,* then hollowed out at the bottom to create a box to collect the resin (Plate 40). The Semelai collect the accumulated oleoresin each week. The hollow is then set on fire for 2–5 minutes, which cleans out the hardened resin left after collection and acts as a stimulant for further exudation. As long as trees are given a several-month resting period, the Semelai maintained that *D. kerrii* trees have been tapped in this manner more or less continuously for 20 years or more without endangering the life of the tree.

Gianno (1990) stated that within the memory of the Semelai of the Malay Peninsula, *Dipterocarpus kerrii* has been the only species tapped for resin on a regular basis for sale. In the past, apparently, a majority of heads of households owned or co-owned many of the trees. Since major logging began in the 1960s, however, the few remaining trees have been tapped for local consumption. Other species tapped occasionally for local use of keruing oil include *D. chartaceus, D. crinitus, D. grandiflorus, D. sublamellatus,* and *Anisoptera megistocarpa.* In the 1980s, however, there was increased interest in collecting keruing oil for export; the volatile components were distilled in Singapore for use primarily as a base for perfume because keruing oil is similar to but less expensive than patchouli oil derived from the grass *Pogostemon cablin*

(Gianno 1986b). The two primary components of keruing oil are the sesqui-terpenes α-gurjunene and alloaromadendrene (Ibrahim 1988).

Because of commercial interest in *Dipterocarpus kerrii* volatiles, accompanied by diminishing stands of these trees, Ibrahim et al. (1987) investigated ways to improve collection of the resin. Even though Gianno (1986b, 1990) reported that the Semelai had used the boxing method without damaging trees, Ibrahim and coworkers thought that newer methods would increase the yield and produce better-quality resin as well as inflict less damage on the trees. They found that bark chipping rather than cutting a large hole in the trunk did less damage. Resin secretion was negligible without a stimulant, however, and sulfuric acid, so commonly used on a large scale in many countries, was preferable to fire in stimulating resin flow. There was an increase in secretion with increased acid concentration up to a point. Although the average amount of resin obtained using the bark-chip method with sulfuric acid was comparable to that obtained by the traditional methods, Ibrahim and coworkers concluded that their tapping method would ensure profitable extraction of the resin with little damage to the tree. They suggested further development of resin tapping technology. Peters (1994) also emphasized that the technology of resin collection from dipterocarps such as *D. kerrii* needs to be improved because local people no longer use the traditional methods that preserved the lives of the trees.

Dipterocarpus alatus is another species native to mainland Southeast Asia whose resin is used by indigenous people for illumination and for waterproofing baskets and boats. Resin from this species is now the main source of cash income for many villagers in Laos, from which it is exported for use in paint, varnishes, and lacquer but particularly as a fixative for perfumes. Requests for resin from businessmen suggest an increase in demand for this crop, which is so important economically to Laotian villagers (Ankarfjärd and Kegl 1998).

Dipterocarpus grandiflorus and *D. gracilis* are among several Philippine species most used for oily resin. Both species are widely distributed canopy forest trees, occurring in abundance where there is a pronounced dry season. The resin is called *baláu* and is used for waterproofing and as a solvent and perfume fixative. *Baláu* has also been collected using the boxing method (Plate 40), and the flow may be more than a kilogram per day. Although resin is usually removed every few days, with firing at intervals to prolong the flow, the same tree can apparently be tapped for a number of years. These species are excellent timber trees and have been used for their wood after resin tapping for a number of years.

In the Philippines, palosapis resin is produced from *Anisoptera thurifera,* a common, widely distributed dipterocarp. Resin from *Anisoptera* is similar to that obtained from *Dipterocarpus* and has similar uses. Tapping of dipterocarps such as *A. thurifera* is an important source of income for Philippine farmers while they wait for their crops to ripen, especially for those farmers whose land is already so depleted of nutrients that they become full-time tappers (Ella and Tongacan 1987). The traditional method of boxing and refreshing by burning at frequent intervals is sufficiently crude that it often results in death of the tree when forestry regulations regarding its use are not followed. Thus, with improper boxing techniques, there has been a rapid depletion of trees for tapping. For sustained resin production, Ella and Tongacan thought that more research is necessary to analyze factors controlling resin yield, including crown size, use of different chemicals to stimulate flow, and so on, that is, a number of factors also considered by Ibrahim et al. (1987) for *D. kerrii.*

Legumes

Although leguminous oleoresins occur worldwide in the tropics, the oily resins of the New World species of *Copaifera* (Fabaceae, or Leguminosae), commonly called oil of copaiba, are the most heavily utilized. Some species of *Daniellia* in Africa produce oily resins, sometimes referred to as African copaiba. *Eperua,* with a New World distribution, and the predominantly Asian *Sindora* also produce oleoresins. Species used frequently in Brazil include *C. langsdorfii,* widely distributed in central and southern Brazil, and *C. guianensis, C. multijuga,* and *C. reticulata* in Amazonia. Commercial varieties of copaiba oil often are known by the names of the towns or regions from which they are exported, for example, Pará and Maranhão. The resin is also obtained from Colombian and Venezuelan species. Resin is collected from the Venezuelan *C. pubiflora* and *C. venezuelana,* but *C. officinalis,* known as Maracaibo copaiba or sometimes as *aceite,* is especially used.

New World species of *Copaifera* have wood resin with a volatile fraction larger than that of the African species and that also contains few labdatriene diterpenoids, which polymerize to form hard copals (Chapter 9). Rather, in some New World *Copaifera* species such as *C. multijuga,* the volatile fraction (predominantly sesquiterpene hydrocarbons) may constitute as much as 97%

of the resin (Langenheim 1981, Cascon and Gilbert 2000). In an analysis by Cascon and Gilbert, the most prominent of these sesquiterpene hydrocarbons were β-caryophyllene (as much as 60% of the resin in some species), α-copaene, α- and β-selinene, β-bisabolene, α-bergamotene, and δ-cadinene. Oxidized sesquiterpenes such as caryophyllene oxide also occur in abundance in *C. guianensis*. The relative proportions of sesquiterpenes to diterpenes varies with species, which results in differing degrees of viscosity. For example, *C. guianensis* has 45% diterpenes, contrasted with only 3–20% in *C. multijuga*. Furthermore, the kinds of diterpenes vary. In *C. multijuga* the bicyclic copalic acid predominates whereas in *C. guianensis* it is the tetracyclic kaur-16-en-19-oic acid. On the other hand, *C. duckei,* related to the commonly harvested *C. reticulata,* has a greater proportion of five sesquiterpene hydrocarbons, constituting 44% of the resin. Thus *C. duckei* falls between *C. multijuga* and *C. guianensis* in sesquiterpene content; its 55% diterpene content is concentrated in the more fluid hardwickiic acid (Figure 1-4) and polyalthic acid. *Copaifera officinalis* also contains high amounts of hardwickiic acid (Cocker et al. 1965). Braga et al. (1998) reported 10 diterpene resin acids and numerous sesquiterpene hydrocarbons and alcohols as well as a new sesquiterpene acid in resin of *C. cearensis*. The anti-inflammatory properties of *C. cearensis* have been supported by in vitro assays (Fernandes et al. 1992).

The widespread and highly variable *Copaifera langsdorfii* is considered the principal source of the more viscous copaiba resin known as the Maranham or Rio de Janeiro variety (Plowden 2001). The more viscous resins may account for their being called copaiba balsam in some cases rather than copaiba oil. Both bicyclic and tetracyclic diterpene acids occur in *C. langsdorfii* wood resin, including polyalthic, (−)-kaur-16-en-19-oic, (−)-16-β-kauren-19-oic, and eperu-8(20)-en-15,18-dioic acids (Ferrari et al. 1971). Leaf resins of *C. langsdorfii* have sesquiterpene hydrocarbons similar to some *Copaifera* wood resins, with large quantities of α- and β-selinene, β-caryophyllene, α-copaene, and δ-cadinene among others.

There is a need to recognize the variability of chemical compositional patterns within and between *Copaifera* species for use of the resins as medicinal and cosmetic products (Cascon and Gilbert 2000). As indicated for other resins used as medicines (Chapters 8 and 10), variation in chemical composition can influence the pharmacological action, toxicity, and hence, quality control of the product.

Copaiba oil has been used medicinally in various ways. Recognized early by Brazilian Indians who treated wounds with it, the resin became known medicinally to Europeans in the 17th century (Pio Corrêa 1931). A Portuguese monk, writing about the natural products of Brazil in 1625, mentioned a drug called cupayba, which was listed in the *London Pharmacopoeia* in 1677. Copaiba oil was included in the *U.S. Pharmacopoeia* from 1820 to 1940, when it was admitted to the *National Formulary*. Long collected from the widely dispersed rain forest trees by rubber extractionists who found they could collect quantities sufficient to sell for medicinals, it has been used especially for recalcitrant inflammatory urinary and pulmonary infections as well as skin ailments such as herpes, eczema, and psoriasis; resins from *C. langsdorfii, C. officinalis,* and *C. reticulata* have been mentioned most in treating these afflictions. Copaiba oil is also used for various ailments involving mucous membranes such as sore throat, bronchitis, and ulcers as well as an antirheumatic, antiseptic, etc. (Trease and Evans 1972, Baile et al. 1988, Duke and Vásquez 1994, Fleury 1997). Ohsaki et al. (1994) found potent antitumor activity of diterpenes in the resin, and Paiva et al. (1998) reported gastroprotective effects of the resin on experimental gastric ulcer models in rats. Both *C. coriacea* and *C. multijuga* are well known for a wide variety of medical uses. *Copaifera* resin also is used in cosmetic preparations, both as a fragrance and odor fixative, in soaps, detergents, lotions, and perfumes as well as in photochemistry to enhance halftones and emphasize shadows (Leung and Foster 1996, Duke and Vásquez 1994).

In Brazil during the late 1970s and through the 1980s, there was interest in using copaiba oil as diesel fuel (Figure 7-8) because the oleoresin of some species of *Copaifera* consists predominantly of hydrocarbons. Since trees are too dispersed to collect sufficient quantities for fuel, at least in the central Amazonian rain forest (0.5–2 trees per hectare), plans were made to grow plantations for tapping (Alencar 1982). The trees are slow-growing primary forest trees, however, and were deemed to take too many years (70 years according to some; Plowden 2001) to reach adequate size for yields to be economically feasible. There has been some research on using *Copaifera* as overstory trees for cacao plantations, thus making it a multiple-purpose tree (Chapter 11). For both fuel and overstory, however, Brazilian research has subsequently lagged, favoring a search for other renewable sources of energy such as sugarcane and cassava.

The earliest methods for harvesting the resin of *Copaifera* involved chopping a large hole or wedge near the base of the tree. Fire was sometimes used to hasten the flow of the resin, similar to the boxing technique for collecting resin from dipterocarps in Southeast Asia. Resin is generally now collected from holes bored into the trunk (Figure 7-9). Holes are sometimes bored at two different heights, with resin flowing out the lower hole.

Assessing the yield of *Copaifera* resin is difficult. Numerous anecdotal reports have individual trees yielding 40–60 liters at a single time. Such reports have led numerous proponents, such as Laird (1995), to suggest industrial utilization of *Copaifera* resin. On the other hand, Amerindian folklore indicates that obtaining resin from any tree is far from certain. Harvesters are supposed to keep from looking up into a tree before attempting a harvest, or the resin will be sucked into the upper trunk (Shanley et al. 1998).

Figure 7-8. Cartoon similar to ones that appeared in Brazilian newspapers and magazines during the 1980s, depicting scientists' suggestions for potential use of *Copaifera* resin for diesel fuel. Finding alternative, renewable resources for fuel is particularly important in petroleum-poor countries, and *Copaifera* may again receive consideration in the future.

There is such folklore about harvesting other tropical resin-producing trees (Chapters 8 and 10), emphasizing the extreme genetic variability of resin production in some tropical species as well as possible seasonal variation in synthesis. In other research, yield varies as does the composition of the resin (Plowden 2001). Thus Plowden emphasized that defining an average harvest of *Copaifera* resin is a challenging task, even for trees of a particular species in the same region.

The study of *Copaifera multijuga* in central Amazonia by Alencar (1982) was the first scientific analysis of resin production in Amazonian members of the genus. In contrast to reports of 40–60 liters, the largest yield was 3.5 liters, and the average first harvest only 0.2 liters per tree, with declines in suc-

Figure 7-9. Copaiba oil being obtained from *Copaifera* in eastern Amazonia as part of an experiment for possible use by the Tembé Indians. Courtesy of C. Plowden.

cessive harvests. Plowden (2001) found even lower average yields in several unidentified *Copaifera* species in eastern Amazonia. His study showed that the highest yields come from trees with trunks in the upper to middle size ranges (45–65 cm diameter breast high) and that resin flows only from the heartwood. Some species of *Copaifera* initially produce resin in the sapwood, in canals that form concentric rings near the vascular cambium from which they differentiate (Chapter 3). As indicated for stump wood in pines, it is assumed that the resin saturates the dead cells of the heartwood as it develops. The middle-sized trunk range has been found to yield the most resin in other tropical genera (e.g., Ella and Tongacan 1992). However, Plowden thought that the uncertainty of finding sizable amounts of resin in some native Amazonian trees, even in this optimal size range, indicates that physical stress or microbial attack is necessary to stimulate resin production in sapwood, promoting lysigeny of canals to form large resin-filled cavities in the heartwood. Studies of resin production in other species of *Copaifera* are needed to test Plowden's hypothesis as to how resin becomes available for harvest. Further research is crucial if hopes for development of an Amazonian industry (as suggested by Laird 1995 and others) are to be realized.

In Africa, the fluid resin from *Daniellia oliveri* is called African copaiba or wood oil. Since it ultimately hardens in the air, it also is referred to as ilorin (or illurin) balsam. *Daniellia oliveri* is a large tree characteristic of savanna forests from Senegal to Cameroon, including parts of Nigeria, and the resin is a chief product of Nigeria. Indigenous people tap the tree and use the resin to fumigate huts, scent garments, and kill vermin. The resin contains a high proportion of sesquiterpenes and the diterpene daniellic acid (illurinic acid), and is produced in large canals arranged in concentric circles in the wood, as in *Copaifera* (Moens 1955). *Daniellia pynaertii,* a large tree in periodically inundated areas in the Congo, Cameroon, and southern Nigeria, produces a resin similar to that of *D. oliveri* (Léonard 1950).

Another caesalpinioid oily resin is produced by *Eperua.* Resin from *E. oleifera,* often called *copáiba-jacaré* in Brazil, is used in the preparation of lacquers in Amazonia. Another species, *E. purpurea,* is the dominant tree of the yevaro forests in unflooded sandy soils in Venezuelan Amazonia. The wood is often used in building rural bridges because of its remarkable resistance to insect attack, which Janzen (1974) attributed to the resin. Medina and de Santis (1981) identified the major constituent of the resin of *E. purpurea* as

(±)-labd-8(20),13-dien-15-oic acid, which occurred along with another lab-dane acid and its corresponding alcohol. The chemical composition of *E. purpurea* is somewhat different from *E. falcata*, which contains a mixture of diterpene acids related to eperuic acid, and large quantities of oleyl alcohol (Blake and Jones 1963). Oleyl alcohol, a fatty acid derivative, and the pre-dominance of nonpolymerizing labdane diterpenes distinguish *Eperua* from most other genera of tribe Detarieae. The oleoresin from *Prioria copaifera*, the Central American cativo tree, is characterized by the diterpene cativic acid (Grant and Zeiss 1954). It is called cativo gum and is often used by bees to make nests (Chapter 5).

Resinous wood oils are also obtained from two species of the primarily Asian genus *Sindora*. In the Philippines, *S. inermis* yields kayu-galu oil, and *S. supa*, supa oil. In addition to local uses, the resin of *S. inermis* was exported by Chinese traders to Singapore.

CHAPTER 8

Fragrant and Medicinal Balsams

Balsams are not as fluid as oleoresins but are relatively soft and initially malleable. Although some researchers (e.g., Arthur Anderson 1955, Leung and Foster 1996) consider true balsams restricted to phenolic resins that consist primarily of cinnamic and benzoic acids, I prefer the broader concept, which includes terpenoid resins as well as phenolic ones. In all cases, however, balsams are noted for the fragrance of their volatile compounds. A gradient in the degree of fluidity or solidity as well as fragrance, from mildly to highly scented, characterizes balsams. Because of their initial malleability they have often been used as salves for wounds, and because of their aroma they have been used in perfumery and cosmetics, and burned as incense.

Majno (1975) emphasized that it is difficult to grasp the importance of aromas in human history and that "the use of tree sap in treating injuries must be older than man's oldest documents." Very early, fragrant resins became a basic necessity. They masked the foul smell of wounds as well as aided in their healing. Majno questioned why balsams came to be used as balms and noted that in the language of the Old Testament, the surgical use of balsam has an overtone of salvation—there was no better proof that exudate from wounded trees was the treatment par excellence for human wounds. Majno also pointed out the analogy of "whatever fills the gashes in plants should also heal gashes in people" for the use of resin in Native American medicine.

Although indigenous people have used any locally available fragrant resin as incense, this has been particularly true of balsams derived from members of the family Burseraceae. Similarly, in most societies, the burning of incense has religious significance, often considered a gift to God or gods.

Some balsams are named using a qualifier (e.g., Canada or Peru balsam) whereas other resins with the properties of balsams have names such as elemi, storax, styrax, and for the classical incenses, frankincense and myrrh.

The Balsams

Conifer Balsams

Only a few conifers produce balsam. One of the most well known is Canada balsam, obtained from balsam fir (*Abies balsamea*, Pinaceae), a conifer widely distributed in northern North America (Plate 2 and Figure 2-1). It is called Canada balsam because the area of greatest commercial collection is in one of the Canadian provinces, Quebec. More rarely, when the resin is particularly fluid, with large quantities of volatile components, it may be called Canadian turpentine. The constitutive terpenoid resin is collected from blisters in the bark (Figure 8-1), often by puncturing them with the pointed spout

Figure 8-1. Resin being collected from bark cysts or blisters for a chemosystematic study of *Abies grandis*. Similar techniques are used for commercial collecting of Canada balsam from *A. balsamea*. Courtesy of E. Zavarin.

of a can into which the balsam flows. The resin usually contains approximately 20% volatile mono- and sesquiterpenes and 80% nonvolatile diterpenes, which form a hard transparent film when dried. The film's high refractive index, approximately that of glass, makes it useful for microscopic work and as a cement for lenses. Although this resin has been partially supplanted by synthetic preparations, some researchers still prefer natural balsam. The medicinal value of Canada balsam was recognized as early as 1607 (Hill 1952). Northern tribes of Native Americans have used the resin as an antiseptic salve for sores, cuts, and burns as well as internally for colds, consumption, menstrual irregularity, as a laxative, and for other ills (Vogel 1970, Moerman 1998). It has also been used as an irritant and stimulant as well as a component of collodion, material used as a substitute for adhesive plaster in surgery for coating wounds and burns. In fact, Canada balsam was included in the medicine chest of the Lewis and Clark expedition as a wound dressing. It was an official medicine of the *U.S. Pharmacopoeia* from 1820 to 1916.

Several other conifers produce resin similar to Canada balsam. For example, somewhat similar balsam-type terpenoid resins are obtained from Douglas fir *(Pseudotsuga menziesii)*, sometimes called Oregon balsam, and hemlock *(Tsuga canadensis)*. The European firs *Abies alba* and *A. pectinata* yield Alsatian or Strasbourg turpentine, the *olio d'abete* (formerly called *olio d'abezzo*) of the Italians (R. White and Kirby 2001), and it has a diterpenoid composition similar to that of *A. balsamea* (Trease and Evans 1972). Both resins contain the abietane and pimarane resin acids common to pinaceous resins (Figure 1-4) but also a large proportion of a labdane alcohol, *cis*-abienol. This compound, with its conjugated diene system, may be partly responsible for the setting characteristics of the resin, that is, producing a hard film when it dries (Mills and White 1994). The β-phellandrene content of the monoterpenes has also been considered important in determining the setting characteristics.

Leguminous Balsams

One of the so-called true balsams is obtained from a tropical tree, *Myroxylon* (Fabaceae, or Leguminosae). Its name is derived from the Greek *myron*, sweet oil or perfume, and *xylon*, wood, which is fragrant because of the resin. Resin from *Myroxylon* has been important historically as a medicine and as an export item used for various purposes (Hale 1911). Although there are divergent opinions regarding the number of species in the genus (the only resin-producing one in subfamily Papilionoideae), two have been commonly referred to as sources of phenolic balsams: Tolu balsam from *M. balsamum* and Peru balsam from *M. pereirae*. These species are so similar morphologically that some researchers regard them as a single species, *M. balsamum* (Record and Hess 1943, Arbeláez 1956, Mors and Rizzini 1966). The resins, however, differ considerably, and thus are considered as derived from two varieties of the single species, varieties *balsamum* and *pereirae*; I follow that taxonomy here.

Myroxylon balsamum var. *balsamum* (syn. *M. toluifera*) is widely distributed in South America, common throughout forested regions of Brazil and Argentina; however, most of the resin entering commerce came from Colombia and Venezuela (Record and Hess 1943). In fact, the name Tolu balsam is derived from a small town near Cartagena, Colombia. The tree also occurs abundantly in the Magdalena and Cauca Valleys and northwestern parts of the mountains in Colombia. In Brazil, however, it is called *balsamo* or *oleo vermelho*. This balsam was once used by the Incas for embalming, and indigenous Colombians continue to use it to treat wounds and stop bleeding, especially when bleeding is profuse. The resin is more fluid than Peru balsam, perhaps because of the greater quantities of phellandrene and alcohols such as guaiacol and creosol, and therefore flows readily from gashes cut in the bark. The fragrant odor of the resin has led to its use in the perfumery industry, particularly for its delicate hyacinth-like scents, which blend well with floral and oriental compounds. Medicinally, it is used as an ingredient in cough mixtures (Tolu syrups) and for its antiseptic properties in topical preparations. Tolu balsam trees have been successfully cultivated in Sri Lanka and India as well as Sumatra and some of the West Indies (Smith et al. 1992).

Myroxylon balsamum var. *pereirae* (syn. *M. pereirae*) grows mainly on the Pacific side of southern Mexico through most of Central America (Figure 8-2), where the tree has sometimes been planted as shade for coffee. Because of the long-term exploitation of trees that grow abundantly in a strip of land on the Pacific coast originally in Guatemala but now in El Salvador, the area has been called the Balsam Coast. The earliest mention of Peru balsam is in the writing of Nicolás Monardes in 1565. Monardes noted that *balsamo* was produced by a tree in New Spain that the Indians called *xilo*. The resin was called Peru balsam because in the early days it was taken to Callao, Peru, for shipment to Spain. Subsequently, it has been shipped directly to consuming countries but the name persists. It remains an important export from El Salva-

Figure 8-2. *Myroxylon balsamum* var. *pereirae*, the source of Peru balsam, in Pacific coastal forests of Central America. *Myroxylon* is the only leguminous resin-producing genus of subfamily Papilionoideae. The abundant wind-dispersed fruit (below, left of the flower) aids in dispersal and spontaneous establishment in various sites.

dor (Coppen 1995). The resin was sufficiently important in religious contexts during the 16th century that Popes Pius IV and V declared that it could be used as a substitute for the consecrated sacramental oil used in baptism, ordination, and extreme unction. Their edicts emphasized that the destruction of balsam trees was sacrilegious.

Because *Myroxylon balsamum* var. *pereirae* does not exude resin naturally to any extent, the usual Indian method of obtaining the resin was to beat the trunk to loosen the bark (Howes 1949). The outer portions of the bark were removed, and exposed surfaces of the sapwood were covered with a cloth to absorb the exudate. These surfaces were then scorched with a torch to induce flow, and the saturated cloths were boiled and pressed to separate the balsam. This was known as the *panol* (cloth) or *trupp* (rag) collecting process (Duke 1985). The most abundant constituents of Peru balsam are peruresinotannol along with benzoic and cinnamic acids and their esters. Important volatiles include cinnamyl cinnamate (or styracin), benzyl cinnamate (or cinnamein), and benzyl benzoate (Tyler et al. 1988, Wren 1988). Wahlbert et al. (1971), however, listed numerous mono-, sesqui-, and triterpenoid compounds (e.g., α-pinene, *para*-cymene, α- and β-bourbonene, caryophyllene, and α- and γ-muurolene) as well as the predominant phenolic compounds. The characteristic pleasant odor is reminiscent of cinnamon when fresh and more like vanilla when aged, probably because of the presence of vanillin and an alcohol, nerolidol. Balsam from both varieties of *Myroxylon balsamum* is sometimes adulterated with components of *Copaifera* or *Pinus* resin before use.

Formerly credited with medicinal value, the resin was once listed in the *U.S. Pharmacopoeia*. It is now used primarily for its aroma in ointments, cosmetics, perfumes, and soaps. It was purchased by the perfumery industry in the United Kingdom in considerable amounts through 1960 (Adamson 1971). However, it is still used locally in some regions to treat asthma, catarrh, rheumatism, diarrhea, and hemorrhoidal pain but more commonly as an antiseptic dressing for ulcerated surfaces (Duke 1985). Phenolic esters such as the benzyl and cinnamyl cinnamates may be responsible for the antiseptic action (Coppen 1995). Also, it is still used as a parasiticide for ringworm (Wren 1998). Peru balsam was used in 1967 in a study of wound healing in a U.S. Army hospital (Majno 1975). Because little residual scarring is left on the skin after application, it has been suggested that the Peru balsam may help reduce scar tissue formation and aid in healing (Lewis and Elvin-Lewis 1977).

Peru balsam trees have been introduced into southern Florida, Sri Lanka, India, and western central Africa, where it is cultivated on a limited scale. The Singapore Botanic Gardens served as a springboard for introducing this resin-producing tree into tropical Asia, where it apparently tolerates alkaline and nutrient-poor soils. Apart from the few introductions into the Old World, Peru balsam is little cultivated, mainly as a shade tree and in backyards (Smith et al. 1992). Thus most balsam comes from wild trees, although some may be tended and in some cases planted to increase their density in the forest, particularly along the Balsam Coast. Peru balsam trees are also cultivated in some of the states of Mexico and, once established, can tolerate heavy tapping and live for a century (Linares and Bye 1987). Smith et al. (1992) predicted that with selected stock, Peru balsam tree cultivation in tropical America could become viable economically and ecologically (Chapter 11).

Storax and Styrax

Although storax and styrax have their own well-known resin names, they are grouped here with *Myroxylon* as true balsams because of their predominantly phenolic composition. Storax is interesting in also having a significant terpenoid complement, particularly when it occurs in leaves. Storax is the name generally used for resin from *Liquidambar* and *Altingia*, traditionally considered members of the family Hamamelidaceae but now placed in a separate family, Altingiaceae (Chapter 2). On the other hand, styrax is resin from *Styrax* (Styracaceae). Storax and styrax are sometimes used synonymously in the literature, however, which unfortunately can result in confusion regarding the botanical source of the resin. Resin from *Styrax* is also called gum benjamin, and the most commonly used resin from Asian species is known as benzoin.

Storax

The term storax is from the Arabic *assitirax*, meaning a sweet-smelling exudation. Of the five species of *Liquidambar*, three *(L. orientalis, L. styraciflua,* and *L. formosana)* are well known for their balsamic exudations. Levant storax is obtained from *L. orientalis*, a medium-sized tree that forms forests in Anatolia, in the southwestern coastal region of Turkey. Since the resin is only produced as a result of wounding, the usual method of collecting it has been to severely bruise the cambial area of the bark in early summer. The yellowish fragrant resin that diffuses through the inner bark is removed in fall and either mechanically pressed out or extracted in boiling water. The volatile content is very low (only 0.05%), 90% being styrene, with the monoterpenes α- and β-pinene primarily constituting the remainder; most of a saponified sample consisted of phenolic compounds such as cinnamyl alcohol (30%) and cinnamic acid (40%), and isomers of 2,3-butanediol and 3-phenylpropanol also occurred in considerable amounts (Hafizoglu et al. 1996). Additionally, three triterpenoid acids—oleanonic and 3-epioleanolic acids (Huneck 1963), and liquidambronovic acid—have been identified.

Ancient Egyptians used Levant storax for incense as well as for impregnating burial gauze to preserve mummified corpses better. The resin has been used in folk medicine since classical times, for uses similar to those of the predominantly phenolic balsams from the leguminous *Myroxylon*. Most Levant storax used to be imported from both Turkey and Italy (Pratt and Youngken 1951), but Turkey is now the only source for international trade (Coppen 1995).

American storax, or sweet gum, comes from *Liquidambar styraciflua* (Figure 2-8) of the southeastern United States and midelevation forests of Mexico (Plate 13), Guatemala, and Honduras; it is similar to resin from *L. orientalis*. The tree, however, appears to develop resin more spontaneously than *L. orientalis*, often exuding into internal cavities between the wood and bark, which may be found as large excrescences on the outside of the bark, which are tapped. The resin has been called *xochicotzatl* by Mexican Indians and was an article of tribute paid to Aztec rulers (A. Peterson and Peterson 1992). Several of the tributary provinces exploited by the Aztecs were located in extensive cloud forests where *Liquidambar* resin was an important forest product. The resin was used in medicines, smoked with tobacco, burned as incense, and perhaps formed into figurines used in sacrifices (Alcorn 1984). Apparently, when the Spanish conqueror of Mexico, Cortés, first met the Aztec emperor Montezuma II in 1519, Montezuma inhaled smoke from a tube that held *Liquidambar* resin mixed with tobacco (Díaz del Castillo 1956).

Native Americans chewed *Liquidambar* resin as a tooth preservative as well as to cure fevers and dysentery. It was also used as a salve for wounds, sores, and ulcers (Moerman 1998). The resin was called either *copal* or *copalm* by the Spanish and French, who thought it produced healing comparable to that of balm of Gilead from *Commiphora opobalsamum* or balsams of Tolu and Peru from *Myroxylon* (Vogel 1970). It was even reported that animals cured their wounds by rubbing against a tree where the resin exuded. *Liquid-ambar styraciflua* was included in the *U.S. Pharmacopoeia* in 1925 (as American storax), as a stimulant, expectorant, and antiseptic, and remains on the list.

Cinnamyl alcohol (storesin) constitutes 30–50% of the resin of *Liquid-ambar styraciflua,* with cinnamic acid, free and esterified as cinnamyl cinnamate (styracin) and phenylpropyl cinnamate, about 18%. Vanillin, borneol, and bornyl acetate are present as well as varying amounts of styrene. Styrene is employed in the chemical synthesis of rubber and the plastics Lustron, Styron, and Victron. Commercial U.S. supplies of American storax during the 1960s and 1970s were purchased mainly from Honduras, with smaller amounts from Guatemala and Nicaragua; this trend continues (Coppen 1995). Because American storax often contains more styrene, which has a gasoline-like odor, it is considered less desirable than Levant storax. However, a large quantity is now produced in the United States for the tobacco industry, where it is used to flavor cigarettes (Tyler et al. 1988). Polymerization of styrene enables preservation of *Liquidambar* resin as amber (Chapter 4).

The chemical composition of leaf resin of *Liquidambar styraciflua* is quite different from that exuded from the trunk. Differences in terpenoid composition in different organs are usual and may be related to the ecological roles of these secondary compounds (Chapter 5). However, the difference in *Liquidambar* is particularly interesting because of the predominance of terpenoids in one organ (leaves) but phenolics in the other (trunk). Leaves were analyzed to establish the compounds responsible for their characteristic strong aroma: 36 terpenoid components were identified by GC-MS, with monoterpenes (terpinen-4-ol, α-pinene, sabinene, α-terpineol, and α- and γ-terpinene) the major constituents, constituting 92% of the volatiles (Wyllie and Brophy 1989). In fact, the resin is considered to be a rich source of terpinen-4-ol, which is of pharmacological interest. Of the remaining components, sesquiterpenes (δ-cadinene, β-caryophyllene, and germacrene D) are the most abundant. This is a terpenoid composition similar to that of *Melaleuca alternifolia* (the source of Australian tea tree oil), whose use in folk medicine has led to a number of clinical investigations. Tea tree oil has rapid bactericidal action against a wide range of organisms and exhibits few significant side effects. Although leaf resin from *L. styraciflua* has not been bioassayed, Wyllie and Brophy indicated that the similarity of its terpenoid composition to tea tree oil suggests similar bactericidal properties. The role of variation of the composition in the leaf terpenoids in relation to herbivory under different forest conditions has been studied ecologically in Mexico (Sánchez-Ramos et al. 1999).

Liquidambar branch buds also secrete resin, which can be obtained by boiling the branches in water. Buds are typically covered with resin from late fall through spring, similar to young branches of *Populus* and *Betula* (Chapter 3). *Liquidambar styraciflua* branches have been collected locally in Guatemala for use as a type of deodorant, sweetening the smell of bodies and clothes (Tripplett 1999). Whether the composition of these external resins is more similar to the terpenoids secreted internally in the leaves or whether they also contain phenolic compounds is not known.

Liquidambar formosana, native to the warmer parts of southeastern China, Taiwan, Vietnam, and Indochina, is also cultivated in Japan. In China, resin is collected from trees 20 years or more old. The resin contains cinnamic acid and esters, cinnamic alcohol, borneol, and a variety of volatile terpenes such as α- and β-pinene and camphene (Ivanov et al. 1969). It is very aromatic when burned as incense, yet its taste is bitter and pungent. It is made into pills for internal use or ground into powder for wounds. It also is prescribed for bleeding, boils, carbuncles, and toothaches. Although the resin is used locally, it apparently has not entered international trade (Coppen 1995).

Altingia, which is distributed from Assam, India, and southern China to the Malay Peninsula, Sumatra, and Java, is the source of liquid storax (Burkill 1966). The Sanskrit name for the resin is *rasa mala;* it has been used since A.D. 700 by Hindus who have lived on Java. Two species, *A. excelsa* and *A. gracilipes,* produce resin, but it is the scented resin of *A. excelsa,* a tall rain forest tree of Southeast Asia, that was exported to China. There, it was used as a tonic and stimulant, particularly for chest complaints (L. Perry 1980). It was sometimes used as a substitute for benzoin from *Styrax,* or storax from *Liquidambar.* Interestingly, Burkill indicated that meliponid bees "perhaps get most of the *Altingia* resin, and by breaking open their nests an impure storax is obtained."

Asian *Styrax*

Resin is the chief product from *Styrax,* from the tree or shrub species of eastern and southeastern Asia (Burkill 1966). Unlike that of most resin-producing trees, the wood is soft and used only for firewood following death of the plant from excessive resin tapping. Styrax, benzoin, and gum benjamin are all names for the resin, but the Asian resin is generally called benzoin. (*Ben* is from the Arabic for fragrant or the Hebrew for branch, and *zoa,* exudation; hence, fragrant or branch exudation.) It is a phenolic resin, fitting into the category of true balsams used by some researchers. Resin is only produced following injury or wounding by tapping, and this has led to reports that some species do not produce resin, which may or may not be true.

There are three Asian resin-producing species *(Styrax benzoin, S. paralleloneurus,* and *S. tonkinensis)* that are important, and several others that yield resin in lesser amounts and quality (Burkill 1966). As indicated in Chapter 2, taxonomic revision of the Old World species of *Styrax* is needed; hence, there is some nomenclatural confusion about the resin-producing species. The best known source of benzoin is *S. benzoin* (Figure 8-3), a relatively tall evergreen tree native to Sumatra, the Malay Peninsula, and western Java. This resin is usually called Sumatra benzoin. Sumatrans first tapped wild trees for resin, but repeated tapping killed many trees. To ensure continued availability of trees, cultivators began to plant *Styrax* seeds with upland rice in their swiddens (a form of slash-and-burn agriculture). Rice plants, followed by weeds after the rice is harvested, provide the shade needed by seedlings. After 7 years the trees can be tapped twice a year and have an economic life of about 8 years (Peltzer 1978). The formation of resin is induced in the wood and bark tissues following injury to the cambium; the first wounds do not produce much resin, but later ones produce more of the fragrant components. The resin forms as yellowish white globular or pear-shaped masses, with the whitest tears the most esteemed. This is similar to *Boswellia,* in which the most desirable frankincense comes from the second or third wounding.

Sumatra benzoin consists principally of coniferyl cinnamate, cinnamyl cinnamate (styracin), and coniferyl benzoate associated with free cinnamic and benzoic acids (Figure 1-8). There are also traces of fragrant benzaldehyde and vanillin, as well as styrene. However, the proportions of these phenolic compounds differ in analyses (Burkill 1966, Duke and Ayensu 1985, Leung and Foster 1996). Benzoin has antiseptic, stimulant, expectorant, and diuretic properties (Tyler et al. 1988). Benzoic acid, now primarily a synthetic product, was initially obtained from Sumatra benzoin. Benzoic acid and its sodium salt are extensively used as a preservative of foods, drinks, fats, and pharmaceutical preparations. Medicinally, benzoic acid is used primarily as an antifungal agent. Because of the pharmacological uses, there has been concern regarding adulteration of commercially available resins. Shah et al.

Figure 8-3. *Styrax benzoin,* source of Sumatra benzoin.

(1971) found market samples of benzoin (the kind not noted) adulterated with colophony (assumed to be pine resin?) whereas Helliwell and Jennings (1983) reported Sumatra benzoin to be adulterated with "gum dammar," assumed by them to be derived from *Shorea*.

The largest number of trees from which benzoin has been obtained occur on Sumatra. In addition to *Styrax benzoin,* the widely distributed *S. paralleloneurus* (syn. *S. sumatranum*) has been commonly used. The resin from this species differs from that of *S. benzoin* in being composed principally of cinnamic acid. This difference, however, could be an artifact of chemical analysis, which was done prior to the availability of modern techniques. Other Sumatran species indicated as producing benzoin include the deciduous *S. ridleyanus* and *S. subpaniculatus* (Burkill 1966). *Styrax serrulatus*, which occurs in northeastern India and parts of the Malay Peninsula, apparently produces an inferior benzoin.

The deciduous *Styrax tonkinensis* is the important resin producer that has been tapped for centuries in what was eastern Siam (now Thailand) and Laos; this resin is often referred to as Siam benzoin. Shroeder (1968) described the cinnamyl compounds from this benzoin. However, Leung and Foster (1996) reported that Siam benzoin is composed primarily of siaresinolic acid and coniferyl benzoate. *Styrax tonkinensis* is a tree that grows to 25 m in height in secondary forests in northern Laos and Vietnam, where it is also grown in plantations for pulp. The species has been introduced into southern parts of China for both resin and pulp production. Pinyopusarerk (1994) described methods of tapping in Laos. A tapper makes a notch into the cambium of the trunk, and the bark is removed. A number of incisions are made, staggered at intervals of 20–30 cm along the trunk. Exuded resin is left on the tree to harden; it may be 4–5 months after tapping that the tears of benzoin are collected. The first tapping is made on trees 3–5 years old in natural forests, and 6–8 years old in regenerated forests (as in *S. benzoin*). Tapping continues for only 6–8 years, less if bark removal has been excessive and permanent damage was done to the tree. In Laos, collection of benzoin is an important cottage industry, widely practiced by people in the highlands. The evergreen *S. benzoides* is the source of benzoin from what was northwestern Siam, and Java, and is tapped by local people during the wet season.

Benzoin has been traded in Asia since the protohistoric period, during which it was a major commodity in the luxury trade (Wolters 1967). There is evidence that benzoin was being traded to China about A.D. 800, where it was used as a fixative for perfume. The Arabs were apparently instrumental in expanding the trade to European ports and are the source for one name used today for the resin. They found benzoin sufficiently similar to frankincense, which they called *luban,* that they referred to the resin from *Styrax* as *luban jawi,* frankincense of Java. Although the Arabs obtained the resin in Java, Burkill (1966) thought that most of the resin came from trees in Sumatra. In taking the resin to Europe, the Arabic *luban jawi* was transformed to *ban jabi,* then to gum benjamin. In the 1700s, even benzoin of the worst quality was valued for medicinal purposes more highly than frankincense.

Benzoin has been popular as an antiseptic, parasiticide, and stimulating expectorant since the time of Marco Polo (1254–1324) (Pratt and Youngken 1951). In the United States, Sumatra benzoin is more customarily used in pharmaceutical preparations than that from other *Styrax* species (Helliwell and Jennings 1983). It has also been used as a carminative, diuretic, and inhalant for bronchitis and laryngitis. Virgin's milk, the milky alcoholic solution of benzoin, has been employed in feminine hygiene. In fact, benzoin has been used as a folk medicine for many ailments (Duke 1985).

Siam benzoin, in contrast to Sumatra benzoin, has apparently been preferred in the flavor and fragrance industries, for example, for alcoholic beverages, baked goods, candies, chewing gum, gelatins, ice cream, puddings, and soft drinks as well as components to fix the fragrance of cosmetics, colognes, lotions, perfumes, and toiletries (Morton 1977, Duke 1985). It is also used as incense and as a preservative of fat (Leung and Foster 1996). Siam benzoin contains predominantly benzoic acid and its esters whereas Sumatra benzoin contains predominantly cinnamic acid and its derivatives (Coppen 1995).

Some scholars believe that the biblical onycha (Exodus 30.34) may have been benzoin (Duke 1983) although others think that it was a *Commiphora* resin (Moldenke and Moldenke 1952). In any event, among its many other uses, benzoin has been burned as incense in religious ceremonies, often in combination with frankincense.

Other Styraxes

Styrax officinalis, a deciduous Mediterranean shrub, produces a resin called styrax-resin or even storax. It was known by the Greeks Theophrastus and

Herodotus and recommended as a remedy by Hippocrates. Storax, as discussed, generally refers to resin from *Liquidambar*. The confusion in names may have come during a time that Southeast Asian styrax (benzoin) was unavailable in Mediterranean countries and resin from *L. orientalis* was substituted for it (Zeybek 1970). Furthermore, controversy has existed as to whether *S. officinalis* even produces resin or whether the resin attributed to it actually comes from *L. orientalis*. To try to solve this mystery, Zeybek tried to induce resin synthesis by wounding *S. officinalis;* he was unsuccessful, which supported the view of Tschirch (1925) that the shrub does not produce resin. However, continued reports of resin from *S. officinalis* led Zeybek to question whether some "races" produce resin, and others not. This may be another example of individual plants producing far less resin than others in a population or not producing resin during certain seasons or only under certain environmental conditions.

Several New World species of *Styrax* produce resin. *Styrax camporum* and *S. ferrugineus*, evergreen small trees or shrubs common in the grasslands of central Brazil, exude resin in the form of tears when the bark is wounded. Called *estoraque* or *beijoeiro*, it has a pleasant odor reminiscent of resin reported to come from *S. officinalis* and is used in popular medicine (Mors and Rizzini 1966). *Styrax argenteus* and *S. warscewiczii* both produce a resin, also called *estoraque*, in Mexico and Mesoamerica. Thus the possible confusion with storax from *Liquidambar* is perpetuated. *Styrax argenteus* is one of the most common species of the genus in this area, but *S. warscewiczii* generally replaces *S. argenteus* above 2000 m (Fritsch 1997). Resin from these species has been used as an incense in some areas (Morton 1981, Tripplett 1999). Therefore, use of the name *estoraque* in Latin America for species of *Styrax* may perpetuate confusion with storax, resin from *Liquidambar*.

Another Latin American species of *Styrax* that produces useful resin is the evergreen tree *S. tessmannii* (syn. *S. guyanensis*). Its fragrant resin is used by Amazonian medicine men along the Río Vaupés in their magical practices (Schultes and Raffauf 1990). The Tikuna Indians of the Río Loretoyacu also use this resin to soothe the pain of tooth decay by packing resin into the tooth cavity, although it is considered toxic if ingested. *Styrax pearcei* is another resin-producing species reported from the Andean region (Howes 1949).

Elemis

In general, resin called elemi is obtained from members of the family Burseraceae, particularly *Canarium, Dacryodes, Protium,* and sometimes *Bursera,* although some from Clusiaceae *(Calophyllum* and *Symphonia)* and Rutaceae *(Amyris)* have been so called. Like other resins grouped as balsams, elemis are more viscous than oleoresins, being semisolid, and very fragrant. Although the volatile fraction may be distilled to produce elemi oil, this is not done routinely, unlike volatile components in oleoresins.

The soft, malleable nature of elemi results partially from the considerable quantity of liquid sesquiterpenes. Triterpenes, another component, sometimes crystallize, producing an opaque white appearance. Most elemis contain large quantities of the triterpenes α- and β-amyrin (originally isolated from *Amyris* resins), and elemi acids of the euphane skeletal group (Chapter 1) were originally isolated from Manila elemi, *Canarium luzonicum.* However, the chemistry of this group of resins has not been systematically studied.

Old World Elemis

Manila elemi, from *Canarium luzonicum,* is one of the best known and is the single largest source of the world's supply of elemi. Among the many species of *Canarium* in the Philippines, *C. luzonicum* is the only one that has been exploited commercially. However, Coppen (1995) mentioned that *C. indicum* (syn. *C. commune*) produces resin that has been used locally. Resin from *Canarium* comprises primarily volatile sesquiterpenes and nonvolatile triterpenes. When fresh, the terpenoid resin is soft and granular, resembling crystallized honey, but later it hardens. Although Manila elemi may be collected throughout the year in Philippine rain forests, in areas that have a dry season, elemi is collected mostly during the wet season when there is greater resin flow. Great quantities of resin exude from the bark of large trees over a period of several months (Brown 1921).

Local people in the Philippines have used *Canarium luzonicum* resin externally as a rubefacient and antirheumatic; it is also used in plasters that are heated and applied to the chest to stop severe coughing. The elemi is often called by its Spanish name, *brea blanca* (white pitch). It has a pleasant resinous odor but burns with a smoky flame when used as incense or for torches. However, the torches give a very bright light and burn a long time (Brown 1921). Manila elemi had been exported from the Philippines in large quantities, particularly for preparing varnishes. The resin was not made directly into varnish but added to various spirit varnishes, making them less brittle and more plastic. Mixing elemi with copal (Chapter 9), used for very hard finishes, made melting the copal much easier as well as giving a paler and more brilliant varnish. Today it is used principally for fragrance purposes, with France the largest single market (Coppen 1995).

Elemi from *Canarium schweinfurthii* (called incense tree in West Africa) resembles Manila elemi but contains a smaller proportion of volatile components. The fragrant resin, which exudes from the bark when the trunk is wounded, is first colorless and then turns sulfur yellow. It burns with the odor of frankincense and has been used locally to fumigate dwellings; it was also mixed with oil to anoint the body. The genus *Canarium* is well represented in Southeast Asia, and many species yield resins that have not been studied in detail. Gianno (1990) listed *C. littorale, C. patentinervium,* and *C. pilosum* as used by the Semelai in peninsular Malaysia. In the Solomon Islands, resin from *C. oleosum* is used on wounds and for itches.

Elemi is also collected from the pantropical genus *Dacryodes.* In the West Indies it comes from *D. excelsa* (syn. *D. hexandra*), which is widespread there in the mountains (Plate 17). Parsons et al. (1991) described the triterpenoid chemistry of resin from *D. normandii* from Gabon. The resin of five species, *D. costata, D. incurvata, D. laxa, D. rostrata,* and *D. rugosa,* has been used by the Semelai in peninsular Malaysia (Gianno 1990). Gianno also mentioned the important use of *Triomma malaccensis,* which was traded overseas even during the protohistoric period in Southeast Asia.

New World Elemis

Elemi exported from Brazil, called hard elemi or Brazilian elemi, is obtained from species of *Protium,* particularly *P. icicariba* and *P. heptaphyllum.* Many of the 100 or more species of *Protium* in the New World tropics occur in Brazil and yield resin; in Amazonia the resin is often locally referred to as *breu.* Other names for resin from *Protium* include *almécega* and *jauaricica.* The most important species commercially is *P. icicariba.* Exudation of resin is accelerated by making incisions in the bark of this tree. The resin is very fragrant and used locally for incense, caulking, and in the manufacture of varnishes (Mors and Rizzini 1966). It is being evaluated for its anti-inflammatory properties (Siani et al. 1999a, b). The other important species, *P. heptaphyllum*, is a dominant tree in some semideciduous forests, where it is called the incense tree (Plate 41). Mors and Rizzini reported that the air becomes fragrant from resin dropping from wounds in the bark and forming masses on the ground. Tacamahaca resin, or elemi from Colombia and Venezuela, also comes from *P. heptaphyllum*.

The major volatile components in leaves of *Protium heptaphyllum* are the monoterpene terpinolene and the sesquiterpenes β-elemene and β-caryophyllene whereas the primary component of fresh stem resin is terpinolene (Zoghbi et al. 1995). On the other hand, Siani et al. (1999) found the primary volatile components of fresh stem resin to be *para*-cymene and α- and γ-terpinene. These results again emphasize the high degree of variability of resin composition in some taxa. Leaf resin from *P. heptaphyllum* has been evaluated for its protective action against *Schistosoma* cercariae (Frischkorn and Frischkorn 1978).

The Tembé and Ka'apor Indians of eastern Amazonia use *breu* as a fire starter, incense, medicine, and for caulking wooden boats. These Indians also rely on the harvest of *breu* as one of their main economic products (Balée 1994). Although the Ka'apor people sometimes wound other burseraceous trees to obtain resin for medicinal or other specialized uses, the Tembé people rely on abundant lumps of naturally exuded *breu* rather than tapping the trees. For commercial purposes, Tembé Indians recognize two major types: *breu branco*, which accumulates in rounded white lumps, and *breu sarara*, which has flatter lumps that often blacken. *Protium pallidum* and *P. trifoliatum* are considered to produce *breu branco* whereas *P. giganteum*, *P. glabrescens*, *P. morii*, *P. polybotryum*, and *Tetragastris panamensis* (Burseraceae) produce various types of *breu sarara*.

Using Indian collection practices, Plowden et al. (2002) studied resin flow and accumulation that occurred naturally. They found that resin flow was stimulated by curculionid weevils *(Sternocoelus),* which chew into the nutrient-rich inner bark. Exudation may be a defensive response of constitutive resin in the phloem or partially induced by the weevils. Unlike bark beetles in conifers, the weevil larvae burrow into the soft resin lumps that form on the outside of the bark, where they shape a protective chamber for their development. This is an interesting adaptation to resin, which usually defends against insects that bore into the plant (Chapter 5). Although resin accumulation varied considerably according to species, and with more resin from larger trees in wetter sites, resin lump size correlated most closely with the number and size of bore holes made by the weevil larvae. Because harvesting the resin often removes lumps containing live larvae, the collection of *breu* may reduce the local population of these weevils, which may affect the sustainability of the *breu* harvest. Thus Plowden and coworkers stressed that the weevil could be considered a "good pest" and wondered how general this relationship might for other elemi species.

Phytochemists are seeking to identify essential oils from *breu* that are marketable to the perfume industry (Zoghbi et al. 1993–95). Plowden et al. (2002) emphasized that if such endeavors succeed, harvesters could earn more money, but the pressure to overharvest would increase. Their study predicted that if resin with enclosed larvae is harvested, it would take at least 4 years for resin levels to build up to initial levels. Therefore, they suggested experimenting with several management practices, including seeding trees with insects, a practice that works well with lac insects responsible for producing natural shellac (K. Sharma et al. 1999). These are important considerations for the success of *breu* as a cash crop and for the sustainability of the forest system (Chapter 11).

Weevil activity resulting in resin exudation has also been reported from other species of *Protium* in Colombian and Peruvian Amazonia (C. Plowden, pers. comm.). Furthermore, resin exudation is sufficiently abundant in these areas, as in Brazilian Amazonia, that the local people harvest the lumps for commercial use rather than tap the trees. Although this evidence is only anecdotal, Plowden hypothesizes that where the weevil is present, harvesting of *Protium* resin depends on insect activity, whereas where the weevil is absent, tapping occurs. On the other hand, great demand for the resin may be another factor, as indicated by *P. copal*, which has a long history of heavy use of its resin and associated tapping procedures to maximize the harvest.

Protium copal produces another important New World elemi. It is a small, subcanopy tree, spanning environments from moist evergreen primary forest to deciduous second-growth forest in southern Mexico, Guatemala, and Belize. Resin from *P. copal* has been called copal, which has become a generic Mayan term in this region for resins used for incense and medicines, including those from *Bursera, Liquidambar.* and *Pinus*. The Maya borrowed the term copal from the Aztec Nahuatl *copalli* but often use *pom* or *copal pom* to refer specifically to burseraceous resin, and in lowland regions specifically to *P. copal* (Chapter 6). Although this usage of copal is important to understand regionally, it is in conflict with well-entrenched commercial usage referring to the many leguminous and araucarian resins that produce hard varnishes (Chapter 9).

Resin from *Protium copal* was used as incense by the Maya prior to the arrival of the Spanish in the Americas (Chapter 6) and was also important in the indigenous materia medica for treating respiratory and various kinds of dental problems and for its vermifuge and antiparasitic properties. Nonetheless, it is best known today for its use as incense in traditional Mayan and Catholic religious ceremonies. The incense has long been used in churches, holy processions, planting ceremonies, and healing rituals in areas where *P. copal* occurs. Tripplett (1999) suggested that *copal pom* to the Maya represents more than just incense. Burning incense was perceived as part of a larger pan-Mayan belief in the transformative actions and processes that occur during ceremonies (Chapter 6).

Protium copal is collected by local people in Guatemala by tapping the bark of the tree. The resin is well known in southern Guatemala, where Mayan traditions are still strong. A number of incisions are made in the trunk, and the resin is collected after about a week. The greatest quantities are collected during the wet season, although some resin is produced during the dry season. The collectors tap only trees with trunks greater than 15 cm in diameter and often at least 10 years old. They have learned to assess the trees' health, letting them rest. The soft, sticky resin is taken from the tree, compressed, and shaped for sale at local markets. About 11 different types of *pom* may be found in southern Guatemalan markets. In some areas, sale of the resin provides supplemental income for farmers (Tripplett 1999).

The resin of *Protium copal* is less familiar in northern than southern Guatemala. Neels (2000), however, expressed the possibility of developing a new, viable industry in resin harvesting in a community forestry concession in the Maya Biosphere Reserve, in the northern part of the country. Previous studies of resin yield in tropical pines and dipterocarps, for example, demonstrate that flow can be stimulated with chemicals (e.g., Messer 1990, Orallo and Veracion 1984). Neels's approach was to assess factors that increase yield without chemicals. Similar to other situations (e.g., Ella and Tongacan 1992), tree size (diameter breast high) and crown condition had the greatest effect on yield. It was assumed that larger tree size is related to a greater number of secretory ducts. Neels's statistical data also tended to corroborate Tripplett's (1999) more qualitative analyses regarding tree size in southern Guatemala. Increased crown size was thought to provide increased biosynthetic capacity through larger photosynthetic area. However, Neels was unable to assess environmental variation quantitatively at different sites. In contrast to Tripplett, Neels did not find a pattern of production associated with wet and dry seasons. Tree-to-tree variation was sufficiently great (which is typical of tropical trees) that she thought the only relatively accurate prediction would be on the basis of yield per hectare. From an ecological perspective, it is interesting that she found trigonid bees removed as much as 25% of the resin during the second half of the tapping season. Also, peccaries removed resin from the cuts for unknown reasons, but perhaps for similar reasons that coatis scrape resin from *Trattinnickia aspera*, another burseraceous tree (Chapter 5).

Bastard elemi and yellow elemi in Brazil are not derived from burseraceous trees but from *Calophyllum brasiliense* and *Symphonia globulifera* (Clusiaceae). Other Clusiaceae, especially *Garcinia*, produce gamboge, a type of dammar (Chapter 9). Various Indian groups in Brazil and Colombia use the resin from various species of *Clusia* to treat different ailments. Resinous floral parts of *Clusia insignis* are used to relieve toothache by natives of the uppermost Río Negro, Colombia, and resin of the large bush, *C. planchoniana*, is applied to gums to treat toothache by Kuripako Indians (Schultes and Raffauf 1990). The resinous bark of *C. opaca* is sun-dried and pulverized by Taiwano Indians. Mixed with the oil of *Oenocarpus bataua* (syn. *Jessenia bataua*), a palm, it is applied to sprains and aching joints. The Makú Indians in Brazil paint resin of *C. opaca* over wounds to hasten healing.

At one time, elemis from Mexico were second in importance commercially to those from the Philippines (Howes 1949). The Mexican resin was not from Burseraceae or Clusiaceae but from species of *Amyris* (Rutaceae). Yucatán elemi was probably derived primarily from *A. plumieri* and *A. elemifera*. It is a hard, yellow, translucent, highly aromatic resin that has been used in place of Canada balsam as a mounting medium in microscopic work. *Amyris plumieri* has also been used in lacquers. Mexican elemis from several other species of *Amyris* have characteristics similar to those of Yucatán elemi. Distilled resin from the wood of *A. balsamifera*, called West Indian sandalwood, occurs in northern South America, Central America, and the West Indies. It also is called candlewood in Cuba. The volatile constituents are primarily hydrocarbon and oxygenated sesquiterpenes, used as incense or as a blender and fixative for soap fragrances (*Kirk-Othmer Encyclopedia of Chemical Technology* 2001).

Other Important Balsams

Resins from three other genera of Burseraceae, *Boswellia, Commiphora,* and *Bursera,* constituting the subfamily Bursereae, are significant as fragrant balsams used in perfumery and particularly as incense. The burning of incense could be considered a universal cultural trait because it has had religious significance throughout history, from the first civilizations through the present (Atchley and Cuthbert 1909, Groom 1981; Chapter 6). The most familiar incenses, the classical incenses, are frankincense *(Boswellia)* and myrrh *(Commiphora)* because of the frequency of their mention in the Bible and the classics. In fact, the words frankincense and myrrh are often used synonymously for the word incense.

The Egyptians, Persians, Sumerians, and Assyrians used incense for mortuary rituals and offerings to their deities before the rise of Greek and Roman civilizations (Atchley and Cuthbert 1909). Some archeologists think that at least myrrh was reaching Egypt from southern Arabia or adjacent areas by 3000 B.C., but organized camel transport probably was not available until about 1000 B.C. The Greeks began to use incense of frankincense and myrrh as a substitute for human and animal sacrifice in the sixth century B.C., and it was adopted by the Romans from at least the beginning of the second century B.C. Incense was also recognized as a purifier and fumigant with which bad smells could be disguised, and it was used for other purposes as well. Frankincense and myrrh were so highly prized during the Greek and Roman periods that they were traded thousands of kilometers from their source in southern Arabia and adjacent areas (Chapter 6). In addition to the classical incenses, resins from *Bursera* and *Protium* (Burseraceae) were commonly used as incense in the New World. Moreover, resins from various species are known for a variety of medical benefits. *Boswellia* and *Commiphora,* especially, have long been used in traditional medicine and are being analyzed in detail, chemically and clinically. Resins of *Bursera* have been used for a variety of purposes throughout Mexico.

Boswellia

Frankincense

Despite the long history of the use of frankincense, taxonomic knowledge of the trees producing the resin remains inadequate (Thulin and Warfa 1987). Chemical knowledge of the constituents and how the chemicals contribute to fragrances and flavor likewise present problems. In fact, Tucker (1986) lamented, "the superficial researcher may encounter many errors in the past literature that have been quoted and requoted, creating vast confusion."

Frankincense (archaic for "choice incense") also is called gum or oil of olibanum. The term olibanum, anglicized from the Arabic *al lubán*, is particularly used in the chemical literature (e.g., *Kirk-Othmer Encyclopedia of Chemical Technology*). The resin is obtained from certain species of *Boswellia* (Figure 8-4; Burseraceae), small trees or shrubs that occur in the arid regions of northeastern Africa and southern Arabia (Chapter 2). Hepper (1969) revised *Boswellia* in southern Arabia and Somalia. He recognized a single Arabian species, *B. sacra*, indicating that *B. carteri* differs mainly in habit, a view that Mona (1978) supported. Small trees of *B. sacra* are often barely taller than the camels that nibble on the branches covered by small succulent leaves (Plate 42). Hepper reported that the best frankincense occurs on trees growing in the narrow strip of Oman's desert plateau that borders the mountains of the former South Yemen, where the plants have ideal soil conditions and tropical sun accompanied by heavy dew from the monsoons. *Boswellia sacra* also occurs in Somalia and on the island of Socotra (Figure 6-11).

Hepper (1969) recognized three other *Boswellia* species in Somalia: *B. carteri*, *B. frereana*, and *B. bhau-dajiana*, the latter poorly known. Tucker (1986) followed Hepper's designations, but Thulin and Warfa (1987) dissented. Thulin and Warfa restudied the taxonomy of frankincense-producing *Boswellia* species in southern Arabia and Somalia with particular interest in possible domestication of the trees in Somalia for resin production. They concluded that there are only two frankincense-producing *Boswellia* species in these areas, *B. sacra* and *B. frereana*, and provided a complete analysis of the issue as it stood in 1987. The Arabic name for *B. sacra* includes numerous variants of *mogar* whereas the Somali name is *mohar*, and the most common name is *mohar madow*. *Boswellia frereana* is distinguished morphologically from *B. sacra*; its common Somali name is *yagar*. Thulin and Warfa indicated that *B. papyrifera* is the closest relative of *B. sacra*; it is called the elephant tree because elephants used it in Ethiopia, Sudan, and East Africa. Tucker also thought that *B. papyrifera* provided the principal source of frankincense in antiquity whereas *B. sacra* was the primary provider during classical times. *Boswellia serrata* in India, and *B. ameero* and *B. socotrana* from the island of Socotra have provided additional sources of frankincense (Rees 1995).

Figure 8-4. Flowering branch of *Boswellia* (see also Plate 42), a frankincense-producing tree, showing resin exuding from the bark.

In ancient times, frankincense could only be harvested by specially appointed families (not true for myrrh). Because the resin was considered to be divine, gatherers were restricted from impure acts (e.g., sex) during harvests (as stated by Pliny the Elder; Groom 1981). Other than that, the method for obtaining frankincense apparently has not changed much in the past several thousand years (Figure 8-5). By making incisions in the bark of the trunk or branches, where the secretory tissue occurs (Chapter 3), the resin exudes and is collected when it hardens sufficiently. Or instead of making incisions, portions of the bark may simply be scraped away with a special tool that has

Figure 8-5. Arabs collecting incense, from André Thevet (1575), *La Cosmographie Universelle*.

both a sharp end and a blunt end (Howes 1949). The sharp end is used for decorticating, and the blunt end for help in harvesting the resin. Pastoral families do most of the harvesting, repeatedly scraping the bark of trees over several months during the dry season to collect the purest grades (figure on Part III page). The first and second scrapings are considered low quality with only the third giving real or pure frankincense. This change in resin may be the result of compounds that are induced by repeated wounding. Two kinds of frankincense are recognized: *zakana* or male frankincense, which has circular, deep yellow or red tears, and *kundura unsa* or female frankincense, pale reddish white and translucent (Rees 1995). However, some *Boswellia* resins have a milky white appearance that is likened by pastoral Somalis to camel's milk. This cloudiness could result from different amounts of polysaccharides in the gum resin. As the resin is easily damaged by rain, it is only collected during the dry season between monsoons. On average, one tree can yield about 1.5 kg of resin each year. After tapping 5–6 years, the trees are rested (Grieve 1959, Hepper 1969).

The odor of frankincense has been described as fresh-balsamic, with a fruity overtone (Tucker 1986). These characteristics make it valuable to perfumers, and it is a minor component blended into modern perfumes and aromatherapy oils (Arctander 1960). α-Pinene is often the predominant monoterpene, and several γ-butyrolactones with strong coumarinic odors are key contributors to the characteristic fragrance of frankincense (*Kirk-Othmer Encyclopedia of Chemical Technology* 2001). The chemistry underlying the fragrance of frankincense, however, is complex because it is a mixture of numerous volatile compounds, with more than 20 monoterpenes and 28 sesquiterpenes (Rees 1995). Because polysaccharides from the walls of the epithelial cells of the secretory structures are incorporated with the terpenoids (Chapter 3), frankincense is considered a gum resin. The gum portion of *Boswellia sacra* resin comprises two polysaccharides: one with galactose and arabinose, the other with galactose and galacturonic acid. Tucker summarized the fairly sizable chemical literature to 1986 but warned that the literature is very confusing with regard to the botanical source of the resins analyzed. Thus variation noted in chemical analyses may be the result of either inherent variation in composition characteristic of most resin-producing species or misidentification of the plant source of the resin sampled.

Even now, frankincense is collected and exported from the northern coast of the Horn of Africa (Figure 6-11). Frankincense is the third most important source of Somalian foreign exports, after livestock and bananas, with 10,000 families depending on it for a livelihood (Farrah 1994). Of several species of *Boswellia* from which resin is collected in Somalia, *B. frereana* provides a superior source of frankincense. The complex mixture of mono- and sesquiterpene hydrocarbons constituting the resin of this species was isolated by GC-MS (Strappaghetti et al. 1982). The monoterpene *para*-cymene was the most abundant compound; other significant compounds included α-pinene, sabinene, myrcene, limonene, α-terpinene, α-cubebene, and cembrene. Both lupane and dammarane-type triterpenoids were reported (Fattorusso et al. 1985). Exemplary of the problems just discussed for *B. sacra*, however, the *Kirk-Othmer Encyclopedia of Chemical Technology* (2001) reported a composition of *B. frereana* dominated by ocyl acetate (50%) with only 5% α-pinene and no *para*-cymene.

Although frankincense is still burned in religious ceremonies, the incense used today in Catholic churches is a mixture of balsams (Grieve 1959): about 66% frankincense, 27% benzoin *(Styrax benzoin)*, and 7% storax *(Liquidambar orientalis)*. This combination is used for liturgical purposes because the odor given off during burning is distinctive, different from the pure frankincense used in the secular perfume industry. The smoke of this mixture is also visible without excessive clouding and thus is ideal for ceremonial purposes.

Medicinals

As mentioned, *Boswellia serrata* is used as Indian frankincense. However, the resin, *salai guggal,* is also well known for its medicinal properties. It is used in the traditional Ayurvedic system of medicine to alleviate inflammatory diseases such as rheumatoid arthritis and gout. Detailed pharmacological studies established that alcoholic extracts of the resin display marked anti-inflammatory activity in mice and rats (Singh and Atal 1986). Mono-, di-, and triterpenes have been reported, but five triterpenoid boswellic acids (ursane skeleton, Figure 1-5; Pardhy and Battacharyya 1978) were demonstrated to be the biologically active principle (Mahajan et al. 1995).

The anti-inflammatory activity of boswellic acids and the anticarcinogenic effect of the alcoholic extract of resin from *Boswellia serrata* led Shao et al. (1998) to assume that boswellic acids might be the biologically active principle responsible for antitumor activity. They isolated four boswellic acids from the resin and found that these triterpenoid acids inhibited in vitro synthesis of DNA, RNA, and protein in human leukemia (HL-60) cells in a dose-dependent manner. G. Singh et al. (1993) proclaimed the anti-inflammatory, antiarthritic, and antipyretic activity of boswellic acids as a basis for calling them "drugs of the future."

A *Boswellia*-based drug called Sallaki is available in India. Systematic pharmacological studies of the boswellic acids by G. Singh and Atal (1986), Safayhi et al. (1991), and Ammon et al. (1991, 1996, 1997) in various in vitro and in vivo assays demonstrated the pivotal role of these acids in the anti-inflammatory activity of Sallaki. One major triterpenoid constituent in *salai guggal,* the resin from *B. serrata,* and other triterpenoids of plant origin had anticomplementary activity (Knaus and Wagner 1996). Pathologically prolonged and sustained activation of the complement system is indicated in a variety of inflammatory afflictions such as rheumatoid arthritis and systemic lupus.

One of the advantages of using *salai guggal* for inflammatory disease is the absence of significant toxic and other side effects commonly found in nonsteroid antirheumatic or corticosteroid treatments. Boswellic acids in *salai guggal* have been shown to possess potent specific inhibitory action in the synthesis of leukotrienes, which in various inflammatory diseases are associated with or thought to be responsible for maintenance of chronic infection. Among the diseases with increased leukotriene formation are chronic polyarthritis, ulcerative colitis, Crohn's disease, and certain brain tumors. Furthermore, the immunomodulatory activity of boswellic acids from *Boswellia serrata* has been shown to be neither cytotoxic nor to cause the immunosuppression produced by most drugs used for the treatment of chronic inflammatory disease (M. Sharma et al. 1996).

Commiphora

Myrrh

Myrrh is collected from several species of the large and very diverse genus *Commiphora* (Chapter 2). *Commiphora* occurs in shrubby, thicket vegetation in southern Arabia (e.g., Yemen and Oman) and northeastern Africa, primarily Somalia, Ethiopia, and the Sudan as well as the island of Socotra (Figure 6-11), but can grow as far east as India. The medical use of myrrh is somewhat similar to that of frankincense because the resins have some of the same terpenoids. Ancient physicians apparently recognized this relationship because the resins were prescribed interchangeably in early Sanskrit texts.

One of the most common species collected for myrrh is *Commiphora myrrha* (syn. *C. molmol*); the resin is also called oil of myrrh, or stacte. *Commiphora erythraea,* the perfumed bdellium or scented myrrh of antiquity (sometimes called opopanax or bisabol myrrh), was noted for its long-lasting scent (Majno 1975, Groom 1981). *Commiphora* trees contain numerous secretory canals in the bark, which may form large interconnected cavities from which resin flows freely upon any wounding. As in *Boswellia,* cell wall polysaccharides become mixed with the terpenoids, producing a gum resin, which hardens into irregularly rounded teardrops or lumps. This hardened resin is dark and bitter, unlike frankincense, which is pale and sweet. It was known by the Arabs as *murr,* meaning bitter. Myrrh never enjoyed the popularity of frankincense throughout the Roman Empire, but the price of myrrh was much higher than frankincense; hence, myrrh became a status symbol. When cremation replaced interment, incense served to mask the odor of burning flesh. Pliny the Elder referred to Arabia's good fortune resulting not only from incense being used to propitiate the gods but "as a luxury of mankind even in the hours of death" (Groom 1981). For example, Nero apparently lavished the equivalent of a year's Arabian production of myrrh on the funeral of his consort, Poppaea.

Myrrh was burned in temples and perfumed the royal mummies of Egypt; when Tutankhamen's tomb was opened in 1922, the aroma of myrrh was still present. In fact, myrrh was used for various purposes in this tomb: mummification, varnish and cement, and personal ornaments made from the resin (Lucas and Harris 1962). Myrrh was used by essentially all the ancient peoples of Asia Minor to anoint and embalm the dead, especially by the celebrated morticians of ancient Egypt (Klein-Rebour 1964). In fact, the word embalm is derived from the Latin *in balsamum,* meaning to preserve in balsam. In the Hippocratic books, myrrh is the favorite resin, prescribed 54 times (Majno 1975). Myrrh was also a main ingredient in the sacred Jewish anointing oils, as specified in the Old Testament. During his crucifixion, Jesus was offered "wine mingled with myrrh" as an anesthetic "but he received it not" (Mark 15.23), but myrrh was used to embalm his body (John 19.39). Hence, frankincense has been associated with Christ's divinity, but myrrh, with his persecution and death (D. Roberts 1998).

Commiphora africana (Figure 2-12) is another source of myrrh from Ethiopia and the Sudan. The resin hardens soon after it exudes from the bark and becomes transparent and wax-like, resembling a pearl. In Tabaré, Niger, and Uguago, Nigeria, the resin is melted with butter, then applied as perfume. In West Africa, the resin is used to repel termites as well as a folk remedy for numerous ailments, especially cancer (Duke 1983).

Myrrh is included in the formulations of many perfumes by leading manufacturers and, like frankincense, is widely used in aromatherapy (Rees 1995). Tucker (1986) described the odor of myrrh as "warm-balsamic, sweet and somewhat spicy—aromatic and pungent when fresh." It is used in perfumes for a spicy base with an oriental character, a woody base, forest notes, pine fragrances, etc. It blends well with geranium, musk, patchouli, spices, and heavier floral bases (Arctander 1960). In flavoring, myrrh produces a biting-burning, somewhat acrid-aromatic taste and is used in mouthwashes and toothpastes, blending well with clove, thyme, mint, and evergreen for gargles and mouth sprays.

Medicinals

Resins from numerous species of *Commiphora* are used medicinally, particularly in Indian, Arabian, and African cultures. Resins from six species that have received study for medical purposes are discussed here. The resin of *C. molmol* (syn. *C. myrrha*, according to some) is commonly used in Arabian medicine, for example, for treatment of some inflammatory conditions, as an antipyretic, antiseptic, stimulant, mouthwash, and for stomach problems, including cancer (Ageel et al. 1987, Tyler et al. 1988). The volatile fraction of the resin consists of commonly occurring monoterpenes such as pinene and limonene, sesquiterpenes such as cadinene, and phenolics such as cinnamaldehyde and eugenol as well as nonvolatile triterpenoids such as commiphoric acids and amyrins (Al-Harbi et al. 1997). The resin provides protection against the effects of several necrotizing agents, which is attributed to its free-radical-scavenging, thyroid-stimulating, and prostaglandin-reducing properties. *Commiphora molmol* resin was demonstrated to have cytotoxic and antitumor activities in mice equivalent to those of the standardly used but cytotoxic drug cyclophosphamide (Qureshi et al. 1993). The resin's nonmutagenic, antioxidative, and cytotoxic potential suggested to the researchers its appropriateness for use in cancer therapy.

Two sesquiterpenes (furanoeudesma-1,3-diene and curzarene) in the resin of *Commiphora molmol* have analgesic effects blocked by naloxone, indicating an interaction with brain opioid mechanisms (Dolara et al. 1996). This could explain the use of myrrh as a painkiller in ancient times, such as when *vinum murratum* (wine with myrrh) was offered to Jesus before his crucifixion. Dolara and coworkers thought that the use of myrrh may have been replaced later by opium derivatives. They also considered the possibility that furanoeudesma-1,3-diene could still have some medical applications, although its action on the central opioid pathways would probably limit its usefulness.

Commiphora opobalsamum (syn. *C. gileadensis*) is a stunted tree and the source of Mecca myrrh or balm of Gilead. The resin resembles classical myrrh *(C. myrrha)* and is considered one of the important plant medicinals of Saudi Arabia (Mossa et al. 1989). The tree, native to Yemen, was thought to have been introduced into Palestine by the queen of the Arabian kingdom of Saba, or Sheba, on her visit to Solomon (Moldenke and Moldenke 1952). The species was later cultivated around Jericho and became an emblem of Palestine, where cultivated plants were protected by guards. As symbols of victory over the Hebrews, Roman soldiers carried branches of the tree back to Rome. Today, the resin is a rare product, used for perfumes with an oriental character and as a special incense; it is also being analyzed for its medicinal properties. In fact, intravenous administration of an aqueous extract of the resin produced immediate dose-dependent hypotensive and bradycardiac (abnormally slow heartbeat) effects (Abdul-Ghani and Amin 1997).

Commiphora wightii (syn. *C. mukul*) is a drought- and salinity-resistant large shrub or small tree of arid areas in Arabia, Pakistan, India, and Bangladesh; the resin is commonly called bdellium or guggal in India, where it is widely used as incense in religious ceremonies, as a fixative in perfumery, and has been studied intensively for medicinal use. Sukh Dev (1988) reviewed the chemical studies of this resin, particularly those of the active guggulsterones. The World Wildlife Fund's (Switzerland) Biological Diversity Campaign of 1989 included the extract of the resin of *C. wightii* (called gugulipid or guggulipid) as one of nine new natural products for medicinal use. Gugulipid was cleared by the Drug Controller of India in 1986 for marketing and was being manufactured and marketed by 1987 in Bombay (Sukh Dev 1988). R. Singh et al. (1974) demonstrated hypolipidemic and antioxidant effects of the resin as an adjunct to the diet of patients with hypercholesterolemia.

Amma et al. (1978) analyzed the effects of the drug on the female reproductive organs of the rat. The resin has anti-inflammatory activity (Duwiejura et al. 1993), and a detailed summary of pharmacological studies confirm Ayurvedic claims of the effectiveness of gugulipid in management of atherosclerosis and hyperlipidemia (Dhawan 1996). Urizar et al. (2002) showed that guggulsterone is a highly effective antagonist of the farnesoid X receptor (FXR), a nuclear hormone receptor activated by bile acids. They proposed that inhibition of FXR activation is the basis of hypolipidemic activity of guggulsterones. P. Sharma and J. Sharma (1996) supported the use of gugulipid in treating other ailments, particularly arthritis and rheumatism as well as many vascular and neurological problems. They also found that the effects of crude extracts of resin are comparable with those of other known compounds that are antiamoebic, for example, against *Entamoeba histolytica.*

Since the trade potential of guggal seemed promising, improved tapping techniques were investigated; guggal had been collected from wild trees by local people. *Commiphora wightii* occurs in desert regions with little conventional agriculture and poor economic conditions for the people living there. Their crude and destructive methods of tapping, and their use of the trees for fuel, led to such depletion of natural populations that the species was listed as threatened. Secretory structures occur in the secondary phloem (bark), thus the traditional method of cutting into the wood was destructive and did not increase resin flow. By applying ethephon, guggal production was enhanced about 22 times over that obtained in controls (Bhatt et al. 1989). Bhatt and coworkers hoped that such improvements in harvesting technique would help make *C. wightii* a remunerative crop for semiarid and arid regions of India.

Compositional variation of guggal, and thus variation in amounts of compounds responsible for hypolipidemic activity, from different agroclimatic regions in India has been analyzed (M. Sharma 1994). Such studies demonstrate the importance of geographic variation in particular components and possible environmental influence on resin composition in the cultivation of plants for medical use as well as for quality control in marketing products.

A free-flowing fragrant resin exudes from injured bark of the small shrub *Commiphora tenuis,* one of the 52 species of the genus occurring in Ethiopia. This resin is widely used by Somali nomads in southeastern Ethiopia for treatment of animal wounds, particularly those of camels. Asres et al. (1998), using GC-MS, reported 42 mono- and sesquiterpenes. Monoterpenes constituted 97% of the volatile fraction, however, with 61% of the resin consisting of α-pinene. Other monoterpenes commonly present were β-pinene, limonene, and β-myrcene (Figure 1-3), and 3-carene, sabinene, and α-thujene. α-Santalene, α-bisabolene, and furanodiene were the most common among the small quantity of sesquiterpenes. The primary triterpene was oleanolic acid acetate.

The major component, T-cadinol, in the resin of scented myrrh *(Commiphora guidottii)* from Ethiopia and Somalia has smooth-muscle-relaxing properties in rat aorta (Claeson et al. 1991). Minor sesquiterpene components with cadinane, guaiane, oplopane, and eudesmane skeletons were effective but less potent than T-cadinol in relaxing smooth muscle (Andersson et al. 1997).

Commiphora merkeri from eastern and southern Africa has been used in folk medicine to treat infections. A triterpene with an oleanane skeleton (Figure 1-5) from the roots of *C. merkeri* possesses anti-inflammatory and analgesic activity (Fourie and Snyckers 1989).

Commiphora erythraea, a close relative of *C. myrrha,* contains furano-sesquiterpenoids that are toxic to the larvae of the African ear tick *Thipicephalus appendiculatus* (Maradufu 1982). Since three tick taxa are of medical or economic significance to humans and domestic animals, an extract of *C. erythraea* was tested on the American dog tick, lone star tick, and deer tick (Carroll et al. 1989). The resin extract had larvicidal and repellent activity against the American dog and lone star ticks; adult deer ticks were also repelled. The repellent qualities, however, were less than those of permethrin, a synthetic pyrethrin-type compound proven effective against ticks.

Bursera

Species of *Bursera* are prominent in tropical dry ecosystems in Mexico (Chapter 2) and the genus is a New World relative of the Old World *Boswellia* and *Commiphora.* The chemistry of *Bursera* is dominated by mono- and sesquiterpenes (Dominguez et al. 1973) but di- and triterpenes also occur. The genus exhibits great variation in the number of these terpenes and their relative concentrations (Chapter 5). For example, some species produce one predominant compound (e.g., β-phellandrene or β-myrcene) whereas most produce a mixture of as many as 25 compounds, including mono-, sesqui-, and diterpenes. Some resins also contain small amounts of phenylpropanoids (McDoniel and Cole 1972, Becerra et al. 2001). In Mexico and Central America, resin from some *Bursera* species are used for incense and various other purposes. In Mexico, the fragrant resin from species such as *B. bipinnata, B. excelsa, B. jorullensis, B. microphylla,* and *B. penicillata* has been used as incense (Martínez 1959, García Ruiz 1981). Stross (1994–96) thought that *B. bipinnata* was the source of the prototypical *copalli* resin used throughout Mesoamerica for incense in religious ceremonies rather than *Protium copal* (Chapter 6). In some cases, splinters of resinous wood are used for incense rather than the resin itself. *Bursera bipinnata* is often accessible because it is grown as a fencerow tree. The resin has been used for preserving meat, gluing, binding pigment, chewing for dental problems, plugging dental caries, and for other medicinal uses (Breedlove and Laughlin 1993).

Trade of resin from some *Bursera* species in Mexico was once significant, with resin used commercially for varnish, glues, and for treating various ailments. Volatile terpenoids in *B. aloexylon* were used in cosmetics, and resin from *B. microphylla* has immunological stimulation similar to that of myrrh. Bianchi et al. (1968, 1969) isolated antitumor agents from *B. microphylla;* McDoniel and Cole (1972) found antitumor activity in *B. schlechtendalii.* Some other commonly used species include *B. aptera, B. glabrifolia, B. odorata,* and *B. trijuga* (Martínez 1959, Morton 1981).

Although gumbo limbo *(Bursera simaruba)* is native to tropical America, it is a common tree widely cultivated elsewhere in the tropics (Plate 18). It is called *B. gummifera* in Florida and the West Indies. The resin has occasionally been used for incense (Stross 1994–96) but primarily as a traditional medicine in the Yucatán region of Mexico for various diseases (Morton 1981) as well as to alleviate the dermatitis caused by the resin of poison wood (*Metopium brownei*, Anacardiaceae). The active principle in this case is picropolygamain (Peraza-Sánchez et al. 1992), previously isolated from the resin of *Commiphora incisa.* The resin of *B. simaruba* includes a variety of triterpenes such as lupeol, epilupeol, epiglutinol, α- and β-amyrin, and a novel lupane type, lup-20(29)-en-3β,23-diol (Peraza-Sánchez et al. 1995).

Ursolic acid, ursonic acids, and α-amyrin occur in *Bursera delpechiana*, a tree cultivated in India for the resin, which is used in the soap, perfume, and cosmetic industries (Syamasundar et al. 1991).

CHAPTER 9

Varnish and Lacquer Resins

Varnish resins such as dammar, sandarac, mastic, acaroid, and hard copal contain fragrant volatile terpenoids, as in the balsams, but have been used for their nonvolatile components, which form hard finishes. Different resins are used for different kinds of varnish. Copal resins produce the hardest and most durable varnishes whereas dammar resins are noted for their more transparent, pale, spirit varnishes. Others arc used for specialty varnishes, such as sandarac for metal and paper, acaroid resins for metal and leather, and mastic to protect oil and watercolor paintings. Resins collected in the liquid state and applied directly to the surface to be varnished without use of any solvent are called lacquers or, sometimes, natural varnishes.

Dammars

Confusion in Terminology and Plant Sources

The term dammar is derived from the Malayan *damar*, a word that originally referred only to a torch made of resin, and then to any kind of resin (Mantell 1950). In the extensive trade of resins for varnish to Europe, resin from the Asian tropics was called dammar. Hence, the term appears in the literature spelled as damar or dammar. Dammar came into use as an English word when the first British traders visited Malaya because the resin was easily obtained in the ports the English frequented (Howes 1949). I generally use the English version, dammar, but also refer to local names in which *damar* is used.

Although members of the Dipterocarpaceae are the primary source of dammar, resin from a number of other plants has been called dammar. This has caused confusion about the botanical origin of resins referred to as dammars in some areas as well as in the literature (Feller 1964). First, local people have commonly used *damar* for resin in general, but particularly that from trees belonging to the families Dipterocarpaceae and Burseraceae (Gianno 1990). Further confusion in terminology resulted from the naming of the genus *Dammara,* a name obviously derived from the plants' production of resin. Two species previously put in the genus *Dammara* are now considered to belong to the genera *Protium* (Burseraceae) and *Shorea* (Dipterocarpaceae) (Whitmore 1980a). Adding to this complexity, a number of species now assigned to the coniferous genus *Agathis* (Araucariaceae), which occur in mixed stands in forests in eastern, drier areas of the Malay Archipelago, were first assigned to the genus *Dammara. Agathis* resin is called *damar* locally whereas in commerce it is called copal.

The semantic confusion surrounding the botanical source of dammars persists in the medical literature. Jost et al. (1989), attempting to determine the plant source of the resin used in a surgical adhesive that causes an allergic eczema, concluded in apparent exasperation, "we must note the extreme confusion existing with these copals *[Agathis]* and damars, confusion which, in a way, reflects the complexity of the world of resins." One solution to this botanical puzzle would be to call dammars known to come from dipterocarps, diptodammars, but this suggestion has not been adopted (Mantell 1950).

Local Use and Export

The use of the dammar, of course, depends on the chemical composition of the resin. Dipterocarp resins contain volatile sesquiterpenes and nonvolatile triterpenes (Chapter 1). Sesquiterpenes predominate in wood oils (Chapter 7) but triterpenes are more significant in the dammars. Dammars from genera such as *Shorea* and *Hopea,* in the tribe Shoreae, largely contain compounds of the tetracyclic dammarane series, and the pentacyclic ursolic acids and corresponding aldehydes (Cheung and Yan 1972, Mills and White 1994). Bisset et al. (1971) studied the sesqui- and triterpenes of 35 species of *Shorea* (e.g., *S. javanica,* Plate 44), the largest genus of Dipterocarpaceae. They found that the association of sesquiterpene hydrocarbons (copaenes, β-caryophyllene, and β-elemene) with the triterpene shoreic acid distinguishes the genus *Shorea* within Dipterocarpaceae. They also reported that sesquiterpene alcohols such as spathulenol, and triterpenes such us β-amyrin, ursolic aldehyde, and dipterocarpol could be used for some subgeneric distinctions.

The age-old use of dammar by the Malays has been for illumination, as discussed for dipterocarp oleoresins in which triterpenoid components are dominated by volatile sesquiterpenes (Chapter 7). To make torches, powdered dammar was mixed with other materials such as bark chips and wood oil, then wrapped in palm leaves into a cylindrical or tapering form 0.6 m or so long. Some kinds of dammar torches burn better than others; in fact, many dark dammars, for which there was little commercial demand, are well suited for making torches. Other forms of illumination have generally superseded the use of these kinds of torches.

As a group, dammars used for varnishes are distinguished from copals by the free solubility of their triterpenoids in turpentine and other organic solvents. Despite their inferior hardness and durability, compared to copals, they have been especially valued in the manufacture of spirit varnishes because of their good solubility, pale colors, and transparency. Mills and White (1994) pointed out that the optical properties of such natural resin films are "almost optimum" and more attractive than those from synthetic resins. Dammar resin was introduced into Europe in 1829 and has been used extensively since 1844. Pale grades of the resin are made into white paper varnishes and white enamels as well as used for wallpaper and indoor decoration, and artwork.

Because of the use of dammar as a varnish for artwork, conservators have found the confusion over the botanical source of the resin particularly frustrating. They also have been interested in the changes that occur during aging of these varnishes. Therefore, Zumbühl et al. (1998) studied the light-induced aging process in dammars by graphite-assisted laser desorption-ionization mass spectrometry. Light-induced aging leads initially to polymerization of the triterpenoids to trimers; further aging leads to decrease in the average molecular weight of components.

The best forms of dammar to reach the overseas market for varnish have been known in the country of origin as *damar mata Kuching* and *damar pĕnak*. In the trade, several kinds of *damar mata Kuching* are recognized. The most esteemed is Batavian dammar, which occurs in large, very pale lumps and comes primarily from western Java and eastern Sumatra. The botanical source of the *damar mata Kuching* that is exported is often obscure and varies in different regions. Howes (1949) stated that the trees that yield most of this kind of dammar are species of *Hopea* (e.g., *H. dryobalanoides*,

H. sangal, and *H. intermedia*). However, *Shorea* is most important in other areas. For example, *S. assamica* subsp. *koodersii* yields the most in the Moluccas (eastern Indonesia), *S. javanica* and *S. retinodes* on Sumatra, and *S. bracteolata* and *S. hypochra* in Malaya (peninsular Malaysia). *Damar pěnak* is derived from a well-known and abundant dipterocarp, *Neobalanocarpus heimii*. The tree, a chief source of timber, is slow-growing and must attain an age of about 50 years before it reaches a size suitable for resin tapping. Before heavy timbering, there were trees more than a meter in girth and probably more than 1000 years old. The upper part of the bole, which is said to produce the best resin, is reached by using ladders.

Traditional methods of tapping, of wild or cultivated trees, involve removing wood from the trunk. Triangular cuts that become circular with age are arranged in vertical rows around the trunk, as shown for *Shorea javanica* in Sumatra (Plate 44). In the *kebun damar* (dammar gardens), in which the trees are grown with other crops, the first cuts are made when trees are about 20 years old. The cut is several centimeters wide at first but is enlarged with every tapping. For higher holes, the tapper climbs the tree, supported by a rattan belt and using the lower holes as footholds (Torquebiau 1984, 1987). Exuded resin is allowed to harden before it is collected; the stalactitic form is *damar pěnak*, and the globular form, often called cat's-eye because of its shape, *damar mata Kuching*. After collecting the resin, the wound is freshened to induce more flow. There is considerable variation in tree-to-tree yield, and resin production is reported to fall markedly when the tree is flowering and fruiting, only reaching previous levels a year later. Chemical stimulants to induce more resin flow have been investigated (Messer 1990). If tapping is carefully done, it apparently causes little damage to the tree.

Numerous other, less-valuable dammars are produced in Malaya (peninsular Malaysia), but Howes (1949) indicated that the botanical source can rarely be determined. Plants mentioned include *Shorea curtisii*, *S. guiso*, *S. lamellata*, *S. multiflora*, *S. ovalis*, *Hopea* spp., and *Canarium* spp. Thus the resin may be from both dipterocarpaceous and burseraceous trees. Furthermore, *damar kelulut* and *damar siput* are collected from insect habitations. *Damar kelulut* is the lining of the nest of meliponid bees that collect resin, often from several kinds of trees, and use it mixed with wax to reinforce the nest (Plate 29). This is a process similar to that used by temperate zone honeybees in farming propolis (Chapters 5 and 9). *Damar siput* occurs as hollow, shell-like exudations (*siput* in Malay means snail) on trunks of *Shorea* trees and is thought to result from larvae of a beetle that lives in the bark (Burkill 1966).

Dammar has been used locally by the Thai in much the same way as in Malaya. Exploited commercially to a certain degree, it is similar to the better grades of *damar mata Kuching* and *damar pěnak*. It is obtained from the dipterocarps *Neobalanocarpus heimii* and *Shorea maxima*, both large timber trees in the mountainous areas of southern Thailand. Other trees in which the resin is collected from natural exudations for local use are *S. obtusa, S. siamensis*, and *Hopea odorata*.

Dipterocarpaceae are well represented in the forests of India, Sri Lanka, and Myanmar (Burma). In some cases, trees yield dammar copiously, which has been collected for local use with only small amounts exported. In Myanmar, *Hopea odorata* and *Shorea obtusa* yield light-colored resins soluble in turpentine that are used for various purposes. One of the best known of the Indian resins is that from sal *(S. robusta)*, a tree that dominates many forests of central and northern India (Figure 9-1). Its distribution represents the northwestern limit of the family Dipterocarpaceae. The resin occurs as stalactites, is a pale creamy color, and somewhat aromatic. Large pieces of *S. robusta* resin in the Darjeeling district have been dug from beneath trees.

Indian white dammar or piney dammar is obtained from *Vateria indica*, a large, evergreen, dipterocarp tree in forests of western peninsular India. On the other hand, the so-called black dammar of India is a burseraceous resin, from *Canarium strictum* and other species of *Canarium*. The resins from these species of *Canarium* are apparently not as aromatic as those considered elemis (Chapter 8) and are used for varnish. *Canarium strictum* is a large deciduous tree that grows in evergreen forests up to 1700 m on the western coast of India. Its dammar is bright, shiny black when adhering to the tree. The common method of obtaining the dammar is to make vertical gashes around the base of the tree and build a fire against it to stimulate the flow of resin. In this manner, resin will exude for many months.

Dammar as a Source of Petroleum

Experimental data support the probable role of dammar resin in petroleum formation (Grantham and Douglas 1980; van Aarssen et al. 1990, 1992, 1994; S. Stout 1995). Resins from various dipterocarp trees were compared to Miocene resin-rich rocks as well as to Miocene oils from the Mahakam

Delta, Indonesia. Stout, reviewing and adding to the experimental data, found that the resins undergo relatively few chemical changes after their exudation-polymerization and early diagenesis (physical and chemical change occurring in sediments during or after the period of deposition until the time of consolidation). The resins contain varying proportions of solvent-extractable primary sesquiterpene hydrocarbons as well as polymerized polycadin-

Figure 9-1. *Shorea robusta,* copious dammar-producing tree of Southeast Asian and Indian forests.

enes (Figure 4-1). The most abundant primary hydrocarbons in both fresh and fossil resins were monomeric and dimeric sesquiterpenoids (cadalenes) with various degrees of saturation. Functionalized triterpenoids, however, with oleanane, ursane, and dammarane skeletons (Figure 1-5) dominated the nonpolymeric solvent extracts of both fresh and fossil resins. With thermal stress, the polycadinene polymeric fraction dissociates, yielding monomeric and dimeric cadinenes, the latter thought to cyclize internally, forming pentacyclic triterpenoids, also known as bicadinanes. The distribution of bicadinane isomers from dipterocarp resins formed under laboratory conditions has been consistent for both fresh resins and immature fossil ones, and demonstrates that the formation of different isomers is temperature dependent and varies with increasing maturity of the resin.

S. Stout (1995) thought that particular functionalized triterpenoids (bicadinanes) as well as the presence of primary sesquiterpene hydrocarbons in both fresh dipterocarp and fossil resins strongly support the role of dipterocarp resin in petroleum formation in numerous parts of Southeast Asia. These compounds occur in petroleum from South Sumatra; Irian Jaya (or West New Guinea); the Indonesian basins of Tarakan, an island in the eastern Celebes Sea, and Arjuna of Central Java; South China Sea; Malay Basin; lower Assam Basin, India; Surma Basin, Bangladesh; central Myanmar Basin; and southeastern Luzon Basin, Philippines—areas in which Dipterocarpaceae are dominant today and during a time in which they have been recorded in oil source strata.

Gamboge

In addition to Dipterocarpaceae and Burseraceae, resin from *Garcinia* (Clusiaceae) is used as a varnish and is classified as a type of dammar. *Garcinia*, a large genus of tropical Asia and southern Africa, produces a gum resin in the bark. Resin has been collected from several species in Thailand, Cambodia, and other parts of Southeast Asia for varnish, a coloring agent, and at one time, medicine, because of antibacterial properties (Duke 1985). *Garcinia hanburyi* is one of the best-known resin-producing species; its resin is most often called gamboge although it is also called gutta Cambodia and gutta gamba (Wren 1988). Gamboge consists mainly of α- and β-garcinolic along with gambogic and neogambogic acids. It is 20–25% gum, similar to acacia gum. The yellow resin is collected from spiral incisions in the bark; after half-set, it is pressed into cakes that are sold. *Rong* is the Thai word for the resin, which is used to make a golden yellow ink in Thailand for locally made books of black paper. Thus McNair (1930) classified gamboge as a pigment resin. It has also been used in China since the 13th century for making a golden spirit varnish known as pear ground for coating metal, the lacquer built up in layers with each layer sprinkled with gold dust. Malays make a pale varnish they call *getah* from the resin of *G. merguensis* (Burkill 1966).

L. Perry (1980) listed 21 species of *Garcinia* that have been used medicinally. Resin was once used as a purgative, but excessive doses can produce fatal gastroenteritis. Indian gamboge is derived from *G. morella*, which is exported to China for external medicinal use. *Garcinia cowa*, *G. dulcis*, *G. griffithii*, *G. nervosa*, and *G. xanthochymus* also have been used for their bark resins (Burkill 1966).

Calophyllum is another genus of Clusiaceae that produces a gamboge-type resin. *Calophyllum inophyllum*, a strand tree ranging from India to eastern Africa and Polynesia, has resin that occurs in the bark or in seeds. In Indochina, powdered resin is used as a tonic for ulcers and infected wounds. *Calophyllum chiapense* resin is used for wounds in Mexico. Resin from the bark of *C. lowei* and *C. palustre* is used for itches and other skin afflictions (Burkill 1966). In peninsular Malaysia and Indonesia, resin from the bark of another clusiaceous tree, *Cratoxylum formosum* (syn. *C. cochinchinense*), is used for itches as well.

Sandarac

Sandarac is a resin obtained from conifers of the family Cupressaceae s.s. African sandarac comes from *Tetraclinis articulata* (syn. *Callitris quadrivalvis;* Figure 9-2) whereas Australian sandarac comes from various species of *Callitris*. *Tetraclinis* is a genus of a single species occurring primarily in the mountains of Algeria and Morocco, where it is also an important timber tree that once grew in extensive forests. The tree can attain 13–17 m in height with a trunk more than a meter in girth. Its dark, fragrant wood has been popular for cabinets and other furniture for many centuries. The bark of the tree contains schizolysigenous ducts full of liquid resin. Resin is collected by local people either from natural exudations or incisions made in the trunk. The exudate gradually solidifies as small pale yellow to orange teardrops, which are picked off when very hard and dry. African sandarac has very little volatile terpene content, consisting of about 70% polymerizable communic acid (Figure 1-4) with sandaracopimaric acid and its 12-acetoxy derivative also major components. Some phenols (e.g., totarol) and other minor labdanoid diterpenoids are present. The resins of some species of *Juniperus*

Figure 9-2. The Mediterranean *Tetraclinis* (below) was once considered to belong to the primarily Australian genus *Callitris* (above). Similarities and differences in the cones and needles may be noted here in *C. preissii* (see also Plate 9) and *T. articulata*.

and *Cupressus* also contain high proportions of polycommunic acids and are thus somewhat similar chemically to African sandarac. In fact, *Juniperus* resins have sometimes been confused with sandarac in the past (Mills and White 1977).

African sandarac resin was used by Mediterranean peoples from ancient times into the Middle Ages for medicinal purposes. In pharmacy, its chief use has been to prepare pill varnishes and sometimes, in alcoholic solutions on cotton wool, as a temporary filling for teeth (Howes 1949). However, it has been used primarily for special varnishes on metal, leather, and paper. It produces a hard, white, spirit varnish that may be brittle if not mixed with other resins, and the gloss varies according to the solvent. The resin was exported largely from the port of Mogador (now Essaouira, Morocco), thus the name Mogador sandarac has often been applied to it.

Australian sandarac is obtained from various species of *Callitris*, a genus primarily restricted to Australia. *Callitris*, commonly known locally as cypress pine, is widely distributed and a prominent feature in many Australian ecosystems (Bowman and Harris 1995). Although *Callitris* ranges from the arid tropics to the relatively high rainfall areas, most species flourish in very drought-prone sites such as savanna woodlands, often dominating the vegetation. There is considerable disagreement regarding the taxonomy of several apparently hybridizing species, including the ones commonly used for resin, such as *C. preissii* and *C. verrucosa*. *Callitris* wood and foliage are rich in resin, contributing to the termite resistance of *C. glauca* and *C. intratropica* (Weissman and Dietrichs 1975). *Callitris* is reported to be resistant to some species of ants (Howes 1949).

Although Australian sandarac is collected from a variety of *Callitris* species, the main sources of the resin used as a substitute for African sandarac *(Tetraclinis)* are black cypress pine *(C. endlicheri,* syn. *C. calcarata)* in southeastern Australia; *C. preissii* (Plate 9 and Figure 9-2), *C. verrucosa,* and their possible hybrids, common in arid regions; and Murray River cypress pine *(C. glauca,* syn. *C. hugelii),* the most widely distributed species. Aborigines also used the resins of one of the few moist forest species, *C. intratropica,* to make glue and for contraceptives (Bowman and Harris 1995). The chemical and physical properties of *Callitris* resin resemble those of African sandarac because it consists largely of a form of polycommunic acid; however, *Callitris* also contains a characteristic component, callitrisic acid (4-epidehydroabietic acid). Australian sandarac often occurs in pieces larger than African sandarac (Barry 1932). Although it is mainly used for purposes similar to the better known African sandarac, it is not exported from Australia.

Mastic

In contrast to sandarac, a conifer resin, mastic is a resin produced by flowering plants, genera of the family Anacardiaceae. Mastic was known and used in the Mediterranean region for centuries, having been mentioned by Theophrastus in 400 B.C. In fact, mastic was known even earlier in the Middle East during the Late Stone Age and Bronze Age (Mabberley 1997). The plant was called the mastick tree in the Bible (Apocrypha, Sus. 1.54); the name mastic is derived from the resin's use as a masticatory. Although resin from other anacard trees has been called mastic, *Pistacia* is the genus most commonly associated with its production.

Varnish

Pistacia lentiscus, with several described varieties, has been the most important producer of mastic. The name of the genus comes from the Persian *pistah,* the pistachio tree *(P. vera); lentiscus* refers to the lens-shaped tissue into which the resin is secreted. *Pistacia lentiscus* is a widely distributed dioecious (separate male and female plants) shrub or small tree, occurring in coastal areas from Portugal and Spain to Greece, Syria, and Israel as well as North Africa, in Morocco and Tunisia. The island of Chios in the Aegean Sea has been the main source of mastic since the Middle Ages; at one time, trade in this mastic was monopolized by Greek rulers. At various times, however, mastic was obtained from other Mediterranean locales such as Cyprus and Sicily, but no regular trade developed from these islands.

The mastic industry is restricted to the southeastern corner of Chios where the small *Pistacia lentiscus* var. *chia* thrives (Davidson 1948). Although the tree grows normally on other parts of Chios as well as adjacent islands, the resin yielded does not have the desirable properties of so-called true mastic. Variety *chia* is also cultivated for mastic, but only the male tree is propagated because the female apparently yields an inferior mastic. By 5–6 years, young trees may produce small amounts of mastic. Twice the amount of resin occurs in a 10-year-old tree, with great increases by 50–60 years.

Collection of mastic on Chios has been limited by law to a 3-month period, 15 July to 15 October, to avoid obtaining inferior winter mastic, known locally as *kokkolyi* (Davidson 1948). Mastic resin occurs primarily in secretory tissue in the bark of stems and branches (Figure 9-3). Its localized presence in the wood, often restricted to vascular rays, suggests that the resin is traumatically induced there (Grundwag and Werker 1976). Resin is tapped by making numerous small vertical incisions in the bark of the stem and main

Figure 9-3. *Pistacia vera* exuding resin from the bark, typical of other mastic resins, as a result of pruning in the orchard at the University of California, Davis. Courtesy of V. S. Polito.

branches; hardened resin is collected 3 weeks or so later. Clean flat stones are placed around the trunk under the branches to catch fallen resin, to prevent its mixing with the soil. Some spontaneous exudation also occurs. The exudation forms tears or cylindrical pieces (Figure 9-3) that are pale yellow, clear, and brittle, breaking with a conchoidal fracture.

The terpenoid resins of *Pistacia* are more similar to those of some Burseraceae or Dipterocarpaceae than they are to resins from other genera of Anacardiaceae, especially the allergenic phenolic ones (Chapter 10). There has been detailed study of the terpenoid chemistry of *P. lentiscus*. The non-volatile fraction of variety *chia* was analyzed and 10 triterpenoid acids (e.g., moronic, oleanonic, masticadienolic, masticadienonic, and oleanolic acids) were identified, some present also in dipterocarp dammar (Papageorgiou et al. 1997). Two novel nortriterpenoids and two representatives of the rare malabaricane and polypodane types were identified (Marner et al. 1991). Other compounds from the neutral fraction include dipterocarpol, lupeol, β-amyrin, β-amyrone, oleanonic aldehyde, and germanicol. Several of these compounds as well as novel tetracyclic triterpenoids were found in galls induced by insects on the leaves of various *Pistacia* species and in the resin exuded from *P. vera* (Ansari et al. 1994).

Mastic from *Pistacia lentiscus* also contains a considerable amount of a polymeric component, 1,4-poly-β-myrcene, the structure of which has been elucidated (van den Berg et al. 1998). The volatile fraction of the resin makes up about 24% of mastic teardrops; α-pinene is the most abundant compound with lesser amounts of β-myrcene, limonene, camphene, and β-pinene. Interestingly, these compounds are prominent in pine resins (Figure 1-3). However, β-myrcene is the only compound with conjugated double bonds that is prone to polymerization. 1,4-Poly-β-myrcene, the first reported naturally occurring polymer of a monoterpene, is an important discovery. Its elucidation also gives support to discoveries of ancient mastic, including large amounts of resin found filling a gap between an Egyptian sarcophagus and its outer case (Lucas and Harris 1962) as well as resin from a Bronze Age shipwreck (Mills and White 1989). The polymer may also have provided the persisting qualities in the preservation of Pleistocene amber suggested to be from *Pistacia* (Chapter 4).

Mastic from *Pistacia lentiscus* has been extensively used in the manufacture of high-grade varnishes of pale color to protect oil and watercolor paintings. For this purpose, it has the great advantage of being easily removed, either by solvents or friction, without damaging the painting (Mills and White 1977). The gelatinous material known to artists as megilp (from McGuilp, the inventor), used in mixing oil colors, consists of a mixture of mastic, turpentine, and linseed oil. Mastic has also been used in lithography for retouching negatives and in cements for precious stones.

Indian or Bombay mastic, obtained from the shrub or small tree *Pistacia khinjuk*, is similar to Chios mastic but darker. Although Greece is probably the only commercial source of mastic (Coppen 1995), *P. chinensis* (Afghanistan to China and the Philippines) and *P. mexicana* (Mexico) are used locally for varnish.

Other Uses

Mastic from *Pistacia lentiscus* (and other species of the genus) has a variety of uses other than varnish. The age-old use of mastic has been as a masticatory to sweeten the breath because it smells sweet itself. *Mastic* in Greek means to chew, and children in the Mediterranean region still buy mastic for chewing gum. It has been used in dentistry to harden gums, alleviate toothache, and fill cavities. It has been used to flavor alcoholic beverages; for example, the Greeks include it in the liqueur mastiche and in the aperitif ouzo. Egyptians employed mastic in addition to the balsams frankincense and myrrh as an embalming agent, and it was sometimes used as incense (Duke 1985). The best grades of mastic, however, the yellowish white translucent tears, have been employed medically as an aromatic astringent. *Pistacia lentiscus* has also been used for innumerable ailments and is frequently cited in cancer-curing folklore (Duke 1983, 1985).

The antimicrobial effect of the volatile fraction of *Pistacia lentiscus* var. *chia* resin has been studied, concentrating on food spoilage and food-borne bacteria (Tassou and Nychas 1995). Addition of mastic to broth cultures inoculated with *Staphylococcus aureus*, *Salmonella enteritidis*, and *Pseudomonas* spp. inhibited growth of these bacteria. The rate of inhibition was greater on gram-positive bacteria (e.g., *Staphylococcus*) than gram-negative bacteria (e.g., *Salmonella*), in agreement with studies by Shelef et al. (1980). In most cases the amount of inoculum and concentration of the resin affected the growth and survival of the bacteria. The search for alternative food additives that prevent growth of food-borne pathogens has been promoted by food manufacturers to decrease excessive use of chemical preservatives, many of which are suspect because of their potential carcinogenic properties or residual toxicity.

Mastic has long been used as a wound adhesive and on surgical tapes used after dermatologic and plastic surgery (Jost et al. 1989). Mastic offers superior adhesive qualities and a lower incidence of postoperative contact dermatitis and subsequent skin discoloration than benzoin (*Styrax* resin; Chapter 8), which is also commonly used (Lesesne 1992).

During an archeological study of a Late Stone Age village in Iran's northern Zagros Mountains, a resin additive was discovered in a pottery jar along with the earliest chemical evidence of wine (McGovern et al. 1996). The jar was produced 5400–5000 B.C., a time when the first permanent human settlements, based on domesticated plants and animals as well as minor crafts such as pottery making, were being established. The jar had a yellowish residue identified as resin from *Pistacia atlantica*, which grows abundantly throughout the Middle East. The jar contained primarily a calcium salt of tartaric acid, which McGovern and coworkers concluded was from grape juice that would ferment to produce wine in the hot climate. They further thought that the addition of *Pistacia* resin served in part to inhibit the growth of *Acetobacter*, which converts wine to vinegar, as well as to mask any offensive taste or odor.

Pistacia atlantica is the source of a viscous liquid resin that was once very important and known as *chios, chio,* Chian turpentine, or even Cyprus balsam. Some mastic found in Egyptian tombs was still soluble after 2000 years. *P. atlantica* is often a large tree whose native range extends from the Canary Islands across North Africa into the Middle East as well as Transcaucasia, Turkey, and eastern Mediterranean islands, where it is planted as a shade tree. In the past, it was sometimes considered a variety of *P. terebinthus* (Mills and White 1994). The resin is still collected on Cyprus, where it is used as a kind of chewing gum.

Pistacia terebinthus, teil tree or turpentine tree (Apocrypha, Wisd. of Jesus Son of Sirach 24.16), is also a source of Chian-type turpentine. Mills and White (1994) thought that this material was the original terebinth or turpentine, the name only later transferred to conifer resin (Chapter 7). The tree grows in the Mediterranean region, and the resin has a pleasant aroma, similar to jessamine, and a mild taste; it was a favorite of the Greeks among the local aromatic resins. The resin that exudes from cuts in the bark solidifies into a transparent mass. Teil tree turpentine probably was a part of the spicery carried into Egypt from Gilead by the Ishmaelites, as mentioned in Genesis 37.25 (Duke 1983). Few resins have a greater repertoire in anticancer folklore; teil tree turpentine has been employed for cancers of the breast, face, lip, liver, medulla, pylorus, rectum, spleen, testicle, tongue, uterus, and vagina. In fact, it is reported to be "one of the most persistent drugs in history" (Majno 1975). Although the resin may have some antiseptic value, Majno indicated that the main effect may have been to make "the patient's malodorous sores smell better." The ancient gesture of spreading pleasantly smelling resins, even perfumes, on a wound was logical in that the most offensively smelling sores (gangrene) are also the most lethal—obliterate the foul smell and, perhaps, the source of that smell. Another reason for using resins on wounds was that they do not decay; thus there was the hope that they might transmit this property to wounded flesh.

Other anacard mastic-like resins include those from the small tropical American genus *Schinus*. Resin from *Schinus* apparently has a higher proportion of insoluble compounds than that from *Pistacia*; in fact, it appears to be a gum resin (Mills and White 1994). *Schinus terebinthifolius,* the Brazilian mastic tree or peppertree, yields a resin used as a Brazilian folk remedy; the resin is reported to be antiseptic, aphrodisiac, astringent, bactericidal, stimulant, tonic, and viricidal. *Schinus molle* (Plate 18) is another Latin American peppertree, also called Peruvian or American mastic tree. The bark, like that of *S. terebinthifolius,* is the source of what is often called aroeira resin. The resin from the fruits contains numerous mono- and sesquiterpenes such as α- and β-phellandrene, α- and β-pinene, limonene, myrcene, camphene, β-spathulene, etc. (Duke 1985). The gum resin is used as a masticatory, for incense, and medicinally for treating tumors and warts, from Mexico to Peru and Brazil. In fact, it has been used for many maladies, being reported as astringent, balsamic, diuretic, purgative, tonic, viricidal, vulnerary, and useful as a collyrium, or eyewash (Duke 1985).

Acaroid Resin

Howes (1949) referred to resins from *Xanthorrhoea* (Xanthorrhoeaceae, formerly included in Liliaceae), the endemic Australian grass tree (Plate 12), as acaroid resin or gum accroides. The genus was first named *Acoroides* (Nelson and Bedford 1993). Subsequently, its medicinal properties were assigned to "resin acoroides." Differences in spelling as well as references to either the plant or the resin have been perpetuated. Mantell (1950) also reported other names such as grass tree gum, black boy gum, and yackha gum.

The resin is produced at the base of the leaves, and sunlight or fire may melt the resin, which collects as globules on the trunk or between old leaf bases. Aborigines used the resin to fix spearheads to shafts, for instance, and Europeans became aware of it as soon as they arrived in Australia. Although all *Xanthorrhoea* species produce resin, Howes (1949) reported that resin in 5 of the 28 species were important for use. Both red and yellow resin is produced, but red resin occurs in most species. *Xanthorrhoea resinifera* (syns. *X. resinosa*, *X. hastilis*) produces yellow resin, *X. arborea*, *X. semiplana* subsp. *tateana* (syn. *X. tateana*) and *X. preissii* produce red resin, and *X. australis* produces a reddish brown resin. Birch and Dahl (1974) analyzed some of the many resin constituents of *X. australis*, *X. preissii*, and *X. resinifera* (as *X. hastilis*). They also summarized the reported occurrences of compounds in the resins of six *Xanthorrhoea* species that had been studied but emphasized that botanical nomenclatural problems plagued the chemical comparisons. However, the red and yellow resins are markedly different. Birch and Dahl reported hydroxy- and methoxyflavanone derivatives and other compounds such as cinnamyl alcohol and *para*-coumaric acid. Two compounds, an anthraquinone (chrysophanic acid) and its oxyphenylene derivative (xanthorrheol), are thought to be artifacts of fire damage.

The collection of commercial amounts of acaroid resins in the past was mainly in South Australia, particularly on Kangaroo Island, where *Xanthorrhoea semiplana* subsp. *tateana* was exploited (Earl 1917). To collect the resin commercially, the plant is trimmed and dead leaf bases with adhering resin are cut away. The leaf bases are transferred to a jigger, a sieve in which the resin is separated from them. It has been customary to cut off the crown of the plant from the base. Even though the process appears excessive, the plant generally recovers from such treatment and sprouts again.

Xanthorrhoea resins differ from other plant varnish resins in that they contain appreciable quantities of free benzoic and cinnamic acids and thus are similar to some balsams (Chapter 8). They have been used primarily as varnish, but because of their rich color and solubility in alcohol, they also have been used for wood stains (e.g., mahogany) and as a varnish for leather. The resin has been used somewhat as a substitute for rosin in paper sizing and in printing inks. Because the red resins contain resin-tannol, yielding picric acid on nitrification, species that contain large amounts of it were used in the manufacture of munitions during World War II in Australia, and in pyrotechnic compositions (Mantell 1950). The relatively rare yellow resin from *X. resinifera* has been used in medicines and glue (Mabberley 1997).

Hard Copals

As discussed in Chapters 6 and 8, the use of the term copal is controversial. The name copal, derived from the Nahuatl *copalli* by the Spanish, referred to all resins used by the Aztecs. Subsequently, the Maya used *copalli* to refer to resins used from *Protium, Bursera, Pinus,* and *Liquidambar,* depending on which resin-producing trees were most abundant in the areas where they lived (Chapter 8). The generalized Aztec meaning for *copalli,* and the subsequent Spanish transformation to copal, led some amber workers, such as Poinar (1992a), to categorize all unfossilized resins worldwide as copals. However, to use the word copal in such a general manner is unwise and only perpetuates confusion. As discussed, in Southeast Asia *damar* is often used for resins from all plant sources by local people. It is helpful to view the perplexing use of names for resins in a historical perspective, recognizing that names such as *copalli* (copal) and *damar* were used by local people before the plant sources were identified by scientists. Furthermore, there is an entrenched, widespread usage in international commerce in which the word copal refers to resin characterized by extreme hardness and relatively high melting point. This definition has been so pervasive in both industry and the literature that it is important to recognize it (Mantell 1950). Unfortunately, there is apparently no record as to how the word copal began to be used for resins that are so different from the Mesoamerican resins. However, I refer to such resins as hard copals to distinguish them from the softer copals associated with incense (e.g., Mayan *copal pom*). Hard copals have also sometimes been referred to as anime or hard anime resins.

The extreme hardness of copal results from polymers of resin acids such as ozoic acid, an enantiomer of communic acid (Figure 1-4) that can polymerize, enabling fossilization (Chapter 4). Hard copals are obtained primarily from various tropical genera of the family Fabaceae (or Leguminosae) subfamily Caesalpinioideae and from the most tropical of all conifers, *Agathis* (Araucariaceae). Copals were once major export items but have now been largely supplanted by synthetic resins (Whitmore 1980b), partially the result of variability in their chemical composition, which makes them less desirable than the more uniform synthetics. This compositional variability may result from copal's being part of the tree's chemical defense (Chapter 5) as well as from local collectors' mixing resins from trees belonging to different genera (as true of dammars).

Hard copals have been used more or less exclusively for the manufacture of hard varnish, in contrast to dammars, which are used for spirit varnishes. Copals were also used in printing inks (those applied to surfaces other than paper), polishes, finishing compositions (mixtures used to finish or coat surfaces), and linoleum. The Congo Basin in Africa yielded the most abundant leguminous copals from a variety of species for world markets, and much of the early economy of northern New Zealand was built on the large accumulations of copal from *Agathis* (Chapter 6).

Leguminous Copals

Eight copal-producing leguminous tree genera occur in tropical Africa and America. Five occur predominantly in West Africa *(Daniellia, Gossweilerodendron, Guibourtia, Oxystigma,* and *Tessmannia),* one in southern Africa *(Colophospermum),* one in East Africa and the Americas *(Hymenaea),* and the other in West Africa and the Americas *(Copaifera).* Of these, *Copaifera, Daniellia, Guibourtia,* and *Hymenaea* (syn. *Trachylobium*) have produced copal of the greatest commercial value.

Léonard (1950) has provided the most extensive description of the copaliers (copal-producing African legumes). Congo copal, the most important in world trade, was derived mainly from *Guibourtia demeusei* (syn. *Copaifera demeusei*) but was also collected from several species of *Tessmannia: T. africana, T. anomala,* and *T. yangambiensis.* Although Léonard reported *Cynometra sessiliflora* as a Congo copal producer, he later admitted that its identity was probably mistaken (Langenheim 1973). In the Belgian Congo (now Democratic Republic of Congo, formerly Zaire), the flooded forests along the rivers were the best collecting areas, although resin was also found in riverbeds and along sandy river beaches during the dry season. Native collectors located pieces of resin by prodding the ground (to depths of 15 cm to

9 m) with iron-tipped sticks. Interestingly from a cultural standpoint, it was the women and children who did the collecting (Mantell 1950) whereas in other cultures, collecting resin is primarily a man's job.

The taxonomic circumscription of the copal-producing genera has been in a state of flux, (e.g., *Copaifera* and *Guibourtia*), leading to confusion in the literature regarding the sources reported for copal. For example, Léonard's (1950) taxonomic revision transferred several species of copiously copal-producing *Copaifera* to *Guibourtia* (e.g., *C. demeusei* to *G. demeusei,* and *C. copallifera* to *G. copallifera*). Léonard did recognize other African species of *Copaifera,* however, including major copal producers (e.g., *C. baumiana, C. mildbraedii, C. religiosa,* and *C. salikunda*). Other controversial taxonomic changes include those of Breteler (1999), who placed the African *Gossweilerodendron* and *Oxystigma* (syn. *Pterygopodium*) in the otherwise New World *Prioria* (Chapter 2). However, more samples are needed before molecular data "can more clearly help to solve these generic delimitations" (Bruneau et al. 2001).

In addition to collection of resin from the ground in the Belgian Congo, fresh resin was collected from living trees where it had accumulated following natural wounding. No systematic tapping of trees took place. Variation in commercial samples was great both in color (white to yellow to brown) and hardness. The chemical constituents of these copals have not been well studied with modern analytical techniques, but several labdane acids (e.g., eperuic, copalic, and ozoic acids) have been reported. Resin obtained from the ground has often been called semifossilized (e.g., Howes 1949, Mantell 1950) although it probably did not attain the age (5000–40,000 years) considered subfossil here (Table 4-1). The salient point, however, is that the copal dug from the soil was much harder than fresh resin taken from the trees and was thus considered superior for industrial use. Copal from *Guibourtia demeusei* was second in hardness only to Zanzibar copal (*Hymenaea verrucosa;* Chapter 6).

The quantity of copal from the Belgian Congo greatly exceeded that from any other area from the turn of the 20th century till about midcentury (Howes 1949). Export increased from 20,000 kg in 1908 to 13 million in 1920 and 18 million in 1935 but declined, with fluctuations, thereafter. Congo resin was still the universal resin of the varnish makers, however, because it is adaptable for making a wide range of varnishes (Mantell et al. 1942). The gloss and depth of the film provided by Congo copal were sufficiently prized that it became the standard for rubbing-varnish resins. Congo copal is insoluble in all organic solvents but becomes soluble in petroleum solvents when thermally processed (resin heated to about 600°F for varying periods of time; Mantell 1950). Manufacturers of synthetic resins fortify their products with these properties by incorporating Congo copal. These copals were sufficiently important commercially during the period that the Belgians controlled the Congo that cultivation was planned (Léonard 1950). During the subsequent political changes, however, there has been no evidence of cultivation of *Guibourtia demeusei*.

Copals had been obtained, though less extensively, from trees in other parts of West Africa and were known by the country or place of origin (e.g., Angola or Sierra Leone). In Sierra Leone, most of the copal collected was probably from *Guibourtia copallifera*, a tree widespread along the coastal region from Guinea to Nigeria (Dalziel 1937). This species was second in importance only to *G. demeusei* as a West African copal producer (Mantell 1950). Fresh resin was obtained by tapping while the resin flowed freely, until the end of the rainy season. As the resin exudes, it forms nodules 5–8 cm in diameter, which were allowed to harden during the dry season before being removed. Because wasteful tapping practices injured many trees, experimental plantations were established in West African coastal countries. However, these slow-growing primary forest trees require 10 years before reaching size suitable for tapping. Therefore, the experiments were abandoned because the economic gain was considered too little for the time taken to establish plantations. *Guibourtia demeusei* is thought to be the source of Cameroon copal, and *G. ehie* the source of large quantities of West African resin in general.

Several species of *Daniellia*, which also yields a more liquid resin that flows freely, are considered to be the source of much of the fresh tapped resin as well as that procured from the ground in parts of West Africa. The best resin-producing species are tall rain forest trees such as *D. ogea* (Plate 15; yields ogea copal), *D. similis* (Accra copal), *D. alsteeniana*, and *D. thurifera*. *Daniellia alsteeniana* produces an excellent copal in the Congo region whereas *D. thurifera* has a wide distribution from Cameroon westward along the coast and may be a major source of West African copal. It is subject to insect attack, leading to abundant resin production (Léonard 1950). Resin from these trees has a pepper-like fragrance and might be considered a balsam in its fresh state. In fact, it has been sold as copaiba balsam, possibly used as a substitute or adulterant for copaiba oil, which is an oleoresin from New World *Copaifera* species (Chapter 7), and it has often been confused with resin from *D. oliveri*, a tree characteristic of savanna forests. On the other hand, *D. thurifera* has also been sold as Sierra Leone frankincense. Local people use it to fumigate huts or scent garments as well as chewing it. Resin from *Daniellia* demonstrates how the ratio of the volatile and nonvolatile fractions may vary, with a gradient from more fluid oleoresins to more viscous balsams and hard copals, even between individuals of the same species.

Gossweilerodendron, as a genus separate from *Prioria,* comprises two West African species. One, *G. balsamiferum,* is one of the largest rain forest trees in Africa and is common in southern Nigeria and Angola. It is best known by its Nigerian name, *agba.* Resin obtained from the wood has been used for illumination by local people (Dalziel 1937). The resin may account for the wood's resistance to fungi and subterranean termites. Apparently, although the resin is generally thick, it may sometimes be relatively fluid, which would make it more like that of *Prioria* (Ekong and Okogun 1967). However, Léonard (1950) and Mabberley (1997) refer to the resin as copal.

East African copal from *Hymenaea verrucosa* (syn. *Trachylobium verrucosum*) was held in especially high esteem because some grades were the hardest of all the copals (Barry 1936). The extreme hardness results from 95% of the trunk's resin being iso-ozoic acid (Martin et al. 1974), with sesquiterpene hydrocarbons providing most of the volatile fraction. The structures of seven other bicyclic diterpenoid resin acids have been elucidated (Hugel et al. 1966). *Hymenaea verrucosa* occurs in coastal evergreen forests of Kenya and Tanzania and offshore on Zanzibar, the Seychelles, and Mauritius as well as eastern Madagascar.

Local people along the coast of East Africa collect *Hymenaea verrucosa* resin from the living tree or the soil. The soil is dug to a depth of more than 1 m and the resin, usually in flat or disk-like pieces, is picked out. The resin can occur in quite large plates and characteristically has a goose-skin-like surface, the origin of which is not understood. The copal from Madagascar is similar to that from along the East African coast but does not display the goose-skin or crazed surface. It is collected primarily from the soil there and can occur in relatively large amounts. The resin from throughout the region is often referred to as fossilized when it is collected from soil in areas in which the tree no longer exists. In Madagascar the name *sandarosy,* derived from

Arabic, is used for the so-called fossilized resin. It has the same root as the French world *sandaraque* (sandarac in English), which is the resin from African *Tetraclinis* (Rakoto-Ratsimamanga et al. 1968). Again, it is doubtful that these accumulations fit the concepts of subfossil or fossil resin (Table 4-1). In fact, some pieces of resin from Madagascar, sold as amber by gem dealers, are only 50 years old (Alex Brown and G. O. Poinar, pers. comm.). However, unsubstantiated Mio-Pliocene and Pleistocene amber from *H. verrucosa* was suggested to have been deposited on Zanzibar in mangrove swamps, similar to the amber in southern Mexico (Chapter 4).

Although it is relatively difficult to extract resin from the fruit pods of *Hymenaea verrucosa*, they contain an appreciable quantity of it: 500 kg of fruits have been estimated to yield 100 kg of copal. Extensive analyses have been made of the resin acids of the pods; they are principally tetra- and pentacyclic diterpenes related to (−)-kaurane (Hugel et al. 1965a–c; Hugel and Ourisson 1966). The resin chemistry of *H. verrucosa* pods is very different from that of New World species of the genus (Figure 5-6). Instead of diterpenoids with tetra- and pentacyclic skeletons, pod resins of the New World species that have been studied have bicyclic skeletons.

A pendant, thought to be made from Zanzibar copal, was found in an ancient palace (2500–2400 B.C.) at Eshnunna (now Tell Asmar, eastern Iraq); it represents the earliest evidence for contact between the part of East Africa in which *Hymenaea verrucosa* occurs and Mesopotamia (C. Meyer et al. 1991). East African resin trade is known from an early date (Figure 6-1) when Arab traders exchanged copal for cotton cloth with Indian traders. The Portuguese developed this trade in the later 18th century. Resin was exported to China, where it was particularly esteemed for varnish (Rakoto-Ratsimamanga et al. 1968). The resin trade from Madagascar was sufficiently important in the later 19th century that the government organized a monopoly over the transactions. The hardness of this resin created keen demand for it and secured high prices, especially before the Congo copal industry was highly developed. For example, in 1989 more than 232,515 kg of resin were exported to Germany alone. At one time, most of the East African copal was shipped from Zanzibar; hence, it was called Zanzibar copal, although later the resin acquired names from other East African ports (Chapter 6). Some of the copal came to Europe on boats from India, which explains why it was sometimes called East Indian, Bombay, or Calcutta copal.

The genus *Hymenaea* is primarily New World, however, extending from Mexico throughout the Caribbean, into Central and South America to southernmost Brazil, and various species are the source of commercially used copals. In some cases, the resin, which is softer than African copals, is referred to as anime. As in Africa, this copal has been known in the trade by the country or area from which it was obtained. Much of the copal has been attributed to *H. courbaril,* probably because of its wide distribution, which essentially coincides with that of the genus in the New World (Chapter 2). Other species produce copal that is used locally, however, especially as cement and medicine. Furthermore, in Amazonia the resin is commonly found in the soil of the periodically inundated forests, where the resin accumulates as it does in flooded Congo forests (Langenheim 1995). In this type of forest, the copal is probably derived from *H. oblongifolia,* the closest New World relative to the African *H. verrucosa* (Langenheim et al. 1974). It is the species from which resin was probably collected in abundance for export as Brazil, Pará, or Demerara copal, especially in areas near the mouth of the Amazon. In upland Amazonian rain forests, *H. intermedia* and *H. parvifolia* as well as *H. courbaril* (Plate 16) are towering trees, emerging above the forest canopy, producing copious amounts of resin referred to as *jutaicica.* Large blocks or pieces of resin, occasionally weighing as much as 3 kg, were dug from the ground around trees in eastern Amazonia (Ducke 1925). The resin is produced in pockets that can be increased to large cavities in the wood (Chapter 3), and the highly viscous resin is often relatively slow-flowing, especially under dry conditions. It can be collected from large exudations in notches or branches of the tree or from injured areas of the trunk as well as from the soil where it has fallen around the base of the tree.

James Duke has referred to *Hymenaea courbaril* as the kerosene tree, based on a report by Peireia (1929) that *Hymenaea* produced an oil that burned like kerosene. This seems unlikely because *Hymenaea* is known to produce only a viscous, predominantly diterpenoid trunk resin (Langenheim 1981). In the leaves, the resin is predominantly fluid, with sesquiterpenes, many of which occur in *Copaifera,* sometimes called the diesel fuel tree (Chapter 7). It would be difficult to extract enough resin from *Hymenaea* leaves to use as kerosene. *Hymenaea,* like many resin-producing trees, produces fine timber that is used in various ways; the bark has been used locally in Amazonia for making canoes that are caulked with the resin.

The structures of four diterpenoid resin acids have been elucidated from trunk resins of three Amazonian *Hymenaea* species: *H. courbaril, H. oblongifolia,* and *H. parvifolia* (Cunningham et al. 1973, 1974; Langenheim 1981). *Hymenaea* trunk resin acids are bicyclic, mostly of the enantio-labdane skeletal series; however, *H. courbaril* has yielded both enantio- and antipodal labdane skeletons. Pod resin in New World *Hymenaea* has only been analyzed from *H. courbaril* from the island of Guadalupe off the western coast of Mexico (Khoo et al. 1973). The diterpene skeletons are bicyclic but more highly rearranged than those of the trunk resins.

Hymenaea courbaril resin is collected by local people from the soil around trees along streams in southern Mexico, where the trees are abundant (Plate 43). Relatively large deposits of Oligo-Miocene amber that occur in Chiapas, the southernmost state of Mexico, have been attributed to *Hymenaea.* Although the amber-producing species has not been definitively determined, there is evidence of some relationship to *H. courbaril,* and these deposits were in coastal estuaries dominated by mangroves (Chapter 4). Both the resin and the amber from this region are used locally as incense, cement, and for carving amulets (Chapter 6).

Araucarian Copals

Copals are obtained from various species of *Agathis* (Araucariaceae), the conifer with the most tropical distribution. Although the resins are generally called copal, those of some species have locally been called *damar. Agathis australis,* from New Zealand, is the best-known copal producer; however, species from Indonesia and nearby islands as well as the Philippines have been important commercial sources (Gonzales and Abejo 1978). In the latter regions it has been called Manila copal because it was exported mostly from Manila. Several names were given to resins with different degrees of hardness, for example, *melengket* for soft ones, *pontianak* for harder ones, and *boea* for the hardest (Mantell 1950).

The confused status of the taxonomy of *Agathis* has led to various species being cited as the source of Manila copal. The most common has been *A. alba (almaciga),* which occurs throughout central Malesia. Whitmore (1980a) considers *A. alba* a synonym of *A. dammara,* however, and de Laubenfels (1989) considers it synonymous with *A. philippinensis / A. celebica.* Both Whitmore and de Laubenfels agree that the copal producers in natural stands in peninsular Malaysia and on Sumatra and Borneo should be called *A. borneensis*. The copal of *A. labillardieri* is also used.

Manila copal has been collected from the ground, from large lumps in the forks of branches of trees, and from tapping the trunk. Local people used an iron-tipped shaft, as done in the Congo, to locate resin in the ground. Very large pieces of copal have been recovered from the soil, and it is thought that some was produced by roots. Manila copal collection has been a significant source of cash income for indigenous Philippine collectors, but despite a ban on timbering the trees, they are more difficult to locate and are overtapped close to villages (Conelly 1985). The economic value of Manila copal, however, has led to research on ways to increase production. All current Manila copal is obtained by tapping, however, by making incisions into the living bark and collecting the exudate (Coppen 1995). Research in Indonesia and the Philippines has shown that thick-barked *Agathis* yields significantly more (almost nine times as much) than thin-barked trees (Soenarno 1987). Improved methods of tapping for sustainable production involve choosing the best size of tree, the kinds of cuts so as not to destroy the cambium, the amount of sulfuric acid to stimulate flow, and the periodicity of tapping (Ella and Tongacan 1992).

Although Manila copal has not been used so extensively since the advent of petroleum-based synthetics, there is still a steady demand for specialized uses such as varnish on food-can labels and color prints (Whitmore 1980b). From 1988 to 1993, the greatest amount of Indonesian copal was exported to Germany, and Philippine copal to Taiwan (Coppen 1995). In addition to export, there has been local demand for *Agathis* resin nearly everywhere the trees occur. In Irian Jaya (or West New Guinea), the principal source of Manila copal in Indonesia, *Agathis* trees are used solely for their resin. Small plantations established for this purpose are controlled for a sustainable non-damaging tapping system.

Different species of *Agathis* have been grown for both resin and timber. For example, *A. dammara* (syn. *A. alba*) has been used in agroforestry ecosystems such as the taungya system on Java. *Agathis* seedlings need shade and are therefore planted in the shade of food crops, then the fast-growing leguminous tree *Leucaena leucocephala* provides shade until the saplings can tolerate sunlight and also controls weed growth and acts as a fertilizer (Whitmore 1980b). When *Agathis* is tapped, yield increases during the first 6 months; it can be stimulated by applying diluted sulfuric acid, however, as done for oleoresins from pines (Chapter 7). Trees respond by producing woody, cankerous growths when too intensively tapped (Plate 45). Both *A. dammara* and *A. macrophylla* are used in forest-enrichment planting in peninsular Malaysia (Chapter 11).

Agathis australis, the kauri pine or just kauri, has been the source of one of the most valuable copals. The kauri is the monarch of the New Zealand forest; its tall, unbuttressed, cylindrical bole is almost without taper (Figure 2-3). The largest trees reach more than 50 m high (30 m before the first branch) and attain more than 10 m in girth. Average trees, however, are much smaller. Today, *A. australis* occurs only in the northernmost lowlands of the North Island. At the beginning of European occupation, there were thousands of hectares of *A. australis* forest. As a result of exploitation for both resin and timber, dense virgin *Agathis* forest is now very restricted, to reserves. *Agathis* together with *Podocarpus* and other southern hemispheric conifers form mixed forests, however, and young kauri trees remain in cutover land.

Like the misnomer kauri "pine," the resin was erroneously called "gum." Although technically a terpenoid resin, discussion of the history of its use is difficult without using the word gum. Therefore, kauri pine gum is sometimes used as an expedient here and certainly is found in the literature. Kauri resin has a distinctive chemical composition; much of it is a copolymer of the diterpene communic acid with communol, which gives it special properties, producing a very hard varnish. Other major components are abietic acid, agathic acid, and sandaracopimaric acid (and its corresponding alcohol) as well as various monoterpenes (Thomas 1969). Other *Agathis* species have several different diterpene acid compositional patterns, which influence their usability. For example, *A. dammara* (syn. *A. alba*) has communic acid and sandaracopimaric acid but large proportions of agathic acid and the acetate of agatholic acid in one compositional type but large amounts of torulosic acid characterizing another compositional type (Mills and White 1994). This kind of diterpene compositional variation is similar to that seen in mono- and sesquiterpenes of other conifers (Chapter 5).

There is variation in the composition of the resin produced in the bark, wood, and leaves of *Agathis australis* (Thomas 1969). Wood resin occurs at the sapwood–heartwood boundary. The resin that is collected commercially is produced primarily from canals in the bark, from which resin exudes abundantly in response to any injury. Descriptive names have been given to resins that exude in different manners from different parts on the plant. That collecting in the forks of branches is known as crutch gum, that accumulating on the bark in long stalactites is candle gum, and that from the root is sugar gum. Pieces weighing 1–2 kg are usual for crutch gum, but lumps of 23–90 kg have been obtained from a single tree (McNeill 1991). In the past, some candle gum was obtained by a method of tapping called bleeding (Figure 9-4). If

Figure 9-4. Tapping of *Agathis australis,* kauri, for resin with a small tomahawk by climbing the large clear bole. The climber returned every 6 months to collect the exuded resin and make fresh incisions. Courtesy of *New Zealand Geographic.*

the cuts do not penetrate the wood and are not too numerous, tapping apparently does little harm because the tree sheds bark, obliterating the wound with time. However, the tendency to overdo the cutting led to regulation of this procedure in New Zealand.

Over time, layers of forest debris covered the resin that fell to the ground, to considerable depths. Enormous amounts of resin accumulated, with some single nuggets weighing more than 270 kg (Figure 9-5; McNeill 1991). It was this accumulation, particularly in peat swamps, that led to the development of the New Zealand gum industry (Chapter 6). In these swamps, where rising water tables slowly waterlogged kauri forest over perhaps thousands of years, there were both kauri logs and resin to depths of more than 8–10 m. In other areas, longer burial led to the formation of amber (Chapter 4). Kauri trees today do not grow in swamps; however, note the importance of swamp-type conditions in the accumulation of large masses of leguminous resin in the Congo Basin, East Africa, and Amazonia, and araucarian resin in New Zealand.

Figure 9-5. Kauri gum export merchants. Note the large pieces of resin to the right. Courtesy of *New Zealand Geographic.*

From the earliest Maori occupation until the first *pakeha* (people other than Maori) arrived, kauri gum could be found lying on the ground on the North Island. Early *pakeha* reported various Maori uses of it (McNeill 1991). Resin taken directly from the tree was boiled until soft, then juice of *puka* (*Meryta sinclairii*, Araliaceae) was added and the gum was chewed communally. Maori medicinal use of chewed gum, however, to treat vomiting, diarrhea, and digestive upset, suggests that early observers were witnessing more pharmacological than social usage. Maori used the resin, which burns hot and bright, as a fire lighter, and these properties were later exploited by Europeans in making matches. Maori burned the resin as torches for illumination; they also used the resin to attract fish, and repel insects in *kumara (Ipomoea chrysorrhiza)* plantations.

The first European report of kauri resin led to a long-standing case of mistaken identity. Captain James Cook collected some resin under mangroves in 1769. Cook concluded that the material came from those plants, and the mangrove was named *Avicennia resinifera*. Although mangroves produce a green aromatic exudate, it was not until 1819 that Samuel Marsden identified the resin that Cook collected as coming from *Agathis*.

There was little interest in the possible use of kauri resin in the early 19th century. For example, when Charles Darwin was shown a kauri tree in New Zealand in 1835, he noted in his journal, "a quantity of resin oozes from the bark which is sold at a penny a pound to Americans but its use is kept a secret." Nonetheless, in 1836, according to local legend, a shipment of kauri resin was sent to London to test for possible commercial importance but was pronounced worthless and dumped into the Thames. A boy playing along the river found a piece of the resin and took it home, however, as a "stone that floats." His father, who was in the varnish trade, had the piece analyzed, the results leading to the beginning of the New Zealand export industry (McNeill 1991).

Between 1850 and 1950, 450 million kilograms of kauri resin were exported from New Zealand, primarily to North America and England. Kauri gum buyers recognized at least 50 different grades, from almost colorless and clear dial gum to almost black swamp gum as well as candle gum (young) and sugar gum (oxidized) among the many kinds. Its primary use was in both oil- and spirit-based varnishes as well as linoleum. For half a century, kauri resin was the province of Auckland's most valuable product, exceeding that of gold, wool, and kauri timber (Chapter 6). In fact, from small beginnings in

1850, there was a 10-fold increase in export until the peak in 1899 (Hayward 1982). Exports remained high until World War I, when they dropped by half as markets in England and Germany closed and the gum diggers departed for war service. At the end of the war, markets returned and the gum industry made a small recovery by using machinery to obtain vast amounts of smaller-sized chip gum. However, exports declined from the 1920s on, with market crises during the Great Depression and World War II, reaching an all-time low in the 1980s as a result of the development of inexpensive synthetic substitutes.

Before the advent of kauri resin, hard-varnish makers used different resins, particularly the African leguminous copals, and sometimes even amber (Mantell et al. 1942). The varnish makers were highly competitive, often using secret formulas. The usual method was to melt the resin and combine it with a heated oil (usually linseed), which was dangerous because most copals have high melting points and mixing them with inflammable oil at high temperature could result in explosion. Furthermore, the completed varnishes were often difficult to apply, took a long time to dry, and gradually darkened with age. Whereas kauri resin had the problems of difficult application and long drying time, it had two advantages over other copal varnish bases: it held its color better and, because of its lower melting point, was much easier and safer to handle during manufacture. By the 20th century, *Agathis* resin was considered the premier varnish resin, especially because other highly valued copals, such as Zanzibar copal, were less abundant.

Although thousands of liters of kauri gum varnish were manufactured in Britain, North America, and Europe, some varnish makers mixed the resin with shellac and other copals to produce a durable French polish that became popular for coating high-quality furniture. Although it made it more difficult to apply, the addition of *Agathis* resin to shellac gave better water resistance and greater elasticity to the polish. By the early 1900s, much furniture in the northern hemisphere was being coated with this polish.

As with dammars, only the palest varieties, considered the highest in quality, were suitable for varnishes. When the linoleum industry developed in 1863, however, lower-quality resin could also be used. Linoleum was made from linoxyn (oxidized product of linseed oil) melted with resins and allowed to cool somewhat. This melted product was mixed warm with cork and pigments and rolled into sheets, thereby producing a tough, durable floor covering. *Agathis* resin was used initially because it was the cheapest copal available, but later, in the 20th century, Congo copal became cheaper.

Synthetics have largely supplanted *Agathis* resin varnishes. The natural resin tends to be used only for specialty items such as polishes for stringed instruments or by furniture restorers trying to match the original polishes on antiques. Kauri resin is often mixed with other resins such as sandarac and mastic, since alone it tends to develop a crazed surface with age, as true of some leguminous copals and ambers. Other, lesser uses for *Agathis* resin included impression molds for false teeth (still used in India) and as a binding agent in phonograph records.

Lacquers and Specialty Varnishes

The terms natural lacquer or natural varnish are used for resins collected in the liquid state and applied directly to the surface to be varnished without the use of a solvent or drying agent. Such lacquers are in everyday use in various parts of Asia but are best known from China and Japan. A unique type of resinous varnish, *barniz de Pasto*, is sometimes called a lacquer and is compared to the traditional oriental lacquers.

Anacard Lacquers

Chinese and Japanese lacquers are derived from *Toxicodendron vernicifluum* (syns. *Rhus verniciflua*, *T. verniciferum*, *R. vernicifera*), a deciduous tree of the family Anacardiaceae that is native to central and western China and widely cultivated by the Chinese (Figure 9-6). The tree was introduced into Japan at a very early time and grows well there. Tapping usually commences when the tree is about 30 cm in diameter and is carried out until the tree is 50–60 years old. Midsummer is considered the best time for tapping because the exudate is very fluid in spring and very thick in fall. After repeated tapping, the tree is allowed to rest several years. The resin is produced in canals in roots, stems, leaves, and immature fruits (Baer 1979). It consists primarily of pentadecylcatecols (Chapter 1), as does the resin from *T. radicans* (poison ivy), *T. vernix* (poison sumac), and *Metopium* (poison wood), but the resin from *T. vernicifluum* includes gum and a small quantity of albuminous matter.

The resin consists of a complex of phenolics called urushiol (Chapter 10), gum, and a small quantity of albuminous matter. Despite fatty acid-like side chains of urushiol, Japanese lacquer dries by a mechanism different from that of drying oils (true oils) because of the phenolic structure of the rest of the molecule (Chapter 1). Although a similar chemistry for other lacquers is assumed, they have not been studied as carefully as Japanese lacquer (Mills and White 1994). Moreover, the albuminous compounds apparently play an important role in the drying of the lacquer. Lacquer has a peculiar property

Figure 9-6. *Toxicodendron vernicifluum*, the deciduous Chinese lacquer tree whose resin is used to produce fine lacquers.

of requiring a damp atmosphere, not a dry one, to dry and harden, which cannot be hastened by heat (Mills and White 1994).

Archeologists have found Chinese lacquer used to coat small objects in the early Late Stone Age (4000 B.C.). According to Mills and White (1994), the techniques had reached an advanced level by the Shang dynasty (1600–1100 B.C.). The art reached its highest level of development during the Ming dynasty (A.D. 1368–1644), however, and the methods employed for very fine work often are still closely guarded secrets. As many as 30 different operations may be used in preparing a single article, with more than 300 different coats of lacquer applied to special pieces. After a groundwork is completed, subsequent coats may be colored or metallic dusts may be used.

The Japanese acquired the art of lacquering from the Chinese and developed it to a high degree of perfection, particularly during the 17th century. The Japanese outstripped their predecessors in excellence of the products, except for lacquer carving (Hill 1952). *Toxicodendron succedaneum* (syn. *Rhus succedanea*) was introduced into Brazil from Japan for lacquer (Mors and Rizzini 1966). The bark of the tree can be tapped in the third year, yielding an extremely caustic resin that is used as a lacquer or varnish, mainly in home industries. The Portuguese name for the lacquer is *charão*.

The anacard genus *Gluta* is the source of Burmese black lacquer, known by the inhabitants of Myanmar (Burma) as *thisi*. It is similar in many ways to Chinese and Japanese lacquer although the lacquerware is quite different. The resin is obtained from *G. usitata* (syn. *Melanorrhoea usitata*), a deciduous tree of tropical Asia. In Myanmar, the tree grows primarily in open forests. Under favorable conditions the tree grows rapidly and may be tapped when only 2–3 years old, although old trees produce more lacquer than young ones. Trees are tapped when in leaf because they do not produce resin when leafless in the dry season. Another species, *G. laccifera*, is also used for lacquer in Southeast Asia. Although the resin from *Gluta* is chemically similar to that from *Toxicodendron*, it has been reputed to cause less dermatitis than *Toxicodendron*.

Barniz de Pasto and Other Rubiaceous Resins

Resin obtained from *Elaeagia*, a shrubby small tree of the family Rubiaceae that grows in the southwestern Colombian Andes at elevations of 2300–2500 m, is used for a unique kind of lacquer. Called *mopa mopa*, the resin has commonly been attributed to *E. utilis* (Arbeláez 1956), but Mora-Osejo (1977) discovered that it originates from *E. pastoensis*. European naturalists who visited Pasto in the 17th–19th centuries described a resin used in producing an exquisite material most often called *barniz de Pasto* but also *mopa mopa de Pasto, barniz de Mocoa,* and *barniz de Condagua* (Kaplan et al. 1999). This material was used to decorate wooden objects, leather, and gourds. Chroniclers of the 16th century described such goods as coming from Pasto through Quito and into Inca centers in what is now Peru.

Elaeagia pastoensis resin was used as a paint binder to decorate wooden *qeros* (ceremonial drinking vessels) during the colonial period in Peru and adjacent areas in the Andes but apparently had also been used for the same purpose at least at the end of the Inca period (Kaplan et al. 1999). Mora-Osejo (1977) documented the techniques of the *barnizadores* in Pasto, reporting that the resin is gathered from buds and young stems of the tree by people other than the artisans and sold to the *barnizadores* as large cakes of resin. This bulk resin is then purified and transformed into a transparent, colorless, elastic mass that is stretched into a thin sheet. For modern folk art, *barniz de Pasto* is colored by powdered pigments introduced into the thin resinous sheet or by an underlayer of metallic sheet, then cut into shapes and applied to the wooden substrate. Today, some *barnizadores* ensure that the thin cut-out shapes of *barniz de Pasto* adhere to the wooden substrate by first coating the wood with oil paint. Apparently, the preparation methods of the *barnizadores* today are similar to those used in the past. However, Kaplan and co-workers indicated that either the use of colored resin or the application of an undercoat of paint is different today. Arbeláez (1956) thought that *barniz de Pasto,* as a unique type of lacquer, could compete with Japanese lacquer in the beauty of the objects created. Today it is mostly used for modern folk and tourist art.

Resin from other species of *Elaeagia* as well as those from several other genera of Rubiaceae have been used for a variety of purposes. For example, some species of *Elaeagia* have been found to be effective against pulmonary tuberculosis. *Coutarea* produces a resin used for incense in Central America. Certain species of the Asian genus *Gardenia* also produce resin freely on buds and young shoots (Chapter 3), and two Indian species, *G. gummifera* and *G. lucida,* are sold in bazaars as dikamali "gum." For dikamali, the shoots and buds are broken off with drops of resin attached. The resin is greenish yellow and strong-smelling. Sometimes it is collected and pressed into cakes to be sold. It has been employed in various ways in native medicine. *Gardenia laccifera* is another species whose resin probably has been used.

Araliaceous Varnishes

Species from three genera of Araliaceae yield resins that were used as a readily photopolymerizable protective varnish for armor and other metalwork in Japan. Resins from *Acanthopanax sciadophylloides (koshiabura), Dendropanax trifidus (kakuremino),* and *Evodiopanax innovans (takanotsume)* were used (Mills and White 1994). Other species of *Dendropanax* produced resins for similar use in Korea and China, known during the T'ang dynasty as *ushitsu* (golden varnish).

Amber Varnish

In the Baltic factories during the early 20th century, trash pieces of amber were dry-distilled to make amber oil and amber varnish (Chapter 6). This varnish dried to a hard finish but was very dark. It was used on Stradivarius violins; besides its dark color, the expense of the varnish was a drawback to its commercial success.

Shellac

Lac, known as shellac in its refined, flake form, is often called resin. It is an insect secretion, however, not a plant resin, and is mentioned here only to clarify its origin. Lac is always formed on plants, which provide the sole source of food for the lac insects, and it does have similarities to some varnish resins from plants.

Both a resinous material and a red or crimson dye are derived from the Indian lac insect *(Laccifera lacca,* syn. *Tachardia lacca),* which belongs to the scale insect family (Coccidae). Many of its relatives produce lac but do not live in sufficiently large colonies to produce material adequate for commercial collection. The larvae of *Laccifera* aggregate on a branch to feed, and the varnish-like secretion of individual larvae becomes a continuous coat beneath which further development and metamorphosis of the insect occur.

The lac insect breeds on a variety of trees but some of the most important include species of *Butea* (Fabaceae), *Ziziphus* (Rhamnaceae), and *Schleichera*

(Sapindaceae). The first reference to the use of lac in varnish making is 1590, when it was stated in the *Ain-i-Akbari* that lac was collected to make varnishes in the palace of Akbar, the Mughal emperor of India. Lac is produced almost entirely in India, which furnishes 97% of the total output, the remainder coming from Myanmar (Burma) and Thailand. The Indians produce an outstanding finish by using lac in the form of sticks colored by the incorporation of pigments. It is a spirit varnish, yielding a tough film with a smooth finish that is capable of a high polish (Barry 1932). Detailed discussion of the formation of lac, the manufacture of shellac, and related subjects has been presented from different perspectives by Barry (1932), Howes (1949), and Mills and White (1994).

CHAPTER 10

Miscellaneous Resins

These resins do not fit easily into the categories of oleoresins, balsams, copals, or varnishes and lacquers. Most have been used medicinally; some have also been used for flavoring *(Humulus)*, as perfume fixitive *(Cistus)*, rubber substitute *(Parthenium)*, or fuel *(Euphorbia)*. Some conifer resins are mentioned, but emphasis is on angiosperms from 14 families, including herbs, shrubs, or trees growing in habitats ranging from dry steppes and deserts through temperate forests to tropical rain forests. The resins are presented by plant families (sometimes by genus) or where possible by the resin's vernacular name.

Umbelliferous Resins

Ammoniacum, asafoetida, and galbanum, produced by *Dorema* and *Ferula* (Apiaceae, or Umbelliferae), have a long tradition of use. These plants are very common on the plains and steppes of Iran and Afghanistan. Shortly after the rainy season starts, the plants send up thick stems from perennial rootstocks. When fully grown they can be 2–3 m high and so abundant that they form a kind of open forest (Plate 46; Hill 1952). Resin is collected from stems and roots. McNair (1930) classified resins from *Dorema* and *Ferula* as tannol resins, that is, predominantly esters of aromatic phenols that behave in some ways like tannins (polyphenols). Such behavior may have ecological implications because tannins can complex with proteins in the digestive system of some herbivores (Chapter 5). The resins also contain small amounts of terpenes, particularly sesquiterpenes, as well as true gums.

Ammoniacum

Dorema ammoniacum, one species in a small genus, produces a resin called gum ammoniacum. The name ammoniacum was reputedly derived from that of the temple of Jupiter Ammon in Libya, where it was commonly collected. The plant is a very large perennial herb, native to central Asia, Iran, and to the north. Resin exudes from punctures in the stem, which can occur from insect attack, and the resin that is usually collected comes from these natural exudations. The medicinal value of the resin was mentioned by Hippocrates in the fifth century B.C. It is still used in Indian and Western medicine and is listed in the *British Pharmacopoeia* as an antispasmodic and expectorant (Chevallier 1996). It is occasionally used for chronic bronchitis and persistent coughs.

Ammoniacum is medicinally similar to asafoetida and galbanum from *Ferula*. In fact, African or Moroccan ammoniacum is thought to be derived from species of *Ferula* (Howes 1950). The ammoniacum referred to by Dioscorides, Pliny the Elder, and later Greek and Latin writers on medicine probably came from *Ferula* rather than *Dorema*.

Asafoetida and Galbanum

Gum resins from species of the large genus *Ferula*, commonly known as giant fennel, have been used since ancient times for medicine, contraception, cooking, perfumery, and incense. About 17 species produce resins of commercial importance, with some used primarily for flavoring and others primarily as medicinals (Raghavan et al. 1974). *Ferula assa-foetida* (Figure 10-1) is one of the most important species from which the resin called asafoetida is obtained. A perennial plant that grows to 2 m high, it is native to Iran, Afghanistan, and Pakistan. When plants are 4 years old, just prior to the flowering stage, the stems are cut off and a series of slices made through the roots. The resin then wells up from canals in the cortex of the roots and is collected when it hardens as tears or masses of varying color (Chevallier 1996). It is one of the rare plants in which resin is tapped from roots. The resin is composed of ferulic acid, umbelliferone, asaresinotannols, sesquiterpene farnesiferols, and coumarins (including foetidin), several disulfides, monoterpenes (α- and β-pinene), a trace of vanillin, and gums composed of galactose-arabinose, rhamnose, and glucuronic acid (Leung and Foster 1996).

The medicinal use of asafoetida can be traced to the seventh century B.C. Hindu medical treatise *Charaka Samhita*, in which it is proclaimed as the best treatment for getting rid of gas and bloating (Chevallier 1996). In India and the Middle East, it has continued to be used for digestive problems. *Ferula jaeschkeana* (Plate 46), a source of asafoetida, has been shown to have contraceptive properties in humans (Farnsworth et al. 1975). Asafoetida resin is also currently used for bronchitis and whooping cough in tablet form as well as for lowering blood pressure and thinning blood. Duke (1985) has presented a long list of medicinal uses covering a wide range of ailments. India is a large importer of asafoetida but also reexports some of the resin to

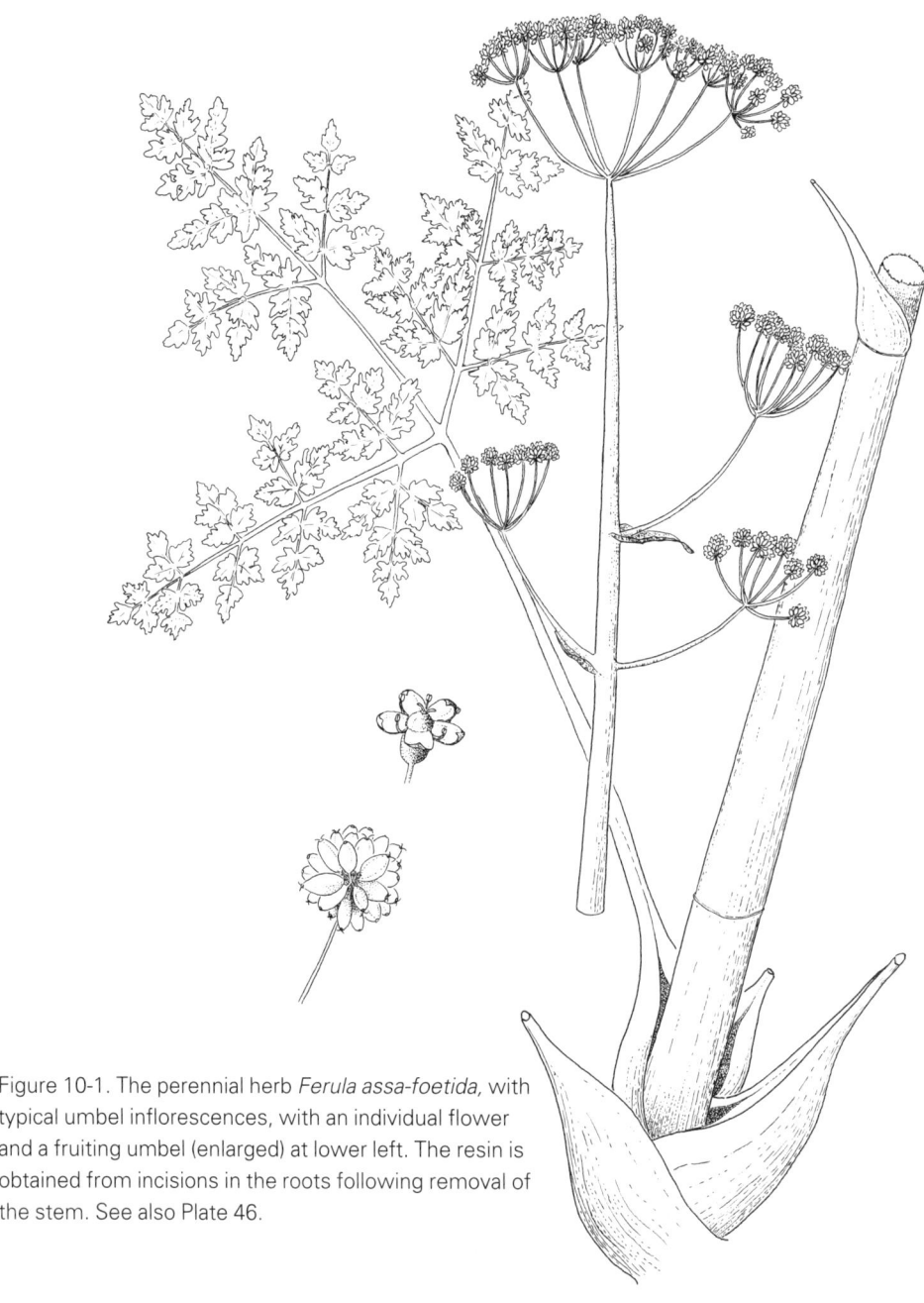

Figure 10-1. The perennial herb *Ferula assa-foetida,* with typical umbel inflorescences, with an individual flower and a fruiting umbel (enlarged) at lower left. The resin is obtained from incisions in the roots following removal of the stem. See also Plate 46.

various Middle Eastern countries. Shivashankar et al. (1972) reviewed the chemical composition of different varieties of asafoetida and the standards in using them for Indian food packers.

Although resin from *Ferula assa-foetida* is most commonly called asafoetida, it is also referred to as devil's dung; the disagreeable odor results from the presence of disulfides, particularly propenyldisulfide. Despite the implications of the name, the resin was the most popular spice in ancient Rome (Chevallier 1996). The aroma is as persistent as garlic, and the taste is even stronger. It is still used as a flavoring, notably in Worcestershire sauce. In Iran, it is rubbed on plates warmed for serving meat. When used as a flavoring, it is generally very dilute and certain objectionable compounds have been removed, or the resin has been blended with diluents. It is also commonly used as a fixative or fragrance component in perfumery. On the other hand, it is used in veterinary practice to repel cats and dogs in situations where they are unwanted (Morton 1977, Duke 1985). Resins called asafoetida are also obtained from *F. foetida* (eastern Iran) and *F. narthex* (Afghanistan) and used as a condiment (called food of the gods in Iran) and more widely as a medicine (Mabberley 1997). Samini and Unger (1979) discussed the provenance and quality of asafoetida-producing *Ferula* species in Afghanistan.

Ferula gummosa (syns. *F. galbaniflua*, *F. rubricaulis*), a perennial plant native to central Asia, is like *F. assa-foetida* in that it exudes a gum resin from the lower part of the stem and root when the stem is cut a few centimeters above the ground; it is collected after it hardens into separate tears or brownish yellow-green masses. The resin is commonly called galbanum and has long been used as medicine, incense, and food flavoring. It is mentioned in the Bible: "Moses, take unto thee sweet spices, stacte, and onycha, and galbanum; . . ." (Exod. 30.34). Today, it is used in perfumes and soaps for its fragrance and it is employed in the manufacture of varnish. In food, it is found in nonalcoholic beverages, baked goods, candies, condiments, gelatins, and puddings (Duke 1983). Medicinally, galbanum is a digestive stimulant, antispasmodic, and reduces flatulence, cramps, and colic. It is also used as an expectorant and in salves for soothing chilblain and healing wounds (Chevallier 1996). Extracts of galbanum apparently have preservative and antimicrobial properties. Its spermaticidal properties may be one reason why galbanum has been used as a contraceptive.

Resin from sumbul, the central Asian *Ferula sumbul,* has been used as a nerve tonic. Sagapenum is a resin that was once collected, probably from *Ferula;* Howes (1950) suggested *F. persica* and *F. szowitziana.* Its use is similar to asafoetida and galbanum.

A resin called opopanax is derived from *Opopanax chironium* in ways similar to asafoetida. The resin was formerly used medicinally but is used presently for scent making (Mabberley 1997). The term opopanax is also apparently used for resin from species of *Commiphora* (Burseraceae), particularly *C. kataf* and *C. erythraea* (Guenther 1950, Vol. 4), which is called bisabol myrrh or sweet myrrh (Chapter 8).

Silphium

In the seventh century B.C., Greek colonists from Thera (now Thíra) founded the coastal city of Cyrene in what is now Libya. According to Theophrastus, the colonists soon discovered a plant that made them famous and some of them very wealthy. They called it *silphion,* later latinized to silphium. Silphium was probably an undescribed species of *Ferula* and should not be confused with the genus *Silphium* (Asteraceae). Extant species that have been compared against coin depictions suggest that if silphium was not *F. tingitana,* then it was a closely related species (Gemmill 1966, Penn 1994). Silphium was also called laser and laserpitium from the names of resin from other species of *Ferula* (Koerper and Kolls 1999). Moreover, the term silphium was apparently used to refer either to the plant or its resin.

Silphium only grew on the dry mountainsides of Cyrenaica, facing the Mediterranean in a band about 200 km long and 55 km wide. Attempts to cultivate it in Syria and Greece were unsuccessful, leaving the city of Cyrene as the sole exporter of the plant and its resins. Silphium became the city's distinctive symbol, with the plant featured on two Cyrenian coins, one showing just the plant and the other showing a seated woman touching the plant with one hand and the other pointing to her genitals (Penn 1994). In Greek art, the god Dionysus, typically representative of love, sensuality, wine, and drama, is often shown holding a silphium plant in his hand.

Koerper and Kolls (1999) listed the myriad medicinal uses for silphium from classical sources. Since it was considered the most therapeutic of all plants in the ancient pharmacopoeia, it was highly prized. Koerper and Kolls, however, raised a controversy as to whether the most important use of silphium was as a contraceptive or aphrodisiac. The generally held view that sil-phium was used as a contraceptive is supported by the writings of Theophras-tus (about 300 B.C.) and Pliny the Elder (about A.D. 77), and more recently by Riddle (1985), Riddle and Estes (1992), and Riddle et al. (1994). Studies of the effects of the resin from three extant *Ferula* species—*F. assa-foetida, F. jaeschkeana* (Plate 46), and *F. orientalis*—have also demonstrated antifertil-ity properties in rats (Prakash et al. 1986, M. Singh et al. 1988). Furthermore, *F. assa-foetida* resin has been reported to act as a contraceptive and abortifa-cient in human tests (Farnsworth et al. 1975). Nonetheless, Koerper and Kolls thought that the imagery and stylized representations on Cyrenian coinage reflects an "imitative principle in early medical practice." Interestingly, Rid-dle et al. (1994) imply that the numismatic motifs support the use of silphium as a contraceptive whereas Koerper and Kolls employ the same motifs to fur-ther their ideas regarding aphrodisiac usage. Other circumstantial evidence for the aphrodisiac view is drawn from the writings of Avicenna, who attrib-uted aphrodisiac qualities to a recognized substitute for Cyrenaic silphium, and to the poetry of Catullus, linking silphium to carnal pleasures. Obvi-ously, the question has not been resolved, but whatever the use of silphium, it was so great that the plant that produced the resin was exterminated.

Silphium resin seems to have become a medium of international exchange similar to bullion, judging from its storage in Rome. It made Cyrene the rich-est city on the African continent until the development of Alexandria, but its downfall was linked to the scarcity of the plant by the first century A.D. Pliny reported that a little later, in Nero's time, "only a single stalk had been found there [in Cyrenaica] within our memory." This careless treatment, including overexploitation of the plant and probably the land as well, could only have occurred if the regulations described earlier by Theophrastus had been aban-doned. At one time, the government had prescribed both the amount and manner in which the resin could be extracted, measures that served to main-tain the semicultivated plant. The collapse of an economy based so much on this plant provides a lesson in how short-term consumer gain can preclude long-term benefits. Export of silphium probably ceased after the beginning of the third century, and by the fifth century the plant was extinct or virtually so (Gemmill 1966). The conquests of Alexander the Great in the East resulted in trade of less-expensive and inferior asafoetida resin into the Greek market.

Convolvulaceous Resins

Among the hundreds of species in the cosmopolitan family Convolvulaceae, only a few have been reported to produce resinous compounds. The resin is a complex of chemicals referred to collectively as jalapine or convolvuline, which occurs in the periderm (thin bark tissue) of tuberous roots. The chemical structures of these complexes have not been completely elucidated, and the taxonomic status of the plants producing them is also uncertain. Furthermore, interchange of the common names for the plants and the resins has created considerable confusion in the literature.

Jalap

The tuberous root of *Ipomoea purga* (Figure 10-2), an evergreen vine, has been used as a laxative or purgative since pre-Hispanic time (Martínez 1969). It is one of several distantly related tuberous New World *Ipomoea* species, including *I. jalapa*, *I. orizabensis*, and *I. stimulans*, that produce a group of valued purgative medicinals known commonly as jalaps (McDonald 1989). Pharmacologically, jalap is considered to be a hydragogous cathartic, an energetic purgative in large doses, and a soft laxative in small doses. Duke (1985) noted that if given in sugar or jelly, it forms a "safe purge" for children; it also can be used as a vermifuge for children.

The tuberous roots of *Ipomoea purga*, called jalapas, were first introduced into Europe from the New World by Spanish explorers in the mid-1500s. Through the late 1600s, unsuccessful attempts were made to cultivate *I. purga* in a number of botanical gardens, especially in England, France, and Germany (Linajes et al. 1994). The British subsequently introduced the species into Jamaica and India, where it was successfully cultivated. Numerous fakes and adulterations resulted from the great demand for jalapa root. Many of the fakes were roots from other members of the Convolvulaceae that were similar. The proliferation of fake jalapas from the mid-1500s to mid-1800s resulted in nomenclatural chaos. Linajes and coworkers emphasized that the confusion resulted from many authors who generated lists of both common and scientific names based on limited morphological descriptions, chemical analyses, and assays of the roots.

All jalapa roots exported by the Spanish were collected from the large natural population of *Ipomoea purga* in the Jalapa–Xico region of central

Veracruz, Mexico. Popular demand for the root led to cultivation there in the mid-1880s. Cultivation was in primary forest vegetation, using shrubs to support the vines, thereby conserving the native plant diversity. This production system apparently was maintained for at least 150 years through generations of local farmers with little modification. Davis and Bye (1982) indicated that

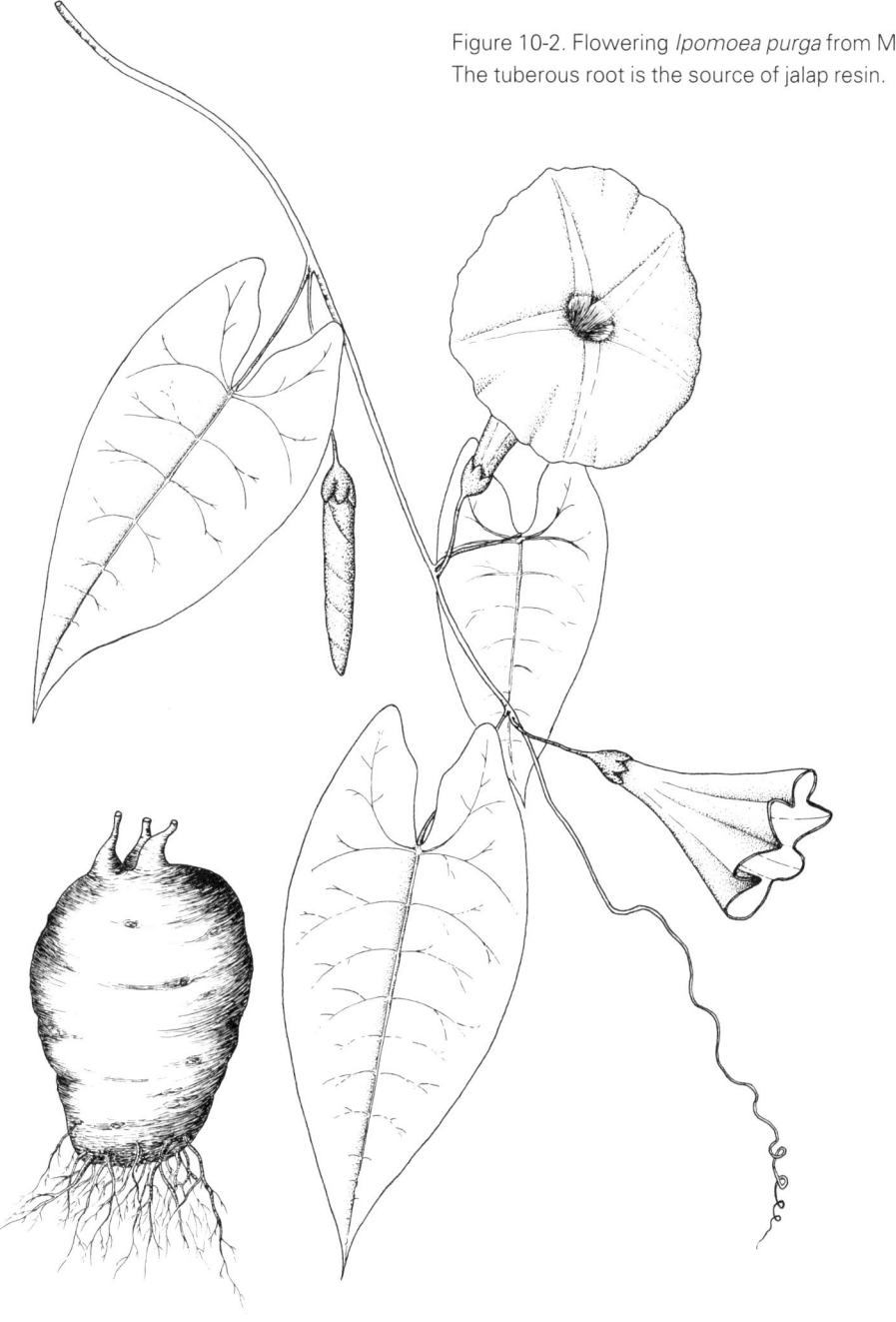

Figure 10-2. Flowering *Ipomoea purga* from Mexico. The tuberous root is the source of jalap resin.

jalapa grows well under management and can be considered intermediate between a wild and a cultivated plant. Even though production of the root has declined, its quality and the demand for laxatives have not, and its export supplements farmers' earnings from their primary crops (Linajes et al. 1994).

Other *Ipomoea* Resins

In addition to the purgative action of the resins from *Ipomoea,* some species have allelopathic potential for suppressing weed growth. In the tropics and temperate zone of Mexico, a multicrop system of agriculture often provides natural control of weeds (Anaya et al. 1990). There, cultivation of species such as *I. tricolor* is common to control weeds. The allelopathic potential of the mixture of resin glycosides in *I. tricolor* was demonstrated against weeds in sugar-cane fields in the southeastern tropical region of Mexico (Pereda-Miranda et al. 1993).

The periderm tissue of sweet potato *(Ipomoea batatas)* is also known to contain allelopathic compounds (J. Peterson and Harrison 1991a). For example, growth of *Cyperus esculentus* and *Medicago sativa* were significantly reduced when plants were grown in field soil that had previously supported sweet potatoes (H. Harrison and Peterson 1986). Decaying sweet potato residues incorporated into the soil significantly inhibited growth of sweet potato vine cuttings, and nodulation in cowpea plants *(Vigna unguiculata)* was negligible under these conditions (Walker and Jenkins 1986). Furthermore, the resin glycoside fraction of periderm tissue of sweet potato produced antibiosis on the diamondback moth, *Plutella xylostella* (J. Peterson et al. 1998). The active glycoside fraction was identical to the one shown to interfere with seed germination of *Abutilon theophrasti, Amaranthus retroflexus,* and *Panicum miliaceum* (J. Peterson and Harrison 1991b).

Scammony

Hippocrates (about 400 B.C.) and Theophrastus (about 300 B.C.) were familiar with the resin known as *scammonium,* or scammony, obtained from the roots of a convolvulaceous plant. Dioscorides (about A.D. 77) described a plant as the source of scammony, probably *Convolvulus sagittifolius* rather than *C. scammonia* (Shellard 1961).

Shellard (1961) stated that it is impossible to say when the roots of *Convolvulus scammonia* were first used as a source of the drug, which is known for its purgative and cholagogic action. The resins obtained from the dried latex and dried root were included in the definition of scammony resin given in the first *British Pharmacopoeia* in 1864. Scammony was defined as the gum resin obtained from the living roots of a plant growing in Syria. Although some changes occurred in subsequent editions of the *British Pharmacopoeia*, in 1914 scammony was excluded after more than 2000 years of use because the resin had been replaced by other drugs. However, scammony root was included as a synonym for ipomoea in the 1948 edition of *British Pharmacopoeia*. The roots of other *Convolvulus* species, for example, larger bindweed *(C. sepium),* which has a worldwide distribution, also produce a resin similar to that of *C. scammonia.*

Hashish and Hops Resin

Cannabis is the source of the resin called hashish when it is extracted from the trichomes (called marijuana if left in the floral bracts), and *Humulus* is the source of hops resin. These genera are the only two in the family Cannabaceae, and their distinctive resins have a long history of human use.

Hashish

Hashish and marijuana resin are produced by the hemp or marijuana plant (*Cannabis sativa*, Figure 10-3). Despite a number of species having been described, most botanists consider the genus to be variable, with one species having two subspecies (Chapter 2). Others, however, still recognize subspecies *indica* as a separate species (e.g., Clarke 1998). If viewed as a single species, *C. sativa* is a taxon that has diversified into numerous ecotypes and cultivated races. It is widely distributed, having spread as a weed over the globe at nearly every latitude that supports plant life; it also grows in a variety of habitats from sea level to more than 2000 m in the Himalaya. Although very adaptable, dispersal of the plant was undeniably facilitated by humans who found it valuable for several purposes.

Cannabis sativa subsp. *indica* yields the greatest quantity of resin, which is produced in trichomes on floral bracts surrounding developing fruits and their subtending leaflets in female plants (Figure 10-3). The trichomes (Figures 3-7 and 3-9) secrete resin containing about 30 cannabinoids and a number of fragrant mono- and sesquiterpenoids. The psychoactive principle is the cannabinoid Δ^1-tetrahydrocannabinol (Δ^1-THC), which can constitute as much as 50% of the resin (Figure 1-8). Most THC is Δ^1, which is also referred to as Δ^9 (Arguell et al. 1984). Pure resin is collected from the trichomes in various ways (Chapter 6).

Figure 10-3. *Cannabis sativa,* the source of hashish and marijuana. Center, flowering apical shoot of a resin-producing female plant (upper left, a seed; upper right, calyx with two pistils; middle right, close-up of a flower; and lower right, habit of the plant); middle left, flower from a male plant.

Greater quantities of resin (along with similar amounts of THC) are commonly reported to be produced in *Cannabis* plants from warm compared to cool temperate areas (Tyler et al. 1988). It seems likely that the greatest quantity of resin would be produced in warm, dry areas, as this generally is the pattern for trichome-produced surface resins in shrubs and perennial herbs; the assumption is that they protect against both herbivory and desiccation (Chapter 5). In experimental studies, Korte (1970) reported that plants from warm climates produce 10 times as much THC as plants from central Europe grown under identical conditions. On the other hand, Bazzaz et al. (1975) found significant differences in THC content between populations but also greater yield under cool rather than warm conditions when grown experimentally. Moreover, Bazzaz and coworkers clearly demonstrated tropical and temperate ecotypes, which can be distinguished not only by THC content but by photosynthetic rate and chlorophyll content.

THC is relatively neutral in fragrance. Volatile terpenes are largely responsible for the fragrance and aroma when hashish is heated and may have led to incense as one of its earliest uses (Chapter 6). The incense fragrance of hashish has been described as amber-scented but possessing a distinctive smell that is superior to musk in its sensual stimulation (Rosenthal 1971). The most common terpenoid components are monoterpenes such as limonene, β-phellandrene, and α- and β-pinene (1–4%) along with sesquiterpenes such as β-caryophyllene (38–46%), *trans*-bergamotene, β-farnesene, and α- and β-selinene. The numerous monoterpenes individually range from only a trace to 3.9%; overall, the sesquiterpenes are more significant, ranging from 1.2 to 46% (Clarke 1998). Thus many of the differences in aroma and flavor of hashish through the geographic range of the plant result from variation in terpene composition rather than that of THC. This geographic variation in terpene composition parallels that found for numerous resin-producing plants in ecological studies (Chapter 5).

Archeological records in Europe and the Middle East indicate the most ancient use of *Cannabis sativa* was as a source of fiber: hemp. Narcotic use of the resin may have preceded all others in some areas, however, because its effects must have assumed an early magico-religious importance (Schultes 1970). *Cannabis* resin was also noted as important in the treatment of various human maladies, including malaria and forgetfulness, in the herbal *Pên-ts'ao ching* attributed to the Chinese emperor Shên-Nung before 2000 B.C. Shên-

Nung further mentioned that it frees the psyche, but later writers warned that it is a liberator of sin. The Chinese pharmacopoeia *Rh-Ya*, compiled in the 15th century B.C., contains the earliest mention of *Cannabis* for shamanistic purposes. Li (1974) pointed out that in ancient China, as in many cultures, medicine had its origin in magic. Medicine men were often "practicing magicians," hence, related to shamanism. By the second century A.D., Chinese physicians mixed it with wine and gave it to patients as an anesthetic before surgery. By the 10th century in Chinese herbals, *Cannabis* resin was being used as a cure for many diseases but especially for its analgesic effects (Li 1974).

Uses of *Cannabis* that were more ritualistic occurred in Asia Minor where it was used as a psychotropic drug by the ancient Scythians. Emboden (1979) concluded that use of *Cannabis* in shamanistic rituals by the Scythians and other people in the region played an important role in both curing the sick and in burial ceremonies.

Cannabis resin has a long history of widespread use for providing relief from a variety of ailments in the Western world. Considerable attention was given to medical use of *Cannabis* during the period 1840–1900 with about 100 papers published on the subject (Tippo and Stern 1977). In fact, marijuana enjoyed fairly wide medical acceptance until about 1900 and was used legally for medicinal and industrial purposes until 1941. At that time, it was removed from the *U.S. Pharmacopoeia and National Formulary*. Once Δ^1-THC was finally isolated in 1965 and control of dosage was possible, the active principle of the resin was found conclusively to have effective medical uses, especially in relieving the symptoms of some diseases and the side effects of drugs used to treat some diseases, rather than curing or preventing disease. For example, it can reduce the pressure in the eyes of glaucoma patients. It has been recommended as an analgesic, an anticonvulsant, and as an aid for many other problems, with the virtue of having extraordinarily low toxicity. There has been increasing use of *Cannabis* resin to ease the suffering of the terminally ill and patients undergoing debilitating cancer treatment such as chemotherapy (Mechoulam 1986a, Grinspoon and Bakaler 1997).

Cannabis as a medical agent has two major drawbacks. First, because most of the therapeutic effects are the result of THC, they also cause medically undesirable psychotropic effects. Until a complete separation by molecular modification is achieved between the therapeutic and cannabimimetic effects, wide use of *Cannabis* as a medical drug cannot be expected (Mechoulam 1986a). Second, even if the psychotropic effects are eliminated, it will be necessary to look for more than one compound in *Cannabis* because all its therapeutic properties cannot be concentrated in one cure-all drug.

For medical purposes, marijuana is generally used rather than hashish. High-quality hashish is so pure and potent (as much as 50% THC), however, that a cancer or AIDS patient need consume only a very small amount to obtain therapeutic effects (Clarke 1998). *Cannabis* is grown especially for medical use in some places, for example, Amsterdam. However, widespread illicit use of marijuana and hashish has led to problems in legalizing it as a prescription drug in many countries (Chapter 6). On the other hand, eight U.S. states, led by California and Arizona in 1986, have voted to legalize the use of medical marijuana. Furthermore, medical investigations have been authorized in California that could conceivably lead to its legalization for certain medical purposes in the United States (Chapter 11).

Hops

Humulus, the genus of the hop plant, comprises three dioecious species of tall, coarse, climbing twiners. Although resin is produced in characteristic trichomes on different parts of the plant (Chapter 3), the largest quantities occur in the female inflorescence. Because most plants grown commercially for the resin develop fruit without fertilization, resin is obtained only from the bracts of the cone-like infructescences, called hops. Resin from one species, *H. lupulus* (Figure 10-4), has been used medicinally in China and Japan as a stomachic, sedative, diuretic, and bitter tonic. An antibiotic, lupulon, with tuberculostatic properties has been reported from the resin (L. Perry 1980).

However, the most common use for resin from *Humulus lupulus* is in making beer. The practice of using hops for flavoring malt beverages (malt alone produces insipid beverages) goes back to mid-eighth century A.D. (Guenther 1952, Vol. 6). At that time, King Pepin the Short, ruler of the Franks (a Germanic-speaking people who lived in present-day western Germany, northern France, and Belgium), donated a hops garden to a monastery; subsequently, a number of German monasteries became well known for their brews. By 1320–1330 hopped beer became popular in Germany and then spread to the Netherlands, France, Sweden, and Bohemia (now the Czech Republic). Hop plant growing began in North America during the mid-1600s but did not become important until 1800. Hops resin not only gives a pleasant taste and aroma to the beer, it adds enzymes to help coagulate unwanted proteins that can make the beer cloudy. The bitter resin, characterized by what is known as the alpha acid fraction, is also somewhat bacteriostatic. Its presence minimizes the growth of infecting lactic bacteria and the bacterium *Pediococcus damnosus* (syn. *Streptococcus damnosus*), whose name suggests the master brewer's reaction to the results of its growth in the beer. The volatile mono- and sesquiterpenes determine the aroma of the resin.

Figure 10-4. *Humulus lupulus* with cone-like infructescences, hops, from which the resin is obtained. When grown commercially, the vines are allowed to climb tall trellises attached to wires strung in long rows.

Twenty-five volatile compounds in hops resin have been identified by GC-MS, but particular aromas are determined by terpene composition, again by the relative proportions of the mixture of mono- and sesquiterpenes (Zupanec et al. 1991). In fact, several chemotypes were discovered during quantification of the terpenes in extracts of 11 hop cultivars (Langezaal et al. 1992). One cultivar was characterized by a large amount of α-humulene (about 40%), a second by β-myrcene (42–58%), and a third by almost equal amounts of these compounds (27% and 25%, respectively). These ratios are similar to those found in some sesquiterpene compositional types in *Hymenaea*, a kind of variation that has been demonstrated as a major defensive property of certain resins (Chapter 5).

Hop leaf resin contains smaller quantities and a different composition of volatile terpenoids than that in the floral bracts (Langezaal 1993). Although 35 terpenoids were identified by GC-MS in leaf resin, the primary components were α-humulene (3–23%) and β-caryophyllene (5–29%). One of the predominant components in floral hops, β-myrcene, occurred at a concentration of 1% or less in leaf resin. Such differences in resin composition between organs has been shown in chemosystematic and ecological studies of numerous resin-producing plants (Chapter 5).

The best-quality hops, with the largest amount of the most desired aromatic content for use in beer making, are grown in the Czech Republic, Poland, and southeastern England, and in the United States, in Oregon and southeastern Washington. The plant is often grown in orchards consisting of rows of female vines trained onto trellises that can be 6 m or more high (Figure 10-4).

Propolis

In temperate climatic regions, bees (primarily honeybees, *Apis mellifera*) collect resin from a variety of plant sources and bring the strongly adhesive substance back to their hives for use in reinforcing the structure (König 1985). The bees masticate the resin with salivary secretions, and the partially digested material is mixed with beeswax. This mixture is referred to as bee glue, or propolis. About 50% of the resin in propolis consists of numerous flavonoid components, and about 10% is volatile terpenoids; beeswax makes up another 25–35% along with about 5% pollen and 5% various other substances (Marcucci 1995, Bankova et al. 2000). Tropical meliponid and trigonid bees also collect resin, but theirs is usually not considered propolis (Chapter 5).

Poplar *(Populus)* bud resins are among the most commonly used for glue, but bees only collect resin from species in the taxonomic section *Aigeiros* (Greenaway et al. 1990). Although there are reports of resin collection from other temperate zone genera such as *Aesculus, Alnus, Betula, Prunus,* and *Salix* (Chapter 3), Wollenweber and Buchmann (1997) pointed out that numerous studies confirm that in central Europe, bees collect resin almost exclusively from poplars, *Populus nigra* in particular. They also reported that the best indicator for propolis is the flavonoid aglycone pattern of the resin. Thus the propolis fingerprint in Europe usually can "hardly or not at all be discerned from that of poplar bud exudate."

The composition of propolis is often variable because of the different seasons in which different taxa are visited by honeybees as well as the inclusion of beeswax and other contaminants. For example, bees may incorporate more wax in the mixture when resin is scarcer or difficult to collect. When and how the beekeeper harvests the propolis further introduces variation into the product. Propolis is collected by commercial beekeepers either by scraping the material from wooden hives or by using specially constructed collection mats; the raw product is then processed to remove beeswax and other impurities. Although standardization is possible in principle, chemical tests have yet to be applied. Lack of quality control is clear in the propolis products marketed in various countries (Tŏth 1985).

The long history of bee domestication has led to human exploitation of propolis for many purposes since early times. Its history in folk medicine is particularly noteworthy. For example, the Greek physician Hippocrates (about 400 B.C.) prescribed it to help heal both internal and external sores and ulcers. Ancient Egyptians depicted propolis-making bees on vases and other ornaments, and used it to alleviate many ailments. The Hebrew word for propolis is *tzori,* and its therapeutic properties are mentioned throughout the Old Testament.

Reports on pharmacological use of propolis appear in the early 1900s (Ghisalberti 1979). However, most research on composition, pharmacology, and commercial use of propolis has been done since the mid-20th century. A wide spectrum of health-promoting and restoring effects has been claimed. In fact, since propolis is reputed to have antiseptic, antimycotic, bacteriostatic, astringent, choleric, spasmolytic, anti-inflammatory, anesthetic, and anti-oxidant properties, Burdock (1998) noted that "the list of preparations and uses is almost endless." Nevertheless, as Tŏth (1985) and Marcucci (1995) emphasized, propolis cannot be considered a remedy for all ills. Modern drug research is based on well-defined preparations that can be administered in a form that allows reproducible results. Some progress in this direction has been made through scientific and medical research since the 1970s, particularly in eastern Europe and eastern Asia, where medical use of propolis is common.

Many studies have shown propolis to be very antimicrobial (Lindenfelser 1967); thus it is often called a natural antibiotic (Brumfitt et al. 1990). Consistent with its antibiotic effects on bacteria, fungi, protozoa, and viruses are its long-known properties of wound healing and tissue repair. Propolis extracts are also noted for producing surface anesthesia. The resin is a good local anesthetic with peripheral action on the mucous membrane of the eye, greater than that of cocaine and with an infiltrative action equal to that of procaine (Ghisalberti 1979). It is used for various diseases of the mucous membrane of the mouth such as ulcers and gingivitis. It is commonly used in topical home remedies and as an ingredient in toothpaste and dental floss (1–5% of the finished product). Additionally, because of the high levels of flavonoids in propolis, it can function as an antioxidant and free radical scavenger in humans (Pascual et al. 1994), and it apparently protects vitamin C from oxidation. Research, including clinical trials, on the numerous posited medical benefits of propolis has been reviewed by Burdock (1998), who further discussed evidence that propolis causes dermatitis in a small percentage of sensitive persons, which, however, is relieved when the skin is no long in contact with the material. Propolis may be found as an unintentional additive in some beeswax and extracted honey.

Allergenic Anacard Resins

Some, but not all, resins from the Anacardiaceae are allergenic. In fact, some anacard resins discussed in Chapter 9, such as those from *Pistacia* and *Schinus*, are called mastic because they are chewed or used in surgical tape. Those resins are primarily composed of terpenoids whereas the dermatitis-causing resins contain irritating phenolic compounds, so irritating to some that they may be considered poisonous.

Most of the allergenic-resin-producing species have traditionally been considered to belong to the large genus *Rhus*. Whether *Toxicodendron* should be considered a genus separate from *Rhus* has long been controversial. Gillis (1971) justified the separation based on the presence of poisonous resins and a suite of distinctive morphological characters; his interpretation is followed here. *Toxicodendron* occurs primarily in North America and eastern Asia. The literature regarding its species is voluminous, in large part because the resins are notoriously allergenic to humans. *Toxicodendron* includes the most widely spread natively occurring taxa in the family Anacardiaceae. They are all dioecious (separate male and female plants), woody, and exhibit noteworthy phenotypic plasticity. Two sections of the genus include resin-producing species: section *Toxicodendron* includes poison ivy and poison oaks whereas section *Venenata* includes poison sumac and lacquer trees (Chapter 9).

From the perspective of dermatologists, the transfer of the poisonous-resin-producing taxa from *Rhus* to *Toxicodendron* introduces nomenclatural confusion because the toxic effects of these plants are rooted in the medical literature as rhus dermatitis (Baer 1979). Certainly, the taxa have a colorful medical history. Reports of the dermatitis during the 19th and 20th centuries are legion, with the numerous major treatments summarized by Rostenburg (1955) and Kligman (1958). Some of the topical treatments cited by Kligman are astounding: "Qualitatively, they range from the preposterous to the fantastic . . . . A brief listing of some of the more interesting agents which have been thought beneficial clearly reveals the profound emotional effects of therapeutic desperation: morphine (topically!), bromine, kerosene, gunpowder (the symbolism here is beautiful), . . . aqua regia (!), buttermilk, cream and marshmallows (!), . . . strychnine (!), etc." Since then, attempts to cure or prevent the dermatitis have moved from topical applications to use of injections, oral prophylaxes, or possibly inducing an immune tolerance in individuals prior to their exposure to the plant (Baer 1979).

Poison Ivy, Poison Oak, and Poison Sumac

Plants of *Toxicodendron* section *Toxicodendron*, to which poison ivy and poison oaks belong, extend from central China and Japan south through Southeast Asia and from southern Canada to western Guatemala and some Caribbean islands. *Toxicodendron radicans* (syns. *Rhus radicans, R. toxicodendron*; Figure 2-13) is the most common species of poison ivy. Gillis (1971) considers *T. radicans* the most common and widespread species of Anacardi-aceae. It ranges from southern Canada through the eastern third of the United States, and throughout Mexico to Guatemala; it also is native to East Asia. In the United States, it often occurs in disturbed, ruderal sites whereas in Japan it grows in old-growth forests. The plant is primarily scandent but may develop into a large shrub if there is nothing to climb. Plants are so variable that numerous species of poison ivy were named. Gillis preferred to recognize the variability by describing nine subspecies of *T. radicans*, however, because he thought there was insufficient genetic difference to justify recognition of separate species. The name poison ivy was given to the plant by John Smith in 1624 in his *Generall Historie of Virginia, New-England, and the Summer Isles,* in which he noted the similarity of its climbing habit to that of English ivy *(Hedera helix)* and its trifoliate leaves to those of Boston ivy *(Parthenocissus tricuspidata)*. He also noted that it caused an "itchynge" and blisters.

One of the closest relatives of *Toxicodendron radicans* is western poison oak *(T. diversilobum,* syn. *Rhus diversiloba).* It occurs along the Pacific coast of North America from Baja California to British Columbia. Poison oak exists as a shrub, vine (Figures 2-13 and 10-5), or rarely a tree over a wide range of habitats; it is adapted to a greater range of rainfall and temperature than any other California shrub. In fact, Jepson (1936) suggested that poison oak exceeded any other California shrub species in number of individuals. The variability of its leaves often makes it difficult to recognize and may result in part from its adaptation to extremes of shade and sun. The high degree of leaf plasticity is reflected in its epithet, *diversilobum* (different lobes), refer-ring to the variability of the lobing of its leaflets. The similarity of its lobed leaves to those of coastal live oak *(Quercus agrifolia),* also extremely variable, is responsible for its vernacular name, poison oak. Like *T. radicans, T. diversi-lobum* also does extremely well in disturbed sites, which increases the possi-bility of human encounters.

In contrast to the variability in habit of *Toxicodendron diversilobum* and *T. radicans,* the eastern poison oak *(T. toxicarium,* syn. *Rhus toxicarium)* is a shrub or subshrub distributed primarily in the sand hills of the southeastern United States, the pine barrens of New Jersey, and other sandy habitats in the upper Mississippi Valley and west Texas.

The New World species of *Toxicodendron* section *Venenata* include *T. vernix* (syn. *Rhus vernix*), the poison sumac, which grows in bogs and swamps from southern Quebec to southern Florida, and *T. striatum* (syn. *R. striata*), an upland tree occurring from Mexico to Colombia. The Asiatic members of section *Venenata* are the lacquer trees, which occur in China, Korea, Japan, the foothills of the Himalaya, and Southeast Asia (Chapter 9).

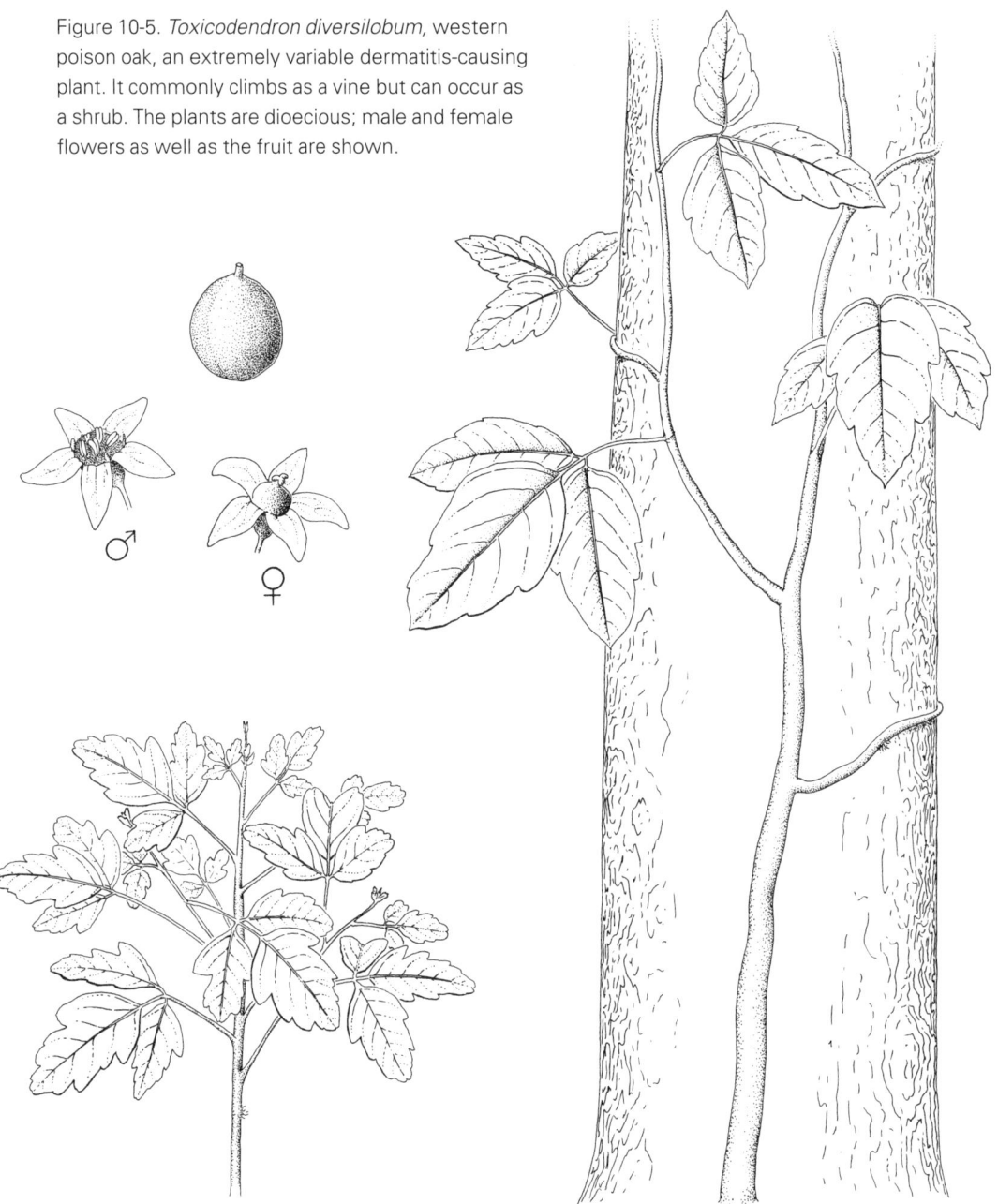

Figure 10-5. *Toxicodendron diversilobum,* western poison oak, an extremely variable dermatitis-causing plant. It commonly climbs as a vine but can occur as a shrub. The plants are dioecious; male and female flowers as well as the fruit are shown.

Plate 1. *Pinus sylvestris,* Scots pine, in northern Finland, the most widespread European pine.

Plate 2. *Abies balsamea,* balsam fir, amid red-leaved sugar maples *(Acer saccharum)* in Ontario, Canada.

Plate 3. The spruce-fir subalpine forest zone, dominated by *Picea engelmannii,* Engelmann spruce, and *Abies lasiocarpa,* subalpine fir, on the western slope of the Colorado Rocky Mountains.

Plate 4. *Pinus jeffreyi,* close relative of *P. ponderosa,* ponderosa pine (see also Plate 34), on the left, and *Pseudotsuga menziesii,* Douglas fir, on the right, in middle elevations on the western slope of the Sierra Nevada, California.

Plate 5. Typical open pinyon pine–juniper *(Pinus edulis–Juniperus monosperma* and *J. scopulorum)* forest in southern Nevada.

Plate 6. Branch of *Cedrus atlantica,* Atlas cedar, with resin exuding from a young cone.

Plate 7. *Pseudolarix amabilis,* golden larch, a possible relative of the Baltic amber tree, in the Morris Arboretum, Philadelphia. Courtesy of B. A. LePage.

Plate 8. *Agathis robusta,* Australian relative of
A. australis, the kauri, with heavy resin exudation
in Queensland, Australia.

Plate 9. *Callitris preissii,* an Australian sandarac
tree, in the University of California, Santa Cruz,
Arboretum.

Plate 10. *Taxodium distichum,* bald cypress, in the Okefenokee Swamp, Florida.

Plate 11. *Metasequoia glyptostroboides,* dawn redwood, the living fossil, characteristically silhouetted in fog in its small remaining native cloud forest habitat, Sichuan province, China. Courtesy of W. B. Gittlen.

Plate 12. Acaroid-resin-producing *Xanthorrhoea australis,* grass tree, in New South Wales, Australia.

Plate 13. *Liquidambar styraciflua,* sweet gum, with red fall foliage in mountain forest in Tamaulipas, Mexico. Courtesy of J. G. Martínez-Avalas.

Plate 14. *Larrea tridentata,* creosote bush, dominant shrub in the Mojave Desert.

Plate 15. Copal-producing *Daniellia ogea* in rain forest of central Ghana.

Plate 16. *Hymenaea courbaril,* copal-producing tree, emergent above the upland rainforest canopy, to the left of the Jeep, in central Amazonia, Brazil.

Plate 17. *Dacryodes excelsa* exuding elemi resin from a cut in the trunk, in Puerto Rican rain forest. Courtesy of M. Lerdau.

Plate 18. *Bursera simaruba* in Mexico, with red peeling back typical of some resin-producing *Bursera* species.

Plate 19. *Commiphora* in open dry forest in northern Kenya.

Plate 20. *Schinus molle,* pepper tree or American mastic tree, in a southern Mexican village.

Plate 21. *Eremophila fraseri,* resin-producing shrub in Western Australian desert. Courtesy of M. Pennacchio.

Plate 22. *Mimulus aurantiacus,* sticky monkey flower, a shrub in central California coastal chaparral.

Plate 23. Cone enclosed in Baltic amber. From Grimaldi (1996) with permission of Harry N. Abrams, Inc.

Plate 24. Eocene fossil *Metasequoia* log with exposed amber-filled cyst, from Axel Heiberg Island, Canada, ×3. Courtesy of B. A. Le Page.

Plate 25. *Hymenaea protera* leaf (length, 5.25 cm) in Dominican amber. Dots are secretory pockets where resin was produced. From Grimaldi (1996) with permission of Harry N. Abrams, Inc.

Plate 26. *Hymenaea protera* petal in Dominican amber. Dots are secretory pockets where resin was produced. Courtesy of F. M. Hueber, Smithsonian Institution.

Plate 27. Section of trunk of *Hymenaea verrucosa* from Zanzibar, Africa, showing almost pure resin between the bark and heartwood. Courtesy of D. Grimaldi with permission of the American Museum of Natural History.

Plate 29. Nest made entirely from resin of the dipterocarp *Anisoptera thurifera* by a megachilid bee, *Chalicodoma pluto,* in the northern Moluccas. Courtesy of A. C. Messer.

Plate 28. Squirt-gun resin response of *Bursera schlectendalii* leaves when resin canals are cut by a chrysomelid beetle *Blepharida* larva (at left). Courtesy of D. L. Venable.

Plate 30. Male flower of *Clusia grandiflora* from the Gran Sabana, Venezuela, showing a bee balling up resin from infertile staminodia surrounded by fertile red stamens. Courtesy of J. Lokvam.

Plate 31. Flowering *Dalechampia tiliifolia* vine in rain forest of Chiapas, Mexico. Courtesy of W. S. Armbruster.

Plate 32. Close-up of leaves of the Argentinian *Larrea cuneifolia,* with exuded resin making the leaves appear shiny. A branch-mimicking grasshopper is camouflaged near a leaf. Courtesy of D. F. Rhoades.

Plate 33. *Betula resinifera*, Alaska paper birch, near Fairbanks; black spruce forest in the background. Courtesy of J. P. Bryant.

Plate 34. Fire in *Pinus ponderosa,* ponderosa pine, forest in northern California, the intensity abetted by the abundance of terpenes in the trees. Courtesy of J. Greenlee.

Plate 35. Amber chest (length, 56.25 cm), 18th century, Danzig, from the Ekaterininsky Palace, St. Petersburg. Courtesy of D. Grimaldi with permission of the American Museum of Natural History.

Plate 36. Early 18th century Chinese amber carving of a carp leaping from the water (height 6.8 cm, not including the base). From Grimaldi (1996) with permission of Harry N. Abrams, Inc.

Plate 37. The Amber Room in the 1930s at the Ekaterininsky Palace, St. Petersburg (Leningrad at the time), before it was disassembled by the Nazis. Courtesy of D. Grimaldi with permission of the American Museum of Natural History.

Plate 38. Stylized amber bear pendant (length, 6.5 cm) from the Middle Stone Age (about 4500 B.C.) of Jutland, Denmark, with permission of the National Museum of Denmark, Copenhagen.

Plate 39. Experimental plantation in Florida to find high-resin-yielding progeny of *Pinus elĺiottii,* slash pine.

Plate 40. Local person collecting keruing oil from *Dipterocarpus kerrii* in Malaysia. Courtesy of R. Gianno with permission of the Connecticut Academy of Arts and Sciences.

Plate 41. *Protium heptaphyllum* being tapped for elemi *(breu)* in Bahia, Brazil. Courtesy of D. C. Daly.

Plate 42. *Boswellia sacra,* source of frankincense, in Arabia. Courtesy of the Royal Botanic Gardens, Kew.

Plate 43. Resin being collected by local people from a large *Hymenaea courbaril*. The trees typically grow along streams in southern Mexico, as shown in the background. See also Plate 16.

Plate 44. Resin being collected from tapped *Shorea javanica* in Sumatra. Courtesy of R. Gianno.

Plate 45. *Agathis dammara* in the Celebes with cankerous growths resulting from overtapping for resin. Courtesy of T. C. Whitmore with permission of the New York Botanical Garden Press.

Plate 46. *Ferula jaeschkeana,* the source of an asafoetida-type resin in Iran. Courtesy of the Royal Botanic Gardens, Kew.

Plate 47. *Daemonorops micracantha* fruit covered with red phenolic resin. Courtesy of J. Dransfield.

Other Poisonous Anacards

Resins from several tropical anacard trees used for fruit and or seed, such as cashew *(Anacardium occidentale),* Australian native cashew *(Semecarpus australiensis),* and mango *(Mangifera indica),* can produce a toxic reaction. *Anacardium* is a tropical American genus that exudes resins used by local people for caulking boats, waterproofing fishnets, and as varnish as well as preventing termite and other insect attack in woodwork (Mabberley 1997). Resin is also collected by euglossine bees for nest building. The fruit of *A. occidentale* consists of a kidney-shaped seed (the so-called cashew nut) surrounded by a hard shell; it subtends the swollen pear-shaped pedicel. The nut inside the shell and the pedicel are not allergenic, but resin from other parts of the plant are mildly so as a result of anacardic acid and cardol (Mabberley 1997). The shell contains compounds closely related to those of Japanese lacquer (Tyman 1979).

Anacardium occidentale was one of the first trees the early Portuguese explorers introduced into other parts of the tropics. Fond of the nut, they brought it to India and other areas in Southeast Asia (Howes 1949). It is now extensively cultivated along the western coast of India and in both East and West Africa as it grows well under dry conditions and in poor soils unsuited for other crops. Before World War II, Indian cashew nutshell liquid (known commercially as CNSL) was resin collected during the nut roasting process (Howes 1949). CNSL was one of the only known, economically significant plant sources of phenols used for various purposes, including plastics, brake linings, clutches, etc. World War II, however, seriously affected the CNSL industry. India today is the primary center for industrial use of CNSL as a baking enamel (Mills and White 1994).

Seeds of some species of *Semecarpus* are also eaten, for example, those of *S. australiensis,* called Australian native cashew. However, contact with the seeds or any part of the tree can cause severe dermatitis in susceptible individuals (Oeirichs et al. 1997). Aborigines leached the seed with water, roasted it, then removed the shell, thus making the nut edible. In attempts to make the nut usable commercially because of the similarity of its flavor to that of *Anacardium occidentale,* Oeirichs and coworkers identified the allergen: urushiol. *Semecarpus australiensis* is also known as tar tree because the resin is black; the exudate from its fruit and that of *S. anacardium* are used for ink and as a black dye for linen.

Mangifera is an Indomalesian genus, but *M. indica* is widely cultivated throughout the tropics for its mango fruit. Eating the fruit can cause a mild dermatitis in highly sensitive individuals. The resin occurs in ducts in the rind of the fruit. These ducts form a longitudinal multilayered network embedded in the outer region of the thick rind (Joel 1980). This network branches and anastomoses in all directions. Joel hypothesized that the complicated duct system confers protection against either vertical penetration of the ovipositor of fruit flies or the movement of their larvae through the rind. Significant differences occur in the duct systems of different mango cultivars. The resistance rendered by a particular duct system depends on the number of ducts, the dimensions of the duct, and the thickness of the duct layer. Thus the characteristics of the duct system may have agricultural significance by providing resistance to fruit pests.

Resins from three other anacard genera of tropical American trees can produce severe contact dermatitis in susceptible individuals (Martínez 1969). *Metopium* (poison wood) comprises three species in Florida, Mexico, and the West Indies, with *M. brownei* and *M. toxiferum* particularly toxic. *Pseudosmodingium* includes seven resin-producing Mexican species, with *P. perniciosum,* as the name suggests, very irritating. *Comocladia* has 20 species, with *C. engleriana* in Mexico noteworthy for its caustic resin.

Chemistry of the Poisonous Resins

The importance of understanding the chemistry of the dermatitis-causing resins of Anacardiaceae is underscored by the study of the active compounds, beginning as early as 1858. However, it was not until 1916 that McNair determined their phenolic catechol nature and possession of unsaturated side chains. Then, Majima (1922) recognized the similarity of the components in poison ivy *(Toxicodendron radicans)* and the lacquer tree *(T. vernicifluum),* calling the complex of closely related derivatives of catechol urushiols after *urushi,* the Japanese name for the plants of the *Toxicodendron* group. Dawson (1954, 1956) elucidated the structures of the side chains (generally 15–17 atoms) in *Toxicodendron* and the variation in different degrees of saturation, with double bonds of either *cis* or *trans* geometry.

The toxic constituents of the resin in poison ivy, poison sumac *(Toxicodendron vernix),* and the lacquer tree are primarily pentadecylcatechols with some heptadecylcatechols. In contrast, in western poison oak *(T. diversilo-*

bum) they are primarily heptadecylcatechols with some pentadecylcatechols (Figure 1-8; Corbett and Billets 1975, Baer 1979).

The resin chemistry of the Burmese lacquer tree (*Gluta*; Chapter 9) is similar to that of *Toxicodendron* except that the phenolic compounds have side chains of 18 carbons with a terminal phenyl group that makes drying very slow. Resin from *Gluta* also apparently causes less severe dermatitis in susceptible individuals than *Toxicodendron*.

Labdanum

Cistus, a genus of perennial evergreen shrubs in the family Cistaceae, occurs in dry scrub and open woodland from the Canary Islands through southern Europe and North Africa to Transcaucasia. The plants are called rockroses because they often frequent rocky places and have blooms that superficially resemble single-flowered roses (Figure 10-6). Leaves of all species are covered with glandular trichomes that secrete resin, which consists mainly of terpenoids but contains some flavonoid aglycones as well (Vogt et al. 1987). Although of Arabic origin, the common Greek name for the resin since antiquity has been *ladan,* and today it is called either labdanum or ladanum. (Ladanum should not be confused with laudanum, an archaic term that refers to certain opium preparations used as powerful analgesics.)

Resin is particularly copious from *Cistus ladanifer,* a shrub of the western Mediterranean region that is an early colonizer following fire. Today, resin is collected by dragging a kind of rake through the plants; the resin sticks to the tines. Pliny the Elder reported that herdsmen obtained the resin from goats' beards where it had collected as the animals moved amid the shrubs (Figure 10-6). In Spain, a leading producer today, labdanum is obtained by boiling twigs collected in spring and early summer, skimming off the resin that comes to the surface. In France, solvent extraction is used, a process that retains more of the fragrant volatile compounds. It is interesting, however, that little is known chemically about the volatile fraction of the resin (Guenther 1952, Vol. 6).

Cistus creticus (syn. *C. incanus* subsp. *creticus)* in the eastern Mediterranean region and on the island of Crete, a smaller shrub than *C. ladanifer,* is the source of a similar aromatic resin. Seven labdane-type diterpenes were identified by GC-MS, and four labdane diterpene esters of malonic acid were found in the resin coating the leaves and stem of *C. creticus* (Demetzos et al.

Figure 10-6. The Mediterranean rockrose *Cistus ladanifer* produces copious amounts of labdanum on leaves and stems. The resin was once brushed from the beards of goats after they had wandered among the shrubs. Flower and fruit are shown above.

1990, 1994). Chinou et al. (1994) reported that these diterpenes have cytotoxic as well as antibacterial and antifungal properties. Demetzos et al. (1999) showed that antistaphylococcal activity is primarily due to the diterpene sclariol. Moldenke and Moldenke (1952) concluded that the biblical labdanum was derived from *C. creticus*, *C. salvifolius*, or *C. villosus*, species that occur in the eastern Mediterranean rather than *C. ladanifer* from the western region. In early times, herdsmen collected the resin of the eastern species by combing the fleece of their flock, where the resin had collected as the animals wandered among the shrubs.

The odor of labdanum varies considerably, according to species and geography. Its fragrance may be of the "heavy oriental type or powerful and sweet with something in common with ambergris, or may have a detrimental ammoniacal odor" (Howes 1950). In fact, labdanum is the nearest approximation to ambergris, from whales, among plant exudates. It is used in the preparation of artificial amber and as a fixative in perfumes, particularly of the following types: carnation, hyacinth, lavender, lily, narcissus, patchouli, rose, verbena, violet, and wallflower. Labdanum also is utilized in fumigating preparations such as pastilles. In perfuming soap, it tends to be most favored for toilet soaps with lavender and sandalwood scents. Labdanum was once apparently used extensively in medicine (Moldenke and Moldenke 1952). It was dissolved in beer for dyspnea (shortness of breath), administered internally by catheters, enemas, and suppositories (Duke 1983). It was also used to prepare plasters and wound dressings as well as used as a stimulant and expectorant for catarrh and asthma (Grieve 1959).

Eleven neoclerodane diterpene acids from the resin of *Cistus populifolius* have been isolated and characterized (Urones 1994). The goal was to obtain highly functionalized diterpenoids from a readily available source that could be used a starting point for synthesis of new antifeedant agents for pests (Chapter 11).

Desert Shrub Surface Resins

Numerous desert shrubs produce resins that cover young leaves and stems (Chapters 2 and 3). It is assumed that the resins provide some protection against desiccation, high-intensity radiation, and herbivory (Chapter 5), but they are also useful to humans.

Myoporaceae

The family Myoporaceae comprises only three genera, two of which are prominent in Australia (Chapter 2). Aborigines considered some species of *Eremophila* and *Myoporum* important for medicinal and ceremonial purposes. *Eremophila* is a very large shrub genus, restricted to Australia, with many species producing copious amounts of resin that coats the leaves. Resins such as those from *E. fraseri* (Plate 21) have also been used as natural adhesives and sealants. As indicated by the name of the genus, from the Greek for "desert loving," *Eremophila* species are adapted to arid or semiarid habitats and are useful in revegetation because of their tolerance of drought, fire, and grazing. The thickness, distribution, and amount of resin that accumulates on leaves vary considerably. Resin constituting 5–10% of the dry biomass of leaves is typical but it can be as much as 20% (Chapter 3). Large quantities of resin can harden to a transparent varnish sufficiently thick to be chipped off in species such as *E. abietina*, *E. fraseri*, *E. ramiflora*, and *E. viscida*.

The large amounts of resin on leaves and terminal branches of Myoporaceae have led to interest in the chemistry of the resins. Hegnauer (1990) noted a striking range of carbon skeletons for the terpenoids, and Ghisalberti (1994) isolated and elucidated the structures of more than 200 compounds from the resins of Myoporaceae. Monoterpenes were not widely represented whereas sesquiterpenes were common, although relatively few species accumulated them in significant quantities. *Myoporum* is characterized by furanoid sesquiterpenes, but *Eremophila* has large quantities of novel diterpenes (Figure 1-6). The furanosesquiterpenes have been extensively studied because of their toxicity to animals that feed on the plants. In fact, Ghisalberti suggested that any species of the family that contains one or more furanosesquiterpenes should be regarded as potentially toxic to higher animals. All the structural classes of diterpenes from *Eremophila* are structurally and stereochemically unique (Chapter 1). The large number of unique terpene structures in *Eremophila* leads to questions about the circumstances under which they evolved (Chapter 2).

Fatty acids also occur in myoporaceous resins. They rarely dominate, but fatty acids and alcohols are important in two species of *Eremophila* (Ghisalberti 1994). Flavonoids (flavones) also often constitute a significant portion of *Eremophila* resins, but their identification has received little study. These additional compounds add to the incredible complexity of *Eremophila* resins.

Apart from limited tests regarding the toxicity of furanosesquiterpenes, little is known about the biological activity of chemicals in *Eremophila*. However, inferences can be made by comparison with similar compounds from other plants that have been analyzed. For example, ngaione (Figure 1-6), a furanosesquiterpene found in numerous *Eremophila* species, is an enantiomer of the sweet potato phytoalexin ipomeanarose (Wilson and Burke 1983). *Eremophila latrobei* produces a diterpene of known antibiotic activity previously isolated from *Capraria biflora* (Scrophulariaceae; Forster et al. 1986). The phenylpropanoid verbascoside, found in undifferentiated callus tissue of several *Eremophila* species, exhibits antihypertensive and analgesic activity and potentiates antitremor L-dopa activity in animals (Dell et al. 1989).

Eremophila alternifolia and *E. longifolia* have featured prominently in the pharmacopoeia of the Aborigines of coastal and central Australia (Richmond and Ghisalberti 1994). Verbascoside and geniposidic acid, which have previously unknown cardioactivity, have been identified from leaf tissue of these species (Pennacchio et al. 1996). These compounds caused significant but opposite changes in activity in rat heart. Verbascoside from *E. alternifolia* leaves significantly increased heart rate, contractile force, and coronary perfusion rate whereas geniposidic acid from *E. longifolia* leaves decreased heart rate, contractile force, and coronary perfusion rate. However, the inhibitory effects were soon followed by stimulatory effects. Pennacchio and coworkers suggested that the stimulatory effect was mediated by verbascoside, which also occurs in *E. longifolia* leaves. It was not clear from the study whether the effects of the two compounds are mediated by the same mechanism or separate ones.

Asteraceae

McLaughlin and Hoffmann (1982) surveyed 195 plant species native to the southwestern United States and northwestern Mexico for potential feedstocks for biocrude production in arid lands. The highest cyclohexane extract (more than 10%) came from resinous shrubs and perennial herbs in Asteraceae subfamily Asteroideae. In fact, many of the most resinous ones were in the tribe Astereae, including *Baccharis*, *Chrysothamnus*, *Grindelia*, *Gutierrezia*, *Haplopappus*, *Lessingia*, and *Xanthocephalum*, often in numerous species (Chapter 2 and Appendix 3). The most highly resinous taxa included *B. sarothroides*, various subspecies of *C. nauseosus*, and *C. paniculatus*, *Gutier-*

rezia microcephala, H. linearifolius, L. germanorum, and *X. gymnosperm-oides.* Resins coat the stems, leaves, and flower heads of these plants, with greatest yields commonly from trichomes of the involucres or leaves (Chapter 3). Maximum total extractables apparently are inversely proportional to plant size; this trade-off may result from the plant's putting more carbon into resin, which defends against herbivores or protects from aridity and high temperatures, than into growth (Chapter 5).

Following their survey, Hoffmann and McLaughlin (1986) focused on study of the perennial *Grindelia* as a potential cash crop. The common name gum plant or gum weed comes from the plants' being tacky or sticky to the touch. The greatest concentration of resin occurs in trichomes of the flower head, with those of the leaves second (Chapter 3). Some species contain various flavonoids along with monoterpenoids such as borneol, terpineol, and α- and β-pinene; the major components isolated from seven *Grindelia* species are a group of 12 labdane diterpene acids, however, with grindelane acids the most abundant (Timmermann et al. 1983).

Resin acid composition varies between individuals and plant parts in *Grindelia* (Timmermann et al. 1987). The quantity of resin varied, 5–18%, within and between *Grindelia* species such as *G. camporum* and *G. squarrosa.* Their diterpene acids have an abietane skeleton and thus are similar to some diterpene acids characterizing pine rosin (Figure 1-4). Therefore, grindelane acids could possibly be used as substitutes for pine rosin in the naval stores industry (Hoffmann et al. 1984). Although the shortage of rosin production in the United States has been somewhat alleviated by imports from China, Hoffmann and McLaughlin (1986) pointed out that a domestic renewable source of rosin substitutes might provide a good chance for *Grindelia* resins to penetrate the market. In fact, they reported that the U.S. market had been able to absorb as much as 45.4 million kilograms of rosin in the recent past. Furthermore, several other applications have been patented for *G. squarrosa* resin in the food, rubber, coatings, textile, and polymer industries (Timmermann and Hoffmann 1985). Hoffmann and McLaughlin predicted that successful development of *G. camporum* as a cash crop could be important to arid agriculture in the Southwest and provide an even greater investment opportunity for the processing sector of the industry (Chapter 11).

Gum plants, including *Grindelia camporum, G. humilis, G. robusta,* and *G. squarrosa,* were used by Native Americans to treat bronchial problems, skin irritations caused by poison ivy, and other complaints (Vogel 1970). Resin from *G. camporum* still provides a valuable remedy for bronchial asthma (Chevallier 1996). Both an antispasmodic and expectorant, it helps relax muscles of the smaller bronchial passages and clear congested mucus. It may also desensitize nerve endings in the bronchial tree and slow the heart rate, both leading to easier breathing. Moreover, this resin has been used to treat whooping cough, hay fever, and cystitis.

Dragon's Blood

The fanciful name dragon's blood has been used for the deep red or ruby red phenolic resins from two different monocots: *Dracaena* (Convallariaceae; Figure 10-7) and *Daemonorops* (Arecaceae). The name also has been used for the exudate from the dicot *Croton* (Euphorbiaceae). Thus, similar to the problems with use of the name dammar (Chapter 7), resins called dragon's blood have different plant origins; the plants also occur on different continents. Moreover, it is difficult to determine the generic origin on a morphological basis from older collections of the resin, such as those in the Royal Botanic Gardens, Kew, Economic Botany Collection (Pearson and Prendergast 2002).

Dragon's blood from *Dracaena cinnabari* was alluded to by Dioscorides, Pliny the Elder, and other ancient writers. Dioscorides referred to it as *kinnabai.* According to Pliny, the name dragon's blood originated from a battle between a dragon-like creature and an elephant that led to the mixing of the blood of the two animals. The resin was considered magical and was esteemed for its alleged medicinal properties by the Greeks, Romans, and Arabs. The resin came from the island of Socotra (Figure 6-11) where it exudes from the plant stem in teardrops and is collected after the rainy season (Balfour 1888).

During Roman times, the Arabs sent the resin to Europe via Bombay or sometimes by way of Zanzibar. Thus one trade name for Socotran dragon's blood was Zanzibar drop. The Arab name, *dam-ul-akh-wain,* is still current in the region. Another species, *D. schizantha,* occurs along the coast of Somalia and is probably the origin of Zanzibar dragon's blood (Barry 1932). In Europe, dragon's blood was formerly used for diarrhea and dysentery and as an astringent in tooth powders. It was a source of mahogany-colored varnish and was used to stain marble.

Resin from the dragon tree (*Dracaena draco*, Figure 2-6), collected from incisions made in the stem of the plant during voyages to the Canary Islands during the 15th century, was one of the valued products of the early explorers. Although the resin of *D. draco* was important to the inhabitants of the Canary Islands, as indicated by drawings in their sepulchral caves, and the early explorers, this resin never became an important article of trade. González et

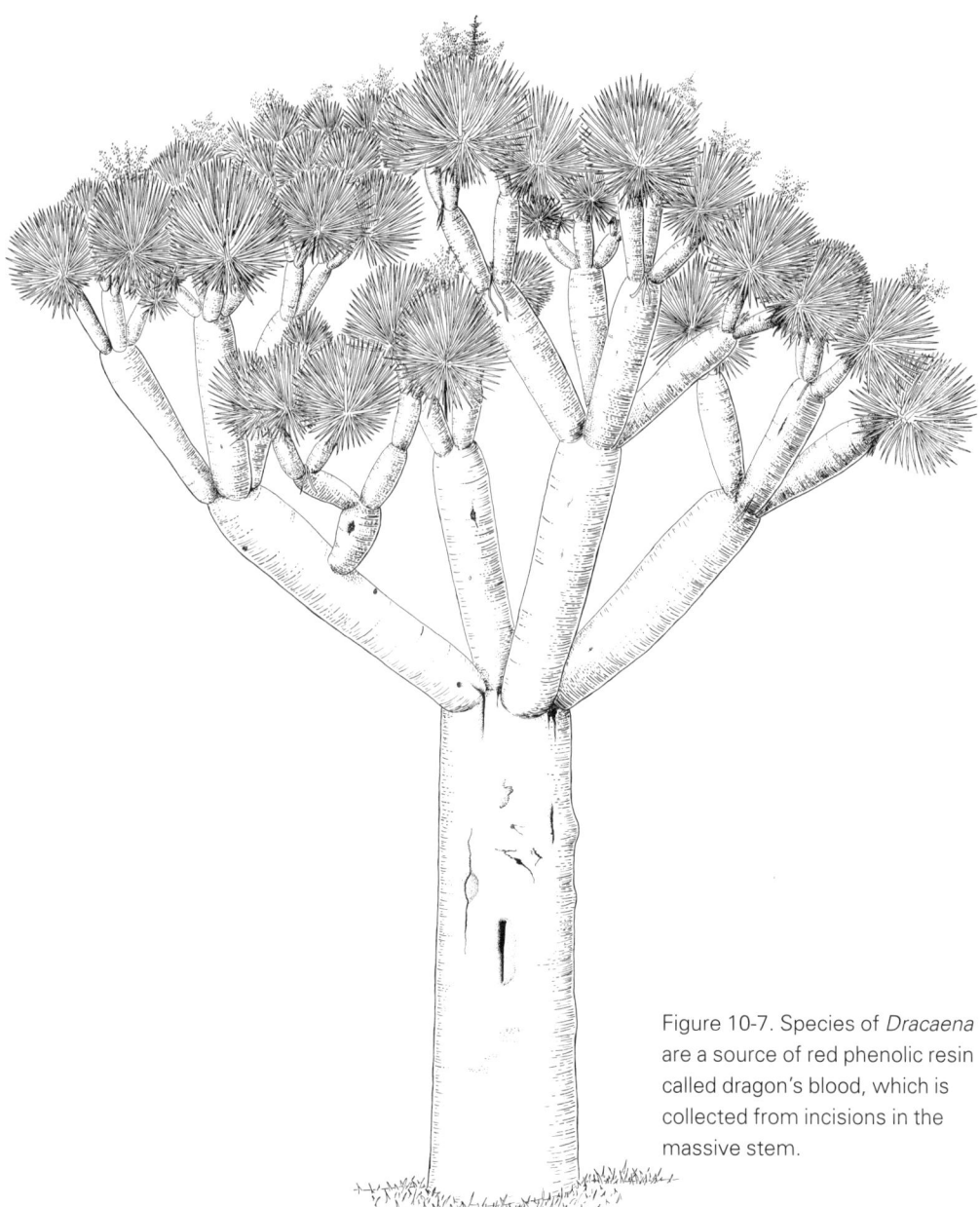

Figure 10-7. Species of *Dracaena* are a source of red phenolic resin called dragon's blood, which is collected from incisions in the massive stem.

al. (2000), however, analyzed the resin, finding 21 phenolic compounds, including a phenylpropanoid, chalcones, flavonoids, and a cinnamyl alcohol.

Resin from *Dracaena cochinchinensis* has been found to be induced by infection of the fungi *Cladosporium* and *Fusarium* (J. Wang et al. 1999). Such fungicidal compounds are called phytoalexins, which in the *Dracaena* resin are hydroxylated and methoxylated flavans.

A group of about nine allied species of lianas in the large Indomalesian rain forest genus *Daemonorops* (noted for rattans; Figure 2-7) has been the primary source of dragon's blood in commerce. This group of palm species occurs primarily on Borneo, Sumatra, and peninsular Malaysia, but those on Sumatra have been the main producers of this dragon's blood. The resin occurs in a brittle layer on the surface of immature fruits about the size of a cherry and covered with imbricated scales (Plate 47). Immature fruits are collected because the resin tends to loosen as the fruits ripen. After the fruits are dried and the resin is dislodged, the resinous powder is softened by heat, then molded into sticks or cakes. The phenolic resin of *D. draco* and related species, including *D. didymophylla*, *D. micracantha*, *D. propinqua*, and *D. rubra*, consists of more than 50% of the resin alcohol dracoresinotannol, which is associated with benzoic and benzoyl acetic acids. Benzoyl acetic ester, dracorsene, dracoalban, and cinnamic acids also occur (Duke 1985).

Arnone et al. (1997) described six new A-types of flavonoid deoxyproanthocyanidins from *Daemonorops* resin. The structures of all phenolic compounds so far isolated from this resin, including the well-known red pigments dracorhin and dracorubin, apparently derive from the oxidation of the two simplest compounds of the resin: 5-methoxyflavon-7-ol and 5-methoxy-6-methylflavan-7-ol. Although the resin of *Daemonorops* is composed primarily of phenolic compounds, Piozzi et al. (1974) reported diterpene resin acids from *D. draco*. Interestingly, these are pimaric and abietic acids characteristic of pine resins.

Duke (1985) listed many uses for *Daemonorops* resin. Since it is astringent, it has been used in collyriums (eyewashes), dentifrices, and mouthwashes, and for diarrhea and dysentary. Roa et al. (1982) demonstrated in vivo antimicrobial activity of the anthocyanidin pigments dracorhodin and dracorubin in *D. draco* against *Staphylococcus aurea*, *Klebsiella pneumoniae*, *Mycobacterium smegmatis*, and *Candida albicans*. It is also used to tint plasters, tinctures, toothpaste, imitation tortoiseshell, and particularly, varnishes, even to stain marble a deep red. The great Italian violin makers of the 18th century varnished their instruments with this dragon's blood. In the Orient, bamboo furniture is stained with the resin.

Pearson and Prendergast (2002) reported that dragon's blood from *Dracaena* and *Daemonorops* can easily be purchased, but most Internet sites do not give either the botanical name or the locality of origin. Although Edwards et al. (1997) were able to distinguish fresh resins of *Dracaena cinnabari* and *Daemonorops draco* chemically using Raman spectroscopy, it is not known how well this technique would aid in analysis of degraded archeological specimens, or for restoration work of medieval paintings where dragon's blood was used as a pigment.

Several Latin American species of *Croton* yield a red exudate commonly called dragon's blood. In some cases it is called a resin whereas in others it is referred to as a latex (Coppen 1995). Species producing *sangre de draco* include *C. draco*, *C. draconoides*, *C. lechleri*, *C. urucurana*, and *C. xalapensis*. They are distributed from Mexico through Central America and parts of South America, including Venezuela, Ecuador, Peru, and Brazil. Similar to that from *Dracaena*, the dragon's blood is collected from incisions made in the stem. Pieters et al. (1992) confirmed the wound-healing effect of crude *Croton* resin with in vivo experiments with rats. A component of this dragon's blood, 3′,4-O-dimethylcedrusin, also improved wound healing in vivo. The effect was greater with the crude resin, however, which Pieters and coworkers attributed to the additional polyphenols that precipitate cell protein and form a crust over the wound.

Other Resins

Podophylloresin

Podophyllin is a mixture of lignan resins (podophylloresin) extracted from the rhizome and root of *Podophyllum peltatum* (Berberidaceae). Called mayapple or American mandrake, *P. peltatum* is a perennial herb that grows in the woodlands of eastern North America. Indian podophylloresin is a similar substance obtained from dried rhizomes and roots of *P. emodi* (syn. *P. hexandrum*), which grows in Tibet and Afghanistan. *Podophyllum emodi* produces somewhat higher amounts of resin than *P. peltatum* and is the main commercial source of podophyllin (Tyler et al. 1988, Moraes et al. 2000). Interestingly, a similar podophyllin also has been found in four species of *Juniperus* (Neish 1965).

The irritant and cathartic effects of podophylloresin are the result of podophyllotoxins. These phytophyllotoxins are the starting material for semisynthesis of the anticancer compounds etoposide and teniposide (Stähelin and Wartburg 1991). Anticarcinogenic research has shown that one or more of these constituents delay the formation of, or completely inactivate, some nuclear and cytological figures during the metaphase of mitosis in some cells of experimental mice. Although the mechanism is not clearly understood, the effect also may be of value in treating venereal warts and some other forms of papillomas. The purgative properties of podophyllin can produce drastic effects.

Because *Podophyllum emodi* was declared in danger of extinction at the 1989 Convention on International Trade in Endangered Species of Fauna and Flora (CITES), the export of bulk amounts of *P. emodi* has been prohibited. Collection of the plant is still allowed for export of podophyllotoxin-enriched resin because of demand for the material to test for new therapies (e.g., Ekstrom et al. 1998, Ajani et al. 1999).

Increasing demand and the compromised availability of source material has prompted researchers to search for alternative sources of podophylloresin. Leaves of *Podophyllum peltatum* can yield greater quantities of podophyllotoxin than the rhizomes of *P. emodi* (Moraes et al. 2000). The use of resin-rich leaves rather than rhizomes provides an excellent opportunity for domestication of the mayapple and its development as a renewable, high-value crop for small farmers in the United States, as suggested by Meijer (1974).

Poplar Bud Resins

Balsam poplar *(Populus balsamifera)* belongs to section *Tacamahaca* of the genus (family Salicaceae). Balsam poplars (a number of subspecies are recognized) are also called *tacamahac,* a name of West Indian origin. Their winter buds contain fragrant resins that coat young leaves when they unfurl from the bud (Figure 10-8). In contrast to bud resins from species of section *Aigeiros,* they lack the flavones and methylated flavonols that characterize propolis. Instead, they contain dihydrochalcones and sesquiterpenols, with the phenylpropanoid cinnamyl cinnamate responsible for much of the strong scent.

Native North Americans used balsam poplar resin medicinally (Vogel 1970, Moerman 1998). They boiled the resinous buds in fat to make a salve to dress wounds or to put up the nostrils for relieving the congestion of head colds or bronchitis. The salve was also used for eczema and persistent sores as well as abating pains and soreness in the limbs. Among white people, the resin was said to be used as an ingredient in nervine plasters, used to warm, irritate, and gently blister. The cottony down from fruits (Figure 10-8) was used by some Native Americans as an absorbent for open sores. Resin is still collected and extracted as a flavor ingredient called populus (Burdock 1995).

Figure 10-8. *Populus balsamifera* subsp. *trichocarpa*, western balsam poplar or black cottonwood, secreting resin from epidermal cells on buds of unfurling leaves (Figure 3-10). Fruiting structure with typical cottony seeds is shown above.

It has a sweet balsamic odor with a slight cinnamon undertone and is used to flavor alcoholic beverages. *Populus balsamifera* and other species of section *Tacamahaca* are not collected by honeybees for propolis.

Guaiac

The resin guaiac occurs primarily in the heartwood of the small genus *Guaiacum* (Zygophyllaceae), a genus of evergreen trees of warm, dry areas in the New World, and is collected from *G. officinale* and *G. sanctum*. The wood of *Guaiacum,* called lignum vitae, is the hardest of commercial timbers. Although small tears of the resin can be collected from exudations along the trunk, guaiac constitutes 18–25% of the heartwood and is obtained in several ways (Howes 1949). A hole is bored in a log, which is burned, causing resin to flow out the hole. Resin may also be obtained by boiling sawdust in seawater or extracting it with solvents. Guaiac is usually marketed in large blocks. Wren (1988) indicated that the phenolic resin consists of several different lignan resin acids (e.g., guaiaretic, guaiacic, and dihydroguaiaretic acids), vanillin, and terpenes (e.g., guaiagutin).

Guaiacum officinale and G. sanctum have been listed at various times in the U.S. Pharmacopoeia, and the resin is permitted as a food additive in the United States and Europe. Guaiac has been used to alleviate skin diseases, gout, and rheumatism (Morton 1981). *Guaiacum officinale* was almost exterminated by local people in the areas where it grew for use in treating venereal disease, and as a result it was introduced for cultivation into Spain in 1501 (Mabberley 1997). In Europe, guaiac was used with mercury for treatment of syphilis, and in lozenges (Plummer's pills) for sore throats, and was included in the *British Herbal Pharmacopoeia* as a treatment for chronic rheumatic conditions in 1990. The resin has also been used in veterinary medicine as an internal antiseptic (Vogel 1970).

Creosote Bush Resin

Resin from another member of the Zygophyllaceae, the desert-dwelling creosote bush (*Larrea tridentata,* Plate 14), has been used by native North Americans for various purposes (Mabry et al. 1977). It was used as an antiseptic, for bronchial and pulmonary complaints, and for treatment of rattlesnake bites as well as for waterproofing baskets and fixing arrowheads to shafts. The resin is a complex mixture of phenolic aglycones, some volatile terpenoids, and a lignan, norhydroguaiaretic acid (NDGA, Figure 1-8), which is a major component. NDGA has been found to be effective in shrinking cancerous tumors and is used as an antioxidant in food and pharmaceutical preservation.

Gharu Wood

Gharu wood (sometimes referred to as *gaharu* or *gahur*), from several species of *Aquilaria* and, occasionally, *Gonystylus* of the large tropical family Thymelaeaceae, is a major resinous forest product in Southeast Asia (Burkill 1966, Gianno 1986b, Chang et al. 1997). Although healthy trees can produce scented wood, it is those species that have resin-saturated wood in the trunk and roots, caused by fungal infection, that are of commercial interest. Only particular species of *Aquilaria* (e.g., *A. agallocha*, *A. beccariana*, *A. hirta*, and *A. malaccensis*) and *G. bancanus*, and then only about one tree in 10 in a population, will have the fungal infection accompanied by resin-saturated wood. *Gonystylus* is considered to produce a lower-quality product (Wollenberg 2001). Healthy wood is light and soft whereas wood saturated with resin becomes dark, hard, and heavy. The highly scented saturated wood is not only used for incense, carvings, and clothing drawers, the resin is extracted for perfumes, cosmetics, and traditional medicine (J. Miller 1969).

During the Southeast Asian protohistoric period (first to 15th centuries A.D.), China was a major importer of gharu wood from the Malay Peninsula and elsewhere, making it one of the oldest woods exported for incense. It also was held in high esteem in northern India and the Middle East in biblical times. Although various names have been used for the heavy resinous wood, gharu wood, derived from the Sanskrit word connoting the wood's heaviness, is the name most used in commerce. However, it has also commonly been referred to as aloe wood or eagle wood. The Bible was the source, through Saxon, of the name aloe wood, from the Hebrew *ahaloth*. Eagle wood has a more complicated transition. The Portuguese seized the trade in the wood from the Chinese in the late 1500s (Burkill 1966). They distorted the Sanskrit into *pao d'aquila*, which the French translated into *bois d'aigle*, and thence the British into eagle wood. Thus eagle wood is the modern equivalent of aloe wood.

Southeast Asian tropical forest dwellers have always collected gharu wood, but there have been fluctuations in export demand for the product. The Semelai on the Malay Peninsula (peninsular Malaysia) provide an example of collectors and the future for such collection. In contrast to collection of keruing oil (Chapter 7) and dammar (Chapter 9), gharu wood collection is replete with ritual and superstition, which Gianno (1990) attributed to great uncertainty in finding trees with good wood. The fungal patches occur in spike-like deposits in various parts of the tree, such as deep in the heartwood, near the bark, in the roots, or in the branches (Jalaluddin 1977). Older trees, however, in which there is evidence of disease and dieback, are generally the best candidates. Luck, however, seems to play a greater role in the collection of gharu wood than other resins. According to Gianno, the "Semelai compare it to playing the national numbers game." The price of gharu wood depends on the strength of the aroma, the size of the pieces of wood, and the degree of blackness.

There was an active period of Semelai collection of gharu wood during the early 20th century, but only in the late 1970s did trade in the wood reappear (Gianno 1986b). The long hiatus in collecting allowed regrowth of many *Aquilaria* trees. Unfortunately, the chainsaws that came with the logging industry allowed Semelai to simply fell the trees. Gianno indicated that many Semelai did so because they thought that the forest was doomed anyway and that they should try to extract as much as they could first.

In the 1970s, interest in Indonesian gharu wood (called gharu there) focused initially on Sumatra, then Kalimantan (southern Borneo), and most recently, Irian Jaya (or West New Guinea; Wollenberg 2001). Singapore, Taiwan, Saudi Arabia, and the United Arab Emirates have been the dominant buyers of Indonesian gharu. A boom in East Kalimantan began in 1989–1990 when collectors tripled or quadrupled their annual income by collecting it, most commonly from *Aquilaria malaccensis* and *A. beccariana*.

The economic value of gharu led Wollenberg (2001) to examine the economic benefits gained by Kenyah swidden farmers in northern central Borneo and the implications for sustainable management of the resource in East Kalimantan. Gharu wood provided the Kenyah villagers their most significant cash income. The high incomes available from its collection, its role as the primary source of cash for most households, and the high economic return relative to the amount of work required indicated how gharu wood was valued locally, which provided economic incentives for continued harvest. However, the incentives did not lead to sustainable management. Harvesting techniques, combing an area thoroughly before moving on, resulted in rapid diminution of resources. The practice of felling trees unnecessarily, by people from outside the Kenyah villages who had little incentive for conservation, also depleted *Aquilaria* populations. These and other conditions encouraged only passive conservation by the villagers. Wollenberg concluded that the economic incentives driving gharu wood harvesting in Kalimantan unfortunately appeared to be on an "economically unstable trajectory." By 1998, villagers were moving northward to collect in neighboring Malaysia. A similar movement had occurred in 1991–1992 from East Kalimantan. In fact, uncontrolled exploitation of *Aquilaria,* together with the wasteful trial-and-error method of searching for resin-soaked wood, have essentially eliminated *Aquilaria* trees from all but inaccessible forests (Peters 1994). The question remaining is whether the decimation has been so complete that there will be no long-term regeneration as there had been in the past in some areas.

On the other hand, in peninsular Malaysia, *Aquilaria hirta,* a small tree in secondary forest, produces incense wood of higher quality than *A. malaccensis.* Gianno (1986b) suggested that *A. hirta* might be considered for cultivation since there are buyers willing to pay high prices for it. As an early-succession forest tree, *A. hirta* probably would mature quickly and thus be able to produce gharu wood relatively readily. Perhaps fungal inoculation would increase the frequency and intensity of occurrence. If successful, such cultivation would provide a valuable cash crop for Southeast Asian people, mitigating some of the problems associated with collecting gharu wood from the forest.

Guayule Resin

Guayule (*Parthenium argentatum,* Figure 10-9), a perennial shrub in the Chihuahuan Desert of northern Mexico and the southwestern United States, is the only species of the genus and of the tribe Heliantheae (of the family Asteraceae) to produce both resin and rubber. The resin contains sesquiterpene phenolic esters (guayulins; Behl et al. 1983), triterpenes (argentatins), methylated flavonoids, and cinnamic acid (Rodriguez 1977, Isman and Rodriguez 1983). Guayule also is the only species of *Parthenium* that produces significant amounts of guayulin A (stems and roots) and guayulin B (stems and leaves); other species are all characterized by large amounts of sesquiterpene lactones. Rodriguez suggested that guayule shifted from the sesquiterpene lactone defense system of other *Parthenium* species to one based on sesquiterpene

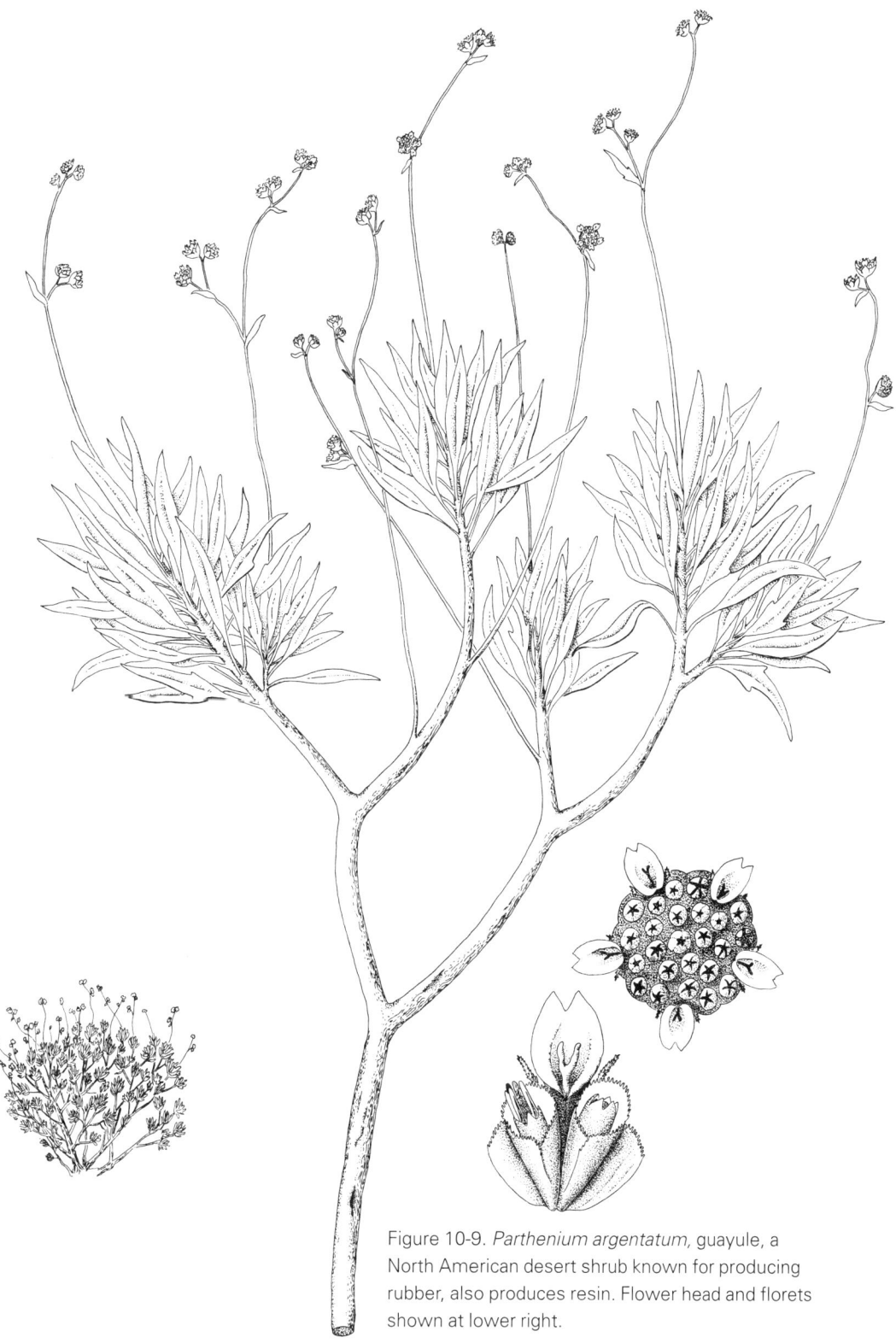

Figure 10-9. *Parthenium argentatum,* guayule, a North American desert shrub known for producing rubber, also produces resin. Flower head and florets shown at lower right.

phenolic esters and aromatic acids. Guayulins A and B are main components (10–15%) of the resin, stored in canals and adjacent cells in the cortex and exuding from the stem at any point of injury.

Understanding the co-occurrence of both resin and rubber has been important in attempts to isolate them so that they can be used industrially. The economic viability of guayule as a domestic source of natural rubber would be enhanced if the resin, a ubiquitous by-product of processing guayule for rubber, could be used to reduce gross production costs (Schloman et al. 1986). The practical value of resin-derived by-products, however, depends on the accessibility and utility of particular components.

Resin yields vary with cultivation site, harvest date, and shrub strain (Schloman et al. 1983). Resin composition varies with shrub strain and processing procedure (as true of cedarwood oil, Chapter 7). Schloman et al. (1986) studied seasonal effects on resin composition in a single guayule strain, using high-performance liquid chromatography. High resin levels are positively correlated with large numbers of leaves. Leaves also contain more resin than woody tissue whereas rubber levels are highest in woody tissue and bark. Also, young stems have higher resin levels than older stems. Triterpenoid levels show little seasonal variation whereas sesquiterpene esters (guayulin A) vary significantly with season. Behl et al. (1983) showed that guayulin A is as potent an allergen as urushiol from poison ivy, thus the potential for allergenicity varies according to guayulin A level. These studies show the complexity of resin production in guayule and reinforce the need for further research for industrial use of the resin as a by-product of rubber processing.

Resins from Latex

Components called resin are contained in the latex of the large cosmopolitan genus *Euphorbia* (Euphorbiaceae). There has been interest in the possible use of resin from some desert species as a source of hydrocarbons for petroleum (Calvin 1979, 1984). The Italians have investigated *E. abyssinica*, abundant in Ethiopia, and the Americans have studied *E. lathyris* and *E. tirucalli* as promising sources. The French have sought hydrocarbons from *E. resinifera* in Morocco, where latex from the cactus-like plant is used by veterinarians. Dried latex of *E. resinifera* is used commercially elsewhere for a well-known drug, gum euphorbium. Furthermore, *E. balsamifera* subsp. *adenensis* is a source of a valuable glue in Arabia.

Conifer Resins

Because of the great importance of pines as a source of naval stores (Chapters 6 and 7), there is a tendency to overlook the use of conifer resins for other purposes. For example, pure heptane, distilled from resin of *Pinus jeffreyi* (Plate 4), was used to develop the octane scale for rating petroleum as a motor fuel (Mirov and Hasbrouck 1976). In addition to the numerous industrial applications for pine resin, there have been medicinal uses. Turpentine (the volatile fraction of the oleoresin; Chapter 7) from pines such as *P. halepensis* in the Mediterranean region and *P. elliottii* in the southeastern United States has been used for catarrh, cough, dysuria (difficult or painful urination), gonorrhea, and rheumatism. Early physicians in America recommended ground pine resin steeped in water, called tar water, for numerous ailments such as smallpox, ulcers, and syphilis. Rosin (the nonvolatile fraction of the oleoresin) from pines such as white pine *(P. strobus)* and pitch pine *(P. rigida)* has been used for abscesses, boils, and cancers. It also has been used for toothache and skin diseases such as psoriasis and ringworm. Confederate surgeon Francis Porter, in a report written in 1863, described the long-leaf pine *(P. palustris)* of the southern United States as "one of God's great gifts to man." He reported that turpentine and pine tar were widely used as a stimulant, astringent, diuretic, and laxative as well as one of the best means of chasing away fleas.

Pinus massoniana and *P. tabulaeformis* are two species mentioned together as used medicinally in China (Stuart 1911). A soup is made of turpentine, water, and onions, which is used in poultices to be warmed and applied as needed for boils, carbuncles, various skin eruptions, old ulcers, and indolent wounds. Also, the resin of *P. merkusii* is used to make an unguent for abscesses.

Spruce resin exudes naturally from various species of *Picea* (Pinaceae) in the United States and Canada, with *P. rubens* the chief source for its use as chewing gum. It is a thin, clear, bitter, sticky oleoresin secreted in blisters in the bark and fissures in the sapwood (Hill 1952). When exposed to air, it hardens; it is collected when semisoft or hard. Before the advent of chewing gum made from the latex of the chicle tree (*Manilkara zapota,* syn. *Achras zapota;* Sapotaceae), about 225,000 kg of spruce resin gum were chewed annually (Hill 1952). The resin softens in the mouth, takes on a reddish color, and has a pleasing resinous taste. Moerman (1998) also mentioned extensive use of resin from other spruce species by Native Americans, for example,

P. engelmannii (Plate 3), *P. glauca,* and *P. sitchensis.* In each case the resin was used as a salve on cuts and wounds. It was often used for rheumatism or arthritis, and as a cold and cough remedy. *Picea engelmannii* resin was even taken with a spoon in a concentrated form directly from bark blisters as a treatment for cancer.

Hemlock pitch from *Tsuga canadensis* (Pinaceae) was listed in the *U.S. Pharmacopoeia* from 1831 to 1894, employed externally in medical plasters as a counterirritant.

A number of cupressaceous species have been used for other than cedarwood oil (Chapter 7). Cedar of Lebanon *(Cedrus libani)* once formed forests in the Holy Land. Now it grows only in scattered areas of Turkey, Lebanon, and Syria but is cultivated in England, France, and the United States. It contains a resin that makes the wood fragrant; Lebanese believe that the fragrance from this tree purifies the air and that smoke of burning resin helps chest problems suggestive of asthma. Resin has also been used for blenorrhagia (a free discharge of mucus, sometimes used synonymously with gonorrhea), bronchitis, and skin eruptions. Medicinally, the resin is considered diuretic and insecticidal. It is also used for burns, cancer, and indurations of the limbs (Duke 1983).

Cupressus sempervirens, an important biblical timber tree, has wood that is highly resistant to decay. The island of Cyprus, where the tree was worshipped, was named for this cypress. The resin is a valuable perfumery item, a source of ambergris- and labdanum-like odors. Cypress resin was a folk remedy for tumors of the eyes, nose, breast, and uterus, and indurations of the liver, spleen, stomach, and testicles (Duke 1983).

The resin from *sabino (Taxodium mucronatum),* the national tree of Mexico, which occurs from southern Mexico to Guatemala, has been used for various medicinal purposes. The tree only produces resin upon injury such as that from fire (Martínez 1959). A distillation product of the wood, called *alquitran* is an effective anesthetic when applied to the tongue. The resin is used for treatment of bronchitis and other chest infections, and for wounds, ulcers, and toothaches among other ailments.

Resin from the leaves and berry-like cones of red cedar *(Juniperus virginiana)* has been widely used by Native Americans for inflammatory rheumatism and for colds (Vogel 1970, Moerman 1998). They inhaled the smoke as well as used it in vapor baths. Moerman also mentioned similar uses of the resin from other North American juniper species (e.g., *J. scopulorum*, Plate 5). Duke (1983) listed several medicinal uses for Greek juniper *(J. excelsa,* syn. *J. procera)* as well. The stimulating resin is used against ulcers, and coning branches are burned as a fumigant for rheumatic pains. Mixed with honey, the resin is used for liver ailments. In Lebanon it is used in liniments and salves, in cough and liver medicines, and in pills and suppositories. Volatiles in the resin from cade *(J. oxycedrus)*, called heath in the Bible, are used in men's fragrances as they give them a woody, or smoky, leathery character. They are also used in food manufacturing to give a smoky flavor to unsmoked fish or meat products. Cade oil (resin) is added to antiseptic soaps and serves as a fragrance in creams, detergents, lotions, and perfumes. It was used for a plethora of purposes in folk remedies (Duke 1983). In Europe, cade oil is taken internally as a vermifuge and applied externally to old wounds and ulcers to promote healing. It has a long history for treating parasitic skin diseases of animals and more recently has been applied to similar human afflictions such as eczema and psoriasis.

The resin of the taxodiaceous tree *Cryptomeria japonica* (Cupressaceae s.l.) is a component of an ointment used in China and Japan. In Japan, resin from the wood of *Chamaecyparis obtusa* (Cupressaceae s.s.) is put in medications to treat gonorrhea. Distilled leaf volatiles from western white cedar (*Thuja occidentalis,* Cupressaceae s.s.) have been used by Native Americans as a stimulant of both heart and uterine muscles. These compounds were listed in the *U.S. Pharmacopoeia* from 1942 to 1950 and used as a soap liniment because of wartime shortage of lavender oil (Vogel 1970, Moerman 1998).

Amber in Medicine

As discussed in Chapter 6, amber has been worn since prehistoric times as a protection against evil, which included various diseases. This lore, attributing guardian properties to amber, has lasted through the present. For example, peasants in the Baltic area wear amber earrings and necklaces to ward off headaches and throat problems; in Mexico, amulets carved to represent various part of the body are worn to bring healing to them.

Since the time of Hippocrates (about 400 B.C.), powdered amber and amber oil have been used in medicines (Rice 1987). The powder was taken as a remedy for stomach problems. The oil, resembling turpentine, was used internally for ailments such as asthma and whooping cough; more frequently, it was used externally as a liniment to be rubbed on the chest.

During the Middle Ages, amber continued to be used for stomach ailments, goiters, and against some poisons. Camillus Leonardus discussed the medical value of amber in his *Speculum Lapidum* of 1502. White amber was so highly prized for its medicinal uses during the rule of the Teutonic Knights in Prussia that the rights for white amber were never sold when amber monopolies were released. This interest in the medicinal use of amber may be related to the Teutonic Order's having originated in association with the German Hospital of St. Mary, only later assuming its military character. By the mid-1500s, Duke Albert had commissioned a study of the medical properties of white amber, and a scientific treatise his court physicians published lent some credibility to its use (Chapter 6).

In China, a syrup of amber, a mixture of "liquid acid—of amber and opium" (Rice 1987), was used as a sedative, anodyne (anything that relieves pain or soothes), and antispasmodic. Powdered amber and amber oil were used as powerful diuretics. Amber fumigation was done by throwing powdered amber on a hot brick.

Amber was still being prescribed by physicians in France and Germany into the late 1800s and early 1900s. This continued popularity was related not only to the belief that it prevented infections but that it acted as a charm against them. Its use as a mouthpiece for cigarettes, cigars, and pipes was originally talismanic (Rice 1987). Amber was used for mouthpieces of hookahs, with their numerous hoses, because it was thought that the amber was germicidal. As late as 1935, amber oil was still reported to be used in pharmaceutical products.

CHAPTER 11

Future Use of Resins

In a lecture to the International Academy of Wood Science on the achievements and the future of forest products research, Hillis (1986) made a charming comparison of the history of forest products to the story of Amber St. Claire, from Kathleen Winsor's 1944 book, *Forever Amber*. A beautiful girl in the early 1600s, Amber spent most of her life in London associating with the highest people and involved with all facets of society. Her beauty and charm, and her place as the king's mistress, brought her attention, but she neither became respectable nor attained a permanent leading position. Hillis was talking about all forest products but his point was clear about resins. Extractive forest products such as resins assumed great importance for particular uses for periods of time throughout history, but large-scale uses have often been replaced by alternatives. So it was that Amber St. Claire achieved high status only for a while, and was replaced. So have the glory periods for resins passed, or is there still a future for their use, and perhaps new uses on the horizon? In other words, what are the prospects, opportunities, and limits?

The usefulness of plant resin and amber to humans has been documented in Part III and depicted in a time line (Figure 6-1). No material was as versatile in the preindustrial world as resin because of its many useful attributes (Gianno 1990). Numerous large-scale industrial uses of resins have been replaced by synthetic products in developed countries, especially from petrochemicals that provide some of the same properties more cheaply and with a more consistently uniform quality in bulk processing. Nonetheless, new industrial uses for various components of resins, particularly as chemical feedstocks, portend continuing use of resins in the future. With decrease in the availability of cheap petrochemical intermediates, increase in energy costs, and the problems presented by hazardous effluents, resins are potentially important as renewable resources. The change in status of resins is suggested by the fact that nontimber forest products (NTFPs) are no longer called minor forest products. Sustainable harvest and trade of NTFPs can be economically and ecologically viable, especially in tropical forests where there are forest-dependent communities of people (Schmincke 1995). Likewise, the resins from desert shrubs, with uses ranging from medicine to fuel, provide a relatively new commercial source.

Traditional Uses

Amber Jewelry and Artwork

Since the Stone Age, humans have created beautiful jewelry and ornaments as well as artwork from amber (Chapter 6), and there is every reason to expect this creative process to continue. The enduring aesthetic value of amber is recognized worldwide. For example, the great Amber Room in the Ekaterininsky Palace in St. Petersburg, lost during World War II, is being tediously reconstructed. The Chinese continue a long history of prizing amber for various artistic creations. Indians and others in Mexico still carve amber for sale. Jewelry made from the abundant amber in the Dominican Republic is second only to Baltic amber in sales in the United States.

In addition to amber, the hard copals from araucarian and leguminous trees (Chapter 9) are made into artwork. Various objects continue to be carved from large lumps of kauri resin in New Zealand, where novices are shown how to use the resin artistically (McNeill 1991). Large amounts of *Hymenaea* resin from extensive deposits in Colombia are obtained through gem dealers for jewelry by both professionals and amateurs. Moreover, professional artists use various kinds of resin as a medium, for example, for molded masks and tiles (Rachel Berwick, pers. comm.).

Lacquer artwork from China, Japan, and other parts of the East remains highly valued (Prendergast et al. 2001). *Barniz de Pasto*, initially used by the Incas, continues to be employed in Peruvian folk craft. Just like amber, there is reason to expect continuation of lacquer artwork in the future.

Incense

Historians have stated that use of incense could be considered a universal cultural trait (Chapter 8). Incense today is not as commercially important as it was during the thousand years of the Arabian incense trade routes (Chapter 6). During this peak period, incense use was associated with traditional or pagan belief systems. Incense continues to be much used today in traditional religious ceremonies of the Maya. Additionally, use of incense has persisted through later ages in various monotheistic religions. For example, in the West, incense is a prominent part of both Greek and Roman Catholic church ceremonies, and in the East, in Hindu and Buddhist temples. Incense thus continues to be an offering or a mark of respect to God (or gods) and the church. Moreover, the aroma of incense smoke still is used to mask unpleasant odors wherever they may occur as well as to be enjoyed for its own sake in ceremonies, processions, etc. Therefore, the various resins that provide aromas when burned probably will continue to serve in their age-old roles throughout the world.

Other Special Uses

In all likelihood, hops resin will be continue to be used in making beer, as it has since the mid-19th century, for its characteristic flavor as well as its antimicrobial properties (Chapter 10). Also, fine pale varnish, such as mastic from *Pistacia*, will probably still be used to protect oil paintings and watercolors. It has the advantage of being easily removed without damaging the painting (Chapter 4). The tradition of using different natural varnishes for stringed instruments and for restoring antique furniture may also be maintained. Although these are minor uses economically, they are testimonies to the often unique properties of resins for specific purposes.

Medicinal Uses

New Therapeutic Uses for Resins

Many resins have been used throughout history in indigenous cultures and have persisted in Eastern herbal medicine for a variety of medical purposes (Chapters 7, 8, and 10). Research has increased on the efficacy of such traditional medicines, employing the scientific methods used to evaluate Western medicines. The marketing of resin as a prescription drug indicates the pharmacological interest in these resins. For example, an extract of the resin of *Commiphora wightii* called gugulipid became commercially available by

1987 for management of atherosclerosis and hyperlipidemia. Both the research on and the marketing of drugs derived from resins point to the possibility of their further medical use in the West as well as in the countries where they have been traditionally used (G. Singh et al. 1993). Since *C. wightii* is now threatened in India, better agronomic approaches for obtaining the resin are being studied (Bhatt 1989). The plant has been overexploited by poor people living in the arid areas where it is native, thus such research has the potential of enhancing production of this increasingly important medicinal plant as well as improving the local economy. Other resin-producing species, long used for their therapeutic effects in traditional medicine, are likely candidates for future studies.

Resins from numerous species of the Australian desert shrub *Eremophila* are being evaluated for a variety of medicinal purposes because of their antihypertensive, analgesic, and cardioactive properties. Research on the pharmacological activity of these plants, such as that of Richmond and Ghisalberti (1994) on the unusual terpenoid chemistry of these resins, is likely to expand in the future.

Podophyllotoxin, from resins in the roots and rhizomes of *Podophyllum*, is the starting material for semisynthesis of anticancer compounds. Most podophyllotoxin is obtained from the Himalayan species, *P. emodi,* and the resin is now so valuable that it has become endangered. With increasing demand for the resin, Moraes et al. (2000) discovered that the leaves of American mayapple *(P. peltatum)* can produce large quantities of podophyllotoxin. Furthermore, because mayapple leaves are renewable without destroying the plant, their use for podophyllotoxin production provides a potential alternative crop for American farmers. Obtaining podophyllotoxins from new sources, as well as studies of population variation in resin composition, will no doubt continue to receive attention.

Propolis

Propolis, resin collected and mixed with wax by honeybees for use in their hives, has a long history of medical use for a wide spectrum of health benefits (Chapter 10). Research regarding its biological, chemical, and medical properties has increased since the 1970s, especially in eastern Europe. Sustained and enhanced interest in understanding the large number of medical benefits proposed for propolis is likely to continue. However, several problems could hinder commercial success. First, Ghisalberti (1979) questioned whether there would be adequate propolis for commercialization, but this problem has abated somewhat with substantial increases in production worldwide. Second, Bankova et al. (2000) emphasized the importance of knowing the botanical origin of propolis, which largely determines the chemical composition of the resin and, hence, the medical properties of propolis. Product variability has increased since resins collected by bees in both the temperate zones and the tropics are now often marketed as propolis. Moreover, variation in the overall chemistry of propolis (resin plus other compounds) affects the nature and quality of the product, which is sold in various forms (e.g., raw propolis, liquid extracts, powders, tablets, and capsules). Bankova and co-workers urged greater research on the human and veterinary use of propolis through cooperation between beekeepers, scientists, and agencies concerned with quality control.

Marijuana

Medical use of *Cannabis* in the United States is receiving accelerated attention with governmental authorization of research on its medical value. Current recommendations by medical organizations in the United States contrast with past governmental prohibitions regarding the use of cannabinoids for medical purposes (Chapters 6 and 10). The American Medical Association, the National Institutes of Health, the Institute of Medicine, and the World Health Organization have recommended further research on marijuana's potential in treating various conditions (Lauerman 2001). In 1997 the editor of the *New England Journal of Medicine* went even further: "Federal authorities should rescind the medical use of marijuana for seriously ill patients and allow physicians to decide which patients to treat. The government should change marijuana status [from] Schedule I drug [illegal, to] . . . Schedule II [medically useful drug] . . . and regulate accordingly." Despite their recommendations, requests by scientists and physicians to conduct such research had been denied until 1999.

To provide rigorous scientific evidence for further consideration, the California legislature passed a bill, signed into law by the governor in 1999, to give the University of California monetary and legal support to study specific ways in which *Cannabis* might be used safely and effectively for medical purposes. The Center for Medicinal Cannabis Research (CMCR) was established with headquarters at the San Diego campus of the University of California, with the San Francisco campus as a major collaborator. Money was allocated to extend over a 3-year period if the CMCR is successful. Establishing such a center was considered to represent a model for close collaboration of federal, state, and academic entities, supporting this kind of research.

The CMCR was to support and fund research to explore the use of marijuana as an alternative treatment for conditions and symptoms defined by the National Academy of Sciences, the Institute of Medicine, and an expert panel of the National Institutes of Health: (1) severe appetite suppression, weight loss, and cachexia resulting from infection by human immunodeficiency virus, (2) chronic pain resulting from specific injuries and diseases such as AIDS, (3) nausea associated with chemotherapy, and (4) muscle spasticity caused by diseases such as multiple sclerosis. Some of the studies that have been approved include the effect of marijuana on neuropathic pain, effects in combination with opioids for cancer pain, mechanisms of cannabinoid analgesia, effects on spasticity, and the effects of repeated treatments on driving ability. The marijuana used in the medical studies has been obtained from the National Institute on Drug Abuse in accordance with the procedures developed by Public Health Services.

Although tetrahydrocannabinol (THC) has been available since the 1980s in a synthetic pill form called Marinol, it is irregularly absorbed by the stomach (Mechoulam 1986a). Thus smoking is still considered an effective way to deliver THC to the blood. Because chronic smoking of marijuana may have negative long-term effects and accurate dosages are difficult to keep consistent, however, there is interest in alternative methods of marijuana administration. Furthermore, a means of disassociating the psychotropic effects of THC from their desired therapeutic effects needs to be considered (Chapter 10). Collaboration with pharmaceutical companies outside the United States, such as in Canada, the Netherlands, and Belgium, where marijuana already is legal for medical use, may prove helpful in creating alternative reliable preparations to smoking.

If the CMCR is successful in obtaining sufficient scientific data supporting the medical use of marijuana, state and federal governments could be obligated to deregulate marijuana for its medical use as a prescription drug. There are those who predict that it is not a question of if, but when, this occurs (Grinspoon and Bakaler 1997). Medical use of *Cannabis* resin is an issue that obviously will receive considerable attention in the United States and other parts of the world.

Industrial Uses

Chemical Feedstocks

As industrialization advanced, the large-scale use of pine resin for traditional naval stores products changed with the development of other, more sophisticated products (Chapter 7). Although pine resins have to compete with functionally equivalent or substitutable chemicals such as petroleum-based synthetic resins, citrus oil terpenes, and vegetable fatty acids, pine resins are and probably will remain important sources for niche markets with particularly high needs per unit volume of material (Hodges 1997). Sterols derived from resins in the paper pulping process provide an interesting example of specialized use, to supply the rapidly growing market for cholesterol-reducing food additions such as the margarine product Benecol. Additionally, there has been a significant increase in use of monoterpenes as precursors in the manufacture of vitamins.

Rosin, however, consisting primarily of diterpene resin acids, has remained the most important pine resin commodity and probably will continue to be so because of its uses as intermediate chemicals in numerous unrelated industries. Some of its many uses are as tackifiers, emulsifiers in synthetic rubber, paper sizing to control water absorptivity, surface coatings, and printing inks (Chapter 7). The pine species producing the rosin is generally not important because there is relatively little variation in the diterpene acid composition. Furthermore, for many purposes, rosins and rosin derivatives are interchangeable because they are mostly used in some chemically modified form.

Interest in turpentine is often focused on monoterpenes because pine resins worldwide yield 10-fold greater amounts of monoterpenes than any other plant (Plocek 1998). The primary monoterpenes in resins of some pines, α- and β-pinene, are used in producing polyterpene tape, and large amounts of β-pinene are involved in synthesis of flavors and fragrances (Chapter 7). Therefore, in contrast to rosin, turpentine from particular species of pines is preferred for certain products. In 1996, aroma products and polyterpene tape accounted for most turpentine use in developed countries whereas pine oil cleaner and solvent uses predominated in developing countries (Figure 7-4). However, Plocek thought that sustained economic development and improved disposable income levels in some developing countries would lead to greater consumption of aroma chemicals. Thus he projected that, long-term, "the impact of these new markets is potentially tremendous" for the aroma industry.

The aroma industry also seeks inexpensive, large-scale resin sources of particular volatile mono- and sesquiterpenes from tropical trees in the families Burseraceae (e.g., *Protium*) and Dipterocarpaceae (e.g., *Dipterocarpus*) as well as aromatic phenolics from the Fabaceae (or Leguminosae; e.g., *Myroxylon*). These resins would provide a cash crop of NTFPs for some indigenous peoples and encourage sustainable forest practices. In fact, Ankarfjärd and Kegl (1998) concluded that use of sesquiterpenes from *Dipterocarpus* resin in perfumes is an "encouraging example of a long-used nontimber product that has found its way into new markets when traditional use has declined in importance."

Resins from arid-land plants are being analyzed for use of particular groups of compounds. Shrubs such as *Grindelia* could provide particular diterpenes, valuable because of their many uses as substitutes for those from pine rosin. Hoffmann and McLaughlin (1986) noted that a renewable source of rosin substitutes, with other uses in the rubber, coatings, textile, and polymer industries (Chapter 10), could have important market possibilities. This is especially true for a crop that could be grown in underutilized arid areas of the United States. For example, *G. camporum* has been investigated because its tolerance to high soil salinity would be a valuable attribute in making cultivation possible in soils degraded by salinization. Ravetta et al. (1997) showed experimentally that there is a slight increase in resin production at higher salinity levels, but they pointed out that more experiments are needed to evaluate this kind of response. Other desert regions of the world as well could potentially benefit in use of such plants.

Other highly functionalized diterpenes are sought from shrubs such as the Mediterranean *Cistus* for initiation of synthesis of new antifeedant pesticides. Furanosesquiterpenes in the Australian shrub *Myoporum* are very toxic to mammals, and their potential value as antifeedants has been touted (Chapter 10). Antimicrobial effects of the volatile fraction of resin from the Mediterranean tree *Pistacia*, against food spoilage and food-borne bacteria, have also been of interest (Chapter 9). This research may become profitable as food manufacturers search for better ways to preserve foods than some of the chemicals commonly used, which have potential carcinogenic properties.

There has been interest in the use of plant chemicals to reduce agricultural feeding damage from birds. Watkins et al. (1999) devised a model to predict if derivatives of cinnamic acid (a common constituent in phenolic resins) could prove effective as bird repellents. They thought that modeling offers a promising way to prospect for new bird repellents, although the full extent of its use as an aid in screening new repellents awaits further experimentation.

Fuel Sources

Resins with a high hydrocarbon content are a potential source of fuel. For example, the predominance of sesquiterpene hydrocarbons in most New World species of *Copaifera* has led to the possibility of their use as diesel fuel (Chapter 7). Efforts have been made to use resins in the latex of different *Euphorbia* species for fuel, but *Copaifera* resin is more attractive because the sesquiterpene hydrocarbons can be used directly in a diesel engine without the further processing necessary for *Euphorbia* latex (Chapter 10). Therefore, genetic engineering was suggested for transferring the gene for production of sesquiterpene hydrocarbons from *Copaifera* to plants such as *E. lathyris* that could be grown in the arid U.S. Southwest (Calvin 1984). Calvin proposed that plants grown for their hydrocarbons, as a substitute for petroleum, should become more important, particularly in areas where land is unsuitable for food production. He challenged the U.S. agricultural community to commit to an "energy agriculture that would have long-term benefits," a situation that has thus far lagged.

Most of the 195 species of shrubs and perennial herbs with high resin content in the arid U.S. Southwest that McLaughlin and Hoffmann (1982) surveyed for biocrude production have not been analyzed in detail. Extensive research on alternate sources of energy, funded during the energy crisis of the 1980s in the United States, was abandoned following a glut of petroleum availability. Perhaps as energy crises reappear, the considerable data collected on plant fuel sources, as well as Calvin's challenge, will be assessed and ultimately come to some fruition. However, much will depend on how the political winds blow in terms of financial support for research and development as well as on the availability other plant sources of energy such as ethanol.

The large deposits of Miocene fossil resin, in which sesqui- and triterpenes contributed to the formation of petroleum over extensive areas of Southeast Asia (Chapters 4 and 9), should continue to be investigated and used. Their presence may stimulate exploration for petroleum from resinous sources in other areas; there are conifer-derived oils in Canada, Australia, and New Zealand, for example (Murray 1998). Additionally, there are compounds (e.g., hydrocarbons) in resins that are sufficiently distinctive structurally that they can be used as biomarkers for petroleum exploration.

Dutta and Schobert (1995) suggested that fossil resins occurring in large quantities in coals, such as those from the Wasatch Plateau in Utah (Chapter 4), could be used as feedstocks, converted to thermally stable compounds for advanced jet fuels. The traditional source has been petroleum containing a large proportion of alkanes, which are thermally unstable at elevated temperatures. Although resin would not meet all the demands from the commercial and military sectors, they could provide some of the needed raw material.

Tropical Forest Management for Resin Use

Sustainable use of all ecosystems is desirable, but in the tropics and subtropics, emphasis has been placed on forestry with the dual goals of sustainability and conservation of biodiversity. Many copiously resin-producing tropical trees are good multiple-purpose trees, ones that can be used for resin, timber, and other purposes. But how can a balance of economic success and sustainable management be developed? There are often different problems, depending on whether the resource is a natural or cultivated forest. Moreover, for resin-producing trees, there are differences in how resin is collected. The basic issue in all cases centers around the economic value of the product and the income the product provides, which motivates people to conserve the resource. An interesting contrast in solving these problems is provided by gharu wood from *Aquilaria* in Malaysia and Kalimantan, or southern Borneo, and resin from *Agathis* from Australasia.

Aquilaria poses particular difficulties because trees are frequently felled to obtain the resin-saturated wood (Chapter 10). Wollenberg (2001) showed that the economic incentives for Kalimantan villagers are complex and that sustainability and conservation are difficult to achieve. Gianno (1986b) reported a similar situation for resin collection from native trees in peninsular Malaysia but suggested the possibility of cultivating a rapidly growing species of *Aquilaria* in plantations and inoculating the tree with the fungus that produces the resin-saturated wood. *Aquilaria* as well as other resin-producing trees provide examples of the need for tropical farmers and professional foresters to cooperate in developing successful management of the resources (Stanley 1991, Smith et al. 1992, Neels 2000). Through better understanding of the historical importance of tree species in the cultures of the local people as well as by working with sensitivity to their needs and constraints, professional foresters and development planners can plan and execute more effective forestry programs.

In the case of *Agathis,* the resin became so economically important for local people in some areas of Australasia that they overexploited the resource. They used such intensive tapping practices (Plate 45) that they killed trees over a large area. Fortunately, these particular problems can be averted by use of resin-tapping systems that do not even damage trees used for timber, as exemplified by *Agathis* in Irian Jaya (or West New Guinea) (Whitmore 1980b). Peters (1994) provided an encouraging thought in this direction: "In theory [resin tapping] probably comes closest to conforming to the ideal of sustainable non-timber forest product extraction." The method must be patterned according to the characteristics of the particular tree species, however, and much research needs to be done.

In another approach, chemical ecologists such as Kursar et al. (1999) use the defensive traits of tropical trees as a means for more effective screening of native plant populations for useful chemicals. For example, some chemicals may be concentrated in either young or mature leaves; hence, collecting does not have to decimate the entire resource. This manner of bioprospecting is a harbinger of future research. Furthermore, although the chemicals may be valuable to people in both developed and developing countries, this approach emphasizes the extracting and assaying of compounds in the source countries in the tropics. Thus the local economy benefits (Capson et al. 1996), providing an important conservation incentive for tropical forests.

Four presently used sustainable management systems that also conserve tropical resin-producing trees provide examples for future practice: (1) collection from native trees in extractive and indigenous reserves, (2) agroforestry, (3) plantations, including reforestation plantations for watershed protection and wildlife sanctuary, and (4) enrichment silviculture in natural ecosystems. The latter three involve some cultivation, which can be valuable in relieving pressure on wild populations, at the same time preserving reservoirs of genetic diversity.

Extractive and Indigenous Reserves

In an attempt to maintain large tracts of Amazonian forest under sustainable use, a type of reserve called an extractive reserve was created (Fearnside 1988). Extractive reserves enjoy legal protection over more than a million hectares of forest traditionally used by 50,000 extractivists in four Brazilian states (Elisabetsky 1996). The extractivists manage the land collectively, removing NTFPs for sale without felling the trees. Although Brazilian extractivists have primarily been rubber tappers since the rubber boom in the late 19th century, they also have collected various Amazonian resins. For example, *Copaifera* resin (copaiba oil) has long been collected for use in medicine and, more recently, cosmetics (Chapter 7) whereas *Hymenaea* copal has been collected for fine hard varnishes (Chapter 9). Extractive reserves have emerged as one of the most promising Amazonian development strategies, combining socially just land use that also reconciles economic development and environmental conservation (Allegretti 1990).

Resins from other Amazonian trees such as *Protium* are still used locally for various purposes, from medicines to caulking boats, in indigenous reserves such as those of the Ka'apor and Tembé Indians. Resin as a potential cash crop for these indigenous people is being investigated by forest ecologists and phytochemists from Brazil's Instituto Nacional de Pesquisas da Amazônia (INPA), particularly for feedstocks for the aroma industry (Chapter 8). Projects regarding NTFPs in indigenous reserves should continue to provide improved sustainable management for increased resin harvests. Export value doubtless will determine what kinds of resins are collected from Amazonian resin-producing trees, as well as the quantities that both extractivists and indigenous peoples will collect. As such, as with gharu wood, economic incentives will need to be assessed in the context of preserving the resource.

Harvesting resin from *Protium copal* in an indigenous community forest concession in Guatemala is also being studied for its viability as an industry (Chapter 8). It would provide income based on domestic use of resin from forests that contain an unusually large population of the trees. However, more research is needed on the chemistry of the resin and possible seasonal effects on its quality and quantity as well as improved harvesting methods

(Neels 2000). Furthermore, Neels concluded that if new markets for the resin were to develop, socioeconomic analyses need to be undertaken. These analyses should be directed toward understanding the effect of increased resin harvest on the local society and how the caretakers of the forest would benefit.

Agroforestry

Agroforestry, as the name implies, is the combination of agriculture and forestry, often the growing of trees amid agricultural crops. Ideas for expanding future tropical agroforestry practices can be taken from several successful examples in Asia and Latin America. Three Asian examples are noteworthy. First, Sumatrans tapped wild trees of *Styrax benzoin* for resin that has been commercially important over the centuries, but repeated tapping killed many trees (Chapter 8). Therefore, cultivators began to plant *Styrax* seeds with upland rice in swiddens, with trees available for tapping after 7 years. A second successful agroforestry example is provided by fast-growing species of *Agathis* (e.g., *A. dammara*) on Java (Whitmore 1980b). *Agathis* seedlings are planted with various food crops, followed by fast-growing trees such as *Leucaena leucocephala* for shade, until the saplings can tolerate full sun and are ready for tapping. *Agathis* may be tapped before it is used for pulpwood at 30 years or for veneer at 50 years (Chapter 9). The *kebun damar* (dammar gardens) of *Shorea javanica* in southern Sumatra provide a third agroforestry example. Rice is grown for 1–2 years, then coffee, pepper, or some other crop is planted together with *Shorea* and other useful trees such as cloves *(Syzygium aromaticum)*. Until the dammar trees reach an age for tapping, other products are harvested to provide cash income for the farmers (Torquebiau 1984, 1987). Coppen (1995) indicated that much is still to be learned about the biology and silviculture of *Shorea javanica*, but valuable knowledge and experience have been gained from research that continues (Plate 44). It is hoped that the successful development of *S. javanica* in Indonesia will encourage the use of other dipterocarps in agroforestry in other countries (Chapter 9).

Studies have considerably refined the understanding of the ways in which Amerindians use the tropical rain forest, which can serve in the establishment of successful agroforestry and forestry practices (Dufour 1990). Anthropologists and ecologists previously thought of swiddens, fallows, and forests as more or less separate entities. Now, however, there is a better understanding of swiddens becoming forests and the degree to which human management is part of the transition. Distinctions between domesticated and wild plants, or natural and managed forest, are also not as clear as previously thought. Research needs to take into account the long history of occupation and use of these tropical forests and recognize that the complex agricultural systems of Amerindians have much to offer in the design of sustainable agro-ecosystems (Ewel 1986).

A different agroforestry approach for resin-producing plants is the farming of jalap *(Ipomoea purga)* in Mexico. Cultivation is done in the tropical forest, using shrubs to support the vines (Chapter 10). Even though production of the root has declined, demand for laxatives has not, and the plant remains important for its export, supplementing farmers' earnings from their primary crops (Linajes et al. 1994).

The need for reforestation in Central America is widespread. It has become critical to find better ways to protect watersheds as well as to increase the kinds of plants for local farmers to grow in managed forest settings. A culture of resin tappers who use native pines in Honduras provides an example of community-based sustainable forest management in a country where deforestation is escalating as a result of population increase. Resin tapping in a community agroforestry cooperative has been successful in safeguarding natural resources that provide income for farmers (Stanley 1991). Stanley thought that long-term success lies in the hands of governments and international donors, however, for support of strong organizations to help manage and replant community forests, and to structure market incentives to make the marketing of the resin economically viable.

Trade in Tolu and Peru balsams (from varieties of *Myroxylon balsamum*) is "still vigorous," and synthetics have failed to displace these resins from a wide variety of uses (Smith et al. 1992). Peru balsam trees (Figure 8-2) are already a part of some agroforestry systems, at the village level and as a shade tree in coffee plantations in Central America. These resin-producing trees tolerate a broad range of environmental conditions; thus little selection or breeding would be needed to obtain material suitable for planting in many parts of the tropics. Smith and coworkers thought that with good seed sources, plantations of these trees in the American tropics could become economically and ecologically viable. Additionally, as more tropical countries develop industrial capacity, the internal demand for such balsams is likely to grow.

Plantations

In a silvicultural study of natural populations of *Copaifera multijuga* in central Amazonia, Alencar (1982) concluded that tapping native trees would provide resin sufficient for medicinal purposes, as had long been done, but was not adequate for use as diesel fuel. Therefore, Alencar suggested that plantations are necessary to obtain enough resin for fuel and various industrial uses (Chapter 7). However, several problems have to be overcome. First, the young plants are slow to reach maturity, when they produce the greatest resin yields, a problem noted for other rain forest resin producers. This becomes an economic liability if the trees are not grown together with other harvestable crops, as in agroforestry. Second, only certain trees are high-yielding, thus there must be selection of seed for plantations. Third, monocultural plantations of Amazonian species are unlikely to succeed without recognizing the potential problems with disease, as demonstrated by rampant disease in rubber plantations, for example. Thus plantation culture methods should take into account both resin yield and compositional variability that may be important to the trees' natural defense system (Chapter 5). Fortunately, some variability in the hydrocarbon fraction of *Copaifera* resin may be of little concern to its use as diesel fuel. However, there has been a shift from interest in the resin as a source of fuel in Brazil to an emphasis on its use for medicines and in the cosmetics industry. Therefore, compositional variation is significant in the use of the resin because some compounds are more sought than others (Chapter 7).

International research assistance for cultivation of tropical pines is much greater than for most other tropical tree crops (Smith et al. 1992). A network approach has helped generate results from a working group on pines in the International Union of Forest Research Organizations. Using a germplasm screening network in 60–70 countries, researchers at the Oxford Forestry Institute have been able to identify promising material for varying environmental conditions in different countries. Thus plantations of different species of pines have been established throughout the tropics. In some places, such as Indonesia, *Pinus caribaea* and *P. merkusii* are tapped before the trees are felled for pulp, and resin is extracted in the sulfate pulping process (Zobel et al. 1987). *Pinus elliottii* is increasingly planted in China to make up for the decreasing availability of native pines for tapping, with some plantations serving a dual role in reforestation projects. Plantations of *P. elliottii* have also been established in subtropical regions of Brazil and Argentina, and plantations of *P. radiata* in Chile, to supply resin for turpentine and rosin. Resins from *P. elliottii* and *P. radiata* are rich in β-pinene, thus they are sought by the aroma industry more than resins from some other species (Plocek 1998). *Pinus caribaea* has been planted in substantial numbers in northern Brazil and Venezuela. Zimbabwe, Kenya, and South Africa are initiating naval stores industries, using these same three species in plantations. All indications point to continued expansion of pine plantations in which the resin would be used.

In some of the smaller developing countries, where pine tree resources are or could become available, use of pine resins has several attractions (Coppen et al. 1984). They not only provide an industry based on a renewable resource, collection by tapping provides labor for local people, and the total investment cost can be relatively low. Despite the existence of a substantial international market, Coppen and coworkers recommended that small developing countries base their operations initially on serving the home market, for example, supplying the need for soap, and paint manufacturing as well as the paper industry. Consumer demand in developing countries is expected to grow with increased industrialization and urbanization, thus there are opportunities to replace imported products with those produced locally. Later, if the industry becomes profitable, it could expand into regional and international markets.

Enrichment Planting

If the economic value of resin increases with greater emphasis on renewable resources, various resin-producing trees could be used as part of enrichment planting, that is, establishment of seedlings or saplings in logged or poorly stocked forests. A serious problem, as indicated, is that copiously resin-producing tropical angiosperms are often primary forest trees that grow very slowly as seedlings and saplings. If the resins are sufficiently valued commercially, however, the ultimate profitability might outweigh the need to wait for maturity in this type of planting. On the other hand, some fast-growing conifers could serve. For example, fast-growing species of *Agathis* (e.g., *A. macrophylla* and *A. dammara*) have been used in enrichment planting in peninsular Malaysia (Whitmore 1980b). Candidate pines have been discussed, although for enrichment planting, native species are preferable to exotic ones (Laird 1995). Traditional home gardens are often a source of good native selections for forest enrichment. Furthermore, *Dipterocarpus alata* in Southeast Asia is an example of a fast-growing native secondary forest tree in which there is commercial interest in the resin (Ankarfjärd and Kegl 1998).

Adams (1987) suggested the possible use of invasive juniper species for cedarwood oil. For example, *Juniperus ashei,* which has invaded millions of hectares of overgrazed grassland, could provide biomass adequate to support expansion of the cedarwood oil industry. This invasiveness is part of native plant succession, as shown by *J. virginiana* (Figure 7-7), but it could be viewed as a natural kind of enrichment planting. If the trees were successfully utilized, the "invasive enrichment" could be accelerated by seeding in more junipers. Although this is an example from temperate zone vegetation, *Juniperus* does occur in some tropical ecosystems. Furthermore, the example suggests the possibility that invasive enrichment of native plants might be applicable even in tropical rain forests.

Enhanced Pest Protection of Resin-Producing Trees

Research has shown that it is possible to manipulate the constitutive and induced resin production in conifers. This has far-reaching implications for enhancing the natural resin protection of trees against pests (Phillips and Croteau 1998, Trapp and Croteau 2001). Some components of resin are under the control of relatively few genes (Chapter 5), suggesting the possibility of using routine transformation, regeneration, and micropropagation techniques, but relatively few studies have considered manipulation of natural chemical defenses against pests. Such research could be directed toward trees from which humans use resins directly (e.g., slash pine, *Pinus elliottii*) as well as those that humans use for other forest products such as timber (e.g., Douglas fir, *Pseudotsuga menziesii*); many resin-producing trees also produce fine timber. Furthermore, such biotechnology could be used for trees in reforestation projects since resistance against major pests such as bark beetles or spruce budworm could have far-reaching consequences for general forest health in some regions.

A major challenge for forest biotechnology is to discover genes that determine desirable chemical defensive characteristics without having to await the results of an overall genome project. For example, the genome sizes of conifers (e.g., pines) are seven times larger than that of humans, putting the problem in perspective (Mann and Plummer 2002). Several genes that control conifer resin composition are known. Molecular techniques make it feasible to isolate and characterize promoters that control components of both constitutive and induced resins. Of course, much remains to be learned about the organization, regulation, and molecular genetics of resin production. Nonetheless, Phillips and Croteau (1998) and Trapp and Croteau (2001) suggested several opportunities for manipulating resin production to improve resistance against forest pests and pathogens, including (1) improving the speed and level of resin response at early stages of attack, (2) increasing concentrations of resin components that are especially toxic by modulating promoter strength and copy number of extant genes, and (3) introducing new defense genes.

Raffa (1989), however, has warned that highly coevolved insects such as bark beetles in conifers might rapidly develop mechanisms to detoxify bioengineered genes. A possible deterrent or delay to the development of this kind of tolerance would be a toxic gene that is only expressed under conditions of severe attack. Ecologically sophisticated strategies could involve altering resin chemistry to disguise the host, thereby confounding the beetles' selection of hosts, or attracting beetle predators or parasites. It might even be possible to alter the stereochemistry of a beetle pheromone precursor, thereby disrupting aggregation or mate selection (Chapters 1 and 5). Interplanting a variety of resistant biotypes could be a powerful biocontrol approach, especially over large areas replanted by the forest products industry.

Although there are apparently no theoretical barriers to the development of first-generation transgenic conifers with enhanced defense capabilities (Phillips and Croteau 1998), there are some challenging biochemical caveats (McCaskill and Croteau 1999). These relate to the myriad of terpenoids produced from a single intermediate (isopentenyl diphosphate) and the complex organization and subtle regulatory features of the biosynthetic pathways (Chapter 1). Furthermore, for all the exciting biotechnological possibilities of altering the resin chemistry of trees, there are issues to be considered from the forest industry's perspective (Robinson 1999). Although such technology could revolutionize the forestry industry, adaptive research cooperatives, which are present in agriculture but not in forestry, will be needed. In implementing and integrating biotechnology into commercial forestry, it is essential to consider the risks, costs, and benefits shared by all stakeholders as well as the business and social systems into which the technology is introduced.

In addition to the problems that must be overcome to make forest biotechnology a commercial reality, legitimate environmental risks must be met (Mann and Plummer 2002). Trapp and Croteau (2001) emphasized the importance of not just eliminating forest pests (by engineering control of resinous defense) but consideration of the broader goals of sustainable forest management. Thus they promoted the incremental exploration of biotechnologically based pest-management strategies "in both type and scale, to permit responsible, stepwise consideration of possible unintended consequences."

Thus prospects for the use of biotechnology to engineer enhanced resin protection of trees is complex, but the door is open to research. It seems likely, however, that this kind of technology will initially be restricted to temperate zone conifers growing in developed countries, or possibly to plantations of exotic resin-producing conifers in developing countries (Chapter 7), whereas protection of tropical angiosperms will depend more on natural ecological and forest management solutions.

Archeology and Anthropology of Resins

Although resins have tended to be neglected in archeological studies, they nonetheless have been considered "archaeographic material of greatest importance" (Dunn 1975). Chemical techniques such as FTIR, GC-MS, and NMR enable information about resins to be obtained from anthropological and archeological objects that was not possible previously. Among the first uses of these techniques on resin was the determination of the sources of Baltic amber (Beck et al. 1966, 1967), and such research will doubtless continue as more archeological sites are discovered. For example, archeologists are pursuing the origin of amber and resin used by Jícan Indians in northern Peru (Shimada et al. 1997) and at other Latin American sites. Another analysis made possible by such chemical technologies was the identification of a resin used as an additive to wine, which supported evidence for the first use of wine in early permanent human settlements (Chapter 9). Other discoveries about how humans used resins seem to be on the horizon because of the diversity of their uses and resin's resistance to decomposition that allows its chemical analysis at a later time.

Because of the pervasive use of resins in indigenous societies, more anthropological studies of resin technology such as those by Gianno (1986b, 1990) in Malaysia, Balée and Daly (1990) in Amazonia, and Ankarfjärd and Kegl (1998) in Laos should be forthcoming (Chapters 8 and 9). The various roles that resins play in Mayan societies in Central America and Mexico are being investigated (Tripplett 1999) as are those of the Tembé Indians in eastern Amazonia (Plowden 2001). Some of these studies also assist these indigenous peoples by improving resin production for cash income (Neels 2000, Plowden 2001).

Plant resins are very likely to continue serving a variety of human needs in ways both similar to and different from those they have throughout history. The use of resins could be increased through research armed with new concepts and advanced techniques of analysis—chemical, evolutionary, ecological, and ethnobotanical—discussed here. The continuing increase in our knowledge of the many aspects of resins and the myriad of resin components points the way to new uses for them. When one considers the possibilities that new technology and knowledge provide, the future for resins may be even richer than their past, like the filling in of an outline of a drawing.

APPENDIX 1

Resin-Producing Conifers

Numbers of species from Enright and Hill (1995) and Mabberley (1997). Subfamilies of Pinaceae from Frankis (1989), subfamilies of Cupressaceae from Gadek et al. (2000). Key: *, useful resin; †, fossil resin; ?, questionably resin-producing.

FAMILIES AND GENERA	SPECIES	DISTRIBUTION
Araucariaceae		
Agathis†*	20	Indochina to New Zealand
Araucaria†	19	New Zealand, Australia, New Caledonia, S. Brazil, Chile
Wollemia	1	Australia
Cephalotaxaceae		
Amentotaxus	4	China, S.E. Asia
Cephalotaxus	6	E. Himalaya to Japan
Cupressaceae s.l.		
(including Taxodiaceae, T)		
Arthrotaxidoideae (T)		
Athrotaxis	3	Australia, Tasmania
Callitroideae		
Actinostrobus	3	S.W. Australia
Austrocedrus	1	temperate S. America
Callitris*	14	Australia, New Caledonia
Diselma	1	Tasmania
Fitzroya	1	Chile
Libocedrus	5	New Caledonia, New Zealand
Neocallitropsis	1	New Caledonia
Papuacedrus	3	New Guinea, Moluccas
Pilgerodendron	1	S. Chile
Widdringtonia	3	tropics & S. Africa
Cunninghamioideae (T)		
Cunninghamia	2	S.E. Asia
Cupressoideae		
Calocedrus	3	S.E. Asia, Panama, Pacific N. America
Chamaecyparis*	7	N. America, Japan, Taiwan
Cupressus*	13	Mediterranean, Sahara, Asia, N. America
Fokienia	1	China, Indochina
Juniperus*	50	N. hemisphere, tropical African mountains, W. Indies

FAMILIES AND GENERA	SPECIES	DISTRIBUTION
Cupressaceae s.l., continued		
Cupressoideae, continued		
Microbiota	1	E. Siberia
Platycladus	1	China, Korea
*Tetraclinis**	1	Spain to Tunis, Malta
*Thuja**	5	China, Japan, N. America
Thujopsis	1	Japan
*Xanthocyparis**	2	Vietnam & U.S. Pacific Northwest
Sequoioideae (T)		
Metasequoia†	1	China
Sequoia	1	California
Sequoiadendron	1	California
Taiwanioideae (T)		
Taiwania	1	S.E. Myanmar (Burma), Yunnan, Taiwan
Taxodioideae (T)		
*Cryptomeria**	1	Japan
Glyptostrobus	1	China
*Taxodium**	2	S.E. United States, Mexico
Pinaceae		
Pinoideae		
*Pinus†**	80–100	N. temperate zone
Piceoideae		
*Picea**	34	N. temperate zone
Laricoideae		
Cathaya	1	W. China
*Larix**	9	Europe, N. Asia, N. America
*Pseudotsuga**	4	N. temperate zone
Abietoideae		
*Abies**	38	N. temperate zone, Central America
*Cedrus**	2–4	Middle East, Algeria, Cyprus, Himalaya
Keteleeria	2	E. Asia, Indochina
Pseudolarix†	1	China
*Tsuga** (including *Nothotsuga*)	14	E. Asia, N. America
Podocarpaceae		
Acmopyle	2	New Caledonia, Fiji
Afrocarpus	6	tropical & S. Africa
Dacrycarpus	9	Myanmar (Burma) to New Zealand
Dacrydium	25	S.E. Asia to New Zealand
Falcatifolium	5	Malesia to New Caledonia
Halocarpus	3	New Zealand
Lagarostrobos	2	Tasmania, New Zealand
Lepidothamnus	3	New Zealand, S. Chile
Microcachrys	1	Tasmania
Microstrobos	2	Australia, Tasmania
Nageia	6	Indomalesia
Parasitaxus	1	New Caledonia
Phyllocladus	5	New Zealand, Tasmania to Philippines
Podocarpus†	93	tropics N. to Himalaya, Japan
Prumnopitys	8	Costa Rica to Venezuela, S. Chile, New Zealand, New Caledonia

FAMILIES AND GENERA	SPECIES	DISTRIBUTION
Retrophyllum	5	New Caledonia, Moluccas to Fiji, S. America
Saxegothaea	1	Andes of Argentina & Chile
Sundacarpus	1	Sumatra, Philippines, N. Queensland, New Ireland
Sciadopityaceae?		
Sciadopitys	1	Japan
Taxaceae?		
Taxus	7	N. temperate zone
Torreya	7	E. Asia, S. United States, California

APPENDIX 2

Resin–Producing Angiosperms

Classification according to the Angiosperm Phylogeny Group (1998); additional data from Cronquist (1981), Takhtajan (1997), and Mabberley (1997).

ORDERS, FAMILIES, AND GENERA	SPECIES	DISTRIBUTION
Basal Group		
Piperales		
Piperaceae		
Piper	2000	tropics
Monocots		
Alismatales		
Araceae		
Monstera	25	American tropics, W. Indies
Philodendron	300–400	American tropics, W. Indies
Asparagales		
Convallariaceae (including		
Dracaenaceae, or Agavacace s.l.)		
Dracaena	60	Old World tropics, Central America, Cuba
Xanthorrhoeaceae (Liliaceae)		
Xanthorrhoea	28	Australia
Commelinoids (subclass Commelinidae)		
Arecales		
Arecaceae (Palmae)		
Daemonorops	114	Indomalesia, especially W. Malesia
Eudicots		
Ranunculales		
Berberidaceae		
Podophyllum	5	Himalaya, E. Asia, E. N. America
Proteales		
Platanaceae		
Platanus	9	N. hemisphere, S.E. Asia
Core Eudicots		
Caryophyllales		
Plumbaginaceae		
Plumbago	24	tropical, warm regions

ORDERS, FAMILIES, AND GENERA	SPECIES	DISTRIBUTION
Saxifragales		
Altingiaceae (Hamamelidaceae)		
Altingia	8	Indomalesia
Liquidambar	5	E. Mediterranean, E Asia, S.E. N. America, Central America
Rosids (subclass Rosidae)		
Order unassigned		
Zygophyllaceae		
Balanites	25	Africa to Myanmar (Burma)
Fagonia	40	warm, dry regions
Guaiacum	6	warm temperate America, W. Indies
Larrea	5	S.W. N. America, S. America
Metharme	1	Chile
Pintoa	1	Chile
Plectrocarpa	3	temperate S. America
Porlieria	6	Mexico, Andes
Sericodes	1	N. Mexico
Eurosids I		
Malpighiales		
Clusiaceae (Guttiferae, Hypericaceae)		
subfamily Calophylloideae		
Calophyllum	187	tropics
Clusia	145	warm American tropics
Garcinia	200	Asian tropics, S. Africa
subfamily Hypericoideae		
Cratoxylum	6	Indomalesia
Moronobea	10	S. American tropics
Symphonia	17	Madagascar, American tropics
Euphorbiaceae		
Bertya	25	Australia, Tasmania
Beyeria	15	Australia
Croton	750	warm tropics
Dalechampia	120	tropics, warm temperate zones, especially America
Euphorbia	2000	cosmopolitan, especially warm
Humiriaceae		
Humiria	4	S. American tropics
Salicaceae		
Populus	35	N. temperate zone
Salix	400	N. temperate zone, Arctic
Fabales		
Fabaceae (Leguminosae)		
subfamily Caesalpinioideae		
Colophospermum	1	African tropics
Copaifera	30?	American tropics, W. Africa
Daniellia	9	W. African tropics
Eperua	15	S. American tropics

ORDERS, FAMILIES, AND GENERA	SPECIES	DISTRIBUTION
Eurosids I, continued		
Fabales, continued		
Fabaceae (Leguminosae), continued		
subfamily Caesalpinioideae, continued		
Gossweilerodendron	2	W. African tropics
Guibourtia	14	W. African tropics
Hymenaea	14	American tropics, E. Africa
Oxystigma	5	W. African tropics
Prioria	1	American tropics
Sindora	18–20	S.E. Asia, Malesia
Tessmannia	11	W. African tropics
subfamily Mimosoideae		
Acacia	1200	tropics & warm Australia
subfamily Papilionoideae		
Myroxylon	1	Central & S. America
Rosales		
Cannabaceae		
Cannabis	1	central Asia, now worldwide
Humulus	3	N. temperate zone
Moraceae		
Artocarpus	50	Indomalesia
Ficus	800	tropics & warm climates, especially Indomalesia to Australia
Morus	12	tropics & warm climates, African tropics
Rhamnaceae		
Rhamnus	125	N. hemisphere to Brazil, S. Africa
Rosaceae		
Prunus	200	temperate zones, tropical mountains
Sorbus	193	N. temperate zone
Fagales		
Betulaceae		
subfamily Betuloideae		
Alnus	25	N. temperate zone, S. to Assam, S.E. Asia, Andes
Betula	35	N. temperate zone
subfamily Coryloideae		
Ostrya	5	N. temperate zone to Central America
Eurosids II		
Brassicales		
Gyrostemonaceae		
Didymotheca (Gyrostemon)	12	W. Australia, Tasmania
Malvales		
Cistaceae		
Cistus	18	Mediterranean, Canaries, Transcaucasia
Dipterocarpaceae		
subfamily Dipterocarpoideae		

ORDERS, FAMILIES, AND GENERA	SPECIES	DISTRIBUTION
tribe Dipterocarpeae		
Anisoptera	11	Indomalesia
Dipterocarpus	69	Indomalesia
Vateria	2	S. India, Sri Lanka
Vatica	65	Indomalesia
tribe Shoreae		
Hopea	102	Indomalesia
Neobalanocarpus	1	Malesia
Parashorea	14	S. China, S.E. Asia, Malesia
Shorea	194	Sri Lanka to S. China, Moluccas, Lesser Sunda Islands
Thymelaeaceae		
Aquilaria	15	Indomalesia
Gonystylus	20	Indomalesia
Sapindales		
Sapindaceae (including Hippocastanaceae)		
Aesculus	13	N. America, S. Europe to S.E. Asia
Diplopeltis	5	Australia
Dodonaea	68	tropics & subtropics, especially Australia
Rutaceae		
Amyris	40	American tropics, W. Indies
Simaroubaceae		
Ailanthus	5	Asia, Australia
Burseraceae		
subfamily Bursereae		
Boswellia	19–20	tropical Africa & Asia
Bursera	100	American tropics
Commiphora	190	warm Africa, Madagascar, Arabia to Sri Lanka, Mexico & S. America
subfamily Canarieae		
Canarium	77	African tropics, Indomalesia
Dacryodes	40	tropical America, Africa, Asia
Santiria	24	W. African tropics, S.E. Asia
subfamily Protieae		
Protium	>120	American tropics, Madagascar to Malesia
Scutinanthe	2	Sri Lanka, Sulawesi (Celebes)
Tetragastris	9	American tropics
Trattinnickia	11	northern S. America
Triomma	1	W. Malaysia
Anacardiaceae		
Anacardium	11	American tropics
Comocladia	20	American tropics
Gluta	30	Indomalesia, Madagascar
Mangifera	40–60	Indomalesia
Metopium	3	Florida, Mexico, W. Indies
Pistacia	9	Mediterranean, Asia & Malesia, S. United States, Mexico, Guatemala

ORDERS, FAMILIES, AND GENERA	SPECIES	DISTRIBUTION
Eurosids II, continued		
Anacardiaceae, continued		
Pseudosmodingium	7	Mexico
Schinus	27	American tropics
Semecarpus	60	Indomalesia to New Caledonia
Tapirira	16	American tropics
Toxicodendron	30	N. America, E. Asia
Asterids (subclass Asteridae)		
Cornales		
Cornaceae (including Mastixaceae)		
Cornus	65	N. temperate zone, S. America
Mastixia	13	Indomalesia
Ericales		
Styracaceae		
Styrax	130	warm Eurasia, Malesia, tropical Americas
Euasterids I		
Order unassigned		
Boraginaceae (including Hydrophyllaceae)		
Eriodictyon	8	S.W. N. America
Halgania	18	Australia
Gentianales		
Rubiaceae		
Burchellia	1	S. Africa
Carphalea	10	American tropics
Cinchona	40	Andes to Costa Rica
Coffea	90	African tropics, Madagascar
Coutarea	7	Mexico to Argentina
Elaeagia	10	Central & S. American tropics, Cuba
Gardenia	60	tropical and warm Old World
Apocynaceae (including Asclepiadaceae)		
Cryptostegia	2	Madagascar
Plumeria	17	American tropics
Lamiales		
Lamiaceae (Labiatae)		
Cyanostegia	5	Australian tropics
Newcastelia	12	Australian tropics
Prostanthera	100	Australia
Myoporaceae		
Bontia	1	W. Indies, S. American tropics
Eremophila	206	Australia, especially W
Myoporum	28	S.W. Pacific, especially Australia
Scrophulariaceae		
Mimulus (including *Diplacus*)	150	cosmopolitan, especially America
Solanales		
Convolvulaceae		
Convolvulus	100	cosmopolitan, especially warm
Ipomoea	650	tropics, warm temperate zones
Solanaceae		
Anthocercis	9	Australia

ORDERS, FAMILIES, AND GENERA	SPECIES	DISTRIBUTION
Euasterids II		
Apiales		
Apiaceae (Umbelliferae)		
Dorema	12	central & S.W. Asia
Ferula	172	Mediterranean to central Asia
Laretia	2	Chilean Andes
Opopanax	3	Balkans to Iran
Thapsia	3	Mediterranean
Araliaceae		
Acanthopanax (Eleutherococcus)	30	E. Asia, Himalaya, Malesia
Aralia	36	Indomalesia, E. Asia, N. America
Dendropanax	60	American tropics, E. Asia, Malesia
Evodiopanax (Gamblea)	4	E. Himalaya to Japan, W. Malesia
Panax	6	tropical E. Asia & N. America
Asterales		
Asteraceae (Compositae)		
subfamily Asteroideae		
tribe Astereae		
Baccharis	400	Americas
Chrysothamnus	12	W. N. America
Grindelia	60	N. & S. America
Gutierrezia	25	N. & S. America
Haplopappus	150	W. N. America
Heterotheca	25	S. & N. America
Lessingia	14	W. N. America
Olearia	100	New Guinea, Australia, New Zealand
Xanthocephalum	20	S. United States to central Mexico
tribe Eupatorieae		
Brickellia	110	W. United States, Mexico, Central America to Argentina
tribe Helenieae, subtribe Madiinae		
Calycadenia	11	W. United States
Hemizonia	30	California, Baja California
Holocarpha	4	California
Holozonia	1	W. United States
Madia	20	Pacific America
tribe Helenieae, subtribe Pectidinae		
Tagetes	50	warm American tropics
tribe Heliantheae		
Balsamorhiza	14	N. America
Flourensia	30	S.W. United States to Argentina
Helianthus	50	N. America
Parthenium	15	N. America, W. Indies
Silphium	23	N. America
Goodeniaceae		
Coopernookia	6	S. Australia
Goodenia	179	Australasia

Skeletons of Characteristic Components of Fossil Resins

Polymeric skeletons characterize fossil resins, and nonpolymeric skeletons characterize resins often preserved within plant fossil structures. Adapted from K. Anderson (1995). For Class II, polycadinene, Anderson and Muntean (2000) indicated that the model of attachment of cadinene monomers (page 487) is inadequate and needs revision.

Polymeric Skeletons

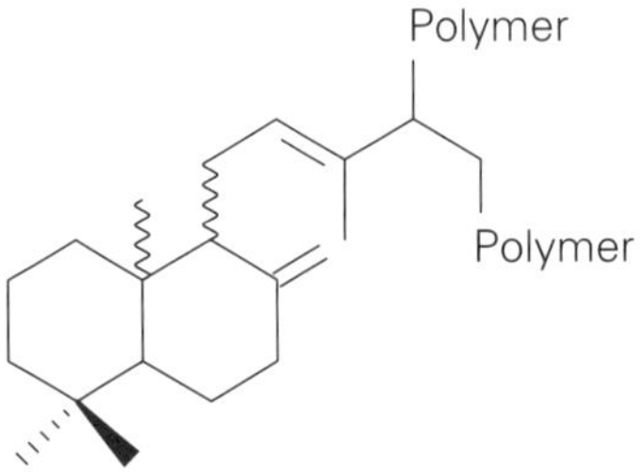

Class I: Polylabdanoid diterpenoids

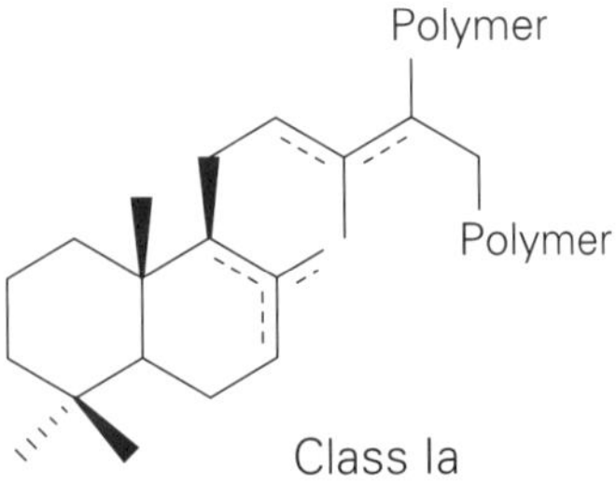

Class Ia

with succinic acid

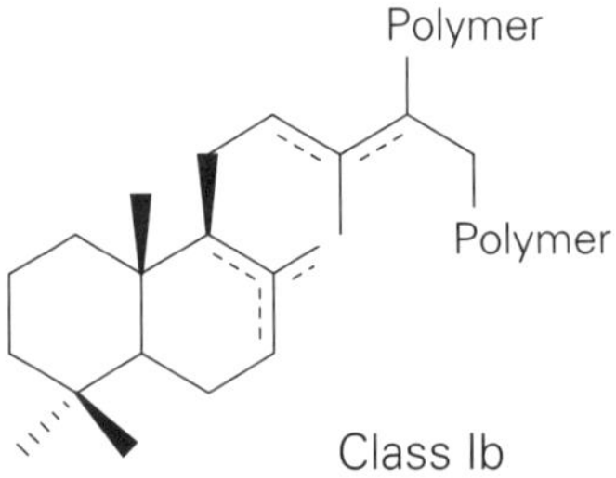

Class Ib

without succinic acid

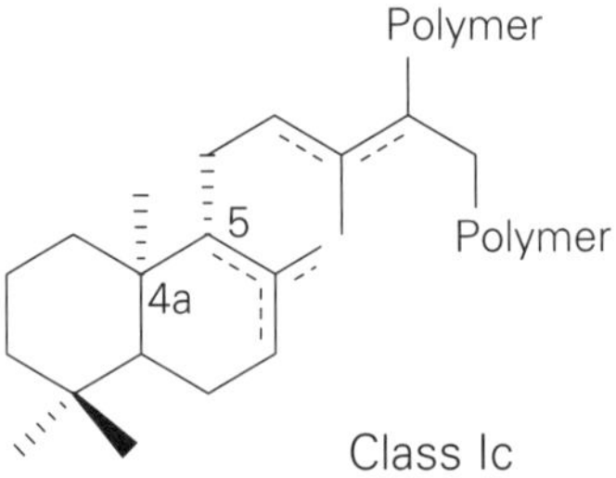

Class Ic

enantiomeric at C-4a and C-5

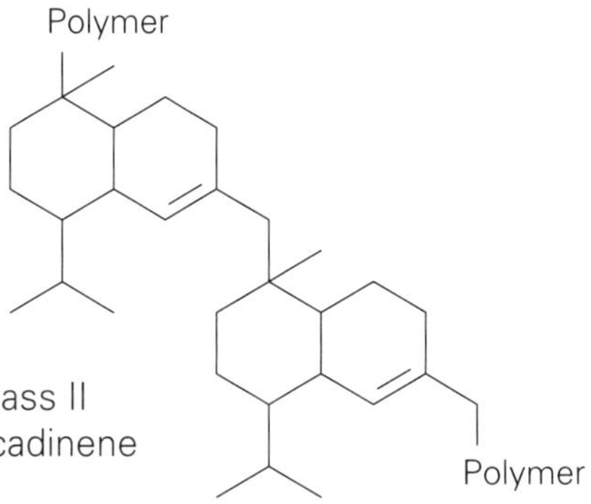

Class II
Polycadinene

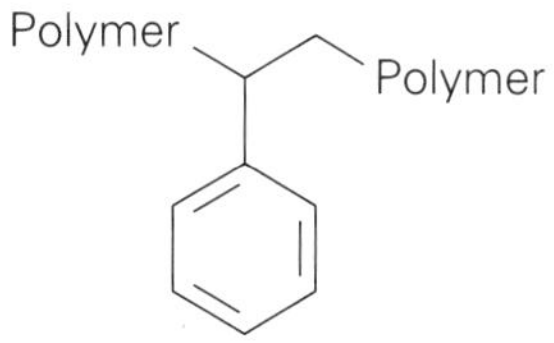

Class III
Polystyrene

Nonpolymeric Skeletons

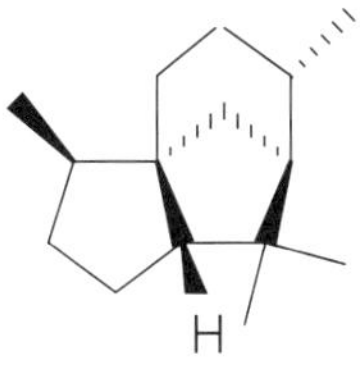

Class IV
Cedrane sesquiterpenoid

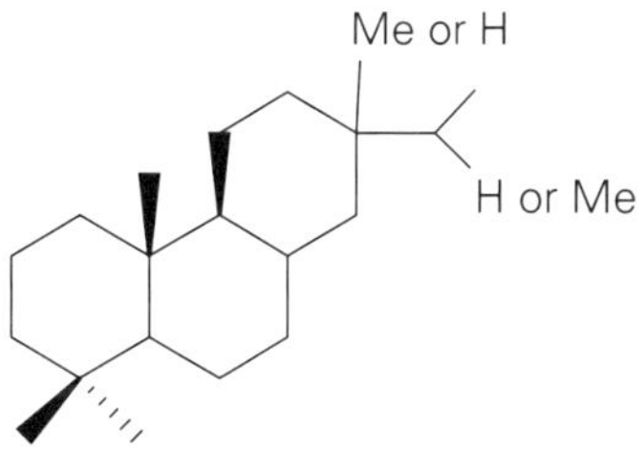

Class V
Abietane / pimarane diterpenoid

APPENDIX 4

Age, Location, and Plant Source of Amber Deposits

Age is given in millions of years ago (Ma) and geologic period or epoch. The botanical source of the resin that fossilized into amber is given by family (or order) and, where postulated, genus. Here, Taxodiaceae is recognized as a family, but its genera are now included in a more inclusive Cupressaceae s.l. (Appendix 1). See text for references. Key: *, questionable date; †, extinct, ?, botanical source suggested but not definite. Locations, by age, of other amber deposits for which no plant source has been proposed are shown in Figure 4-2.

AGE (Ma)	GEOLOGIC AGE	LOCATION OF DEPOSIT	BOTANICAL SOURCE
2.5	Pleistocene	Dominican Republic* Zanzibar* N.E. Angola* Tel-Aviv, Israel*	*Hymenaea* (Fabaceae) *Hymenaea* *Copaifera* (Fabaceae) *Pistacia*? (Anacardiaceae)
	Tertiary	Guayaquil, Ecuador* Medellín, Colombia*	*Protium*? (Burseraceae) *Hymenaea* (Fabaceae)
7	Pliocene	Zanzibar* Victoria, Australia	*Hymenaea* (Fabaceae) *Agathis* (Araucariaceae)
24	Miocene	Rhineland, Germany central Europe Bukit Asam, Sumatra central Sumatra Pará, Brazil* Dominican Republic	*Liquidambar*? (Altingiaceae) Pinaceae? or Araucariaceae?; Burseraceae? *Shorea*? (Dipterocarpaceae) *Shorea*? *Hymenaea* (Fabaceae) *Hymenaea*
29	Oligo-Miocene	Chiapas, Mexico Roxburgh, New Zealand	*Hymenaea* (Fabaceae) *Agathis* (Araucariaceae)
38	Oligocene	Baltic Coast, Europe Rhineland, Germany, coals Devon, England	Pinaceae? or Araucariaceae? *Liquidambar*? (Altingiaceae) Cupressaceae s.s.

		Baltic Coast, Europe
		
		Germany; England
			Pinaceae? or Araucariaceae?;
			  Protium or *Canarium*?
			  (Burseraceae)
			Mastixia (Cornaceae)

AGE (Ma)	GEOLOGIC AGE	LOCATION OF DEPOSIT	BOTANICAL SOURCE
		Rhineland, Germany, coals	*Liquidambar* (Altingiaceae)
		New Zealand	*Agathis* (Araucariaceae)
		Halle, Germany*	Araucariaceae
		Axel Heiberg Island, Canada	*Metasequoia* (Taxodiaceae); *Pinus, Pseudolarix* (Pinaceae)
		South Island, New Zealand	Podocarpaceae
		British Columbia, Canada	*Metasequoia*
		Washington	*Metasequoia*
		Arkansas	Dipterocarpaceae?
54	Eocene	England, southeastern coast	*Protium*? or *Canarium*?
		England, southeastern coast	*Protium*? or *Canarium*? (Burseraceae)
65	Paleocene	Montana	*Liquidambar* (Altingiaceae)
		Wyoming	Taxodiaceae?, Pinaceae?
		Utah; Wyoming	Dipterocarpaceae?
		Atlantic Coastal Plain, U.S.A.	Araucariaceae, *Liquidambar* (Altingiaceae), Taxodiaceae?, Pinaceae?
		Mississippi	Taxodiaceae, Pinaceae
		San Juan Basin, New Mexico	Taxodiaceae
		Arctic Coastal Plain, Alaska	Araucariaceae?, Taxodiaceae?
		Manitoba, Canada*	Araucariaceae?
		New Zealand coals	Araucariaceae
95	Late Cretaceous	Basque Country, Spain	Araucariaceae?
		Mt. Hermon, Lebanon & Syria	Araucariaceae?, Cheirolepidaceae†
		Salzburg, Austria	Araucariaceae?
141	Early Cretaceous	Amman, Jordan	Araucariaceae?
225	Triassic	Arizona	Araucariaceae?
		central Appalachians	*Medullosa, Cordaites*?†
		Montana; Mississippi Valley	Coniferales
310	Late Carboniferous	Northumberland, England	Cordaitales?†

Common Names, Plant Sources, and Uses of Resins

Only family and genus names are given for the plants, and only the primary uses for the resins. More details are provided in the text (use the Indexes).

RESIN NAME	PLANT SOURCE	PRIMARY USE
acaroid resin	*Xanthorrhoea* (Xanthorrhoeaceae)	varnish, medicine
aceite, see copaiba		
almécega	*Protium* (Burseraceae)	incense, medicine
aloe wood, see gharu wood		
alquitran	*Taxodium* (Cupressaceae)	medicine
ammoniacum	*Dorema* (Apiaceae)	medicine
anime, see hard copals		
asafoetida, see galbanum		
baláu	*Dipterocarpus* (Dipterocarpaceae: Dipterocarpeae)	aroma, lacquer
balm of Gilead	*Commiphora* (Burseraceae)	medicine, aroma
balsam		
Canada	*Abies* (Pinaceae)	medicine, mounting medium
Cyprus	*Pistacia* (Anacardiaceae)	chewing gum, varnish
gurjun	*Dipterocarpus* (Dipterocarpaceae: Dipterocarpeae)	illumination
Mecca	*Commiphora* (Burseraceae)	incense
Oregon	*Pseudotsuga* (Pinaceae)	mounting medium
of Peru	*Myroxylon* (Leguminosae: Papilionoideae)	medicine, aroma
of Tolu	*Myroxylon*	medicine, aroma
barniz de Pasto	*Elaeagia* (Rubiaceae)	lacquer
bdellium, see myrrh		
benzoin, see styrax		
boea, see Manila copal		
brea	*Canarium* (Burseraceae)	medicine, incense, varnish
breu branco	*Protium* (Burseraceae)	aroma, medicine
cade	*Juniperus* (Cupressaceae)	medicine, aroma
cedarwood oil	*Juniperus, Cupressus* (Cupressaceae)	aroma
charão	*Toxicodendron* (Anacardiaceae)	lacquer

RESIN NAME	PLANT SOURCE	PRIMARY USE
chio(s)	*Pistacia* (Anacardiaceae)	chewing gum, varnish
colophony, see oleoresin		
copáiba-jacaré	*Eperua* (Fabaceae: Caesalpinioideae)	lacquer
copaiba oil (balsam)	*Copaifera* (Fabaceae: Caesalpinioideae)	medicine, cosmetics
copal, hard		hard varnish, adhesives, linoleum
Congo	*Guibourtia, Tessmannia* (Fabaceae: Caesalpinioideae)	
East African	*Hymenaea* (Fabaceae: Caesalpinioideae)	
kauri	*Agathis* (Araucariaceae)	
Manila	*Agathis*	
South American	*Hymenaea*	
West African	*Copaifera, Daniellia, Gossweilerodendron, Guibourtia* (Fabaceae: Caesalpinioideae)	
copalm	*Liquidambar* (Altingiaceae)	medicine
copal pom	*Bursera, Protium* (Burseraceae), *Pinus* (Pinaceae), *Liquidambar* (Altingiaceae)	incense, medicine
dammar		spirit varnish
Batavian, see *damar mata Kuching*		
black	*Canarium* (Burseraceae)	
damar mata Kuching	*Hopea, Shorea* (Dipterocarpaceae: Shoreae)	
damar pěnak	*Neobalanocarpus* (Dipterocarpaceae: Shoreae)	
piney (Indian white)	*Vateria* (Dipterocarpaceae: Dipterocarpeae)	
dragon's blood	*Dracaena* (Convallariaceae), *Daemonorops* (Arecaceae), *Croton* (Euphorbiaceae)	dye, medicine
eagle wood, see gharu wood		
elemi		aroma, medicine, incense, varnish
African	*Canarium* (Burseraceae)	
bastard	*Calophyllum* (Clusiaceae)	
hard (Brazilian)	*Protium* (Burseraceae)	
Manila	*Canarium*	
yellow	*Symphonia* (Clusiaceae)	
Yucatán	*Amyris* (Rutaceae)	
estoraque, see styrax		
frankincense	*Boswellia* (Burseraceae)	incense, medicine
gaharu or *gahur,* see gharu wood		
galbanum	*Ferula* (Apiaceae)	medicine
gamboge	*Garcinia* (Clusiaceae)	pigmented varnish

RESIN NAME	PLANT SOURCE	PRIMARY USE
getah	*Garcinia* (Clusiaceae), *Altingia* (Altingiaceae)	varnish, medicine
gharu wood	*Aquilaria* (Thymelaeaceae)	incense
gonzetsu (koshiabura)	*Acanthopanax* (Araliaceae)	varnish
guaiac	*Guaiacum* (Zygophyllaceae)	medicine
guggal	*Commiphora* (Burseraceae)	incense, medicine
"gum"		
accroides, see acaroid resin		
benjamin, see styrax		
blackboy, see acaroid resin		
dikamali	*Gardenia* (Rubiaceae)	medicine
euphorbium	*Euphorbia* (Euphorbiaceae)	medicine
grass tree, see acaroid resin		
spruce	*Picea* (Pinaceae)	chewing gum
gutta gamba, see gamboge		
hashish	*Cannabis* (Cannabaceae)	psychoactive drug, medicine
heath, see cade		
hops resin	*Humulus* (Cannabaceae)	beer flavoring
ilorin (illurin)	*Daniellia* (Fabaceae: Caesalpinioideae)	fumigation
jalap	*Convolvulus / Ipomoea* (Convolvulaceae)	medicine
jauaricica, see *almécega*		
jutaicica	*Hymenaea* (Fabaceae: Caesalpinioideae)	varnish, adhesives
kayu-galu oil (supa oil)	*Sindora* (Fabaceae: Caesalpinioideae)	medicine
keruing oil (wood oil)	*Dipterocarpus* (Dipterocarpaceae: Dipterocarpeae)	aroma, lacquer
kinnabai	*Dracaena* (Convallariaceae)	medicine
kokkolyi, see mastic		
labdanum (ladanum)	*Cistus* (Cistaceae)	aroma
marijuana, see hashish		
mastic	*Pistacia, Schinus* (Anacardiaceae)	masticatory, varnish
melengket, see Manila copal		
minyak oil, see keruing oil		
mopa mopa, see *barniz de Pasto*		
myrrh	*Commiphora* (Burseraceae)	incense, medicine
ogea	*Daniellia* (Fabaceae: Caesalpinioideae)	varnish
oleoresin	*Pinus* (Pinaceae)	naval stores (multiple industrial uses)
olibanum, oil of or gum of, see frankincense		
olio d'abete (*olio d'abezzo*), see Alsatian turpentine		
onycha	*Styrax* (Styracaceae), *Commiphora* (Burseraceae)	incense
opopanax	*Opopanax* (Apiaceae), *Commiphora* (Burseraceae)	medicine, aroma

RESIN NAME	PLANT SOURCE	PRIMARY USE
palosapis resin	*Anisoptera* (Dipterocarpaceae: Dipterocarpeae)	aroma
pitch, Burgundy	*Picea* (Pinaceae)	naval stores (multiple industrial uses)
podophylloresin	*Podophyllum* (Berberidaceae)	medicine
pontianak, see Manila copal		
rasa mala	*Altingia* (Altingiaceae)	medicine
rong	*Garcinia* (Clusiaceae)	specialty inks
sagapenum	*Ferula* (Apiaceae)	medicine, food flavoring
salai guggal	*Boswellia* (Burseraceae)	medicine
sandarac		specialty varnish
African	*Tetraclinis* (Cupressaceae)	
Australian	*Callitris* (Cupressaceae)	
sandarosy	*Hymenaea* (Fabaceae: Caesalpinioideae)	varnish
scammony, see jalap		
silphium	*Ferula* (Apiaceae)	medicine, contraceptive
stacte, see myrrh		
storax		medicine
American	*Liquidambar* (Altingiaceae)	
Levant	*Liquidambar*	
liquid	*Altingia* (Altingiaceae)	
styrax	*Styrax* (Styracaceae)	medicine
terebinth	*Pistacia* (Anacardiaceae)	medicine
thisi	*Gluta* (Anacardiaceae)	lacquer
turpentine		
Alsatian	*Abies* (Pinaceae)	medicine
Chian, see *chio*		
Jura, see Burgundy pitch		
Strasbourg, see Alsatian turpentine		
Venice turpentine	*Larix* (Pinaceae)	turpentine
ushitsu	*Dendropanax* (Araliaceae)	specialty varnish
xochicotzatl	*Liquidambar* (Altingiaceae)	incense, medicine

Glossary

adaxial side toward the axis or stem of the plant; versus abaxial, the side away from the axis or stem

aglycone flavonoid compound without a glycosyl (sugar) substituent

alkane synonymous with acyclic, saturated hydrocarbons

ambergris solid, opaque, waxy, ash-colored substance used in perfumery, secreted from sperm whales

amulet object worn on the body because of supposed magical power to protect against injury or evil; talisman

angiosperm plant producing true flowers in which the seeds are enclosed in a true fruit

APG Angiosperm Phylogeny Group

balsam relatively soft and initially malleable resin, generally fragrant; sometimes restricted to phenolic resins of this kind

bark all plant tissues outside the vascular cambium in a woody stem

bituminous coal coal that yields pitch or tar when it burns; soft coal

blister see secretory pocket

boreotropical of a flora comprising tropical genera that was widespread in northerly middle latitudes of North America and Eurasia in the Eocene but that was subsequently extirpated from those areas and is now limited to southerly eastern Asia and Malesia

bract modified, usually reduced leaf-like structure

bracteole small bract that subtends the flower

calyx sepals collectively; the outermost flower whorl, usually green and enclosing the flower bud

cambium embryonic tissue zone (meristem) that runs parallel to the sides of roots and stems; vascular cambium produces secondary xylem and phloem, and cork cambium produces the cork tissues of the bark

canal elongated endogenous (internal) secretory structure or duct

cellulose structural polysaccharide making up major part of the tough secondary walls of plant cells; a polymer of glucose

chemosystematics use of secondary chemicals as characters in plant systematics (the study of biological diversity and its evolutionary history)

clade a branching group in a cladogram, displaying phylogenetic relationships

cladistics a system of arranging taxa by the analysis of primitive and derived characteristics so that the arrangement will reflect presumed phylogenetic relationships

colleter trichome secreting a sticky substance (particularly resin or mucilage) that consists of a multicellular head with or without a stalk

composition of resin, the relative amounts (percentages) of the constituents of a resin

constitutive resin preformed resin; resin synthesized and stored in the plant prior to injury

convergent evolution independent development of similar structures (or chemistry) in organisms that are not directly related; often found in organisms living in similar environments

copal a kind of resin; has been used to refer to all incense resins by Mexican and Central Americans, but commercially has been restricted to leguminous and araucarian plants that produce hard varnishes

cortex primary tissue of a stem or root, bounded externally by the epidermis and internally by the central cylinder of vascular tissue

CP/MAS NMR cross-polarization / magic-angle spinning nuclear magnetic resonance

cuticle waxy or fatty, noncellular layer on the outer wall of epidermal cells, formed of a substance called cutin; an adaptation to preventing desiccation in land plants

cutin waxy substance (polymer primarily of fatty acids) that impregnates the walls of epidermal cells, forming the cuticle

cyst see secretory pocket

dammar (or damar) kind of a primarily varnish resin generally produced by members of the family Dipterocarpaceae; sometimes used for resin similarly derived from other taxa such as Burseraceae

dicot abbreviation for dicotyledon, flowering plant in which the embryo has two cotyledons, leaves are net-veined, and flower parts in fours or fives

dictyosome cellular organelle consisting of stacks of flat vesicles that are broken into a lacework of interconnected tubules, involved in preparation of material to be transported out of the cell; synonymous with Golgi body but occasionally referred to (erroneously) as Golgi apparatus

dioecious plant in which the male and female flowers are separate and borne on different individuals

directional versus **diversifying selection pressure** directional selection for particular characteristics results from pest pressures exerted by dependent organisms, usually in a homogeneous environment, whereas diversifying (multidirectional) selection occurs from pest pressures in a more heterogeneous environment in space and time

diterpene 20-carbon terpenoid compound; often a nonvolatile component in resins

DTMS direct-temperature-resolved mass spectrometry

duct general term for endogenous (internal) secretory structure

elemi a kind of balsam resin, very fragrant and generally derived from members of the family Burseraceae

endogenous secretion material that accumulates in internal structures comprising intercellular spaces surrounded by secretory cells and normally exuding from the plant when it is injured

endoplasmic reticulum flattened membrane network dividing the cytoplasm into compartments and channels; if ribosomes are attached, it is called rough ER; if ribosomes are not present, it is called smooth ER

epithelium cells that create the lining of secretory ducts

ER see endoplasmic reticulum

essential oil volatile mono- and sesquiterpenes occurring together; either volatiles occurring alone, or associated with a nonvolatile fraction in resins

exocytosis process by which a cell releases a large molecule

exogenous secretion material that occurs in epidermal cells; may be discharged directly to the outside or first into a subcuticular space

foraminifera marine protozoans with calcareous shells, often used to date geologic strata

FTIR Fourier transform infrared spectroscopy

GC gas chromatography

GC-MS gas chromatography–mass spectrometry

gland general term for a secretory structure

glycoside plant product whose molecules are composed of an alcohol or alcohol-like group bound to a sugar

gum complex chain of hydrophilic polysaccharides (complex sugars) derived from monosaccharide (simple sugar) moieties; usually thought to be induced by injury to plant tissue

gum resin resin in which carbohydrates from the breakdown of the epithelial cell walls have mixed with the terpenoids during lysigenous development or following damage to the secretory structure; pine gum resin, however, is a term that has been used in the naval stores industry to distinguish resin obtained from tapping the tree from that obtained from stumps, and gum turpentine and gum rosin are the terms most commonly used to refer to the distilled products

gymnosperm plant producing seeds but the seeds not enclosed in a true fruit but rather usually in a woody cone (i.e., a coniferous plant) or in a structure derived from a cone

herbivore, generalist herbivore that consumes numerous host plant taxa, either locally or geographically

herbivore, specialist herbivore that consumes only one or very few host plant taxa

idioblast any cell that differs significantly from neighboring cells, usually by content, size, or structure

Indomalesia plant geographical term for area extending from India through Malesia

induced resin resin synthesized in response to injury

IR infrared spectroscopy

isoprene unit structure of terpenoids

kino a type of gum that contains polyphenols

lacquer, natural resin collected in liquid state and applied directly to a surface to be varnished without the use of a solvent or drying agent

lamella tissue present at the junction where the walls of neighboring cells come into contact

latex (plural, latices) milky or clear plant exudate composed of a complex mixture of substances such as terpenoids, proteins, carbohydrates, phenolics, etc.; occurs in specialized structures called laticifers

laticifer secretory vessels or specialized cells in which latex is produced; laticifers vary in development, anatomy, and distribution

lignan widely distributed family of dimeric substances arising from phenolic coupling of C_6–C_3 units (phenylpropanes)

lignin polymer of phenolic substances impregnating the cellulose framework of secondary plant cell walls; an adaptation for support

lignite soft, brownish black coal in which the texture of the original wood can often be seen

lumen (plural, lumina) central space of trichomes, ducts, or virtually any other enclosed space

lysigeny breakdown of cells that can produce an intercellular space

Ma million(s of) years ago

Malaya former British federation of states in the southern part of the Malay Peninsula; now peninsular or West Malaysia

Malay Archipelago great island group of Indonesia, formerly called the East Indies, lying between the Asian mainland and Australia; also includes the Philippine Archipelago

Malay Peninsula peninsula in Southeast Asia, extending from Singapore to the Isthmus of Kra, including part of Thailand in the north, the southern part constituting peninsular or West Malaysia, a part of the federation of Malaysia

Malesia plant geographical term for the area comprising Indonesia, Malaysia, Brunei, the Philippines, Singapore, Timor, and eastern New Guinea

mastic a masticatory and specialty varnish resin produced primarily by members of the family Anacardiaceae

masticatory substance to be chewed but not swallowed, for example, chewing gum

meristem undifferentiated tissue, the cells of which are capable of active cell division and differentiation into specialized tissue

mitochondrion (plural, mitochondria) membrane-bounded organelle that functions in cellular aerobic respiration and oxidation of food for the production of energy for cellular activities

moiety sharing half or an indefinite part of the whole

monocot abbreviation for monocotyledon, flowering plant in which the embryo has only one cotyledon, leaves are typically parallel-veined, and flower parts are in threes

monophyly all descendants are from a common ancestor

monoterpene 10-carbon terpenoid compound; common volatile component in resins

MS mass spectrometry

mucilage water-soluble complex of acidic and neutral polysaccharide polymers of high molecular weight that occur in various secretory structures and serve multiple roles for plants; can become mixed with resins in surface coatings

natural products same as secondary metabolites but often used commercially to distinguish them from synthetic products produced by humans

naval stores industry based on the processing of pine resins, originally to provide materials for ships

NMR nuclear magnetic resonance

NTFP nontimber forest product

oleoresin relatively fluid terpenoid resin with relatively high proportion of volatile to nonvolatile terpenes compared to other resins

ontogeny (adjective, ontogenetic) life cycle of a single organism; used in some resin studies to distinguish between the development of an organ (e.g., leaf) during the ontogeny or development of the tree (e.g., seedling, sapling, and adult tree).

palisade parenchyma columnar cells, containing many chloroplasts, found just beneath the upper epidermis of leaves

parenchyma most common and unspecialized plant cell type; living, thin-walled, containing only the primary cell wall, usually making up photosynthetic or storage tissue

pedicel stem of an individual flower

periderm outer protective tissue in plants produced by the cork cambium, functionally replacing the epidermis when it is destroyed by secondary growth; it is considered the outer dead portion of the phloem or bark

phenylpropanoid phenolic compound with a six-carbon benzene ring and a three-carbon side chain

phloem conductive tissue composed of living and nonliving tissue; inner portion consists of food-conducting tissue differentiated from vascular cambium, and outer portion consists of corky tissue differentiated from the cork cambium (also called inner and outer bark)

pitch another name for pine resin; also sometimes used for a carbonaceous residue obtained by distillation of substances such as coal tar and petroleum

plasma membrane phospholipid membrane between the cytoplasm of the plant cell and the cell wall; also called cell membrane and plasmalemma

plasmodesma (plural, plasmodesmata) small hole in the primary plant cell wall through which the protoplasm of one cell is in contact with that of adjacent cells

plastids plant cell organelles that function in food manufacture and storage

prenyl indicating the presence of an isoprenoid (terpenoid) group

primary cell wall plant cell wall deposited during the period of cell expansion

primary growth growth that originates in apical meristems of shoots and roots, resulting in an increase in length, as contrasted with secondary growth, increase in diameter

primary plant body part of the plant body arising from apical meristems and their derivative meristematic tissues; composed entirely of primary tissues

propolis a kind of "bee glue," usually consisting of resin accumulated from various plant organs, for the construction of cells or to close off the nest; a term particularly used for honeybees

Py-GC-MS pyrolysis gas chromatography–mass spectrometry

Py-MS pyrolysis mass spectrometry

resinite microscopic particles of fossil resin; distinguished from macroscopic particles, which are amber or fossil resin per se

resinoid for compounds extracted from a resin, usually volatile compounds used for fragrance

rhizome elongated, horizontal, underground stem; may be enlarged for storage or may function in vegetative reproduction

rosin nonvolatile diterpene fraction in pine resin

sandarac specialty varnish resin produced by the cupressaceous genera *Callitris* and *Tetraclinis*

saponification process of combining fatty acids with alkali to make soap

schizogeny formation of an intercellular space separating one parenchyma cell away from another, cellular breakdown not involved; for resins, referable to the formation of secretory structures such as canals and cysts

schizolysigeny formation of an intercellular space, involving both the separation of parenchyma cells from each other and actual breakdown of the cells

secondary growth increase in stem and root diameter made possible by cell division of lateral meristems; secondary growth produces the secondary plant body

secondary metabolites compounds synthesized by the plant that lack any apparent role in primary physiological or metabolic processes but may play various ecological roles in survival of the plant and in ecosystem interactions

secondary plant wall innermost layer of the cell wall, formed in certain cells after cell elongation has ceased; secondary walls have highly organized microfibrillar structure and are often impregnated with lignin

secretory pocket rounded endogenous secretory structure or duct; also called blister or cyst

selection pressure nature and strength of environmental factors tending to select for or against various traits

serotinous cone closed cone, often by resinous coating, until opened by high temperatures such as those caused by fire

sesquiterpene 15-carbon terpenoid compound; often a volatile component in resins

sister group closest group of relatives in a phylogenetic relationship

s.l. sensu lato, in the broad sense

Southeast Asia region of Asia bounded roughly by India (west), China (north), and the Pacific Ocean (east), becoming popular after World War II and replacing phrases such as East Indies, Indochina, and the Malay Peninsula, which formerly designated all or part of the area; Southeast Asia includes the Indochina Peninsula, Malay Peninsula, and the Indonesian and Philippine Archipelagoes; it includes 10 countries: Brunei, Cambodia, Indonesia, Laos, Malaysia, Myanmar (Burma), the Philippines, Singapore, Thailand, and Vietnam

sp. (plural, spp.) species

spirit varnish resin solution that dries by evaporation of a solvent, with or without application of heat

s.s. sensu stricto, in the strict sense

stipule leaf-like appendage on either side of the basal part of a leaf, often scale- or spine-like

stoma (plural, stomata) microscopic opening in the plant epidermis, surrounded by guard cells and serving for gaseous exchange

storax kind of balsam resin derived from *Liquidambar;* sometimes used for the balsam styrax (not referring to the plant *Styrax* but the balsam from it), causing confusion regarding the botanical source of the resin

styrax kind of balsam resin derived from *Styrax;* sometimes the resin styrax is confused with storax

suberin specialized lipid that impregnates the walls of cork cells

subsp. subspecies

swidden type of slash-and-burn agriculture in which fields are cleared by cutting the forest and burning the vegetation; common type of shifting cultivation used in the tropics today

syn(s). synonym(s)

synthase enzyme that catalyzes the formation of the basic skeleton of compounds such as terpenoids; sometimes called cyclase when the reaction products are cyclic

talisman anything bearing engraved symbols; thought to bring good luck or keep away evil, to have magical power or charm; amulet

tar product of destructive distillation of pine resin, often as by-product of charcoal made from coniferous wood

taxon (plural, taxa) natural group of organisms at any taxonomic rank

THC tetrahydrocannabinol

trichome outgrowth of the epidermis, such as a hair (sometimes glandular)

triterpene 30-carbon terpenoid compound; nonvolatile component in resins

turpentine volatile mono- and sesquiterpene fraction of pine resin

turpentining process of obtaining pine resin by tapping the tree

var. variety

wax complex lipid-soluble mixture whose common components are alcoholic esters of fatty acids and straight-chain alkanes; important constituent of the cuticle, acting as a protective coating on the epidermis of exterior plant walls

wood, early portion of the annual ring formed in wood of temperate zone trees during spring, characterized by large cells and thin walls; also called spring wood

wood, late portion of an annual ring formed in wood of temperate zone trees during summer and fall, characterized by small-diameter cells with thick walls; also called summer wood

xylem complex vascular tissue through which most of the water and minerals are conducted through the plant; constituted primarily of nonliving cells (also called wood)

References

The scope of *Plant Resins* precludes an exhaustive bibliography. To enhance the readability of the text, the cited references emphasize review articles and those of major importance. However, some additional references of historic interest or ancillary to topics discussed in the text, but of possible interest, are included here.

Abdul-Ghani, A.-S., and R. Amin. 1997. Effect of aqueous extract of *Commiphora opobalsamum* on blood pressure and heart rate in rats. Journal of Ethnopharmacology 57: 219–222.

Abel, E. 1980. Marihuana—the First Twelve Thousand Years. Plenum Press, New York.

Abercrombie, T. J. 1985. Arabia's frankincense trail. National Geographic 1985(Oct.): 474–512.

Adams, R. P. 1987. Investigation of *Juniperus* species of the United States for new sources of cedarwood oil. Economic Botany 41: 48–54.

Adams, R. P. 1991. Cedarwood oil—analyses and properties. In: Modern Methods of Plant Analysis, pp. 159–173, eds. H. Linskens and J. F. Jackson. Springer-Verlag, New York.

Adams, R. P., E. von Rudloff, and L. Hogge. 1983. Chemosystematic studies of the western North American junipers based on their volatile oils. Biochemical Systematics and Ecology 11: 189–193.

Adams, R. P., C. A. McDaniel, and F. L. Carter. 1988. Termiticidal activities in the heartwood, bark / sapwood and leaves of *Juniperus* species from the United States. Biochemical Systematics and Ecology 16: 453–456.

Adamson, A. D. 1971. Oleoresins: production and markets with particular reference to the United Kingdom. Report G56. Tropical Products Institute, London.

Ageel, A. M., J. S. Mossa, M. Tariq, M. A. al-Yahya, and M. S. al-Said. 1987. Plants Used in Saudi Folk Medicine. King Saud University Press, Riyadh.

Ahmed, S. I., F. Asad, and S. S. Husain. 1969. Anatomical studies of *Commiphora mukul* Engl. and the localization of gum, resins and tannins. Hamdard Medical Digest 12: 492–495.

Ajani, J. A., P. F. Mansfield, and P. Dumas. 1999. Oral etoposide for patients with metastatic gastric adenocarcinoma. Cancer Journal from Scientific American 5: 112–114.

Alcorn, J. B. 1984. Huastec Mayan Ethnobotany. University of Texas Press, Austin.

Alencar, J. C. 1982. Estudos silviculturais de uma população natural de *Copaifera multijuga* Hayne—Leguminosae, na Amazônica central. 2—produção de óleo-resina. Acta Amazônica 12: 75–89.

Alencar, J. C. 1984. Estudos silviculturais de um população natural de *Copaifera multijuga* Hayne (Leguminosae) na Amazônica central. 3. Distribução espacial da regeneração natural pré-existente. Acta Amazônica 14: 255–279.

Al-Harbi, M. M., S. Oureshi, M. Raza, M. M. Ahmed, M. Afzal, and A. H. Shah. 1997. Gastric antiulcer and cytoprotective effect of *Commiphora molmol* in rats. Journal of Ethnopharmacology 55: 141–150.

Allegretti, M. H. 1990. Extractive reserves: an alternative for reconciling development and environmental conservation in Amazonia. In: Alternatives to Deforestation: Steps Toward Sustainable Use of the Amazonian Rain Forest, pp. 252–262, ed. A. B. Anderson. Columbia University Press, New York.

Allen, O. N., and E. K. Allen. 1981. The Leguminosae: a Source Book of Characteristics, Uses and Nodulation. University of Wisconsin Press, Madison.

Alonso, J., A. Barron, J. C. Corral, J. Grimalt, J. F. López, X. Martínez-Delclos, V. Peñalver, and P. R. Trincao. 2000. A new fossil resin with biological inclusions in Lower Cretaceous deposits from Álava (northern Spain, Basque-Cantabrian Basin). Journal of Paleontology 74: 158–178.

Alverson, W. S., K. G. Karol, D. A. Baum, M. W. Chase, S. M. Swenson, R. McCourt, and K. J. Sytsma. 1998. Circumscription of the Malvales and relationships to other Rosidae: evidence from *rbc*L sequence data. American Journal of Botany 85: 876–887.

Amma, M. K., N. Malhotra, R. K. Suri, O. P. Arya, H. M. Dani, and K. Sareen. 1978. Effect of the oleoresin of gum guggul *(Commiphora mukul)* on the reproductive organs of the female rat. Indian Journal of Experimental Biology 16: 1021–1023.

Ammon, H. P. T. 1996. *Salai guggal—Boswellia serrata:* from herbal medicine to a specific inhibitor of leukotriene biosynthesis. Phytomedicine 3: 67–70.

Ammon, H. P. T., T. Mack, G. B. Singh, and H. H. Safayhi. 1991. Inhibition of leukotriene B_4 formation in rat peritoneal neutrophils by an ethanolic extract of the gum resin exudate of *Boswellia serrata*. Planta Medica 57: 203–207.

Ammon, H. P. T., H. H. Safayhi, T. Mack, and J. Sabieraj. 1997. Mechanism of anti-inflammatory actions of curcumine and boswellic acids. Journal of Ethnopharmacology 38: 113–119.

Anaya, A. L., M. R. Calera, R. Mata, and R. Pereda-Miranda. 1990. Allelopathic potential of compounds isolated from *Ipomoea tricolor* Cav. (Convolvulaceae). Journal of Chemical Ecology 16: 2145–2152.

Ander, K. 1941. Die Insektenfauna des baltischen Bernstein nebst damit verknüpften zoogeographischen Problemen. Acta Universitatis Lundensis, Nova Series, Sectio 2 (Lunds Universitets Årsskrift, Ny Följd, Andra Afdelningen) 38: 3–82.

Anderson, Anthony B. 1990. Extraction and forest management by rural inhabitants in the Amazon Estuary. In: Alternatives to Deforestation: Steps Toward Sustainable Use of the Amazonian Rain Forest, pp. 65–85, ed. A. B. Anderson. Columbia University Press, New York.

Anderson, Arthur B. 1955. Recovery and utilization of tree extractives. Economic Botany 9: 108–140.

Anderson, D. M. W. 1985. Gums and resins, and factors influencing their economic development. In: Plants for Arid Lands, pp. 343–356, eds. G. E. Wickins, J. R. Goodin, and D. V. Field. George Allen and Unwin, London.

Anderson, D. M. W., and A. C. Munro. 1969. Gum exudates from the genus *Araucaria*. Carbohydrate Research 11: 43–51.

Anderson, K. B. 1994. The nature and fate of natural resins in the geosphere. IV. Middle and Upper Cretaceous amber from the Taimyr Peninsula, Siberia—evidence for a new structural subclass of resinite. Organic Geochemistry 21: 209–212.

Anderson, K. B. 1995. New evidence concerning the structure, composition and maturation of Class I (polylabdanoid) resinites. In: Amber, Resinite and Fossil Resins, pp. 105–129, eds. K. B. Anderson and J. C. Crelling. Symposium Series 617. American Chemical Society, Washington, D.C.

Anderson, K. B. 1997. The nature and fate of natural resins in the geosphere—VII. A radiocarbon (^{14}C) age scale for description of immature natural resins: an invitation to scientific debate. Organic Geochemistry 25: 251–253.

Anderson, K. B., and R. E. Botto. 1993. The nature and fate of natural resins in the geosphere. III. Re-evaluation of the structure and composition of Highgate copalite and glessite. Organic Geochemistry 20: 1027–1038.

Anderson, K. B., and B. A. LePage. 1995. Analysis of fossil resins from Axel Heiberg Island, Canadian Arctic. In: Amber, Resinite and Fossil Resins, pp. 170–192, eds. K. B. Anderson and J. C. Crelling. Symposium Series 617. American Chemical Society, Washington, D.C.

Anderson, K. B., and J. V. Muntean. 2000. The nature and fate of natural resins in the geosphere. Part X. Structural characteristics of the macromolecular constituents of modern dammar resin and Class II ambers. Geochemical Transactions 1.

Anderson, K. B., E. E. Winans, and R. E. Botto. 1992. The nature and fate of natural resins in the geosphere. II. Identification, classification, and nomenclature of resinites. Organic Geochemistry 18: 829–841.

Andersson, M., O. Bergendorff, R. Shan, P. Zygmunt, and O. Sterner. 1997. Minor components with smooth muscle relaxing properties from scented myrrh *(Commiphora guidottii)*. Planta Medica 63: 251–254.

Andrée, K. 1937. Der Bernstein und Seine Bedeutung in Natur- und Gesteswissenschaften: Kunst und Kunstgewerbe. Technik, Industrie und Handel. Gräfe und Unzer, Königsberg.

Andrée, K. 1951. Der Bernstein, das Bersteinland und Sein Leben. Kosmos, Stuttgart.

Angelopoulou, D., C. Demetzos, and D. Perdetzoglou. 2002. Diurnal and seasonal variation from *Cistus monspeliensis* L. leaves. Biochemical Systematics and Ecology 30: 189–204.

Angiosperm Phylogeny Group. 1998. An ordinal classification for the families of flowering plants. Annals of the Missouri Botanical Garden 85: 531–553.

Ankarfjärd, R., and M. Kegl. 1998. Tapping oleoresin from *Dipterocarpus alatus* (Dipterocarpaceae) in a Lao village. Economic Botany 52: 7–14.

Ansari, S. H., M. Ali, and J. S. Qadry. 1994. New tetracyclic triterpenoids from *Pistacia integerrima* galls. Pharmazie 49: 356–357.

Arbeláez, E. P. 1956. Plantas Utiles de Colombia. Libraria Colombiana, Bogotá.

Archer, B. I., and Audley, B. G. 1973. Rubber, gutta percha and chicle. In: Phytochemistry, Vol. 2, Organic Metabolites, pp. 310–343, ed. L. P. Miller. Van Nostrand Reinhold, New York.

Arctander, S. 1960. Perfume and Flavor Materials of Natural Origin. Arctander, Elizabeth, New Jersey.

Aregullin, M., M. E. Gompper, and E. Rodriguez. 2002. Triterpenes and a sesquiterpene lactone in the resin of *Trattinnickia aspera* (Burseraceae). Biochemical Systematics and Ecology 30: 187–188.

Argurell, S., W. L. Dewey, and R. E. Wilette (eds.). 1984. The Cannabinoids: Chemical, Pharmalogic and Therapeutic Aspects. Academic Press, Orlando.

Armbruster, W. S. 1984. The role of resin in angiosperm pollination: ecological and chemical considerations. American Journal of Botany 71: 1149–1160.

Armbruster, W. S. 1992. Phylogeny and evolution of plant–animal interactions. BioScience 42: 12–20.

Armbruster, W. S. 1993. Evolution of plant pollination systems: hypotheses and tests with the Neotropical vine *Dalechampia*. Evolution 47: 1480–1505.

Armbruster, W. S. 1996. Exaptation, adaptation and homoplasy: evolution of ecological traits in *Dalechampia* vines. In: Homoplasy: the Recurrence of Similarity in Evolution, pp. 227–243, eds. M. J. Sanderson and L. Hufford. Academic Press, San Diego.

Armbruster, W. S. 1997. Exaptations link evolution of plant–herbivore and plant pollinator interactions: a phylogenetic inquiry. Ecology 78: 1661–1672.

Armbruster, W. S., J. J. Howard, T. P. Clausen, E. M. Debevec, J. C. Loquvam, M. Matsaki, B. Cerendolo, and F. Andel. 1997. Do biochemical exaptations link evolution of plant defense and pollination systems? Historical hypotheses and experimental tests with *Dalechampia* vines. American Naturalist 149: 461–484.

Arnone, A. G. Nasini, O. U. de Pava, and L. Merlini. 1997. Constituents of dragon's blood. 5. Dracoflavans B_1, B_2, C_1, C_2, D_1, D_2, new A-type deoxyproanthocyanidins. Journal of Natural Products 60: 971–975.

Arrhenius, S. P., and J. H. Langenheim. 1983. Inhibitory effects of *Hymenaea* and *Copaifera* leaf resins on the leaf fungus *Pestalotia subcuticularis*. Biochemical Systematics and Ecology 11: 361–366.

Arrhenius, S. P., and J. H. Langenheim. 1986. The association of *Pestalotia* on the leguminous tree genera *Hymenaea* and *Copaifera* in the Neotropics. Mycologia 78: 673–676.

Ash, S. R. 1970. *Pagiophyllum simpsonii*, a new conifer from the Chinle Formation (Upper Triassic) of Arizona. Journal of Paleontology 44: 945–952.

Ash, S. R. 1972. Late Triassic plants from the Chinle Formation in northeastern Arizona. Palaeontology 15: 598–618.

Ash, S. R. 1987. Petrified Forest National Park, Arizona. In: Centennial Field Guide, Rocky Mountain Section, Vol. 2, pp. 405–410, ed. S. S. Bues. Geological Society of America.

Ashton, P. S. 1982. Dipterocarpaceae. Flora Malesiana, Series 1, 9: 241–242.

Asres, K., A. Tei, G. Moges, F. Sporer, and M. Wink. 1998. Terpenoid composition of the wound-induced bark exudate of *Commiphora tenuis* from Ethiopia. Planta Medica 64: 473–475.

Assad, Y. O. H., B. Torto, A. Hassanali, P. G. N. Njagi, N. H. H. Bashir, and H. Mahamat. 1997. Seasonal variation in the essential oil composition of *Commiphora quadricincta* and its effect on the maturation of immature adults of the desert locus, *Schistocera gregaria*. Phytochemistry 44: 833–841.

Atchley, E. G., and F. Cuthbert. 1909. A History of the Use of Incense in Divine Worship. Longmans Green, London.

Austin, J. J., A. J. Ross, A. B. Smith, R. A. Fortey, and R. H. Thomas. 1997. Problem of reproducibility—Does geologically ancient DNA survive in amber-preserved insects? Proceedings of the Royal Society, London (Biological Science) 264: 467–474.

Aycke, J. C. 1835. Fragmente zur Natureschichte des Bernsteins. Danzig.

Azar, D. 2000. Les ambres Mésozöique du Liban. Thèse de Docteur en Sciences de l'Université Paris-Sud U.F.R. Scientifique d'Orsay.

Babu, A. M., and A. R. S. Menon. 1990. Distribution of gum and gum-resin ducts in the plant body: certain familiar features and their significance. Flora 184: 257–261.

Bachofen-Echt, A. 1949. Der Bernstein und seine Einschlüsse. Wien.

Baer, H. 1979. The poisonous Anacardiaceae. In: Toxic Plants, pp. 161–170, ed. A. D. Kinghorn. Columbia University Press, New York.

Baldwin, I. T., and C. A. Preston. 1999. The eco-physiological complexity of plant responses to insect herbivores. Planta 208: 137–145.

Balée, W. L. 1994. Footprints of the Forest. Ka'apor Ethnobotany—the Historical Ecology of Plant Utilization of an Amazonian People. Columbia University Press, New York.

Balée, W. L., and D. C. Daly. 1990. Resin classification by the Ka'apor Indians. Advances in Economic Botany 8: 24–34.

Balfour, I. B. 1888. Botany of Socotra. Transactions of the Royal Society of Edinburgh 31.

Balls, E. K. 1965. Early Uses of California Plants. University of California Press, Berkeley.

Balser, C. S. 1960. Notes on resin in aboriginal Central America. Akten des 34. Internationalen Amerikanisten Kongress, pp. 374–380. Wien.

Bandaranayake, W. M., S. P. Gunasekera, S. Karunanayaki, S. Sothecswaran, and M. U. S. Sultanbawa. 1975. Terpenes of *Dipterocarpus* and *Doona* species. Phytochemistry 14: 2043–2048.

Bankova, V., R. Christov, C. Marcucci, and S. Popov. 1998. Constituents of Brazilian geopropolis. Zeitschrift für Naturforschung 53C: 402–406.

Bankova, V. S., S. L. de Castro, and M. C. Marcucci. 2000. Propolis: recent advances in chemistry and plant origin. Apidologie 31: 3–15.

Bannan, M. W. 1936. Vertical resin ducts in the secondary wood of the Abietineae. New Phytologist 35: 12–47.

Baradat, P., C. Bernard-Dagan, C. Fillon, A. Marpeau, and G. Pauly. 1972. Les terpènes du pin maritime: aspects biologiques et génétiques. II. Hérédité de la teneur en monoterpènes. Annales des Sciences Forestières 29: 307–334.

Baradat, P., C. Bernard-Dagan, G. Pauly, and C. Zimmerman-Fillon. 1975. Les terpènes du pin maritime: aspects biologiques et génétiques. III. Hérédité de la teneur en myrcène. Annales des Sciences Forestières 32: 29–54.

Baradat, P., A. Marpeau, and C. Bernard-Dagan. 1978. Variation of terpenes within and between populations of maritime pine. In: Biochemical Genetics of Forest Trees, pp. 151–169, ed. D. Rudin. Umeå, Sweden.

Barkley, F. A. 1937. A monographic study of *Rhus* and its immediate allies in North and Central America, including the West Indies. Annals of the Missouri Botanical Garden 24: 265–498.

Barnes, R. D. 1988. Tropical forest genetics at the Oxford Forestry Institute. Commonwealth Forestry Review 67: 231–241.

Barnola, L. F., M. Hasegawa, and H. Cedeño. 1994. Monoterpene and sesquiterpene variation in *Pinus caribaea* needles and its relationship to *Atta laevigata* herbivory. Biochemical Systematics and Ecology 22: 437–445.

Barry, H. T. 1932. Natural Varnish Resins. Ernest Bonn, London.

Barry, H. T. 1936. The future of natural resins. Journal of the Oil and Colour Chemists Association 19: 75–95.

Barthel, M., and H. Hetzer. 1982. Bernstcin-Inklusen aus dem Miozën des Bitterfeler Raumes. Zeitschrift für Angewandte Geologie 28: 314–336.

Basile, A. C., J. A. Sertie, P. C. D. Freitas, and A. C. Zanini. 1988. Anti-inflammatory activity of oleoresin from Brazilian *Copaifera*. Journal of Ethnopharmacology 22: 101–109.

Bazzaz, F. A., D. Dusek, D. S. Seigler, and A. W. Haney. 1975. Photosynthesis and cannabinoid content of temperate and tropical populations of *Cannabis sativa*. Biochemical Systematics and Ecology 3: 15–18.

Becerra, J. X. 1994. Squirt-gun defense in *Bursera* and the chrysomelid counterploy. Ecology 75: 1991–1996.

Becerra, J. X. 1997. Insects on plants: macroevolutionary chemical trends in host use. Science 276: 253–256.

Becerra, J. X., and D. L. Venable. 1990. Rapid-terpene-bath and "squirt-gun" defense in *Bursera schlechtendalii* and the counterploy of chrysomelid beetles. Biotropica 22: 320–323.

Becerra, J. X., and D. L. Venable. 1999a. Macroevolution of insect–plant associations. The relevance of host biogeography to host affiliation. Proceedings of the National Academy of Sciences U.S.A. 96: 12,626–12,631.

Becerra, J. X., and D. L. Venable. 1999b. Nuclear ribosomal DNA phylogeny and its implications for evolutionary trends in Mexican *Bursera* (Burseraceae). American Journal of Botany 86: 1047–1057.

Becerra, J. X., D. L. Venable, P. H. Evans, and W. S. Bowers. 2001. Interactions between chemical and mechanical defenses in the plant genus *Bursera* and their implications for herbivores. American Zoologist 41: 856–876.

Beck, C. W. 1965. The origin of amber found at Gough's Cave, Cheddar, Somerset. Proceedings, University of Bristol Spelaeological Society 10: 272–275.

Beck, C. W. 1970. Amber in archeology. Archeology 23: 7–11.

Beck, C. W. 1974. Archeological chemistry. In: Science and Archeology, pp. 234–236, ed. R. Brill. MIT Press, Cambridge, Massachusetts.

Beck, C. W. 1986. Spectroscopic investigations of amber. Applied Spectroscopy Reviews 22: 57–110.

Beck, C. W. 1999. The chemistry of amber. [In: Proceedings of the World Congress on Amber Inclusions, eds. J. Alonso, J. C. Corral, and R. López.] Estudios del Museo de Ciencias Naturales de Álava 14: 33–48.

Beck, C. W., and S. Shennan. 1991. Amber in Prehistoric Britain. Oxbow Monographs and Oxford Books, Oxford.

Beck, C. W., M. Gerving, and E. Wilbur. 1966. The provenience of archeological amber artifacts. An annotated bibliography. Part 1. Art and Archeological Technical Abstracts 6(2): 215–302.

Beck, C. W., M. Gerving, and E. Wilbur. 1967. The provenience of archeological amber artifacts. An annotated bibliography. Part 2. Art and Archeological Technical Abstracts 6(3): 201–280.

Bedford, D. J. 1986. *Xanthorrhoea*. Flora of Australia 46: 148–169.

Behl, H. M., B. Marchand, and E. Rodriguez. 1983. Inheritance of sesquiterpenoid phenolic acid esters (guayulins) in F_1 hybrids of *Parthenium* (Asteraceae). Zeitschrift für Naturforschung 38C: 494–496.

Bellman, E. 1913. Amber. Mining Journal 10: 29.

Bellman, E. 1913. Recovery and Treatment of Amber at Palmnicken (East Prussia). Mining Journal 102: 122.

Benayoun, J., and A. Fahn. 1979. Intracellular transport and elimination of resin from epithelial duct-cells of *Pinus halepensis*. Annals of Botany 43: 179–181.

Berenbaum, M. R. 1981. Patterns of furanocoumerin production and insect herbivory in a population of wild parsnip (*Pastinaca sativa* L.). Oecologia 49: 236–244.

Berenbaum, M. R. 2002. Postgenic chemical ecology: from genetic code to ecological interactions. Journal of Chemical Ecology 28: 873–896.

Berenbaum, M. R., and A. R. Zangerl. 1996. Phytochemical diveristy: Adaptation or random variation? In: Phytochemical Diversity and Redundancy in Ecological Interactions, pp. 1–24, eds. J. T. Romero, J. A. Saunders, and P. Barbosa. Recent Advances in Phytochemistry 30. Plenum Press, New York.

Berggen, W. A., and J. A. H. van Couvering. 1974. The late Neogene. Palaeogeography Palaeoclimate and Palaeoecology 16: 1–216.

Bernard-Dagan, C. 1988. Seasonal variations in energy sources and biosynthesis of terpenes and maritime pine. In: Mechanisms of Woody Plant Defenses Against Insects, pp. 93–116, eds. W. J. Mattson, J. Levieux, and C. Bernard-Dagan. Springer-Verlag, New York.

Bernard-Dagan, C., C. Fillon, G. Pauly, P. Baradat, and G. Illy. 1971. Les terpènes du pin maritime. I. Variabilité de la composition monoterpènique dans un individu, entre individus et entre provenances. Annales des Sciences Forestières 223–258.

Bernard-Dagan, C., G. Pauly, G. Marpeau, A. Gleizes, J. P. Carde, and P. Baradat. 1982. Control and compartmentation of terpene biosynthesis in leaves of *Pinus pinaster*. Physiologie Vegetale 20: 775–795.

Berner, R. A., and G. P. Landis. 1988. Gas bubbles in fossil amber as possible indications of the major gas composition of ancient air. Science 239: 1406–09.

Berryman, A. A. 1972. Resistance of conifers to invasion by bark beetle–fungus associations. BioScience 22: 598–602.

Berryman, A. A. 1989. Adaptive pathways in scolytid–fungus associations. In: Insect–Fungus Interactions, pp. 145–159, eds. N. Wilding, N. M. Collins, P. M. Hammond, and J. F. Webber. Academic Press, New York.

Berryman, A. A., K. F. Raffa, J. A. Millstein, and N. C. Stenseth. 1989. Interactive dynamics of bark beetle aggregation and conifer defense rates. Oikos 56: 256–263.

Bevan, C. W. L., D. E. U. Ekong, and J. I. Okogun. 1966. West African timbers. Ozic acid, a new diterpene acid from *Daniellia ogea*. Chemical Communication 1966: 44–45.

Bhargava, A. K., and C. S. Chauan. 1968. Antibacterial activity of some essential oils. Indian Journal of Pharmacology 30: 151–152.

Bhatt, J. R. 1987. Development and structure of primary secretory ducts in the stem of *Commiphora wightii* (Burseraceae). Annals of Botany 10: 405–416.

Bhatt, J. R., M. N. B. Nair, and H. Y. M. Ram. 1989. Enchancement of oleo-gum resin in *Commiphora wightii* by improved tapping technique. Current Science 58: 349–357.

Bianchi, E., M. E. Caldwell, and J. R. Cole. 1968. Antitumor agents from *Bursera microphylla* (Burseraceae). I. Isolation and characterization of deoxypodophyllotoxin. Journal of Pharmaceutical Chemistry 57: 696–697.

Bianchi, E., K. Sheth, and J. B. Cole. 1969. Antitumor agents from *Bursera fragroides* (Burseraceae) (β-peltatin-α-methylether and 5′-desmothoxy-β-peltatin-α-methylether). Tetrahedron Letters 32: 2759–2762.

Bianchini, F., and F. Corbetta. 1977. Health Plants of the World. Newsweek Books, New York.

Bierhorst, D. W. 1971. Morphology of Vascular Plants. Macmillan, New York.

Billing, H. J. 1944. Congo copal. Oil and Colour Trades Journal 3: 666–668.

Birch, A. J., and C. J. Dahl. 1974. Some constituents of the resins of *Xanthorrhoea preissii, australis* and *hastilis*. Australian Journal of Chemistry 27: 331–344.

Birks, J. S., and P. J. Kanowski. 1988. Interpretation of composition of coniferous resin. Silvae Genetica 37: 29–39.

Bisset, N. G., M. A. Diaz, C. Ehret, G. Ourisson, M. Palmade, F. Patil, P. Pesnelle, and J. Streith. 1966. Étude chimio-taxonomiques dans la famille des Dipterocarpacées—II. Constituants du genre *Dipterocarpus* Gaertn. f. Essai de classification chimio-taxonomique. Phytochemistry 5: 865–880.

Bisset, N. G., V. Chavanel, J.-P. Lanz, and R. E. Wolff. 1971. Constituants sesquiterpèniques et triterpèniques des résines du genre *Shorea*. Phytochemistry 10: 2451–2463.

Biswas, C., and B. M. Johri. 1997. The Gymnosperms. Springer-Verlag, Berlin.

Bittrich, V., and C. E. Amaral. 1996. Flower morphology and pollination biology of some *Clusia* species from the Gran Sabana (Venezuela). Kew Bulletin 51: 681–694.

Bittrich, V., and C. E. Amaral. 1997. Floral biology of some *Clusia* species from central Amazonia. Kew Bulletin 52: 617–635.

Blake, S., and G. Jones. 1963. Extractives from *Eperua falcata*. The petrol-soluble constituents. Journal of the Chemical Society 1963: 430–433.

Blanche, C. A., P. L. Lorio, R. A. Sommers, J. D. Hodges, and T. E. Nebeker. 1992. Seasonal cambial growth and development of loblolly pine: xylem formation, inner bark chemistry, resin ducts and resin flow. Forest Ecology and Management 49: 151–165.

Blanchette, R. A. 1992. Anatomical responses of xylem to injury and invasion by fungi. In: Defense Mechanisms of Woody Plants Against Fungi, pp. 76–95, eds. R. A. Blanchette and A. R. Biggs. Springer-Verlag, Berlin.

Blom, F. 1959. Historical notes relating to the pre-Columbian amber trade from Chiapas. Mitteilungen aus dem Museum Volkerkunde Hamburg 25: 24–27.

Bock, W. 1954. *Primaraucaria*, a new araucarian genus from the Virginia Triassic. Journal of Paleontology 28: 32–42.

Boelens, H. M., D. Ruke, and H. G. Haring. 1982. Studies of some basalmics in perfumery. Perfumer and Flavorist 6: 7–14.

Boersig, M. R. 1981. The chemistry and ecology of tarweed glandular exudate. M.S. thesis, University of California, Davis.

Boersig, M. R., and R. F. Norris. 1988. Tarweed glands: their value for survival. Fremontia 1988(Jan.): 21–24.

Bogle, A. L. 1986. The floral morphology and vascular anatomy of the Hamamelidaceae: subfamily Liquidambaroideae. Annals of the Missouri Botanical Garden 73: 325–347.

Bohlmann, J., C. L. Steele, and R. Croteau. 1997. Monoterpene synthases from grand fir *(Abies grandis):* cDNA isolation, characterization, and functional expression of myrcene synthase, (−)-(4*S*)-limonene synthase and (−)-(1*S*,5*S*)-pinene synthase. Journal of Biological Chemistry 272: 21,784–21,792.

Bohlmann, J., J. Crock, R. Jetter, and R. Croteau. 1998a. Terpenoid-based defenses in conifers: cDNA cloning, characterization, and functional expression of wound-inducible *(E)*-α-bisabolene synthesis from grand fir *(Abies grandis).* Proceedings of the National Academy of Sciences U.S.A. 95: 6756–6761.

Bohlmann, J., G. Meyer-Gauen, and R. Croteau. 1998b. Plant terpenoid synthases: molecular biology and phylogenetic analysis. Proceedings of the National Academy of Sciences U.S.A. 95: 4126–4133.

Borden, J. H. 1984. Semiochemical mediated aggregation and dispersion in Coeloptera. In: Insect Communication, pp. 123–149, ed. T. Lewis. Academic Press, New York.

Bowden, B. F., and B. Reynolds. 1982. The chromatographic analysis of ethnographic resins. Newsletter 18, Australian Institute of Aboriginal Studies, Canberra.

Bowen, R. L., Jr. 1958. Ancient trade routes in South Arabia. In: Archeological Discoveries in South Arabia, pp. 35–42, eds. R. L. Bowen and F. P. Albright. Johns Hopkins Press, Baltimore.

Bowen, R. L., Jr., and F. P. Albright. 1958. Archeological Discoveries in South Arabia. Johns Hopkins University Press, Baltimore, Maryland.

Bowman, D. M. J. S., and S. Harris. 1995. Conifers of Australia's dry forests and open woodlands. In: Ecology of the Southern Conifers, pp. 252–270, eds. N. J. Enright and R. S. Hill. Smithsonian Institution Press, Washington, D.C.

Brackman, W., K. Spaargaren, J. P. C. M. van Dongen, P. A. Couperus, and F. Bakker. 1984. Origin and structure of fossil resin from an Indonesian Miocene coal. Geochimica et Cosmochimica Acta 48: 2483–2487.

Braga, W. F., C. M. Rezende, O. A. C. Antunes, and A. C. Pinto. 1998. Terpenoids from *Copaiba [Copaifera] cearensis.* Phytochemistry 49: 263–264.

Bratt, L. C. 1979. Wood-derived chemicals: trends in production in the U.S. Pulp and Paper. 1979(June).

Breedlove, D. E., and R. M. Laughlin. 1993. The Flowering of Man: a Tzotzil Botany of Zincantán. Smithsonian Contributions to Anthropology 35.

Brenan, J. P. M. 1965. The geographic relationship of the genera of the Leguminosae in tropical Africa. Webbia 19: 545–578.

Breteler, E. J. 1999. A revision of *Prioria,* including *Gossweilerodendron, Kingiodendron, Oxystigma,* and *Pterygopodium* (Leguminosae–Caesalpiniodeae–Detarieae) with emphasis on Africa. Wageningen Agricultural University Papers 99: 1–61.

Bridges, J. R. 1987. Effect of terpenoid compounds on growth of symbiotic fungi associated with the southern pine beetle. Phytopathology 77: 83–85.

Brieskorn, C. J., and P. Noble. 1982. The terpenes of the essential oil of myrrh. In: Aromatic Plants: Basic and Applied Aspects, pp. 221–226, eds. N. Margaris, A. Koedan, and D. Voukou. Martinus Nijhoff, The Hague.

Brown, W. H. 1921. Minor Products of Philippine Forests. Department of Agriculture and Bureau of Forestry of the Philippine Islands, Bulletin 22, Vol. 2. Manila.

Brumfitt, W., J. M. T. Hamilton-Miller, and I. Franklin. 1990. Antibiotic activity of natural products: I. Propolis. Microbios 62: 19–22.

Bruneau, A., F. Forest, P. S. Herendeen, B. B. Klitgaard, and G. P. Lewis. 2001. Phylogenetic relationships in the Caesalpiniodeae (Leguminosae) as inferred from chloroplast *trn*L intron sequences. Systematic Botany 26: 487–514.

Brunsfeld, S. J., P. S. Soltis, D. E. Soltis, P. A. Gadek, C. J. Quinn, D. E. Strenge, and T. A. Ranker. 1994. Phylogenetic relationships among genera of Taxodiaceae and Cupressaceae: evidence from *rbc*L sequences. Systematic Botany 14: 253–262.

Bryant, J. P. 1981. Phytochemical deterrence of snowshoe hare browsing by adventitious shoots of four Alaskan trees. Science 313: 889–890.

Bryant, J. P., G. D. Wieland, P. B. Reichardt, V. E. Lewis, and M. L. McCarthy. 1983. Pinosylvin methyl ether deters snowshoe hare feeding on green alder. Science 222: 1023–1025.

Bryant, J. P., and R. Julkunen-Tiitto. 1995. Ontogenetic development of chemical defense by seedling resin birch: energy cost of defense production. Journal of Chemical Ecology 21: 883–896.

Bryant, J. P., and P. J. Kuropat. 1980. Selection of winter forage by subarctic browsing vertebrates: the role of plant chemistry. Annual Review of Ecology and Systematics 11: 261–285.

Bryant, J. P., F. S. Chapin III, and D. R. Klein. 1983. Carbon-nutrient balance of boreal plants in relation to vertebrate herbivory. Oikos 40: 357–368.

Bryant, J. P., F. S. Chapin, P. B. Reichardt, and T. P. Clausen. 1985. Adaptation to resource availability as a determinant of chemical defense strategies in woody plants. In: Chemically Mediated Interactions Between Plants and Other Organisms, pp. 219–237, eds. G. A. Cooper-Driver, T. Swain, and E. E. Conn. Plenum Press, New York.

Bryant, J. P., J. Tahvanainen, M. Sulkinoja, R. Julkunen-Titto, P. B. Reichardt, and T. Green. 1989. Biogeographic evidence for the evolution of chemical defense by boreal birch and willow against mammalian browsing. American Naturalist 134: 20–34.

Bryant, J. P., P. J. Kuropat, P. B. Reichardt, and T. P. Clausen. 1991. Controls over allocation of resources by woody plants to chemical antiherbivore defense. In: Plant Defense Against Mammalian Herbivory, pp. 83–102, eds. R. T. Palo and C. T. Robbins. CRC Press, Boca Raton.

Bryant, J. P., P. B. Reichardt, T. P. Clausen, F. O. Provenza, and P. J. Kuropat. 1992. Woody plant mammal interactions. In: Herbivores: Their Interactions with Secondary Plant Metabolites, Vol. 2, pp. 344–371, eds. G. A. Rosenthal and M. R. Berenbaum. Academic Press, New York.

Bryant, J. P., R. K. Swihart, P. B. Reichardt, and L. Newton. 1994. Biogeography of woody plant chemical defense against snowshoe hare browsing: comparison of Alaska and eastern North America. Oikos 70: 385–395.

Buckley, R. A. 1932. The damars of the Malay Peninsula. Malayan Forest Records 11.

Buffum, W. A. 1897. The Tears of the Heliades or Amber as a Gem. Sampson Low, and Marston and Company, London.

Burbott, A. J., and W. D. Loomis. 1969. Evidence for metabolic turnover in peppermint. Plant Physiology 44: 173–179.

Burburlis, I. E., and B. Winkel-Shirley. 1999. Interactions among enzymes of the *Arabidopsis* flavonoid biosynthetic pathway. Proceedings of the National Academy of Sciences U.S.A. 96: 12,929–12,934.

Burdock, G. A. 1995. Fenaroli's Handbook of Flavor Ingredients, ed. 3, Vol. 1, p. 350. CRC Press, Boca Raton.

Burdock, G. A. 1998. Review of the biological properties and toxicity of bee propolis (propolis). Food and Chemical Toxicology 36: 347–363.

Burkill, I. H. 1966. A Dictionary of the Economic Products of the Malay Peninsula, Vols. 1–2. Crown Agents for the Colony, London.

Burman, E. 1987. The Assassins. Crucible, Wellingborough, England.

Burnham, R. J., and A. Graham. 1999. The history of Neotropical vegetation: new developments and status. Annals of the Missouri Botanical Garden 86: 546–589.

Butler, C. B. 1998. Treasures of the Longleaf Pines. Naval Stores. Tarkel Publishing, Shalimar, Florida.

Button, D. K. 1984. Evidence for a terpene-based food chain in the Gulf of Alaska. Applied Environmental Microbiology 48: 1004–1011.

Byers, J. A., B. S. Lanne, and J. Lofquist. 1989. Host tree unsuitability recognized by pine shoot beetles in flight. Experientia 45: 489–492.

Calvin, M. 1979. Petroleum plantations for fuel and materials. BioScience 29: 533–538.

Calvin, M. 1983. New sources for fuel and materials. Science 219: 24–26.

Calvin, M. 1984. Renewable fuels for the future. Journal of Applied Biochemistry 6: 3–18.

Cano, R. J., and M. K. Borucki. 1995. Revival and identification of bacterial spores in 25–40 million-year-old Dominican amber. Science 268: 1060–1064.

Cano, R. J., H. N. Poinar, D. W. Roubik, and G. O. Poinar, Jr. 1992. Isolation and partial isolation of DNA from *Proplebeia dominicana* (Hymenoptera: Apidae) in 25–40 million year old amber. Medical Science Research 20: 619–622.

Cano, R. J., H. N. Poinar, N. J. Pieniazek, A. Acra, and G. O. Poinar, Jr. 1993. Amplification and sequencing of DNA from a 120–135 million-year-old weevil. Nature 363: 536–538.

Capson, T. L., P. D. Coley, and T. A. Kursar. 1996. A new paradigm for drug discovery in tropical rainforests. Nature Biotechnology 14: 1200–1201.

Carde, J. P. 1984. Leucoplasts: a distinct kind of organelle lacking typical 70S ribosomes and free thylakoids. European Journal of Cell Biology 34: 18–26.

Carde, J. P., C. Bernard-Dagan, and M. Gleizes. 1980. Membrane systems involved in synthesis and transport of monoterpene hydrocarbons in pine leaves. In: Biogenesis and Function of Plant Lipids, pp. 441–444, eds. P. Mazliak, P. Benveniste, C. Costes, and R. Douce. Elsevier, Amsterdam.

Carmago, J. M. F., D. Grimaldi, and S. R. M. Pedro. 2000. The extinct fauna of stingless bees (Hymenoptera: Apidae: Meliponini) in Dominican amber: two new species and descriptions of the male *Proplebeia dominicana* (Willie and Chandler). American Museum Novitiates 3293: 1–24.

Carmen, R. M., D. E. Cowley, and R. A. Marty. 1970. Diterpenoids XXV. Dundathic acid and polycommunic acid. Australian Journal of Chemistry 23: 1656–1665.

Carnmalm, B. 1954. Constitution of resin phenols and their biogenic relations. XVIII. Acta Chimica Scandanavia 8: 1827–1829.

Carroll, G. L. 1988. Fungal endophytes in stems and leaves: from latent pathogens to mutualistic symbiont. Ecology 69: 2–9.

Carroll, J. F., A. Maradufu, and J. D. Warthen, Jr. 1989. An extract of *Commiphora erythraea*: a repellent and toxicant against ticks. Entomología Experimentalis et Applicata 53: 111–116.

Cascon, V., and B. Gilbert. 2000. Characterization of the chemical composition of oleoresins of *Copaifera guianensis* Desf., *Copaifera duckei* Dwyer and *Copaifera multijuga* Hayne. Phytochemisry 55: 773–778.

Cascon, V., L. M. Rocha, S. R. M. Lima, B. Gilbert, and M. A. L. Louras. 1997. Evaluation of the chemical composition and of the toxicity of oleoresins by *Copaifera* spp. Bolletino Chimico Farmaceutico 136: 216.

Caspary, R., and R. Klebs. 1907. Die Flora des Bernsteins und underer fossiler Harze des ost preussischen Tertiärs. Abhandlungen der Königlich Preussischen Geologischen Landesanstalt, n.f. 4: 1–182.

Cates, R. G. 1996. The role of mixtures and variation in production of terpenoids in conifer–insect–pathogen interactions. In: Phytochemical Diversity and Redundancy in Ecological Interactions, pp. 179–216, eds. J. T. Romeo, J. A. Saunders, and P. Barbosa. Plenum Press, New York.

Cates, R. G., and H. Alexander. 1982. Host resistance and susceptibility. In: Bark Beetles in North American Conifers: Evolution and Ecology, pp. 212–263, eds. J. Mitton and K. Sturgeon. University of Texas Press, Austin.

Cates, R. G., and R. A. Redak. 1988. Variation in terpene chemistry of Douglas fir and its relationship to western spruce budworm success. In: Chemical Mediation of Coevolution, pp. 317–344, ed. K. C. Spencer. Academic Press, New York.

Cates, R. G., R. A. Redak., and C. B. Henderson. 1983. Patterns in defensive natural product chemistry: Douglas fir and western spruce budworm interactions. In: Plant Resistance to Insects, pp. 3–20, ed. P. A. Hedin. Symposium Series 208. American Chemical Society, Washington, D.C.

Cates, R. G., J. Zou, and C. Carlson. 1991. The role of variation in Douglas-fir foliage quality in silvicultural management of western spruce budworm. In: Interior Douglas Fir: The Species and Its Management, pp. 115–128, eds. D. Baumgartner and J. Lotan. Washington State University, Pullman.

Chameides, W. L., C. W. Lindsay, J. Richardson, and C. S. Kiang. 1988. The role of biogenic hydrocarbons in urban photochemical smog: Atlanta as a case study. Science 24: 1473–1475.

Chang, I. T., Y. S. Ng, and A. A. Kadir. 1997. A review on *agur (gaharu)* producing *Aquilaria* species. Journal of Tropical Forest Products 2: 272–285.

Chappell, J. 1995. Biochemistry and molecular biology of the isoprenoid biosynthetic pathways in plants. Annual Review of Plant Physiology and Plant Molecular Biology 46: 521–547.

Chararas, C., R. Revolon, M. Feinberg, and C. Ducauze. 1982. Preference of certain

Scolytidae for different conifers. A statistical approach. Journal of Chemical Ecology 8: 1093–1109.

Charon, J., J. Launay, and E. Vindt. 1986. Ontogénèse des canaux sécréteurs d'origine primaire dans le bourgeon de pin maritime. Canadian Journal of Botany 64: 2955–2964.

Charon, J., J. Launay, and J. P. Carde. 1987. Spatial organization and volume density of leucoplasts in pine secretory cells. Protoplasma 138: 45–53.

Charrière-Ladreix, Y. 1973. Etude de la sécrétion flavonoidique de bourgeons de *Populus nigra* L. Cinetique de phénomène secretoire: ultrastructure et évolution du tissue glandulaire. Journal de Microscopie 17: 299–316.

Charrière-Ladreix, Y. 1975. La secretion lipophile des bourgeons d'*Aesculus hippocastanum* L. modifications ultrastructurales des trichomes au cours du processus glandulaire. Journal de Microscopie et de Biologie Cellulaire 24: 75–90.

Chaudhari, D. C., C. L. Goel, B. D. Joshi, D. N. Uniyal, A. Prasad, and R. S. Bist. 1992. Correlation between diameter and resin yield by rill method of tapping. Indian Forester 118: 456–459.

Chaudhari, D. C., S. K. Sheel, D. N. Uniyal, and R. S. Bist. 1996. Supplementing resin production by retapping healed up channels. Indian Forester 122: 746–750.

Chaw, S.-h., A. Zharkikh, H. M. Sung, T. C. Lau, and W. H. Li. 1997. Molecular phylogeny of extant gymnosperms and seed plant evolution: analysis of nuclear 18S rRNA sequences. Molecular Biology and Evolution 14: 56–68.

Chen, Z. D., S. R. Manchester, and H.-y. Sun. 1999. Phylogeny and evolution of the Betulaceae as inferred from DNA sequences, morphology and paleobotany. American Journal of Botany 86: 1168–1181.

Cheniclet, C. 1987. Effects of wounding and fungus inoculation on terpene producing systems of maritime pine. Journal of Experimental Botany 38: 1557–1572.

Cheniclet, C., and J. P. Carde. 1985. Presence of leucoplasts in secretory cells and monoterpenes in the essential oil: a correlative study. Israel Journal of Botany 34: 219–238.

Cheung, H. T., and Y. C. Yan. 1972. Constituents of Dipterocarpaceae resins. IV. Triterpenes of *Shorea accuminata* and *S. resina-nigra*. Australian Journal of Chemistry 25: 2003–2012.

Chevallier, A. 1996. The Encyclopedia of Medicinal Plants. D.K. Publishing, New York.

Chhibber, H. L. 1934. The Mineral Resources of Burma. Macmillan and Co., London.

Chinou, I., C. Demetzos, C. Harvala, C. Roussakis, and J. Verbist. 1994. Cytotoxic and antibacterial labdane-type diterpenes from the aerial parts of *Cistus incanus* subsp. *creticus*. Planta Medica 60: 34–36.

Christiansen, E., P. Krokene, A. A. Berryman, V. R. Franceshi, T. Krekling, E. Lieutier, A. Lönneborg, and H. Solheim. 1999. Mechanical injury and fungal infection induce acquired resistance in Norway spruce. Tree Physiology 19: 399–403.

Claeson, P., R. Anderson, and G. Samuelson. 1991. T-cadinol: a pharmacologically active constituent of scented myrrh: introductory pharmacological characterization and high field proton NMR and carbon-13 NMR data. Planta Medica 57: 352–356.

Clark, A. M., J. C. McChesney, and R. P. Adams. 1990. Antimicrobial properties of the heartwood, bark / sapwood and leaves of *Juniperus* species. Phytotherapy Research 4: 15–19.

Clarke, R. C. 1981. Marijuana Botany. And/Or Press, Berkeley, California.

Clarke, R. C. 1998. Hashish! Red Eye Press, Los Angeles.

Clausen, T. P., J. P. Bryant, and P. B. Reichardt. 1986. Defense of winter dormant green alder against snowshoe hares. Journal of Chemical Ecology 12: 2117–2131.

Clavijero, F. J. 1971. Historia Antigua de México. Editorial Porrua, México.

Clement-Westerhof, J. A. 1987. Aspects of Permian paleobotany. VII. The Majonicaceae, a new family of late Permian conifers. Review of Paleobotany and Palynology 52: 375–402.

Clifford, D. J., and P. G. Hatcher. 1995. Maturation of Class 1b (polylabdanol) resinites. In: Amber, Resinite and Fossil Resins, pp. 92–104, eds. K. B. Anderson and J. C. Crelling. Symposium Series 617. American Chemical Society, Washington, D.C.

Cobb, F. W., Jr., M. Krstic, E. Zavarin, and H. W. Barber. 1968. Introductory effects of volatile oleoresin components on *Fomes annosus* and four *Ceratocystis* species. Phytopathology 58: 1327–1335.

Cocker, O., A. L. Moore, and A. C. Pratt. 1965. Dextrorotary hardwickiic acid. An extractive of *Copaifera officinalis*. Tetrahedron Letters 24: 1983–1988.

Cohen, S. B. (ed.). 1998. The Columbia Gazetteer of the World, 3 vols. Columbia University Press, New York.

Coley, P. D., and T. M. Aide. 1991. Comparison of herbivory and plant defenses in temperate and tropical broadleaved forests. In: Plant–Animal Interactions: Evolutionary Ecology in Tropical and Temperate Regions, pp. 25–49, eds. P. W. Price, T. M. Lewinsohn, G. W. Fernandes, and W. W. Benson. John Wiley and Sons, New York.

Coley, P. D., J. P. Bryant, and F. S. Chapin III. 1985. Resource availability and plant antiherbivore defense. Science 230: 895–899.

Collier, T. J. 1977. Feasibility of petrochemical substitution by oleoresin. Proceedings of Annual Meeting of Lightwood Research Coordinating Council.

Collinson, M., and J. J. Hooker. 1987. Vegetational and mammalian faunal changes in the early Tertiary of southern England. In: Origins of Angiosperms and Their Biological Consequences, pp. 259–304, eds. E. M. Friis, W. G. Chaloner, and P. R. Crane. Cambridge University Press, Cambridge.

Comerford, S. C. 1996. Medicinal plants of two Mayan healers from San Andrés, Petén, Guatemala. Economic Botany 50: 327–336.

Conelly, W. T. 1985. Copal and rattan collecting in the Philippines. Economic Botany 39: 39–46.

Connell, J. H. 1971. On the role of natural enemies in preventing competitive exclusion of some marine animals and in rain forest trees. In: Dynamics of Populations, pp. 298–312, eds. P. J. den Boer and G. R. Gradwell. Center for Agricultural Publishing and Documentation, Wageningen, Netherlands.

Conwentz, H. W. 1886. Die Flora des Bernsteins, Band 2, Die Angiospermen des Bernsteins. Danzig.

Conwentz, H. W. 1890. Monographie der baltschen Bernsteinbäume Danzig. Commissions-Verlag W. Engelmann, Leipzig.

Cook, J. 1770. The Natural History of Lac, Amber, and Myrrh; with a Plain Account of the Many Excellent Virtues These Three Medicinal Substances are Naturally Possessed of, and Well Adapted for the Cure of Various Diseases Incident to the Human Body. . . . Woodfall, . . . , London.

Coppen, J. J. W. 1995. Gums, Resins and Latexes of Plant Origin. Non-wood Forest Products 6. Food and Agriculture Organization of the United Nations, Rome.

Coppen, J. J. W., and G. A. Hone. 1995. Gum naval stores: turpentine and rosin from pine resin. Natural Resources Institute. Food and Agriculture Organization of the United Nations, Rome.

Coppen, J. J. W., P. Greenhalgh, and A. E. Smith. 1984. Gum naval stores: an industrial profile of turpentine and resin production from pine resin. Report G87. Tropical Development and Research Institute, London.

Corbett, M. D., and S. Billets. 1975. Characterization of poison oak urushiol. Journal of Pharmaceutical Sciences 64: 1715–1718.

Cowan, R. S. 1975. A monograph of the genus *Eperua* (Leguminosae: Caesalpinioideae). Smithsonian Contributions to Botany 28: 1–45.

Cowan, R. S., and R. M. Polhill. 1980. Detarieae. In: Advances in Legume Systematics, Part 1, pp. 117–134, eds. R. M. Polhill and P. H. Raven. Royal Botanic Gardens, Kew.

Coyne, J. F., and W. B. Critchfield. 1974. Identity and terpene composition of Honduran pines attacked by the bark beetle, *Dendroctonus frontalis* (Scolytidae). Turrialba 24: 327–331.

Craig, J. C., C. W. Waller, S. Billets, and M. A. Elsohly. 1978. New GLC analysis of urushiol congeners in different parts of poison ivy, *Toxicodendron radicans*. Journal of Pharmaceutical Sciences 67: 483–485.

Crane, P. R., E. M. Friis, and K. R. Pedersen. 1995. The origin and early diversification of angiosperms. Nature 374: 27–33.

Crankshaw, D. R., and J. H. Langenheim. 1981. Variation in terpenes and phenolics through leaf development in *Hymenaea* and its possible significance to herbivory. Biochemical Systematics and Ecology 9: 115–124.

Crelling, J. C. 1995. The petrology of resinite in American coals. In: Amber, Resinite and Fossil Resins, pp. 218–233, eds. K. B. Anderson and J. C. Crelling. Symposium Series 617. American Chemical Society, Washington, D.C.

Crepet, W. L. 1999. Early bloomers. Natural History 108: 40–41.

Crepet, W. L., and K. C. Nixon. 1998. Fossil Clusiaceae from the late Cretaceous (Turonian) of New Jersey and implications regarding the history of bee pollination. American Journal of Botany 85: 1122–1133.

Cronquist, A. 1981. An Integrated System of Classification of Flowering Plants. Columbia University Press, New York.

Cronquist, A. 1988. The Evolution and Classification of Flowering Plants, ed. 2. Columbia University Press, New York.

Croteau, R. 1988. Catabolism of monoterpenes in essential oil plants. In: Flavors and Fragrances: a World Perspective, pp. 65–84, eds. B. D. Moorherjee and B. J. Willis. Elsevier, Amsterdam.

Croteau, R. 1998. The discovery of terpenes. In: Discoveries in Plant Biology, Vol. 1, pp. 329–343, eds. S.-d. Kung and S.-f. Yang. World Scientific, Singapore.

Croteau, R., and M. A. Johnson. 1984. Biosynthesis of terpenoids in glandular trichomes. In: Biology and Chemistry of Plant Trichomes, pp. 133–185, eds. E. Rodriguez, P. L. Healey, and I. Mehta. Plenum Press, New York.

Croteau, R., and C. Martinkus. 1979. Metabolism of monoterpenes: demonstration of (+)-neomenthyl-β-*d*-glucoside as a major metabolite of (−)-menthone in peppermint *(Mentha piperita).* Plant Physiology 64: 169–175.

Croteau, R., and V. K. Sood. 1985. Metabolism of monoterpenes: evidence for the function of monoterpene catabolism in peppermint *(Mentha piperita)* rhizomes. Plant Physiology 77: 801–806.

Croteau, R., S. Gurkewitz, M. A. Johnson, and H. J. Fisk. 1987. Biochemistry of oleoresinosis: monoterpene and diterpene biosynthesis in lodgepole pine saplings infected with *Ceratocystis clavigera* or treated with carbohydrate elicitors. Plant Physiology 85: 1123–1128.

Cunningham, A., S. S. Martin, and J. H. Langenheim. 1973. Resin acids from two Amazonian species of *Hymenaea.* Phytochemistry 12: 633–635.

Cunningham, A., S. S. Martin, and J. H. Langenheim. 1974. Labd-13-en 8-ol-15-oic acid in trunk resin of Amazonian *Hymenaea courbaril.* Phytochemistry 13: 294–295.

Cunningham, A., I. D. Gay, A. C. Oehlschlager, and J. H. Langenheim. 1983. ^{13}C NMR and IR analyses of structure, aging and botanical origin of Dominican and Mexican ambers. Phytochemistry 22: 965–968.

Cunningham, A., P. R. West, G. S. Hammond, and J. H. Langenheim. 1987. The existence and photochemical initiation of free radicals in *Hymenaea* trunk resins. Phytochemistry 16: 1142–1143.

Curtis, J. D., and N. R. Lersten. 1974. Morphology, seasonal variation and function of resin glands on buds and leaves of *Populus deltoides* (Salicaceae). American Journal of Botany 61: 835–845.

Curtis, J. D., and N. R. Lersten. 1990. Internal secretory structures of *Hypericum* (Clusiaceae): *H. perforatum* L. and *H. balearicum* L. New Phytologist 114: 571–580.

Czechowski, F., B. R. T. Simoneit, M. Sachabinski, J. Chojan, and S. Wolowiec. 1996. Physicochemical structural characterization of ambers from deposits in Poland. Applied Geochemistry 11: 811–834.

Czeczott, H. 1961. The flora of the Baltic amber and its age. Prace Muzeum Ziemi 4: 119–145.

Daly, D. 1987. A taxonomic revision of *Protium* (Burseraceae) in eastern Amazonia and the Guianas. Ph.D. dissertation, City University of New York, New York.

Daly, D. 1989. Studies in the Neotropical Burseraceae. II. Generic limits in New World Protieae and Canarieae. Brittonia 41: 17–27.

Dalziel, J. M. 1937. The Useful Plants of West Tropical Africa. Crown Agents for the Colonies, London.

Davidson, D. F. D. 1948. Report on the gum mastic industry in Chios. Bulletin of the Imperial Institute 46: 184–191.

Davis, T., and R. A. Bye. 1982. Ethnobotany and progressive domestication of *Jaltomata* (Solanaceae) in Mexico and Central America. Economic Botany 36: 225–241.

Dawson, C. R. 1954. The toxic principle of poison ivy and related plants. Record of Chemical Progress 15: 38–59.

Dawson, C. R. 1956. The chemistry of poison ivy. Transactions of the New York Academy of Sciences, II, 18: 427–433.

Dayanandan, S., P. S. Ashton, S. N. Williams, and R. B. Primack. 1999. Phylogeny of the tropical tree family Dipterocarpaceae based on nucleotide sequences of the chloroplast *rbc*L gene. American Journal of Botany 86: 1182–1190.

de Angelis, J. D., T. E. Nebeker, and J. D. Hodges. 1986. Influence of tree age and growth rate on radial duct systems in loblolly pine *(Pinus taeda)*. Canadian Journal of Botany 64: 1046–1049.

de Beer, J. H., and M. McDermott. 1989. The Economic Value of Non-timber Forest Products in Southeast Asia. Netherlands Committee for IUCN, Amsterdam.

de Cordemoy, H. J. 1911. Les Plantes à Gommes et à Résines. Octave Doin et Fils, Paris.

de Foresta, H., and G. Michon. 1994. Agroforests in Sumatra—where ecology meets economy. Agroforestry Today 6: 12–13.

de Laubenfels, D. J. 1989. *Agathis*. Flora Malesiana, Series 1, 10: 429–442.

Dell, B. 1975. Geographical differences in leaf resin components of *Eremophila fraseri* F. Muell. (Myoporaceae). Australian Journal of Botany 23: 889–897.

Dell, B. 1977. Distribution and function of resins and glandular hairs in Western Australian plants. Journal of the Royal Society of Western Australia 59: 119–123.

Dell, B., and A. J. McComb. 1974. Resin production and glandular hairs in *Beyeria viscosa* (Labill.) Miq. (Euphorbiaceae). Australian Journal of Botany 22: 195–210.

Dell, B., and A. J. McComb. 1975. Glandular hairs, resin production and habitat of *Newcastelia viscida* E. Pritzel (Dicrastylidaceae). Australian Journal of Botany 23: 373–390.

Dell, B., and A. J. McComb. 1977. Glandular hair formation and resin secretion in *Eremophila fraseri* F. Muell. (Myoporaceae). Protoplasma 92: 71–86.

Dell, B., and A. J. McComb. 1978. Plant resins: their formation, secretion, and possible functions. Advances in Botanical Research 16: 277–316.

Dell, B., C. L. Elsegood, and E. L. Ghisalberti. 1989. Production of verbascoside in callus tissues of *Eremophila* spp. Phytochemistry 28: 1871–1872.

Delle Monache, F. F., E. Corio, I. L. d'Albuquerque, and G. B. Marini-Bettolo. 1969. Diterpenes from *Copaifera multijuga* Hayne. Annali di Chimica 59: 539–551.

Delorme, L., and F. Lieutier. 1990. Monoterpene composition of the preformed and induced resins of Scots pine, and their effect on bark beetles and associated fungi. European Journal of Forest Pathology 20: 304–316.

Demetzos, C., C. Harvala, and S. M. Philianos. 1990. A new labdane-type diterpene and other compounds from the leaves of *Cistus incanus* ssp. *creticus*. Journal of Natural Products 53: 1365–1368.

Demetzos, C., S. Mitaku, A. L. Skaltsounis, M. C. Harvala, and F. Libot. 1994. Diterpene esters of malonic acid from the resin "ladano" of *Cistus creticus*. Phytochemistry 35: 979–981.

Demetzos, C., H. Katerinopoulos, A. Kouvarakis, N. Stratigakis, A. Loukis, C. Ekonomakis, V. Spiliotis, and J. Tsaknis. 1997. Composition and antimicrobial activity of the essential oil of *Cistus creticus* subsp. *eriocephalus*. Planta Medica 63: 477–479.

Demetzos, C., B. Stahl, T. Anastassaki, M. Gazouli, L. Tzouvelekis, and H. Rallis. 1999. Chemical analysis and antimicrobial activity of the resin ladano, of its essential oil and of isolated compounds. Planta Medica 65: 76–78.

de Navarro, I. M. 1925. Prehistoric routes between northern Europe and Italy defined by amber trade. Geographic Journal 66.

Denissova, G. A. 1975. Attempt of classification of terpenoid-containing conceptacles in plants [in Russian]. Botanicheskii Zhurnal S.S.S.R. 60: 1698–1706.

Denno, R. F., and M. S. McClure (eds.). 1983. Variable Plants and Herbivores in Natural and Managed Systems. Academic Press, New York.

de Oliveira, D. M. A., A. M. Porto, V. Bittrich, I. Vencato, A. J. Marsaioli, and A. J. Tidsskrift. 1996. Floral resins of *Clusia* spp.: chemical composition and biological function. Tetrahedron Letters 37: 6427–6430.

Derfer, J. M. 1978. Turpentine as a source of perfume and flavor materials. Perfumer and Flavorist 3: 45.

DeSalle, R., J. Gatesy, W. Wheeler, and D. A. Grimaldi. 1992. DNA sequences from a fossil termite in Oligo-Miocene amber and their phylogenetic implications. Science 257: 1933–1936.

de Santis, V., and J. D. Medina. 1981. Chrystalline racemic labd-8(20)13-dien-15-oic acid in the trunk resin of *Eperua purpurea*. Journal of Natural Products 44: 370–372.

Dev, Sukh. 1988. Ayurveda and modern drug development. Proceedings of the Indian National Science Academy 54A: 12–42.

Devon, T. K., and A. I. Scott. 1972. Handbook of Naturally Occurring Compounds, Vol. 2, Terpenes. Academic Press, New York.

Dhawan, B. N. 1996. A standardized *Commiphora wightii* preparation for management of hyperlipidemic disorders. In: Medicinal Resources of the Tropical Forest, pp. 278–283, eds. M. J. Balick, E. Elisabetsky, and S. A. Laird. Columbia University Press, New York.

Díaz del Castillo, B. 1956. The Discovery and Conquest of Mexico, 1517–1521. Farrar, Straus and Cudahy, New York.

Di Castri, F. 1981. Mediterranean-type shrublands of the world. In: Ecosystems of the World, Vol. 11, pp. 1–52, eds. F. di Castri, D. W. Goodall, and R. L. Specht. Elsevier, Amsterdam.

Dieterich, K. 1901. Analysis of Resins, Balsams and Gum Resins, Their Chemistry and Pharmacognosis: for the Use of the Scientific and Technical Research Chemist, transl. C. Salter Scott. Greenwood and Sons, London.

Dilcher, D. L., P. S. Herendeen, and F. Hueber. 1992. Fossil *Acacia* flowers with attached anther glands from Dominican Republic amber. In: Advances in Legume Systematics, Part 4, Fossil Record, pp. 33–42, eds. P. S. Herendeen and D. L. Dilcher. Royal Botanic Gardens, Kew.

Dioscorides, P. De Materia Medica, 1968 ed. R. T. Gunther. Hafner, New York.

Dixon, W. N., and T. L. Payne. 1980. Attraction of entomophagous and associated insects of the southern pine beetle to beetle and host-tree-produced volatiles. Journal of the Georgia Entomological Society 15: 378–389.

Dörr, A. 1973. Frankincense, myrrh and opopanax. Dragoco Report 20: 100–102.

Dolara, P., C. Lucera, C. Ghelardini, C. Monserrat, S. Aioill, F. Luceri, M. Lodovici, S. Menichetti, and M. N. Romanelli. 1996. Analgesic effects of myrrh. Nature 379: 29–29.

Dominguez, X. A., J. Rzedowski, M. Gutiérrez, and M. E. Gómez. 1973. A phytochemical survey of 21 species of the genus *Bursera* (Burseraceae) natives of Mexico. Revista Latinoamericano de Química 4: 108–121.

Donoghue, M. J., J. A. Doyle, J. Gauthier, A. G. Kluge, and T. Rowe. 1989. The importance of fossils in phylogeny reconstruction. Annual Review of Ecology and Systematics 20: 431–460.

Douce, R., M. Neuberger, P. Bligny, and G. Pauly. 1978. Effects of β-pinene on the oxidative properties of purified intact plant mitochondria. In: Plant Mitochondria, pp. 207–214, eds. G. Ducet and C. Lance. Elsevier / North Holland Biomedical Press, Amsterdam.

Downie, S. R., D. S. Katz-Downie, and M. W. Watson. 2000. A phylogeny of the flowering plant family Apiaceae based on chloroplast DNA *rpl16* and *rpoC1* intron sequence: towards a suprageneric classification of subfamily Apioideae. American Journal of Botany 87: 273–292.

Doyle, J. A. 1998. Phylogeny of vascular plants. Annual Review of Ecology and Systematics 29: 567–599.

Dressler, R. L. 1982. Biology of the orchid bees (Euglossini). Annual Review of Ecology and Systematics 13: 373–394.

Drew, J. 1989. History. In: Naval Stores: Production-Chemistry-Utilization, pp. 3–38, eds. D. E. Zinkel and J. Russell. Pulp Chemical Association, New York.

Drew, J., and M. Propst (eds.). 1981. Tall Oil. Pulp Chemical Association, New York.

Ducke, A. 1925. As leguminosas do estado do Pará. Archivos do Jardim Botanico do Rio de Janeiro 4: 263.

Dufour, D. L. 1990. Use of tropical rainforests by native Amazonians. BioScience 40: 652–659.

Duke, J. A. 1980. Handbook of Legumes of World Economic Importance. Plenum, New York.

Duke, J. A. 1983. Medicinal Plants of the Bible. Trado-Medic Books, New York.

Duke, J. A. 1985. CRC Handbook of Medicinal Herbs. CRC Press, Boca Raton.

Duke, J. A. 1987. CRC Handbook of Agricultural Energy Potential of Developing Countries. CRC Press, Boca Raton.

Duke, J. A., and E. S. Ayensu. 1985. Medicinal Plants of China. Reference Publications, Algonac, Michigan.

Duke, J. A., and J. L. Cellier. 1993. CRC Handbook of Alternative Cash Crops. CRC Press, Boca Raton.

Duke, J. A., and R. Vásquez. 1994. Amazonian Ethnobotanical Dictionary. CRC Press, Boca Raton.

Duncan, D. P. 1989. Tall oil fatty acids. In: Naval Stores: Production-Chemistry-Utilization, pp. 346–439, eds. D. E. Zinkel and J. Russell. Pulp Chemicals Association, New York.

Dunn, F. L. 1964. Excavations at Gua Kechil, Pahang. Journal of the Malaysian Branch of the Royal Asiatic Society 37: 87–124.

Dunn, F. L. 1975. Rain-forest Collectors and Traders: a Study of Resource Utilization in Modern and Ancient Malaya. Monographs 5. Malaysia Branch, Royal Asiatic Society, Kuala Lumpur.

Dupuis, G. 1979. In vitro lymphocyte transformation by urushiol-protein conjugates. British Journal of Dermatology 101: 617–624.

Dussourd, D. E., and R. F. Denno. 1991. Deactivation of plant defense: correspondence between insect behavior and secretory canal architecture. Ecology 72: 1383–1396.

Dutta, R., and H. H. Schobert. 1996. Dammar resin: a chemical model for reactions of Utah resinite. In: Amber, Resinite and Fossil Resins, pp. 263–278, eds. K. B. Anderson and J. C. Crelling. Symposium Series 617. American Chemical Society, Washington, D.C.

Duwiejura, M., I. J. Zeitlin, P. G. Waterman, J. Chapman, G. J. Mhango, and G. J. Provan. 1993. Anti-inflammatory activity of resins from some species of the plant family Burseraceae. Planta Medica 59: 12–16.

Dwyer, J. D. 1951. The Central American, West Indian, and South American species of *Copaifera* (Caesalpiniaceae). Brittonia 7: 143–172.

Earl, J. C. 1917. Grass Trees: an Investigation of Their Economic Products. Department of Chemistry Bulletin Number 6. South Australia.

Eckenwalder, J. E. 1976. Re-evaluation of Cupressaceae and Taxodiaceae: a proposed merger. Madroño 23: 237–300.

Ehrlich, P. R., and R. H. Raven. 1964. Butterflies and plants: a study in co-evolution. Evolution 18: 586–608.

Ekong, D. E. U., and J. J. Okogun. 1969. West African timbers. Part XXV. Diterpenoids of *Gossweilerodendron balsamiferum* Harms. Journal of the Chemical Society, D, 1969: 2153–2156.

Ekstrom, K., K. Hoffman, T. Linne, B. Eriksoon, and B. Glimelius. 1998. Single-dose etoposide in advanced pancreatic and biliary cancer; a phase II study. Oncology Reports 5: 931–934.

Elisabetsky, E. 1996. Community ethnobotany: setting foundations for an informed decision on trading rainforest resources. In: Medicinal Resources of the Tropical Forest, pp. 402–407, eds. M. J. Balik, E. Elisabetsky, and S. A. Laird. Columbia University Press, New York.

Ella, A. B., and A. L. Tongacan. 1987. Tapping of palosapis [*Anisoptera thurifera* (Blanco) Blume subsp. *thurifera*]. Forest Products Research and Development Institute Journal 16: 15–26.

Ella, A. B., and A. L. Tongacan. 1992. Techniques in tapping *almaciga* [*Agathis philippinensis* Warb.] for sustained productivity of the tree: the Philippine experience. Forest Products Research and Development Institute Journal 21: 73–79.

Emboden, W. 1979. Narcotic Plants: Hallucinogens, Stimulants, Inebriants and Hypnotics, Their Origins and Uses. Collier Books, New York.

Endler, J. S. 1986. Natural Selection in the Wild. Princeton University Press, Princeton, New Jersey.

English, S., W. Greenaway, and F. R. Whatley. 1991. Bud exudate composition of *Populus tremuloides*. Canadian Journal of Botany 69: 2291–2295.

Enright, N. J., and R. S. Hill. 1995. Ecology of the Southern Conifers. Smithsonian Institution Press, Washington, D.C.

Erbilgin, W., and K. F. Raffa. 2000. Opposing effects of host monoterpenes on responses by two sympatric species of bark beetles to their aggregation pheromones. Journal of Chemical Ecology 26: 2527–2548.

Erdtman, H. 1959. Conifer chemistry and taxonomy of conifers. In: Biochemistry of Wood, pp. 1–28, eds. K. Krazl and G. Billek, Pergamon Press, Oxford.

Ernest, K. A. 1994. Resistance of creosote bush to mammalian herbivory: temporary consistency and browsing-induced changes. Ecology 75: 1684–1692.

Esau, K. 1977. Anatomy of Seed Plants. John Wiley, New York.

Espinosa-García, F. J., and J. H. Langenheim. 1991a. The effect of some leaf essential oil phenotypes in coastal redwood on the growth of several fungi with endophytic stages. Biochemical Systematics and Ecology 19: 629–642.

Espinosa-García, F. J., and J. H. Langenheim. 1991b. Effects of sabinene and γ-terpinene from coastal redwood leaves acting singly or in mixtures on the growth of some of their fungus endophytes. Biochemical Systematics and Ecology 19: 643–650.

Espinosa-García, F. J., P. Saldivar-García, and J. H. Langenheim. 1993. Dose-dependent effects in vitro of essential oils on growth of two endophytic fungi in coastal redwood leaves. Biochemical Systematics and Ecology 21: 185–194.

Espinosa-García, F. J., R. J. Rollinger, and J. H. Langenheim. 1996. Coastal redwood leaf endophytes: their occurrence, interactions and responses to host volatile terpenoids. In: Endophyte Fungi in Grasses and Woody Plants. Systematics, Ecology, and Evolution, pp. 101–120, eds. S. C. Redlin and L. M. Carris. American Pathological Society, St. Paul, Minnesota.

Estell, R. E., E. L. Fredrickson, D. M. Anderson, K. M. Harstad, and M. D. Remmenga. 1998. Relationship of tarbush leaf surface terpene profile with livestock herbivory. Journal of Chemical Ecology 24: 1–12.

Evans, P. H., J. X. Becerra, D. L. Venable, and W. S. Bowers. 2000. Chemical analysis of squirt-gun defense in Bursera and counter defense by chrysomelid beetles. Journal of Chemical Ecology 26: 745–754.

Evans, W. C. 1996. Trease and Evans' Pharmacognosy. W. B. Saunders, London.

Everett, T. H. 1981. The New York Botanical Garden Illustrated Encyclopedia of Horticulture, Vol. 4. Garland Publishing, New York.

Ewel, J. J. 1986. Designing agricultural ecosystems for the humid tropics. Annual Review of Ecology and Systematics 17: 245–271.

Fahn, A. 1979. Secretory Tissues in Plants. Academic Press, New York.

Fahn, A. 1988. Secretory tissues in vascular plants. New Phytologist 108: 229–257.

Fahn, A., and J. Benayoun. 1976. Ultrastructure of resin ducts in Pinus halepensis: development, possible sites of resin synthesis, and mode of its elimination from the protoplast. Annals of Botany 40: 857–863.

Fahn, A., and R. F. Evert. 1974. Ultrastructure of the secretory ducts of Rhus glabra L. American Journal of Botany 61: 1–14.

Fahn, A., and C. Shimony. 1996. Glandular trichomes of Fagonia L. (Zygophyllaceae) species: structure, development and secreted materials. Annals of Botany 77: 25–34.

Fahn, A., and C. Shimony. 1998. Ultrastructure and secretion of secretory cells of two species of Fagonia L. (Zygophyllaceae). Annals of Botany 81: 557–565.

Fahn, A., and E. Werker. 1972. Anatomical mechanisms of seed dispersal. In: Seed Biology, Vol. 1, pp. 151–221, ed. T. T. Kozlowski. Academic Press, New York.

Fahn, A., and E. Zamski. 1970. The influence of pressure, wind, wounding and growth substances on the rate of resin duct formation in *Pinus halepensis* wood. Israel Journal of Botany 19: 429–446.

Fahn, A., E. Werker, and P. Ben-Tzur. 1979. Seasonal effects of wounding and growth substances on development of traumatic resin ducts in *Cedrus libani*. New Phytologist 82: 537–544.

Fail, G. L. 1990. Ultrastructural studies of leaf / fungus interactions and secondary chemical production in *Hymenaea courbaril,* a tropical leguminous tree. Ph.D. Dissertation, University of California, Santa Cruz.

Fail, G. L., and J. H. Langenheim. 1990. Infection process of *Pestalotia subcuticularis* on leaves of *Hymenaea courbaril*. Phytopathology 80: 1259–1265.

Farjon, A., N. T. Hiep, D. K. Harder, P. K. Loc, and L. Averyanov. 2002. A new genus and species in Cupressaceae (Coniferales) from northern Vietnam, *Xanthocyparis vietnamensis*. Novon 12: 179–189.

Farnsworth, N. A., A. S. Bingel, G. H. Cordell, F. A. Crane, and H. H. S. Fong. 1975. Potential value of plants as a source of new antifertility agents. Journal of Pharmaceutical Sciences. Part 1, 64: 535–598; Part 2, 64: 717–754.

Farrah, A. Y. 1994. The Milk of the Boswellian Forests: Frankincense Production Among the Pastoral Somali. University of Uppsala Press, Uppsala.

Farrell, B. D., D. E. Dussourd, and C. Mitter. 1991. Escalation of plant defense: Do latex and resin canals spur plant diversification? American Naturalist 138: 881–900.

Fattorusso, E., C. Santacroce, and E. Xaasan. 1985. Dammarane triterpenes from the resin of *Boswellia freerana* [i.e., *frereana*]. Phytochemistry 24: 1033–1036.

Favorito, E., and K. Baty. 1995. The silphium connection. Celator 9: 6–8.

Fearnside, P. M. 1988. Extractive reserves in Brazilian Amazonia. BioScience 39: 387–393.

Feller, R. L. 1954. Dammar and mastic infrared analysis. Science 120: 1069–1070.

Feller, R. L. 1964. What's in a name? Dammar or serendipity in the library. Crucible 49: 8.

Fenaroli, G. 2002. Fenaroli's Handbook of Flavor Ingredients, ed. 4, ed. G. A. Burdock. CRC Press, Boca Raton.

Fernandes, R. M., N. A. Pereira, and L. G. Paulo. 1992. Anti-inflammatory activity of copaiba balsam (*Copaifera cearensis* Huber). Revista Brasileira de Farmácia 73: 53–56.

Ferrari, M., U. M. Pagnoni, F. Pelizooni, V. Lukes, and G. Ferrari. 1971. Terpenoids from *Copaifera langsdorfii*. Phytochemistry 10: 905–907.

Firn, R. D., and C. G. Jones. 1996. An explanation of secondary product "redundancy." In: Phytochemical Diversity and Redundancy in Ecological Interactions, eds. J. J. Romero, J. A. Saunders, and P. Barbosa. Recent Advances in Phytochemistry 30: 295–312.

Fleury, M. 1997. On medicinal role of copahu balsam. Acta Botanica Gallica 144: 473–479.

Florin, R. 1951. Evolution in *Cordaites* and conifers. Acta Horti Bergiani 15: 79–109.

Forestales Instituto Nacional de Investigaciones. 1980. Situación de la industria resinera en México, Secretaria de Agricultura y Recursos Hidraulicos, Subsecretaria Forestal y de la Fauna, Instituto Nacional de Investagaciones Forestales.

Forget, P. M. 1992. Regeneration ecology of *Eperua grandiflora* (Caesalpiniaceae), a large-seeded tree in French Guiana. Biotropica 24: 146–156.

Forrest, G. I. 1980. Genotypic variations among native Scotch pine *(Pinus sylvestris)* populations in Scotland (U.K.) based on monoterpene analysis. Forestry 53: 101–128.

Forster, P. G., E. L. Ghisalberti, P. R. Jeffries, V. M. Poletti, and N. J. Whiteside. 1986. Serrulatane diterpenes from *Eremophila* spp. Phytochemistry 25: 1377–1383.

Foster, G. M. 1956. Resin coated pottery in the Philippines. American Anthropologist 58: 732–733.

Fourie, T., and F. O. Snyckers. 1989. A pentacyclic triterpene with anti-inflammatory and analgesic activity from the roots of *Commiphora merkeri*. Journal of Natural Products 52: 1129–1131.

Foxworthy, F. W. 1922. Minor forest products of the Malay Peninsula. Malayan Forest Records 2: 151–217.

Franceschi, V. R., R. Kokense, T. Krekling, and E. Christensen. 2000. Phloem parenchyma cells are involved in local and distant defense responses to fungal inoculation or bark beetle attack in Norway spruce (Pinaceae). American Journal of Botany 87: 314–326.

Frankis, M. P. 1989. Generic inter-relationship in Pinaceae. Notes from the Royal Botanic Garden Edinburgh 45: 527–548.

Freise, F. W. 1934. Brazilian labdanum. Perfumery and Essential Oil Record 25: 135.

French, J. C. 1987. Systematic survey of resin canals in roots of the Araceae. Botanical Gazette 148: 360–371.

Frischkorn, C. G. B., and H. E. Frischkorn. 1978. Cercaricidal activity of some essential oils of plants from Brazil. Naturwissenschaften 65: 480–483.

Fritsch, P. W. 1997. A revision of *Styrax* (Styracaceae) for western Texas, Mexico and Mesoamerica. Annals of the Missouri Botanical Garden 84: 705–761.

Fritsch, P. W. 1999. Phylogeny of *Styrax* based on morphological characters, with implications for biogeography and infrageneric classification. Systematic Biology 24: 355–378.

Fritsch, P. W. 2001. Phylogeny and biogeography of the flowering plant genus *Styrax* (Styracaceae) based on chloroplast DNA restriction sites and DNA sequences of the internal transcribed spacer region. Molecular Phylogenetics and Evolution 19: 387–408.

Frondel, J. W. 1967a. X-ray diffraction study of some fossil and modern resins. Science 155: 1411–1413.

Frondel, J. W. 1967b. X-ray diffraction study of fossil elemis. Nature 215: 1360–1361.

Frost, S. H., and R. L. Langenheim. 1974. Cenozoic Reef Biofacies: Tertiary Larger Foraminifera and Scleractinian Corals from Chiapas, Mexico. Northern Illinois University Press, DeKalb.

Fulling, E. H. 1953. Dragon's blood. Economic Botany 7: 227.

Fundter, J. M. 1982. Names for Dipterocarp Timbers and Trees from Asia. PUDOC, Wageningen.

Gadek, P. A., E. S. Fernando, C. J. Quinn, S. B. Hoot, T. Terrazas, H. C. Sheahan, and M. W. Chase. 1996. Sapindales: molecular delimitation and infra-ordinal groups. American Journal of Botany 83: 802–811.

Gadek, P. A., D. L. Alpers, M. M. Heslewood, and C. J. Quinn. 2000. Relationships within Cupressaceae sensu lato: a combined morphological and molecular approach. American Journal of Botany 87: 1044–1057.

Gallagher, W. B., D. C. Parris, and E. S. Spamer. 1986. Paleontology, biostratigraphy and depositional environment of the Cretaceous–Tertiary transition in the New Jersey coastal plain. Mosasaur, Journal of Delaware Valley Paleontological Society 3: 1–35.

Gambel, T. 1921. Naval Stores History, Production, Distribution and Consumption. Review Publishing and Printing Company, Savannah, Georgia.

Gambliel, H. A., and R. G. Cates. 1995. Terpene changes due to maturation and canopy level in Douglas-fir *(Pseudotsuga menziesii)* flush needle oil. Biochemical Systematics and Ecology 33: 469–476.

García Rivas, H. 1988. Plantas Medicinales de México: Descripción y Usos. Panorama Editorial, México.

García Ruiz, J. F. 1981. La cervelle du ceil: ethnologie du copal au Mexique. Forklore Americano 32: 93–126.

Gemmill, C. L. 1966. Silphium. Bulletin of the History of Medicine 40: 295–313.

Gershenzon, J. 1994a. The cost of plant chemical defense against herbivory: a biochemical perspective. In: Insect Plant Interactions, Vol. 5, pp. 105–173, ed. E. A. Bernays. CRC Press, Boca Raton.

Gershenzon, J. 1994b. The metabolic costs of terpenoid accumulation in higher plants. Journal of Chemical Ecology 20: 1281–1328

Gershenzon, J., and R. Croteau. 1990. Regulation of monoterpene biosynthesis in higher plants. In: Biochemistry of the Mevalonic Acid Pathway to Terpenoids, pp. 99–160, eds. G. H. N. Towers and H. A. Stafford. Plenum Press, New York.

Gershenzon, J., and R. Croteau. 1991. Terpenoids. In: Herbivores, Their Interactions with Secondary Metabolites, Vol. 1, The Chemical Participants, pp. 165–219, eds. G. A. Rosenthal and M. R. Berenbaum. Academic Press, New York.

Gershenzon, J., and R. Croteau. 1993. Terpenoid biosynthesis: the basic pathway and formation of monoterpenes, sesquiterpenes and diterpenes. In: Lipid Metabolism in Plants, pp. 339–388, ed. T. S. Moore, Jr. CRC Press, Boca Raton.

Gershenzon, J., and T. J. Mabry. 1983. Secondary metabolites and the higher classification of angiosperms. Nordic Journal of Botany 3: 5–34.

Gershenzon, J., D. McCaskill, J. I. M. Rajaonarivony, C. Mihaliak, F. Karp, and R. Croteau. 1992. Isolation of secretory cells from plant glandular trichomes and their use in biosynthetic studies of monoterpenes and other glandular products. Analytical Biochemistry 200: 130–138.

Gershenzon, J., G. J. Murtagh, and R. Croteau. 1993. Absence of rapid turnover in several diverse species of terpene accumulating plants. Oecologia 96: 583–592.

Gershenzon, J., M. McConkey, and R. Croteau. 2000. Regulation of monoterpene accumulation in leaves of peppermint (*Mentha ×piperita* L.). Plant Physiology 122: 205–213.

Ghisalberti, E. L. 1979. Propolis: a review. Bee World 60: 59–84.

Ghisalberti, E. L. 1994. The phytochemistry of the Myoporaceae. Phytochemistry 35: 7–33.

Gianno, R. 1986a. Resin classification among the Semelai of Tasek Bera, Pahong, Malaysia. Economic Botany 40: 186–200.

Gianno, R. 1986b. The exploitation of resinous products in a lowland Malaysian forest. Wallaceana 1986(Mar.): 3–6.

Gianno, R. 1990. Semelai Culture and Resin Technology. Connecticut Academy of Arts and Sciences, New Haven.

Gianno, R., D. W. von Endt, W. D. Erhardt, K. M. Kochummen, and W. Hopwood. 1987. The identification of insular Southeast Asian resins and other plant exudates for archeological and ethnological application. In: Recent Advances in the Conservation and Analysis of Artifacts, pp. 229–298, comp. J. Black. Institute of Archeology, London.

Gibbs, R. D. 1974. Chemotaxonomy of Flowering Plants, 4 vols. McGill–Queen's University Press, Montreal.

Gibson, A. J., and C. T. Mason. 1927. The resin industry of India. Indian Forester 53: 379–385.

Gijzen, M., E. Lewinsohn, and R. Croteau. 1992. Antigenic cross-reactivity among monoterpene cyclases from grand fir and induction of these enzymes upon stem wounding. Archives of Biochemistry and Biophysics 294: 670–674.

Gilbert, B., J. I. P. Ferreira, M. B. S. Almeida, E. S. Carvalho, V. Cascon, and L. M. Rocha. 1997. The official use of medicinal plants in public health. Ciência e Cultura 49: 339–344.

Gillett, J. B. 1980. *Commiphora* (Burseraceae) in South America and its relationship to *Bursera*. Kew Bulletin 34: 569–589.

Gilliland, M. G., M. R. Appleton, and J. van Studen. 1988. Gland cells in resin canal epithelia in guayule *(Parthenium argentatum)* in relation to resin and rubber production. Annals of Botany 61: 55–64.

Gillis, W. 1971. The systematics and ecology of poison-ivy and poison-oaks. Rhodora 73: 72–159, 161–237, 370–443, 465–540.

Gimbutas, M. 1963. The Balts. Frederick A. Praeger, New York.

Gimlette, J. D., and H. W. Thomson. 1939. A Dictionary of Malayan Medicine. Oxford.

Gittlen, W. 1998a. Discovered Alive: the Story of the Chinese Redwood. Pierside Publication, Berkeley, California.

Gittlen, W. 1998b. Reunion with a Chinese redwood. California Wild 1998 (Fall): 28–32.

Gleizes, M., J. P. Carde, G. Pauly, and C. Bernard-Dagan. 1980a. In vivo formation of sesquiterpene hydrocarbons in the endoplasmic reticulum of pine. Plant Science Letters 20: 79–90.

Gleizes, M., G. Pauly, C. Bernard-Dagan, and R. Jacques. 1980b. Effects of light on terpene hydrocarbon synthesis in *Pinus pinaster*. Physiologia Plantarum 50: 16–20.

Gleizes, M., C. Bernard-Dagan, J. P. Carde, G. Pauly, and A. Marpeau. 1982. Function of plastids in terpene biosynthesis. In: Biochemistry and Metabolism of Plant Lipids., pp. 511–514, eds. J. F. G. M. Wintermans and P. J. C. Kurper. Elsevier / North Holland Medical Press.

Gleizes, M., G. Pauly, J. P. Carde, A. Marpeau, and C. Bernard-Dagan. 1983. Monoterpene hydrocarbon biosynthesis by isolated leucoplasts of *Citrofortunella mitis*. Planta 159: 373–381.

Glover, P. M. 1937. Lac cultivation in India. Indian Lac Research Institute, Namkum, Ranchi, Bihar.

Goeppert, H. R. 1836. Fossile Pflanzenreste des Eisensandes von Aachen. Nova Acta Physico-Medica Akademie Caesareae Leopoldino-Carolinae Naturae Curiosorum 1836: 19–150.

Goeppert, H. R., and G. C. Berendt. 1845. Die Bernstein und die in ihm befindlichen Planzenreste der Vorwelt, Band 1, Abtheilung 1. Berlin.

Goeppert, H. R., and A. Menge. 1883. Die Flora des Bernsteins und Ihre Beziehungen zur Flora der Tertiärformation und der Gegenwart, Band 1. Danzig.

Gómez-Vásquez, B. G., and E. M. Engleman. 1984. Bark anatomy of *Bursera longipes* and *Bursera copallifera*. International Association of Wood Anatomists Bulletin 5: 335–340.

Gomppers, M. E., and A. M. Hoylman. 1993. Grooming with *Trattinnickia* resin: possible pharmaceutical use by coatis in Panama. Journal of Tropical Ecology 9: 533–540.

Gonzales, E. V., and F. G. Abejo. 1978. Properties of Manila copal *(almaciga)* resin from 15 different localities in the Philippines. Forpride Digest 7: 68–69.

González, A. G., F. León, L. Sánchez-Pinto, J. L. Padrón, and J. Bermejo. 2000. Phenolic compounds of dragon's blood from *Dracaena draco*. Journal of Natural Products 63: 1297–1299.

Gossen, G. H. 1974. Chamulas in the World of the Sun: Time and Space in a Maya Oral Tradition. Harvard University Press, Cambridge.

Gough, L., and J. S. Mills. 1972. The composition of succinite (Baltic amber). Nature 239: 527–528.

Gould, S. J., and E. S. Vrba. 1982. Exaptation—a missing term in the science of form. Paleontology 8: 4–15.

Gower, S. T., and J. H. Richards. 1990. Larches: deciduous conifers in an evergreen world. BioScience 40: 818–826.

Grabowska, J. 1983. Polish Amber. Interpress Publishers, Warsaw.

Graham, A. 1999. Late Cretaceous and Cenozoic History of North American Vegetation. Oxford University Press, New York.

Graham, A. 2003. Historical phytogeography of the Greater Antilles. In: Flora of the Greater Antilles, Vol. 1, ed. T. Zanoni. New York Botanical Garden Press, Bronx.

Grant, F. W., and H. H. Zeiss. 1954. The structure of cativic acid. Journal of the American Chemical Society 76: 5001–5002.

Grantham, P. J., and A. G. Douglas. 1980. The nature and origin of sesquiterpenoids in some Tertiary fossil resins. Geochimica et Cosmochimica 44: 1801–1810.

Grayum, M. H. 1996. Revision of *Philodendron* subgenus *Pteromischum* (Araceae) for Pacific and Caribbean Tropical America. Systematic Botany Monographs 47: 31–34.

Greenaway, W., T. Scaysbrook, and F. R. Whatley. 1990. The composition and plant origins of propolis: a report of the work at Oxford. Bee World 71: 107–118.

Greenfield, M. D., T. E. Shelley, and K. R. Downum. 1987. Variation in host-plant quality: implications for territoriality in a desert grasshopper. Ecology 68: 828–838.

Greenway, P. J. 1941. Gum, resinous and mucilaginous plants in East Africa. East African Agricultural Journal 6: 241–250.

Gregoire, J.-C., D. Couillien, R. Krebber, A. König Winfred, H. Meyer, and W. Francke 1992. Orientation of *Rhizophagus grandis* (Coleoptera: Rhizophagidae) to oxygenated monoterpenes in a species-specific predator–prey relationship. Chemoecology 3: 14–18.

Grieve, M. 1959. A Modern Herbal: the Medicinal, Culinary, Cosmetic and Economic Properties, Cultivation and Folk-Lore of Herbs, Grasses, Fungi, Shrubs & Trees, with

All Their Modern Scientific Uses, 2 vols. Hafner, New York (reprinted 1971, Dover, New York).

Grimaldi, D. A. 1995. The age of Dominican amber. In: Amber, Resinite and Fossil Resins, pp. 203–217, eds. K. B. Anderson and J. C. Crelling. Symposium Series 617. American Chemical Society, Washington, D.C.

Grimaldi, D. A. 1996. Amber: Window to the Past. Harry N. Abrams, with the American Museum of Natural History, New York.

Grimaldi, D., C. W. Beck, and J. J. Boon. 1989. Occurrence, chemical characteristics and paleontology of the fossil resins from New Jersey. American Museum Novitiates 2948: 1–27.

Grimaldi, D. A., A. Shedrinsky, A. Ross, and N. S. Baer. 1991. Forgeries of fossils in amber: history, identification, and case studies. Curator 37: 251–274.

Grimaldi, D. A., E. Bonwich, M. Dellanay, and S. Doberstern. 1994. Electron microscopic studies of mummified tissues in amber fossils. American Museum Novitiates 3097: 1–31.

Grimaldi, D., A. Shedrinsky, and T. P. Wampler. 2000. A remarkable deposit of fossiliferous amber from the Upper Cretaceous (Turonian) of New Jersey. In: Studies of Fossils in Amber, with Particular Reference to the Cretaceous of New Jersey, pp. 1–76, ed. D. Grimaldi. Backhuys, Leiden.

Grimaldi, D. A., J. A. Lillegraven, T. W. Wampler, D. Bookwalter, and A. Shedrinsky. 2001. Amber from Upper Cretaceous through Paleocene strata of the Hanna Basin, Wyoming, with implications for taphonomy of fossil resins. Rocky Mountain Geology 35: 163–204.

Grimalt, J. O., B. R. T. Simoneit, P. G. Hatcher, and A. Nissenbaum. 1988. The molecular composition of ambers. Organic Geochemistry 13: 677–90.

Grinspoon, L. 1977. Marihuana Reconsidered. Harvard University Press, Cambridge.

Grinspoon, L., and J. B. Bakaler. 1997. Marihuana: the Forbidden Medicine. Yale University Press, New Haven, Connecticut.

Groom, N. 1981. Frankincense and Myrrh: a Study of the Arabian Incense Trade. Longman, New York.

Grove, D. 1984. Chalcazingo: Excavations on the Olmec Frontier. Thames and Hudson, London.

Grun, B. 1991. The Timetables of History, ed. 3. Simon and Schuster, New York.

Grundwag, M., and E. Werker. 1976. Comparative wood anatomy as an aid to identification of Pistacia L. species. Israel Journal of Botany 25: 152–167.

Guariguata, M. R., and G. S. Gilbert. 1996. Interspecific variation in rates of trunk wound closure in a Panamanian lowland forest. Biotropica 28: 23–29.

Gudynas, P., and S. Pinkus. 1967. The Palanga Museum of Amber. Mintis Books, Vilnius, Lithuania.

Guenther, E. 1948–52. The Essential Oils, Vols. 1–6. Van Nostrand, New York (reprinted 1972–76, Krieger, New York).

Guignard, L. 1982. L'appareil sécréteur de Copaifera. Bulletin de la Société Botanique de France 39: 253.

Gunn, C. R., and J. V. Dennis. 1976. World Guide to Tropical Drift Seeds and Fruits. Quadrangle / New York Times Book Company, New York.

Gutiérrez, G., and A. Marin. 1998. The most ancient DNA recovered from an amber-preserved specimen may not be as ancient as it seems. Molecular Biology and Evolution 15: 926–929.

Haeuser, J., R. Lombard, F. Lederer, and G. Ourisson. 1961. Isolement et structure d'un noveau diterpène: l'acide daniellique. Tetrahedron 12: 205–214.

Haeuser, J., S. F. Hall, A. C. Oehlschlager, and G. Ourisson. 1970. The structure and stereochemisry of oliveric acid. Tetrahedron 26: 1461–1465.

Hafizoglu, H., M. Reunanen, and A. Istek. 1996. Chemical constituents of balsam from *Liquidambar orientale*. Holzforschung 50: 116–117.

Hale, A. 1911. Balsam of Peru, a Central American contribution to the pharmacopoeia. Bulletin of the Pan American Union 1911: 880–881.

Hall, P., and K. Bawa. 1993. Methods to assess the impact of extraction of non-timber tropical forest products on plant populations. Economic Botany 47: 234–237.

Halos, S. C. 1983. Factors affecting quality and quantity of *almaciga* resin. National Research Council of the Phillipines Research Bulletin 38: 70–113.

Halpine, S. M. 1995. Trace amino acid composition of natural resins. In: Ambers, Resinite and Fossil Resins, pp. 234–254, eds. K. B. Anderson and J. C. Crelling. Symposium Series 617. American Chemical Society, Washington, D.C.

Hammond, C. T., and P. G. Mahlberg. 1973. Morphology of glandular hairs of *Cannabis sativa* from scanning electron microscopy. American Journal of Botany 60: 524–528.

Hammond, C. T., and P. G. Mahlberg. 1977. Morphogenesis of capitate glandular hairs of *Cannabis sativa* (Cannabaceae). American Journal of Botany 64: 1023–1031.

Han, K. P., and D. E. Lincoln. 1994. The evolution of carbon allocation to plant secondary metabolites: a genetic analysis of cost in *Diplacus aurantiacus*. Evolution 48: 1550–1563.

Han, K. P., and D. E. Lincoln. 1997. The impact of plasticity and maternal effect on evolution of leaf resin production in *Diplacus aurantiacus*. Evolutionary Ecology 11: 471–484.

Hanes, C. S. 1927. Resin canals in seedling conifers. Journal of the Linnean Society, Botany 47: 613–636.

Hanover, J. W. 1966a. Genetics of terpenes I. Gene control of monoterpene levels of *Pinus monticola*. Heredity 21: 73–84.

Hanover, J. W. 1966b. Enviromental variation in monoterpenes in *Pinus monticola*. Phytochemistry 5: 713–717.

Hanover, J. W. 1971. Genetics of terpenes II. Genetic variances and interrelationships of monoterpene concentration in *Pinus monticola*. Heredity 27: 237–245.

Hanover, J. W. 1992. Applications of terpene analysis in forest genetics. New Forests 6: 158–178.

Harborne, J. B. 1980. Flavonoids. In: Secondary Plant Products, pp. 329–402, eds. E. A. Bell and B. V. Charlwood. Springer-Verlag, Berlin.

Harborne, J. B. 1991. Recent advances in the ecological chemistry of plants. In: Ecological Chemistry and Biochemistry of Plant Terpenoids, pp. 399–426, eds. J. B. Harborne and F. A. Tomás-Barberán. Clarendon Press, Oxford.

Hare, J. D. 2002. Geographic and genetic variation in the leaf surface resin components of *Mimulus aurantiacus* from southern California. Biochemical Systematics and Ecology 30: 281–296.

Harkins, K. J., and R. A. Linley. 1973. Determinations of balsamic acids and esters. Analyst 98: 819–822.

Harrison, H. F., and J. K. Peterson. 1986. Allelopathic effects of sweet potatoes *(Ipomoea batatas)* on yellow nutsedge *(Cyperus esculentus)* and alfalfa *(Medicago sativa)*. Weed Science 34: 623–627.

Harrison, I. T., and T. Nakano. 1962. Os sesquiterpenos da resinas da jutaicica. Anais Associação Brasiliera de Química 21: 23–29.

Hart, J. A. 1987. A cladistic analysis of conifers: preliminary results. Journal of the Arnold Arboretum 68: 269–307.

Hart, T. B., J. A. Hart, and P. G. Murphy. 1989. Monodominant and species-rich forests of the humid tropics: causes for their co-occurrence. American Naturalist 133: 613–633.

Hartmann, P. J. 1677. Succini Prussici Physica et Civilis Historia. Francofurti.

Hartwell, J. L. 1967–71. Plants Used Against Cancer. A Survey. Lloydia 30–34.

Hays, J. T., and W. Cottle. 1989. New sources. In: Naval Stores: Production-Chemistry-Utilization, pp. 200–221, eds. D. E. Zinkel and J. Russell. Pulp Chemicals Association, New York.

Hayward, B. W. 1982. Kauri Gum and the Gumdiggers: a Pictorial History of the Kauri Gum Industry in New Zealand. Lodestar Press, Auckland.

Hegnauer, R. 1990. Chemotaxonomie der Pflanzen, Band 9. Birkhäuser, Basel.

Helliwell, K., and P. Jennings. 1983. A critical evaluation of commercial Sumatra benzoins. Journal of Pharmacy and Pharmacology 35 (Supplement).

Henderson, J. S. 1997. The World of the Ancient Maya, ed. 2. Cornell University Press, Ithaca, New York.

Hepper, F. N. 1969. Arabian and African frankincense trees. Journal of Egyptian Archeology 55: 66–72.

Herendeen, P. S., and B. F. Jacobs. 2001. Fossil legumes from the Eocene of Tanzania. Botany 2001. Botanical Society of America.

Herendeen, P. S., W. I. Crepet, and D. L. Dilcher. 1992. The fossil history of the Leguminosae: phylogenetic and biogeographic implications. In: Advances in Legume Systematics, Part 4, Fossil Record, pp. 303–316, eds. P. S. Herendeen and D. L. Dilcher. Royal Botanic Gardens, Kew.

Heringer, E. P., and M. B. Ferreirá. 1975. Árvores uteis du região geã econômica da Distrito Federal. Dendrologia. O gênero *Hymenaea*. Cerrado 7: 27–32.

Herms, D. A., and W. J. Mattson. 1992. The dilemma of plants: to grow or to defend. The Quarterly Review of Biology 67: 283–335.

Herms, D. A., and K. A. Raffa. 1995. Effects of stresses on the expression of plant resistance to herbivores: an overview. Proceedings of the World Congress of the International Union of Forestry Research Organization, Tampera, Finland.

Herodotus. The Histories, 1954 transl. A. de Selincourt. Penguin Press, London.

Heywood, V. H., J. B. Harborne, and B. L. Turner (eds.). 1977. The Biology and Chemistry of the Compositae. Academic Press, London.

Hill, A. F. 1952. Economic Botany. McGraw Hill, New York.

Hillis, W. E. 1986. Forever amber: a story of the secondary wood components. Wood Science Technology 20: 203–227.

Hillis, W. E. 1987. Heartwood and Tree Exudates. Springer-Verlag, New York.

Hills, E. S. 1957. Fossiliferous Tertiary resin from Allendale, Victoria. Proceedings of the Royal Society of Victoria 69: 15–19.

Hintikka, W. 1970. Selective effect of terpenes on growth of symbiotic fungi associated with the southern pine beetle. Karstenia 11: 28–32.

Hippocrates. Ancient Medicine; . . . , 8 vols. 1923–95 transl. W. H. S. Jones, E. T. Withington, and P. Potter. Loeb Classical Library, Harvard University Press, Cambridge.

Hodges, A. 1997. International market analysis. Naval Stores Review 1997: 7–21.

Hoffmann, J. J., and S. P. McLaughlin. 1986. *Grindelia camporum:* a potential cash crop for the arid Southwest. Economic Botany 40: 162–169.

Hoffmann, J. J., B. E. Kingsolver, S. P. McLaughlin, and B. N. Timmermann. 1984. Production of resins by arid-adapted Astereae. In: Phytochemical Adaptations to Stress, pp. 251–271, eds. B. N. Timmermann, C. Steelink, and F. A. Loewus. Plenum Press, New York.

Hollick, A., and E. C. Jeffrey. 1909. Studies of Cretaceous remains from Kreischerville, New York. Memoirs of the New York Botanical Garden 3: 1–138.

Hong, Y.-Ch. 1981. Eocene Fossil Diptera (Insecta) in Amber of Fushun Coalfield. Geological Publishing House, Beijing.

Hoots, C. 1993. Retracing the incense route. Mosaic 1993(Feb.): 43–45.

Horntveldt, R. 1988. Resistance of *Picea abies* to *Ips typographus:* the response to monthly innoculations with *Ophiostoma polonicum,* a beetle transmitted blue-stain fungus. Scandinavian Journal of Forest Research 3: 107–114.

Howard, J. J. 1985. Observations on resin collecting by six interacting species of stingless bees (Apidae: Meliponinae). Journal of the Kansas Entomological Society 58: 337–345.

Howard, J. J., J. Coxin, Jr., and D. F. Wiemer. 1988. Toxicity of terpenoid deterrents to the leaf-cutting ant *Atta cephalotes* and its mutualistic fungus. Journal of Chemical Ecology 14: 59–69.

Howes, F. N. 1929. Tapping Peru-balsam. Kew Bulletin 1929: 327–329.

Howes, F. N. 1949. Vegetable Gums and Resins. Chronica Botanica, Waltham, Massachusetts.

Howes, F. N. 1950. Age-old resins of the Mediterranean region and their uses. Economic Botany 4: 307–316.

Hrazdina, G., and R. A. Jensen. 1984. Metabolic pathways as enzyme complexes: evidence for the synthesis of phenylpropanoids and flavonoids on membrane-associated enzyme complexes. Archives of Biochemistry and Biophysics 237: 88–100.

Hubbell, S., D. F. Weimer, and A. Adejare. 1983. An antifungal terpenoid defends a Neotropical tree *(Hymenaea)* against attack by a fungus-growing ant *(Atta).* Oecologia 60: 321–327.

Hueber, F. M., and J. H. Langenheim. 1986. Dominican amber tree had African ancestors. Geotimes 31: 8–10.

Hugel, G., and G. Ourisson. 1965. Diterpènes de *Trachylobium.* Structure et stéréochimie de l'acide zanzibarique. Bulletin de Société Chimique de France 1965: 2903–2908.

Hugel, G., L. Lods, J. M. Mellor, O. W. Theobald, and G. Ourisson. 1965a. Diterpènes de *Trachylobium*. I. Introduction générale. Isolement de kauranol et de huit diterpènes. Bulletin de Société Chimique de France 1965: 2882–2887.

Hugel, G., L. Lods, J. M. Mellow, D. W. Theobald, and G. Ourisson. 1965b. Diterpènes de *Trachylobium*. II. Structure des diterpènes têtra- et pentacyclique de *Trachylobium*. Bulletin de Société Chimique de France 1965: 2888–2894.

Hugel, G., L. Lods, J. M. Mellor, and G. Ourisson. 1965c. Diterpènes de *Trachylobium*. III. Reaction des dérevés trachylobanique. Bulletin de Société Chimique de France 1965: 2894–2902.

Hugel, G., A. C. Oehschlager, and G. Ourisson. 1966. The structure and stereochemistry of diterpenes from *Trachylobium verrucosum* Oliv. Tetrahedron Supplement 8, Part 1: 203–216.

Huneck, S. 1963. Triterpenes of the balsam of *Liquidambar orientalis* Miller (storax). Tetrahedron 19: 479–482.

Hurd, P. D., R. F. Smith, and J. W. Durham. 1962. The fossiliferous amber of Chiapas, Mexico. Ciencia 21: 107–118.

Ibrahim, J. 1988. The essential oil of *Dipterocarpus kerrii*. Journal of Tropical Forest Science 1: 11–15.

Ibrahim, J., A. Abu Said, and A. Abdul Rashid. 1987. Tapping of oleoresin from *Dipterocarpus kerrii*. Malaysian Forester 50: 343–353.

Ikeda, T., F. Matsumura, and D. M. Benjamin. 1977. Mechanisms of feeding discrimination between mature and juvenile foliage by two species of pine sawflies. Journal of Chemical Ecology 3: 677–694.

Isman, M. B., and E. Rodriguez. 1983. Larval growth inhibitors from species of *Parthenium* (Asteraceae). Phytochemistry 22: 2709–2713.

Iturralde-Vinent, M. A., and R. O. E. MacPhee. 1996. Age and paleogeographical origin of Dominican amber. Science 273: 1850–1852.

Iturralde-Vinent, M., and E. Hartstein. 1998. Miocene amber and lignitic deposits in Puerto Rico. Caribbean Journal of Science 34: 308–312.

Ivanov, C. P., L. K. Yankov, and P. T. T. Tho. 1969. On the composition of the essential oil from resin of *Liquidambar formosana* H. Rivista Italiana EPPOS 51: 380–384.

Jacobs, M. 1974. Botanical panorama of the Malesian Archipelago (vascular plants). In: Resources of Humid Tropical Asia (Natural Resources Research 12), pp. 263–294. UNESCO, Paris.

Jain, K. K. 1976. Evolution of wood structures in Pinaceae. Israel Journal of Botany 25: 28–33.

Jalaluddin, M. 1977. A useful pathological condition of wood. Economic Botany 31: 222–224.

Jantan, I. 1988. The essential oil of *Dipterocarpus kerrii*. Journal of Tropical Forest Science 1: 10–15.

Janzen, D. H. 1970. Herbivores and the number of species in tropical forests. American Naturalist 104: 501–528.

Janzen, D. H. 1974. Tropical blackwater rivers, animals and mast fruiting by the Dipterocarpaceae. Biotropica 6: 69–103.

Janzen, D. H. 1975. Behavior of *Hymenaea courbaril* when its predispersal seed predator is absent. Science 189: 145–147.

Janzen, D. H. 1981. Patterns of herbivory in a tropical deciduous forest. Biotropica 13: 221–281.

Jarvis, B. B., and J. D. Miller. 1996. Natural products, complexity and evolution. In: Phytochemical Diversity and Redundancy in Ecological Interactions, eds. J. J. Romero, J. A. Saunders, and P. Barbosa. Recent Advances in Phytochemistry 30: 265–293.

Jarzembowski, E. A. 1999. British amber: a little known resource. [In: Proceeding of the World Congress in Amber Inclusions, eds. J. Alonso, J. C. Corral, and R. López] Estudios del Museo de Ciencias Naturales de Álava 14: 133–140.

Jayasingh, D. B., and F. E. Freeman. 1980. The comparative population dynamics of eight solitary bees and wasps (Aculeata; Apocrita; Hymenoptera) trap nested in Jamaica. Biotropica 12: 214–219.

Jefferies, P. R. 1979. Studies on the resin chemistry of some Western Australian plants. Journal of the Royal Society of Western Australia 62: 95–107.

Jeffrey, E. C. 1905. The comparative anatomy and phylogeny of the Coniferales. Pt. 2. The Abietineae. Bulletin of the Boston Society of Natural History 6: 1–37.

Jepson, W. L. 1936. A Flora of California, Vol. 2. Cunningham, Curtis, and Welch, San Francisco.

Joel, D. M. 1980. Resin ducts in the mango fruit: a defense system. Journal of Experimental Botany 31: 1707–1718.

Johnson, D. 1952. Indian Hemp—a Social Menace. Christopher Johnson, London.

Johnson, H. B. 1975. Plant pubescence: an ecological perspective. Botanical Review 41: 233–258.

Johnson, K. S., F. A. Eischen, and D. E. Giannasi. 1994. Chemical composition of North American bee propolis and biological activity towards larvae of the greater waxmoth (Lepidoptera: Pyralidae). Journal of Chemical Ecology 20: 1783–1792.

Johnson, N. D. 1983. Flavonoid aglycones from *Eriodictyon californicum* (Hydrophyllaceae) resin. Biochemical Systematics and Ecology 11: 211–215.

Johnson, N. D., C. C. Chu, P. R. Ehrlich, and H. A. Mooney. 1984. The seasonal dynamics of leaf resin, nitrogen and herbivore damage in *Eriodictyon californicum* and their parallels in *Diplacus aurantiacus*. Oecologia 61: 398–402.

Jones, W. G., K. D. Hill, and M. J. Allen. 1995. *Wollemia nobilis*, a new living Australian genus and species in the Araucariaceae. Telopea 6: 173–176.

Jost, T., Y. Sell, and J. Foussereau. 1989. Contact allergy to Manila resin. Nomenclature and physico-chemistry of Manila, kauri, damar, and copal resins. Contact Dermatitis 21: 228–238.

Judd, W. S., C. S. Campbell, E. A. Kellogg, and P. F. Stevens. 1999. Plant Systematics: a Phylogenetic Approach. Sinauer Associates, Sunderland, Massachusetts.

Jung, V. W., E. Knobloch, and Z. Kvaček. 1971. Makrofloristiche Untersuchungen in Braunkohlentertiär der Oberpfalz. Mitteilungen der Bayerischen Staatssammlung für Paläontologie und Historische Geologie 11: 223–249.

Kainer, K. A., and M. L. Duryea. 1992. Tapping women's knowledge: plant resource use in extractive reserves, Acre, Brazil. Economic Botany 46: 408–425.

Kaplan, E., E. Pearlstein, E. Howe, and J. Levinson. 1999. Análisis técnico de qeros pintados de los períodos Inca y Colonial. Íconos 2: 30–38.

Kar, A., and M. K. Menon. 1969. Analgesic effect of the gum resin of *Bursera serrata*. Life Sciences 8: 1023–1028.

Karban, R., and I. T. Baldwin. 1997. Induced Responses to Herbivory. University of Chicago Press, Chicago.

Katoh, S., and R. Croteau. 1998. Individual variation in constitutive and induced monoterpenes in biosynthesis in grand fir. Phytochemistry 47: 577–582.

Kelch, D. G. 1998. Phylogeny of the Podocarpaceae: comparison of evidence from morphology and 18S rDNA. American Journal of Botany 85: 986–996.

Kelly, M. J., and A. E. Rohl. 1989. Pine oil and miscellaneous uses. In: Naval Stores: Production-Chemistry-Utilization, pp. 560–572, eds. D. E. Zinkel and J. Russell. Pulp Chemical Association, New York.

Kelley, S. T., and B. D. Farrell. 1998. Is specialization a dead end? The phylogeny of host use in *Dendroctonus* bark beetles (Scolytidae). Evolution 52: 1731–1743.

Khalid, S. A. 1985. Chemistry of the Burseraceae In: Chemistry and Chemical Taxonomy of the Rutales, pp. 281–299, eds. P. G. Waterman and M. F. Grundon. Academic Press, New York.

Khoo, S. F., A. C. Oehlschlager, and G. Ourisson. 1973. Structure and stereochemistry of diterpenes of *Hymenaea courbaril* (Caesalpinioideae) pod resin. Tetrahedron 29: 3379–3388.

Kikuzawa, K. 1982. Leaf survival and evolution in Betulaceae. Annals of Botany 50: 345–354.

Kimmens, A. C. (ed.). 1977. Tales of Hashish. William Morrow, New York.

Kimura, M. 1983. The Neutral Theory of Molecular Evolution. Cambridge University Press, Cambridge.

Kirk, J. 1868. On the copal of Zanzibar. Journal of the Linnean Society of Botany 11: 1–4.

Kirk-Othmer Encyclopedia of Chemical Technology. 2001. John Wiley Interscience, New York.

Klebs, R. 1880. Der Bernstein. Königsberg.

Kleinig, H. 1989. The role of plastids in isoprenoid biosynthesis. Annual Review of Plant Physiology and Plant Molecular Biology 40: 39–59.

Klein-Rebour, F. 1964. L'Égypte et ses parfums. France & Parfums 7: 137–140.

Klepzig, K. D., E. L. Kruger, E. B. Smalley, and K. F. Raffa. 1995. Effects of biotic and abiotic stress on induced accumulation of terpenes and phenolics in red pines inoculated with bark beetle-vectored fungus. Journal of Chemical Ecology 21: 601–626.

Kligman, A. M. 1958. Poison ivy *(Rhus)* dermatitis. American Medical Association Archives of Dermatology 77: 149–180.

Knaus, H., and H. Wagner. 1996. Effects of *Boswellia serrata* and other triterpenic acids on the complement system. Phytomedicine 3: 77–81.

König, B. 1985. Plant sources of propolis. Bee World 66: 136–139.

Koerper, H., and A. L. Kolls. 1999. The silphium motif adorning ancient Libyan coinage: marketing a medicinal plant. Economic Botany 53: 133–143.

Korte, E. 1970. Recent results in hashish chemistry. In: The Botany and Chemistry of *Cannabis*, pp. 119–135, eds. C. R. B. Joyce and S. H. Curry. Churchill, London.

Kostiashowa, Z. 1999. Amber crisis in Sambia: today and 100 years ago. [In: Proceedings of the World Congress on Amber Inclusions, eds. J. Alonso, J. C. Corral, and R. López.] Estudios del Museo de Ciencias Naturales de Álava 14: 255–260.

Krause, K. 1909. Über harzsercenierende Drüsen an den Nebenblätten von Rubaceen. Berichte der Deutschen Botanischen Gesellschaft 27: 446–452.

Krischkorn, C. G. B., and H. E. Krischkorn. 1978. Cercaricidal activity of some essential oils of plants from Brazil. Naturwissenchaften 65: 480–483.

Krüssmann, G. 1985. Manual of Cultivated Conifers. Timber Press, Portland, Oregon.

Krumbiegel, G., and B. Krumbiegel. 1994. Bernstein—Fossile Harzc aus der Welt: Geschichte, Harze, Vorkommen, Gewinnung, Inklusen. Goldschneck-Verlag, Weinstadt.

Kursar, T. A., T. L. Capson, P. D. Coley, D. G. Corley, M. B. Gupta, L. A. Harrison, E. Ortega, and D. Windsor. 1999. Ecologically guided bioprospecting in Panama. Pharmaceutical Biology 37 (Supplement): 1–14.

Kusumi, J., Y. Tsumura, H. Yoshimaru, and H. Tachida. 2000. Phylogenetic relationships in Taxodiaceae and Cupressaceae sensu stricto based on *mat*K gene, *chl*L gene, *trn*L–*trn*F IGS region, and *trn*L intron sequences. American Journal of Botany 87: 1480–1488.

Labandeira, C. C. 1998. The early history of arthropod and vascular plant associations. Annual Reviews of Earth and Planetary Science 26: 329–377.

Labandeira, C. C., B. A. LePage, and A. H. Johnson. 2001. A *Dendroctonus* bark engraving (Coleoptera: Scolytidae) from middle Eocene *Larix* (Coniferales: Pinaceae): Early or delayed colonization? American Journal of Botany 88: 2026–2039.

Laird, S. 1995. The natural management of tropical forests for timber and non-timber products. Oxford Forestry Institute Occasional Papers 49.

Lam, H. J. 1932. The Burseraceae of the Malay Archipelago and Peninsula. Bulletin du Jardin Botanique de Buitenzorg, série 3, 12: 286–297.

Lambert, J. B., and J. S. Frye. 1982. Carbon functionalities in amber. Science 217: 55–57.

Lambert, J. B., J. S. Frye, and G. O. Poinar, Jr. 1985. Amber from the Dominican Republic: analysis of nuclear magnetic resonance spectroscopy. Archaeometry 27: 43–51.

Lambert, J. B., J. S. Frye, T. A. Lee, Jr., C. J. Welch, and G. O. Poinar, Jr. 1989. Analysis of Mexican amber by carbon-13 NMR spectroscopy. Archaeological Chemistry 4: 381–388.

Lambert, J. B., S. C. Johnson, and G. O. Poinar, Jr. 1996. Nuclear magnetic resonance characterization of Cretaceous amber. Archaeometry 38: 325–335.

Lambert, J. B., C. E. Shawl, G. O. Poinar, Jr., J. A. Santiago-Blay. 1999. Classification of modern resins by solid state nuclear magnetic resonance spectroscopy. Bioorganic Chemistry 27: 409–423.

Landis, G. P., and L. W. Snee. 1991. ^{40}Ar / ^{39}Ar systematics and argon diffusion in amber: implications for ancient earth atmospheres. Paleogeography Paleoclimatology Paleoecology (Global and Planetary Change Section) 97: 63–67.

Lange, B. M., M. R. Wildung, D. McCaskill, and R. Croteau. 1998. A family of transketolases that direct isoprenoid biosynthesis via a mevalonate-independent pathway. Proceedings of the National Academy of Sciences U.S.A. 95: 2100–2104.

Lange, B. M., T. Rujan, W. Martin, and R. Croteau. 2000. Isoprenoid biosynthesis: the evolution of two ancient and distinct pathways across genomes. Proceedings of the National Academy of Sciences U.S.A. 97: 13,172–13,177.

Langenheim, J. H. 1964. Present status of botanical studies of amber. Harvard Botanical Museum Leaflets 20: 225–287.

Langenheim, J. H. 1966. Botanical source for amber from Chiapas, Mexico. Ciencia 24: 201–211.

Langenheim, J. H. 1967. Preliminary investigations of *Hymenaea courbaril* as a resin producer. Journal of the Arnold Arboretum 48: 203–229.

Langenheim, J. H. 1969. Amber: a botanical inquiry. Science 163: 1157–1169.

Langenheim, J. H. 1972. Botanical origin of fossil resin and its relation to forest history in northeastern Angola. Publicações Culturais, Companhia de Diamantes de Angola 85: 15–36.

Langenheim, J. H. 1973. Leguminous resin-producing trees in South America and Africa. In: Tropical Forest Ecosystems in Africa and South America: a Comparative Review, pp. 89–104, eds. B. J. Meggers, E. S. Ayensu, and W. D. Duckworth. Smithsonian Press, Washington, D.C.

Langenheim, J. H. 1975. Role of the tropics in the evolution of resin-producing trees. Proceedings XII International Botanical Congress, Leningrad.

Langenheim, J. H. 1981. Terpenoids in the Leguminosae. In: Advances in Legume Systematics, Part 2, pp. 627–655, eds. R. M. Polhill and P. H. Raven. Royal Botanic Gardens, Kew.

Langenheim, J. H. 1984. The roles of plant secondary chemicals in wet tropical ecosystems. In: Physiological Ecology of Plants in the Wet Tropics, pp. 189–208, eds. E. Medina, H. A. Mooney, and C. Vázquez-Yánes. Dr. W. Junk Publishers, The Hague.

Langenheim, J. H. 1990. Plant resins. American Scientist 70: 16–24.

Langenheim, J. H. 1994. Higher plant terpenoids: a phytocentric overview of their ecological roles. Journal of Chemical Ecology 20: 1223–1280.

Langenheim, J. H. 1995. Biology of amber-producing trees: focus on case studies of *Hymenaea* and *Agathis*. In: Amber, Resinite and Fossil Resin, pp. 1–31, eds. K. B. Anderson and J. C. Crelling. Symposium Series 617. American Chemical Society, Washington, D.C.

Langenheim, J. H. 2001. Contribuciones de los estudios a largo plazo a la teoría de la defensa química: perspectivos con árboles resinosos de zonas templadas y tropicales. In: Relacciones Químicas Entre Organismos: Aspectos Básicos y Perspectivas de su Aplicación, pp. 251–304, eds. A. L. Anaya, F. J. Espinosa-García, and R. Cruz-Ortega. Instituto de Ecología, UNAM (Universidad Nacional Autónoma de México) y Plaza y Valdéz, S.A. de C. V. Mexico.

Langenheim, J. H., and C. S. Balser. 1975. Botanical origin of resin objects from aboriginal Costa Rica. Vinculos 1: 72–82.

Langenheim, J. H., and A. Bartlett. 1971. Interpretation of pollen in amber from a study of pollen in present-day coniferous resins. Bulletin of the Torrey Botanical Club 98: 127–140.

Langenheim, J. H., and C. W. Beck. 1965. Infrared spectra as a means of determining botanical source of amber. Science 149: 52–55.

Langenheim, J. H., and C. W. Beck. 1968. Catalogue of infrared spectra of amber. Part I. North and South America. Harvard Botanical Museum Leaflets 22: 65–120.

Langenheim, J. H., and Y. T. Lee. 1974. Reinstatement of the genus *Hymenaea* L. (Leguminosae: Caesalpinioideae) in Africa. Brittonia 26: 3–21.

Langenheim, J. H., and W. H. Stubblebine. 1983. Variation in leaf resin composition between parent tree and progeny in *Hymenaea:* implications for herbivory in the humid tropics. Biochemical Systematics and Ecology 11: 97–106.

Langenheim, J. H., B. Hackner, and A. Bartlett. 1967. Mangrove pollen at the depositional site of Oligo-Miocene amber from Chiapas, Mexico. Harvard Botanical Museum Leaflets 21: 289–324.

Langenheim, J. H., Y. T. Lee, and S. S. Martin. 1973. An evolutionary and ecological perspective of Amazonian Hylaea species of *Hymenaea* (Leguminosae: Caesalpinioideae). Acta Amazônica 3: 5–38.

Langenheim, J. H., W. H. Stubblebine, C. F. Foster, and J. C. Nascimento. 1977. Estudos comparativos da variabilidade na composição de resina da folha a entre árvore parental e progênie de espécies selecionados de *Hymenaea*. I. Comparação de populações Amazônica e Venezuelanas. Acta Amazônica 7: 335–354.

Langenheim, J. H., C. E. Foster, D. E. Lincoln, and W. H. Stubblebine. 1978. Implications of variation in resin composition among organs, tissues and populations in the tropical legume *Hymenaea*. Biochemical Systematics and Ecology 6: 299–313.

Langenheim, J. H., C. E. Foster, and R. B. McGinley. 1980. Inhibitory effects of different quantitative compositions of *Hymenaea* leaf resins on a generalist herbivore *Spodoptera exigua*. Biochemical Systematics and Ecology 8: 385–396.

Langenheim, J. H., S. Arrhenius, and J. C. Nascimento. 1981. Relationship of light intensity to leaf resin composition and yield in the tropical leguminous genera *Hymenaea* and *Copaifera*. Biochemical Systematics and Ecology 9: 27–37.

Langenheim, J. H., D. E. Lincoln, W. H. Stubblebine, and A. C. Gabrielli. 1982. Evolutionary implications of leaf resin pocket patterns in the tropical tree *Hymenaea* (Caesalpinioideae: Leguminosae). American Journal of Botany 69: 595–607.

Langenheim, J. H., C. L. Convis, C. A. Macedo, and W. H. Stubblebine. 1986a. *Hymenaea* and *Copaifera* leaf sesquiterpenes in relation to lepidopteran herbivory in southeastern Brazil. Biochemical Systematics and Ecology 14: 41–49.

Langenheim, J. H., C. A. Macedo, M. K. Ross, and W. H. Stubblebine. 1986b. Leaf development in the tropical legumnous tree *Copaifera* in relation to microlepidopteran herbivory. Biochemical Systematics and Ecology 14: 51–59.

Langenheim, R. L., C. T. Smiley, and J. Gray. 1960. Cretaceous amber from the Arctic Coastal Plain of Alaska. Bulletin of the Geological Society of America 71: 1345–1356.

Langezaal, C. R. 1993. A pharmacognostical study of hop, *Humulus lupulus* L. Pharmacy World and Science 15: 178–179.

Langezaal, C. R., A. Chandra, and J. J. C. Scheffer. 1992. Antimicrobial screening of essential oils and extracts of some *Humulus lupulus* cultivars. Pharmaceutisch Weekblad (Scientific Edition) 14: 353–356.

Lapinjoki, S. P., H. A. Elo, and H. T. Taipale. 1991. Development and structure of resin glands on tissues of *Betula pendula* Roth during growth. New Phytology 117: 210–223.

Larson, J. A. 1980. The Boreal Ecosystem. Academic Press, New York.

Larsson, S. G. 1975. Palaeobiology and mode of burial of the insects of the Lower Eocene Mo-clay of Denmark. Bulletin of the Geological Society of Denmark 24: 193–209.

Larsson, S. G. 1978. Baltic Amber—a Paleobiological Study. Scandinavian Science Press, Klampenburg, Denmark.

Latta, R. G., and V. B. Linhart. 1997. Path analysis of natural selection on plant chemistry: the xylem resin of ponderosa pine. Oecologia 109: 251–258.

Latta, R. G., Y. B. Linhart, L. Lundquist, and M. A. Snyder. 2000. Patterns of monoterpene variation within individual trees in ponderosa pine. Journal of Chemical Ecology 26: 1341–1357.

Lauerman, J. F. 2001. Ma$R_x$ijuana? Harvard Magazine 2001(Mar.): 26–27.

Laufer, B. 1907. Historical Jottings on Amber in Asia. Memoirs of the American Anthropological Association 1. Lancaster, Pennsylvania.

Lavie, P. 1973. Sur l'origine de la propolis. Revue Française d'Apiculture 305: 19–21.

Lawrence, B. M. 1985. A review of the world production of essential oils (1984). Perfumer and Flavorist 10: 1–16.

Leather, S. R. 1987. Pine monoterpenes stimulate oviposition in the pine beauty moth, *Panolis flammea*. Entomología Experimentalis et Applicata 43: 295–297.

Lee, Y. T., and J. H. Langenheim. 1975. Systematics of the genus *Hymenaea* L. (Leguminosae, Caesalpinioideae, Detarieae). University of California Publications in Botany 69.

Leenhouts, P. W. 1955. The genus *Canarium* in the Pacific. B. P. Bishop Museum Bulletin 216.

Leenhouts, P. W. 1959. Burseraceae. Flora Malesiana, Series 1, 5: 209–296.

Le Maitre, D. C. 1998. Pines in cultivation: a global view. In: Ecology and Biogeography of *Pinus,* pp. 407–431, ed. D. M. Richardson. Cambridge University Press, New York.

Léonard, J. 1950. Étude Botanique des Copaliers en Congo Belge. Publications de l'Institut National pour l'Étude Agronomique du Congo Belge, Série Scientifique 45: 1–158.

Léonard, J. 1957. Genera des Cynometreae et des Amherstieae Africaines. Mémoires de l'Academie Royale de Belgique, Classe des Sciences, 30: 1–314.

Léonard, J. 1996. Les délimitations des genres chez les Caesalpinioideae africaines (Detarieae et Amherstieae) (1957–1994). In: The Biodiversity of African Plants, pp. 443–455, eds. L. J. G. van der Maesen, X. M. van der Burgt, and J. M. van Medenbach de Rooy. Kluwer Academic Publishers, Dordrecht, Netherlands.

LePage, B. A., and J. F. Basinger. 1995. Evolutionary history of the genus *Pseudolarix* Gordon (Pinaceae). International Journal of Plant Sciences 156: 910–950.

Lerdau, M., and J. Gershenzon. 1997. Allocation theory and chemical defense. In: Plant Resource Allocation, eds. F. A. Bazzaz and J. Grace. Academic Press, San Diego.

Lerdau, M., M. Litvak, and R. Monson. 1994. Plant chemical defense: monoterpenes and the growth–differentiation balance hypothesis. Trends in Ecology and Evolution 9: 58–61.

Lerdau, M., A. Guenther, and R. Monson. 1997a. Plant production and emission of volatile organic compounds. BioScience 47: 373–383.

Lerdau, M., M. Litvak, P. Palmer, and R. Monson. 1997b. Controls over monoterpene emissions from boreal forest conifers. Tree Physiology 17: 563–567.

Lesesne, C. B. 1992. The postoperative use of wound adhesives. Gum mastic versus benzoin. Journal of Dermatologic Surgery and Oncology 18: 990.

Leung, A. Y., and S. Foster. 1996. Encyclopedia of Common Natural Ingredients Used in Food, Drugs, and Cosmetics, ed. 2. John Wiley and Sons, New York.

Lewinsohn, E., M. Gizen, and R. Croteau. 1991a. Defense mechanisms of conifers. Differences in constitutive and wound induced monoterpene biosynthesis among species. Plant Physiology 96: 44–49.

Lewinsohn, E., M. Gizen, T. J. Savage, and R. Croteau. 1991b. Defense mechanisms of conifers. Relationship of monoterpene cyclase activity to anatomical specialization and oleoresin monoterpene content. Plant Physiology 96: 38–43.

Lewinsohn, T. M. 1991. The geographical distribution of plant latex. Chemoecology 2: 64–68.

Lewis, W. H., and M. P. F. Elvin-Lewis. 1977. Medical Botany: Plants Affecting Man's Health. John Wiley and Sons, New York.

Ley, W. 1951. Dragons in Amber. Viking Press, New York.

Li, H.-l. 1953. A reclassification of *Libocedrus* and Cupressaceae. Journal of the Arnold Arboretum 34: 17–35.

Li, H.-l. 1974. An archeological and historical account of *Cannabis* in China. Economic Botany 20: 437–448.

Lichtenthaler, H. K. 1999. The 1-deoxy-*d*-xylulose 5-phosphate pathway of isoprenoid biosynthesis in plants. Annual Review of Plant Physiology and Plant Molecular Biology 50: 47–65.

Lichtenthaler, H. K., J. Schwender, A. Disch, and M. Rohmer. 1997. Biosynthesis of isoprenoids in higher plant chloroplasts proceeds via a mevalonate-independent pathway. Federation of European Biochemical Societies Letters 400: 271–274.

Linajes, A., V. Rico-Gray, and G. Carrión. 1994. Traditional production system of the root of jalapa, *Ipomoea purga* (Convolulaceae) in central Veracruz, Mexico. Economic Botany 48: 84–89.

Linares, E., and R. A. Bye. 1987. A study of four medicinal plant complexes of Mexico and adjacent U.S.A. Journal of Ethnopharmacology 19: 153–184.

Lincoln, D. E. 1980. Leaf resin flavonoids of *Diplacus aurantiacus*. Biochemical Systematics and Ecology 8: 397–400.

Lincoln, D. E. 1985. Host plant protein and phenolic resin effects on larval growth and survival of a butterfly. Journal of Chemical Ecology 11: 1459–1467.

Lincoln, D. E., and D. Couvet. 1989. The effect of carbon supply on allocation to allelochemicals and caterpillar consumption of peppermint. Oecologia 78: 112–114.

Lincoln, D. E., and M. D. Walla. 1986. New flavonoids from *Diplacus aurantiacus* leaf resin. Biochemical Systematics and Ecology 14: 1459–1467.

Lincoln, D. E., T. W. Newton, P. R. Ehrlich, and K. W. Williams. 1982. Coevolution of the checkerspot butterfly *Euphydryas chalcedona* and its larval food plant *Diplacus aurantiacus*: larval response to protein and leaf resin. Oecologia 52: 216–223.

Lindenfelser, L. A. 1967. Antimicrobial activity of propolis. American Bee Journal 107: 90–92, 130–131.

Linhart, Y. B. 1988. Ecological and evolutionary studies of ponderosa pine in the Rocky Mountains. In: Ponderosa Pine: the Species and Its Management, pp. 77–89, eds. D. M. Baumgartner and J. E. Lotan. Washington State University Cooperative Extension, Pullman.

Linhart, Y. B. 1989. Interactions between genetic and ecologic patchiness in forest trees and their dependent species. In: The Evolutionary Ecology of Plants, pp. 394–430, eds. J. H. Bock and Y. B. Linhart. Westview Press, Boulder, Colorado.

Linhart, Y. B. 1991. Disease, parasitism and herbivory: multidimensional challenges in plant evolution. Trends in Ecology and Evolution 6: 392–396.

Linhart, Y. B., and M. C. Grant. 1996. Evolutionary significance of local genetic differentiation in plants. Annual Review of Ecology and Systematics 27: 237–277.

Linhart, Y. B., M. A. Snyder, and S. A. Hubeck. 1989. The influence of animals on genetic variability within ponderosa pine studies, illustrated by the effects of Abert's squirrel and porcupine. In: Multiresource Management of Ponderosa Pine Forests, pp. 141–148, eds. A. Tede et al. U.S. Department of Agriculture Forest Service General Technical Report RM-184.

Litvak, M. E., and R. K. Monson. 1998. Patterns of induced and constitutive monoterpene production in conifer needles in relation to insect herbivory. Oecologia 114: 531–540.

Litvak, M. E., S. Madronich, and R. K. Monson. 1999. Herbivore-induced monoterpene emissions from coniferous forests: potential impact on local tropospheric chemistry. Ecological Applications 9: 1147–1159.

Litwin, R. J., and S. R. Ash. 1991. First early Mesozoic amber in the western hemisphere. Geology 19: 273–276.

Lokvam, J., and J. F. Braddock. 1999. Anti-bacterial function in sexually dimorphic pollinator rewards of *Clusia grandiflora* (Clusiaceae). Oecologia 119: 534–540.

Lokvam, J., J. F. Braddock, P. B. Reichardt, and T. P. Clausen. 2000. Two polyisoprenylated benzophenones from the trunk latex of *Clusia grandiflora* (Clusiaceae). Phytochemistry 55: 29–34.

Loomis, W. D., and R. Croteau. 1973. Biochemistry and physiology of lower terpenoids. In: Terpenoids, Structure, Biogenesis and Distribution, pp. 147–185, eds. V. C. Runeckles and T. J. Mabry. Recent Advances in Phytochemistry 6.

Lorio, P. L., Jr. 1986. Growth–differentiation balance: a basis for understanding southern pine beetle–tree interactions. Forest Ecology Management 14: 259–273.

Lorio, P. L., Jr. 1988. Growth and differentiation balance relationships in pines affect their resistance to bark beetles (Coleoptera: Scolytidae). In: Mechanisms of Woody Plant Defenses Against Insects, pp. 73–92, eds. W. J. Mattson, J. Levieux, C. Bernard-Dagan. Springer-Verlag, New York.

Lorio, P. L., Jr., and J. D. Hodges. 1968. Microsite effects on oleoresin exudation pressure of large loblolly pines. Ecology 49: 1207–1210.

Lorio, P. L., Jr., and R. A. Sommers. 1986. Evidence for competition for photosynthesis between growth processes and oleoresin synthesis in *Pinus taeda* L. Tree Physiology 2: 301–306.

Low, C. K., and M. A. Abdul Razak. 1985. Experimental tapping of pine oleoresin. Malaysian Forester 48: 248–253.

Lucas, A., and J. R. Harris. 1962. Ancient Egyptian Materials and Industries, ed. 4. Edward Arnold, London.

Lyons, P. C., R. B. Finkelman, C. L. Thompson, F. W. Brown, and P. G. Hatcher. 1982. Properties, origin and nomenclature of rodlets of the inertinite maceral group in coals of the central Appalachian Basin, U.S.A. International Journal of Coal Geology 1: 313–346.

Lyons, P. C., P. G. Hatcher, J. A. Minkin, C. L. Thompson, R. R. Larson, Z. H. Brown, and R. W. Pheifer. 1984. Resin rodlets in shale and coal (Lower Cretaceous), Baltimore Canyon Trough. International Journal of Coal Geology 3: 257–278.

Mabberley, D. J. 1997. The Plant-book, ed. 2. University of Cambridge Press, Cambridge.

Mabry, T. J., D. Difeo, Jr., M. Sakakibara, C. F. Bonsted, Jr., and D. Siegler. 1977. The natural products chemistry of *Larrea*. In: The Biology and Chemistry of the Creosote Bush *(Larrea)* in the New World Deserts, pp. 115–134, eds. T. J. Mabry, J. Hunziker, and D. R. DiFeo, Jr. U.S. I.B.P. Synthesis Series 6. Dowden, Hutchinson and Ross, Stroudsburg, Pennsylvania.

McAlpine, J. F., and J. E. H. Martin. 1969. Canadian amber—a paleontological treasure chest. Canadian Entomologist 101: 819–838.

McCaskill, D., and R. Croteau. 1998. Some caveats for bioengineering terpenoid metabolism in plants. Trends in Biotechnology 16: 349–355.

McCaskill, D., and R. Croteau. 1999. Strategies for bioengineering the development and metabolism of glandular tissues in plants. Nature Biotechnology 17: 31–36.

Macedo, C. A., and J. H. Langenheim. 1989a. A further investigation of leaf sesquiterpene variation in relation to herbivory in two Brazilian populations of *Copaifera langsdorfii*. Biochemical Systematics and Ecology 17: 207–216.

Macedo, C. A., and J. H. Langenheim. 1989b. Microlepidopteran herbivory in relation to leaf sesquiterpenes in *Copaifera langsdorfii* adult trees and their seedling progeny in a Brazilian woodland. Biochemical Systematics and Ecology 17: 217–224.

Macedo, C. A., and J. H. Langenheim. 1989c. Intra- and interplant leaf sesquiterpene variability in *Copaifera langsdorfii*: relation to microlepidopteran herbivory. Biochemical Systematics and Ecology 17: 551–557.

McConkey, M. E., J. Gershenzon, and R. Croteau. 2000. Developmental regulation of monoterpene biosynthesis in the glandular trichomes of peppermint. Plant Physiology 122: 215–223.

McCully, M. E., M. W. Shane, A. N. Baker, C. X. Huang, L. E. C. Ling, and M. J. Cunny. 2000. The reliability of cyroSEM for observation and quantification of xylem embolisms and quantitative analysis of xylem sap in situ. Journal of Microscopy 199: 24–33.

McDonald, J. A. 1989. Neotypification of *Ipomoea jalapa* (Convolulaceae). Taxon 38: 135–138.

McDoniel, P. B., and J. R. Cole. 1972. Antitumor activity of *Bursera schlechtendalli* (Burseraceae): isolation and structure determination of two new lignans. Journal of Pharmaceutical Science 61: 1992–1994.

McGarvey, D. J., and R. Croteau. 1995. Terpenoid metabolism. Plant Cell 7: 1015–1026.

McGovern, P. E., D. L. Glusker, L. J. Exner, and M. M. Volgt. 1996. Neolithic resinated wine. Nature 381: 480–481.

Macht, D. I., and H. F. Bryan. 1935. A contribution to the pharmacology of myrrh, krameria and eriodictyon. American Journal of Pharmacy 107: 500–511.

McLaughlin, S. P., and J. J. Hoffmann. 1982. A survey of biocrude-producing plants from the Southwest. Economic Botany 36: 323–339.

McLaughlin, S. P., B. E. Kingsolver, and J. J. Hoffmann. 1983. Biocrude production in arid lands. Economic Botany 37: 150–158.

McNair, J. B. 1916. The poisonous principle of poison oak. Journal of the American Chemical Society 38: 1417–1421.

McNair, J. B. 1930. Gum, tannin, and resin in relation to specificity, environment, and function. American Journal of Botany 17: 187–196.

McNeill, J. 1991. Northland's buried treasure. New Zealand Geographic 10: 17–45.

McReynolds, R. D., S. V. Kossuth, and R. W. Clements. 1989. Gum naval stores methodology. In: Naval Stores: Production-Chemistry-Utilization, pp. 83–122, eds. D. E. Zinkel and J. Russell. Pulp Chemicals Association, New York.

Maeda, E. 1977. Scanning electron microscope studies on lupulin glands in *Humulus lupulus* L. Japanese Journal of Crop Science 46: 249–253.

Mahajan, B., S. C. Taneja, V. K. Sethi, and K. L. Dhar. 1995. Two triterpenoids from *Boswellia serrata* gum resin. Phytochemistry 39: 453–455.

Mahlberg, P. G., C. T. Hammond, J. C. Turner, and J. K. Hemphill. 1984. Structure, development and composition of glandular trichomes of *Cannabis sativa* L. In: Biology and Chemistry of Plant Trichomes, pp. 23–56, eds. E. Rodriguez, P. L. Healey, and I. Mehta. Plenum Books, New York.

Maiden, J. H. 1989. The Useful Native Plants of Australia. Trubner, London.

Majima, R. 1922. Über der Hauptbestandteil des Japan-Lacks VIII. Mitteilung: Stellung der Doppelbindungen in der Seitenkette des Urushiols und Beweisführung, dass das Urushiol eine Mischung ist. Berichte der Deutschen Chemischen Gesellschaft 95: 172–191.

Majno, G. 1975. The Healing Hand. Man and Wound in the Ancient World. Harvard University Press, Cambridge.

Malone, J. J. 1964. Trees and Politics: the Naval Stores and Forest Policy in Colonial New England, 1691–1775. University of Washington Press, Seattle.

Manguro, L. O. M., K. M. Mukonyi, and J. K. Githiomi. 1996. Bisabolenes and furanosesquiterpenes of Kenyan *Commiphora kua* resin. Planta Medica 62: 84–85.

Manheim, B. S., and T. W. Mulroy. 1978. Triterpenoids in epicuticular waxes of *Dudleya* species. Phytochemistry 17: 1799–1800.

Mann, C. C., and M. L. Plummer. 2002. Forest biotech edges out of the lab. Science 295: 1626–1629.

Mantell, C. L. 1950. The natural hard resins—their botany, sources and utilization. Economic Botany 4: 203–242.

Mantell, C. L., C. W. Kopf, J. L. Curtis, and E. M. Rogers. 1942. The Technology of Natural Resins. John Wiley and Sons, New York.

Mapes, G., and G. W. Rothwell. 1991. Structure and relationships among primitive conifers. Neues Jahrbuch für Geologie und Paläontologie—Abhandlungen 183: 269–287.

Maplestone, R. A., M. J. Stone, and D. H. Williams. 1992. The evolutionary role of secondary metabolites—a review. Gene 115: 151–157.

Maradufu, A. 1982. Furanosesquiterpenoids of *Commiphora erythraea* and *C. myrrha*. Phytochemistry 21: 677–680.

Marcucci, M. C. 1995. Propolis: chemical composition, biological properties and therapeutic activity. Apidologie 26: 83–89.

Marner, F.-J., A. Freyer, and J. Lex. 1991. Triterpenoids from gum mastic, the resin of *Pistacia lentiscus*. Phytochemistry 30: 3709–3712.

Marpeau, A., P. Baradat, and C. Bernard-Dagan. 1975. Les terpènes du pin maritime: aspects biologiques et genetiques. IV. Hérédité de teneur en deux sesquiterpènes: le longifolène et le caryophyllène. Annales des Sciences Forestières 32: 185–203.

Martin, J. T., and B. E. Juniper. 1970. The Cuticles of Plants. St. Martins Press, New York.

Martin, S. S., and J. H. Langenheim. 1974. Enantio-8(17),13(16),14-labdatrien-18-oic acid from trunk resin of Kenyan *Hymenaea verrucosa*. Phytochemistry 13: 523–525.

Martin, S. S., J. H. Langenheim, and E. Zavarin. 1974. Quantitative variation in leaf pocket composition in *Hymenaea courbaril*. Biochemical Systematics and Ecology 3: 760–787.

Martin, S. S., J. H. Langenheim, and E. Zavarin. 1976. Quantitative leaf resin composition in *Hymenaea* (Leguminosae). Biochemical Systematics and Ecology 4: 181–191.

Martínez, M. 1959. Plantas Utiles de la Flora Mexicana. Ediciones Botas, México.

Martínez, M. 1969. Las Plantas Medicinales de México. Ediciones Botas, México.

Martos, I., M. Cossentini, F. Ferreres, and F. A. Tomás-Barberán. 1997. Flavonoid composition of Tunisian honey and propolis. Journal of Agriculture and Food Chemistry 54: 2824–2829.

Mathiou, J. P., and G. Ourisson. 1958. Triterpenoids. Pergamon Press, London.

Mattson, W. J., S. S. Slocum, and C. N. Koller. 1983. Spruce budworm (*Choristoneura fumiferana*) performance in relation to foliar chemistry of its host plants. In: Forest Defoliator–Host Interactions: a Comparison Between Gypsy Moth and Spruce Budworms, pp. 55–65. U.S. Department of Agriculture Forest Service General Technical Report NE-85.

Mauseth, J. D. 1988. Plant Anatomy. Benjamin-Cummings, Menlo Park, California.

Mechoulam, R. 1973. Marijuana. Academic Press, New York.

Mechoulam, R. (ed.). 1986a. Cannabinoids as Therapeutic Agents. CRC Press, Boca Raton.

Mechoulam, R. 1986b. The pharmacohistory of *Cannabis sativa*. In: Cannabinoid as Therapeutic Agents, pp. 1–19, ed. R. Mechoulam. CRC Press, Boca Raton.

Medina, J. D., and V. de Santis. 1981. Constituents of the trunk resin of *Eperua purpurea*. Planta Medica 43: 202–206.

Meiggs, R. 1982. Trees and Timber in the Ancient Mediterranean World. Clarendon Press, Oxford.

Meijer, W. 1974. *Podophyllum peltatum*—mayapple, a potential new cash-crop of eastern North America. Economic Botany 28: 69–72.

Meinzer, F. L., C. S. Wisdom, A. González-Coloma, P. W. Rundel, and L. M. Shultz. 1990. Effects of leaf resin on stomatal behaviour and gas exchange of *Larrea tridentata* (DC.) Cov. Functional Ecology 4: 579–584.

Melchior, H., and E. Werdermann. 1954–64. Bakterien bis Gymnospermen. In: Syllabus der Pflanzenfamilien, Band 1, ed. 12, pp. 1–367, ed. A. Engler. Gebrüder Borntraeger, Berlin.

Mesquita, R. de C. G., and C. H. Franciscon. 1995. Flower visitors of *Clusia nemarosa* G. F. W. Meyer (Clusiaceae) in Amazonian white-sand campina. Biotropica 27: 254–257.

Messer, A. C. 1984. *Chalicodoma pluto*: the world's largest bee rediscovered living communally in termite nests (Hymenoptera: Megachilidae). Journal of the Kansas Entomological Society 57: 165–168.

Messer, A. C. 1985. Fresh dipterocarp resins gathered by megachilid bees inhibit growth of pollen-associated fungi. Biotropica 17: 175–176.

Messer, A. C. 1990. Traditional and chemical techniques for stimulation of *Shorea javanica* (Dipterocarpaceae) resin exudation in Sumatra. Economic Botany 44: 463–469.

Metcalfe, C. R. 1967. Distribution of latex in the plant kingdom. Economic Botany 21: 115–125.

Metcalfe, C. R., and L. Chalk. 1983. Anatomy of the Dicotyledons, Vol. 2. Clarendon Press, Oxford.

Meuzelaar, H. L. C., H. Huaying, R. Lo, and J. P. Dworzanski. 1991. Chemical composition and origin of fossil resins from Utah Wasatch Plateau coal. Journal of Fuels Processing Technology 28: 119–134.

Meyer, C., J. M. Todd, and C. W. Beck. 1991. From Zanzibar to Zabros: a copal pendant from Eshnunna. Journal of Near Eastern Studies 50: 289–298.

Meyer, G. A., and M. E. Montgomery. 1987. Relationships between leaf age and the food quality of cottonwood foliage for the gypsy moth, *Lymantria dispar*. Oecologia 72: 527–532.

Meyer, M. W., and W. H. Karazov. 1989. Antiherbivore chemistry of *Larrea tridentata*: effects on woodrat *(Neotoma lepida)* feeding and nutrition. Ecology 70: 953–961

Meyer, M. W., and W. H. Karazov. 1991. Chemical aspects of herbivory in arid and semi-arid habitats. In: Plant Defenses Against Mammalian Herbivory, pp. 167–187, eds. R. T. Palo and C. T. Robbins. CRC Press, Boca Raton.

Michener, C. D. 1974. The Social Behavior of Bees. Belknap Press, Cambridge, Massachusetts.

Michener, C. D. 2000. The Bees of the World. Johns Hopkins University Press, Baltimore.

Mihaliak, C. A., and D. E. Lincoln. 1985. Growth pattern and carbon allocation to volatile leaf terpenes under nitrogen limiting conditions in *Heterotheca subaxillaris*, Asteraceae. Oecologia 66: 423–426.

Mihaliak, C. A., J. Gershenzon, and R. Croteau. 1991. Lack of rapid monoterpene turnover in rooted plants: implications for theories of plant chemical defense. Oecologia 87: 373–376.

Milburn, M. 1984. Dragon's blood in East and West Africa, Arabia and the Canary Islands. Africa 39: 486–493.

Millar, C. I. 1993. Impact of the Eocene on the evolution of *Pinus* L. Annals of the Missouri Botanical Garden 80: 471–498.

Millar, C. I. 1998. Early evolution of pines. In: Ecology and Biogeography of *Pinus*, pp. 69–91, ed. D. M. Richardson. Cambridge University Press, New York.

Miller, C. N. 1976. Early evolution in the Pinaceae. Review of Paleobotany and Palynology 21: 101–117.

Miller, C. N. 1977. Mesozoic conifers. Botanical Review 43: 217–280.

Miller, C. N. 1988. The origin of modern conifer families. In: Origin and Evolution of Gymnosperms, pp. 448–487, ed. C. B. Beck. Columbia University Press, New York.

Miller, C. N. 1999. Implications of fossil conifers for the phylogenetic relationships of living families. Botanical Review 65: 239–277.

Miller, J. I. 1969. The Spice Trade of the Roman Empire: 29 B.C. to A.D. 641. Clarendon Press, Oxford.

Mills, J. S. 1973. Identity of daniellic acid with illurinic acid. Phytochemistry 12: 2479–2480.

Mills, J. S., and A. E. A. Werner. 1955. The chemistry of dammar resin. Journal of the Chemical Society 1955: 3132–3140.

Mills, J. S., and R. White. 1977. Natural resins of art and archaeology: their sources, chemistry and identification. Studies in Conservation 22: 12–31.

Mills, J. S., and R. White. 1989. The identity of resins from the late Bronze Age shipwreck at Ulu Burun (Kas). Archeometry 31: 37–44.

Mills, J. S., and R. White. 1994. Natural resins and lacquers. In: The Organic Chemistry of Museum Objects, ed. 2, pp. 95–128. Butterworth Heinemann, London.

Mills, J. S., R. White, and L. Gough. 1984. The chemical composition of Baltic amber. Chemical Geology 47: 15–39.

Miranda, F. 1963. Two plants from the amber of the Simajovel, Chiapas, Mexico area. Journal of Paleontology 37: 611–614.

Mirov, N. T. 1967. The Genus *Pinus*. Ronald Press, New York.

Mirov, N. T., and J. Hasbrouck. 1976. The Story of Pines. Indiana University Press, Bloomington.

Misra, R., R. C. Pandey, and Sukh Dev. 1964. The chemistry of the oleoresin from *Hardwickia pinnata*: a series of new diterpenoids. Tetrahedron Letters 49: 3751–3759.

Mitchell-Olds, T., J. Gershenzon, I. T. Baldwin, and W. Boland. 1998. Chemical ecology in the molecular era. Trends in Plant Science 3: 362–365.

Mitton, J. B. 1997. Selection in Natural Populations. Oxford University Press, Oxford.

Moens, P. 1955. Les formations sécrétrices des copaliers congolais. Cellule 57: 35–79.

Moerman, D. E. 1998. Native American Ethnobotany. Timber Press, Portland, Oregon.

Moffat, A. S. 1996. Plant biotechnology: moving forest trees into the modern genetics era. Science 271: 760–761.

Moldenke, H. N., and A. L. Moldenke 1952. Plants of the Bible. Ronald Press, New York.

Momberg, F., R. K. Puri, and T. Jessup. 2000. Exploitation of *gaharu* and conservation efforts in Kayan Mintaraug National Park, East Kalimantan, Indonesia. In: People, Plants and Justice, pp. 259–284, ed. C. Zerner. Columbia University Press, New York.

Mooney, H. A., and W. A. Emboden. 1968. The relationship of terpene composition, morphology, and distribution of populations of *Bursera microphylla* (Burseraceae). Brittonia 20: 44–51.

Mooney, H. A., P. R. Ehrlich, D. E. Lincoln, and K. S. Williams. 1980. Environmental controls on the seasonality of a drought deciduous shrub, *Diplacus aurantiacus,* and its predator, the checkerspot butterfly, *Euphydryas chalcedona.* Oecologia 45: 143–146.

Moraes, R. M., C. Burnandt, Jr., M. Ganzera, X. Li, I. Khan, and C. Chanel. 2000. The American mayapple revisited—*Podophyllum peltatum*—Still a potential cash crop? Economic Botany 54: 471–476.

Mora-Osejo, L. E. 1977. El barniz de Pasto. Caldesia 11: 5–32.

Mors, W. B., and C. I. Rizzini. 1966. Useful Plants of Brazil. Holden Day, San Francisco.

Morton, J. F. 1977. Major Medicinal Plants: Botany, Culture, and Uses. Charles C. Thomas, Springfield, Illinois.

Morton, J. F. 1981. Atlas of Medicinal Plants of Middle America. Charles C. Thomas, Springfield, Illinois.

Mosini, V., and R. Sampri. 1985. Correlation between Baltic amber and *Pinus* resins. Phytochemistry 24: 859–861.

Mossa, A. J., M. A. al-Yahya, I. A. al-Mehal, and M. Tariq. 1983. Phytochemical and biological screening of Saudi medicinal plants. Fitoterapia 54: 147–152.

Müller, W. W. 1976. Notes on the use of frankincense in South Arabia. Proceedings of the 9th Seminar, London (1975) 6: 124–136. Seminar for Arabian Studies, London.

Muller, J. 1970. Palynological evidence on early differentiation of angiosperms. Biological Review 45: 417–450.

Muller, J. 1981. Fossil pollen records of extant angiosperms. Botanical Review 47: 1–142.

Murray, A. P., D. Padley, D. M. McKirdy, W. E. Booth, and R. E. Summons. 1994. Oceanic transport of fossil dammar resin: the chemistry of coastal resinites from South Australia. Geochimica et Cosmochimica Acta 58: 3049–3059.

Murray, A. P., D. Edwards, J. M. Hope, C. J. Boreham, W. E. Booth, R. A. Alexander, and R. E. Summons. 1998. Carbon isotope biogeochemistry of plant resins and derived hydrocarbons. Organic Biochemistry 29: 1199–1214.

Mustoe, G. E. 1985. Eocene amber from the Pacific Coast of North America. Geological Society of America Bulletin 96: 1530–1536.

Mutton, D. B. 1962. Wood resins. In: Wood Extractives, pp. 331–363, ed. W. E. Hillis. Academic Press, London.

Nagy, N. E., V. R. Franceschi, H. Solheim, T. Krekling, and E. Christiansen. 2000. Wound-induced traumatic resin duct development in stems of Norway spruce (Pinaceae): anatomy and cytochemical traits. American Journal of Botany 87: 302–313.

Nascimento, J. C., and J. H. Langenheim. 1986. Leaf sesquiterpenes and phenolics in *Copaifera multijuga* (Leguminosae) on contrasting soil types in a central Amazonian rain forest. Biochemical Systematics and Ecology 14: 615–624.

National Research Council. 1976. Renewable resources for industrial materials. Report of Committee on Renewable Resources for Industrial Materials. Washington, D.C.

Nebeker, T. E., R. F. Schmitz, and R. A. Tisdale. 1995. Comparison of oleoresin flow in relation to wound size, growth rates, and disease status of lodgepole pine. Canadian Journal of Botany 73: 370–375.

Neels, S. 2000. Yield, sustainable harvest and cultural uses of resin from the copal tree (*Protium copal*: Burseraceae) in the Carmelita Community Forest Concession, Petén, Guatemala. M.S. thesis, University of British Columbia, Vancouver.

Neish, A. C. 1965. Coumarins, phenylpropanes and lignin. In: Plant Biochemistry, pp. 581–617, eds. J. Bonner and J. E. Varner. Academic Press, New York.

Nelson, E. C., and D. J. Bedford. 1993. The names of the Australian grass-tree: *Xanthorrhoea* Sm. and *Acoroides* C. Kite (Xanthorrhoeaceae). Botanical Journal of the Linnean Society 112: 95–105.

Neve, R. A. 1991. Hops. Chapman and Hall, London.

New Catholic Encyclopedia. 1967. Vols. 14–15. McGraw Hill, New York.

Newman, J. D., and J. Chappell. 1999. Isoprenoid biosynthesis in plants: carbon partitioning within the cytoplasmic pathway. Critical Reviews in Biochemistry and Molecular Biology 34: 95–106.

Nikolov, N., and H. Helmisaari. 1992. Silvics of the circumpolar boreal forest trees. In: A Systems Analysis of the Global Boreal Forest, pp. 13–48, eds. H. Shugart, R. Leemans, and G. Bonan. Cambridge University Press, New York.

Nissenbaum, A., and A. Horowitz. 1992. The Levantine amber belt. Journal of African Earth Science 14: 295–300.

Noda, N., M. Ono, K. Miyahara, I. Kawasaki, and M. Okabe. 1987. Resin glycosides I. Isolation and structural elucidation of orizabin—I, II, III and IV, genuine resin glycosides from the root of *Ipomoea orizabensis*. Tetrahedron 43: 3889–3902.

Obermann, H. 1977. Differences in chemistry and odor of incense resins. Dragoco Report 24: 260–265.

Oeirichs, P. B., J. K. Macleod, A. A. Seawright, and J. C. Ng. 1977. Isolation and characterization of urushiol components from the Australian native cashew *(Semecarpus australiensis)*. Natural Toxins 5: 96–98.

Ohshaki, A., L. T. Yan, I. Shigeru, M. Edatsugi, D. Iwata, and Y. Komoda. 1994. The isolation and in vivo potent antitumor activity of clerodane diterpenoid from the oleoresin of the Brazilian medicinal plant *Copaifera langsdorfii* Desf. Bioorganic and Medicinal Chemistry Letters 4: 2889–2892.

O'Neil, P. E., A. Smith, and P. E. Heckelman (eds.). 2001. Merck Index: an Encyclopedia of Chemicals, Drugs and Biologicals, ed. 13. Merck, Whitehouse Station, New Jersey.

Orallo, C. A., and V. P. Veracion. 1984. Comparison of four methods of tapping Benguet pine (*Pinus kesiya* Royle ex Gordon) for oleoresin production in Benguet. Sylvatrop Forest Research Journal 9: 55–64.

Oriel, J. D. 1998. Gonorrhoea and the balsams. Sexually Transmitted Infections 74: 127–127.

Otto, A., and V. Wilde. 2001. Sesqui-, di-, and triterpenoids as chemosystematic markers in recent conifers. Botanical Review 67: 141–238.

Otto, A., B. R. T. Simoneit, V. Wilde, L. Kunzmann, and W. Püttman. 2002a. Terpenoid composition of three fossil resins from Cretaceous and Tertiary conifers. Review of Palaeobotany 120: 203–215.

Otto, A., J. D. White, and B. R. T. Simoneit. 2002b. Natural product terpenoids in Eocene and Miocene conifer fossils. Science 297: 1543–1545.

Paclt, J. 1953. On a new subfossil liptobiolite from the Plain of Sharon in Israel. Israel Exploration Journal 3: 242–245.

Page, C. W. 1990. Coniferophytina. In: The Families and Genera of Vascular Plants, ed. K. Kubtizki, Vol. 1, Pteridophytes and Gymnosperms, pp. 282–361, eds. K. U. Kramer and P. S. Green. Springer-Verlag, Berlin.

Paine, T. D., F. M. Stephen, and R. G. Cates. 1993. Within- and among-tree variation of the induced response of loblolly pine to fungus associated with *Dendroctonus frontalis* Zimmerman (Coleoptera: Scolytidae) and sterile wounding. Canadian Entomology 125: 65–71.

Paiva, L. A. F., V. S. N. Rao, N. V. Gremosa, and E. R. Silveira. 1998. Gastroprotective effect of *Copaifera langsdorfii* oleoresin on experimental gastric ulcer models in rats. Journal of Ethnopharmacology 62: 73–78.

Palo, R. T. 1984. Distribution of birch (*Betula* spp.), willow (*Salix* spp.) and poplar (*Populus* spp.) secondary metabolites and their potential role as chemical defense against herbivores. Journal of Chemical Ecology 10: 499–520.

Papageorgiou, V. P., M. N. Bakola-Christianopoulou, K. K. Apazidou, and E. E. Psarros.

1996. Gas chromatographic–mass spectroscopic analysis of the acidic triterpenic fraction of mastic gum. Journal of Chromatography, A, 769: 263–273.

Pardhy, R. S., and S. C. Bhattacharyya. 1978. Tetracyclic triterpene acids from the resin of *Boswellia serrata* Roxb. Indian Journal of Chemistry 16B: 174–175.

Parker, L. R. 1976. The paleoecology of the fluvial coal-forming swamps and associated floodplain environments in the Blackhawk formation (Upper Cretaceous) of central Utah. Brigham Young University Geology Series 22: 99–116.

Parry, E. J. 1920. Gums and Resins: Their Occurrence, Properties and Uses. Sir Isaac Pitman and Sons, London.

Parsons, I. C., A. I. Gray, C. Lavaud, G. Massiot, and P. G. Waterman. 1991. Seco ring-A triterpene acids from the resin of *Dacryodes normandii*. Phytochemistry 30: 1221–1223.

Pascual, C., R. González, and R. G. Torricella. 1994. Scavenging action of propolis extract against oxygen radicals. Journal of Ethnopharmacology 41: 9–13.

Paseshnichenko, V. A. 1995. Regulation of terpenoid biosynthesis in plants and its relation to the biosynthesis of phenolic compounds. Russian Journal of Plant Physiology 42: 699–714.

Pate, D. W. 1994. Chemical ecology of *Cannabis*. Journal of the International Hemp Association 1: 29–36.

Pearson, J., and H. D. V. Prendergast. 2002. *Daemonorops, Dracaena* and other dragon's blood. Economic Botany 55: 474–477.

Peireia, H. 1929. Pequena contribuição para um dicionario das plantas uteis do estado de São Paulo. Typographia Brasil de Rothschild and Co., São Paulo.

Peltzer, K. J. 1978. Swidden cultivation in Southeast Asia: historical, ecological and economic perspectives. In: Farmers in the Forest, pp. 273–286, eds. P. Kundstader, E. C. Chapman, and S. Subhasri. University Press of Hawaii, Honolulu.

Penhallow, D. P. 1907. A Manual of the North American Gymnosperms. Athenaeum Press, Boston.

Penn, R. G. 1994. Medicine on Ancient Greek and Roman Coins. Seaby, B. T. Batsford, London.

Pennacchio, M., Y. M. Syah, E. L. Ghisalberti, and E. Alexander. 1996. Cardioactive compounds from *Eremophila* species. Journal of Ethnopharmacology 53: 21–27.

Penny, J. S. 1947. Studies of the conifers of the Magothy flora. American Journal of Botany 34: 281–295.

Peraza-Sánchez, S. R., and L. M. Peña-Rodríguez. 1992. Isolation of picropolygamain from the resin of *Bursera simaruba*. Journal of Natural Products 55: 1768–1771.

Peraza-Sánchez, S. R., N. E. Salazar-Aquilar, and L. M. Peña-García. 1995. A new triterpene from the resin of *Bursera simaruba*. Journal of Natural Products 58: 271–274.

Pereda-Miranda, R., R. Mata, A. L. Anaya, D. B. Wickramaratne, J. M. Pezzuto, and A. D. Kinghorn. 1993. Tricolorin A, major phytogrowth inhibitor from *Ipomoea tricolor*. Journal of Natural Products 56: 571–582.

Pernet, R. 1972. Phytochemie des Burseracées. Lloydia 35: 280–287.

Perry, J. P., Jr. 1991. The Pines of Mexico and Central America. Timber Press, Portland, Oregon.

Perry, L. M. 1980. Medicinal plants of East and Southeast Asia: Attributed Properties and Uses. MIT Press, Cambridge, Massachusetts.

Peters, C. M. 1994. Sustainable harvest of non-timber forest plant resources in tropical moist forest: an ecological primer. Biodiversity Support Program of the World Wildlife Fund, the Nature Conservancy, and the World Resources Institute, Washington, D.C.

Peterson, A. A., and A. T. Peterson. 1992. Aztec exploitation of cloud forests: tributes of *Liquidambar* resin and quetzal feathers. Global Ecology and Biogeography Letters 2: 165–173.

Peterson, J. K., and H. F. Harrison, Jr. 1991a. Differential inhibition of seed germination by sweet potato *(Ipomoea batatas)* root periderm extracts. Weed Science 39: 119–123.

Peterson, J. K., and H. F. Harrison, Jr. 1991b. Isolation of a substance from sweet potato *(Ipomoea batatas)* periderm tissue that inhibits seed germination. Journal of Chemical Ecology 17: 943–957.

Peterson, J. K., H. F. Harrison, Jr., and A. E. Muckenfuss. 1998. Sweet potato [*Ipomoea batatas* (L.) Lam.] resin glycosides: evidence of antibiosis effects in the diamondback moth *Plutella xylostella* L. (Lepidoptera: Plutellidae). Allelopathy Journal 5: 43–52.

Phillips. M. A., and R. B. Croteau. 1999. Resin-based defenses in conifers. Trends in Plant Science 4: 184–190.

Pieters, L., T. de Bruyne, G. Mei, G. Lemèire, D. van den Berghe, and A. J. Vlietinck. 1992. In vitro and in vivo biological activity of South American dragon's blood and its constituents. Planta Medica 58: 582–583.

Pimenov, M. G., and M. V. Leonov. 1993. The Genera of the Umbelliferae. Royal Botanic Gardens, Kew.

Pimentel, D., and A. C. Bellotti. 1976. Parasitic–host population systems and genetic stability. American Naturalist 110: 877–888.

Pindell, J. H., and S. F. Barrett. 1990. Geologic evolution of the Caribbean region: a plate-tectonic perspective. In: The Geology of North America, Vol. H, The Caribbean Region, pp. 405–432, eds. G. Dengo and J. E. Case. Geological Society of America, Boulder.

Pinyopusarerk, K. 1994. *Styrax tonkinensis:* taxonomy, ecology, silviculture and uses. Australian Centre for International Agricultural Research Technical Report 31. Canberra.

Pio Corrêa, M. 1931. Dictionário das Plantas Úteis do Brasil. Ministério da Agricultura, Rio de Janeiro.

Pio Corrêa, M. 1984. Dicionário das Plantas Úteis para Estratégias de Preservaçao e Desenvolvimento da Amazônia: Fatos e Perspectivas, Vol. 1, pp. 215–220. INPA (Instituto Nacional de Pesquisas da Amazônia), Manaus, Brazil.

Piozzi, F., S. Passannanti, and M. P. Paternostro. 1974. Diterpenoid resin acids of *Daemonorops draco*. Phytochemistry 13: 2231–2233.

Pliny (the Elder). Natural History, transl. H. Rackham, W. H. S. Jones, and D. E. Eichholz, 1967. Loeb Classical Library, Harvard University Press.

Plocek, T. 1998. Turpentine: a global perspective. Perfumer and Flavorist 23: 1–5.

Plowden, C. 2001. The ecology, management, and marketing of non-timber forest products in the Alto Rio Guamá Reserve (eastern Brazilian Amazon). Ph.D. dissertation, Pennsylvania State University, State College.

Plowden, C., C. Uhl, and F. de A. Oliveira. 2002. Breu resin harvest by Tembé Indians and its dependence on a bark-boring beetle. In: Ethnobiology and Biocultural Diversity, pp. 365–380, eds. J. R. Stepp, F. S. Wyndham, and R. K. Zarger. University of Georgia Press, Athens.

Poinar, G. O., Jr. 1991. *Hymenaea protera* sp. n. (Leguminosae, Caesalpinioideae) from Dominican amber has African affinities. Experientia 47: 1075–1082.

Poinar, G. O., Jr. 1992a. Life in Amber. Stanford University Press, Palo Alto.

Poinar, G. O., Jr. 1992b. Fossil evidence of resin utilization by insects. Biotropica 24: 466–468.

Poinar, G. O., Jr. 1999. Cenozoic fauna and flora in amber. [In: Proceedings of the World Congress on Amber Inclusions, eds. J. Alonso, J. C. Corral, and R. López.] Estudios del Museo de Ciencias Naturales de Alava 14: 151–154.

Poinar, G. O., Jr., and J. Haverkamp. 1985. Use of pyrolysis mass spectrometry in the identification of amber samples. Journal of Baltic Studies 16: 210–221.

Poinar, G. O., Jr., and R. Hess. 1982. Ultrastructure of 40-million-year-old insect tissue. Science 215: 1241–1242.

Poinar, G. O., Jr., and R. Poinar. 1999. The Amber Forest. Princeton University Press, Princeton, New Jersey.

Poinar, G. O., Jr., B. M. Waggoner, and U. C. Bauer. 1993. Terrestrial soft-bodied protists and other microorganisms in Triassic amber. Science 259: 222–224.

Polhill, R. M. 1994. Complete synopsis of legume genera. In: Phytochemical Dictionary of the Leguminosae, Vol. 1, Plants and Their Constituents, pp. xliv–liv, eds. F. H. Bisby, J. Buckingham, and J. B. Harborne. Chapman and Hall, London.

Pollard, A. M., and C. Heron. 1996. Archaeological Chemistry. Royal Society of Chemistry, Cambridge, England.

Powell, J. S., and K. F. Raffa. 1999. Sources of variation in concentration and composition of foliar monoterpenes in tamarack *(Larix laricina)* seedlings: role of nutrient availability, time of season and plant architecture. Journal of Chemical Ecology 25: 1771–1797.

Prakash, A. O., V. Saxena, R. K. Tewari, S. Mathur, A. Gupta, and S. Sharma. 1986. Anti-implantation activity of some native plants in rats. Acta Europea Fertilitatis 24: 19–24.

Prasad, R. S., and Sukh Dev. 1976. Chemistry of Ayurvedic crude drugs—IV. Guggulu (resin from *Commiphora mukul*) absolute stereochemistry of mukulol. Tetrahedron 32: 1437–1441.

Pratt, R., and H. W. Youngken. 1951. Pharmacognosy. J. B. Lippincott, London.

Prendergast, H. D. V., H. Jaeschke, and N. Rumball. 2001. Lacquer Legacy at Kew: the Japanese Collection of John J. Quinn. Royal Botanic Gardens, Kew.

Price, R. A. 1995. Familial and generic classification of the conifers. American Journal of Botany 82, Supplement 110.

Price, R. A., and J. M. Lowenstein. 1989. An immunological comparison of the Sciadopityaceae, Taxodiaceae and Cupressaceae. Systematic Botany 14: 141–149.

Price, R. A., A. Liston, and S. H. Strauss. 1998. Phylogeny and systematics of *Pinus*. In: Ecology and Biogeography of *Pinus*, pp. 49–68, ed. D. M. Richardson. Cambridge University Press, New York.

Prockat, F. 1932. Amber mining in Germany. Engineering and Mining Journal 129: 305–307.

Proefke, M. L., K. L. Rinehart, M. Raheel, S. H. Ambrose, and S. U. Wissenbaum. 1992. Probing the mysteries of ancient Egypt: chemical analysis of Roman period Egyptian mummy. Analytical Chemistry 64: 105–111.

Proietti, G., G. Strappaghetti, and S. Corsano. 1981. Triterpenes of *Boswellia frereana*. Planta Medica 41: 417–418.

Qureshi, S., M. M. al-Harbi, M. M. Ahmed, M. Raza, A. B. Giangreco, and A. H. Shah. 1993. Evaluation of the genotoxic, cytotoxic and antitumor properties of *Commiphora molmol* using normal and Ehrlich ascites carcinoma cell-bearing Swiss albino mice. Cancer Chemotherapy Pharmacology 33: 130–138.

Raatikainen, O. J., H. T. Taipale, A. Pelttari, and S. P. Lapinjoki. 1992. An electron microscope study of resin production and secretion by glands of seedlings of *Betula pendula* Roth. New Phytologist 122: 537–543.

Rachmilevitz, T., and D. J. Joel. 1976. Ultrastructure of the calyx glands of *Plumbago capensis* Thunb. in relation to the process of secretion. Israel Journal of Botany 25: 127–139.

Raffa, K. F. 1989. Genetic engineering of trees to enhance resistance to insects. Evaluating the risks of biotype evolution and secondary pest outbreak. BioScience 39: 524–534.

Raffa, K. F. 1991. Induced defensive reactions in conifer–bark beetle systems. In: Phytochemical Induction by Herbivores, pp. 245–276, eds. D. W. Tallamy and H. J. Raupp. Academic Press, New York.

Raffa, K. F. 2001. Mixed messages across multiple trophic levels: the ecology of bark beetle chemical communication defenses. Chemoecology 7: 49–65.

Raffa, K. F., and A. A. Berryman. 1982. Accumulation of monoterpenes and associated volatiles following inoculation of grand fir with a fungus transmitted by the fir engraver, *Scolytus ventralis* (Coleoptera: Scolytidae). Canadian Entomologist 114: 797–810.

Raffa, K. F., and A. A. Berryman. 1983. The role of host plant resistance in the colonization behavior and ecology of bark beetles (Coleoptera: Scolytidae). Ecological Monographs 53: 27–49.

Raffa, K. F., and A. A. Berryman. 1987. Interacting selective pressures in conifer–bark beetle systems: A basis for reciprocal adaptations? American Naturalist 129: 234–262.

Raffa, K. F., and K. D. Klepzig. 1992. Tree defense mechanisms against fungi associated with insects. In: Defense Mechanisms of Woody Plants Against Fungi, pp. 354–390, eds. R. A. Blanchette and A. C. Briggs. Springer-Verlag, Berlin.

Raffa, K. F., A. A. Berryman, J. Simasko, W. Teal, and B. L. Wong. 1985. Effects of grand fir monoterpenes on the fir engraver beetle (Coleoptera: Scolytidae) and its symbiotic fungus. Environmental Entomology 4: 552–556.

Raffa, K. F., T. W. Phillips, and S. M. Salom. 1993. Strategies and mechanisms of host colonization by bark beetles. In: Beetle–Pathogen Interactions in Conifer Forests, pp. 103–128, eds. T. D. Schowalter and G. M. Filip. Academic Press, San Diego.

Raffa, K. F., S. C. Krause, and P. Reich. 1998. Long-term influence of defoliation on *Pinus resinosa* suitability to insect herbivores feeding on diverse plant parts. Ecology 79: 2352–2364.

Raghavan, B., K. O. Abraham, M. L. Shankaranarayana, L. V. L. Sastry, and C. P. Natarajan. 1974. Asafoetida. II. Chemical composition and physicochemical properties. The Flavour Industry 5: 179–181.

Rakoto-Ratsimamanga, A., P. Boiteau, and M. Mouton. 1968. Éléments de pharmacopée Malagasy. Notice 36. Andrakadtaka. Bulletin de Madagasgar 271: 1091–1102.

Ramírez, B. W., and L. D. Gómez. 1978. Production of nectar and gums by flowers of *Monstera deliciosa* (Araceae) and some species of *Clusia* (Guttiferae) collected by New World *Trigona* bees. Brenesia 14–15: 407–412.

Ramos, G. S., and E. M. Engleman. 1982. Estudio de los canales resiniferos de la corteza de *Bursera copallifera* and *Bursera grandifolia*. Boletín de la Sociedad Botánica de México 42: 41–54.

Rane, K. K., and T. A. Tattan. 1987. Pathogenicity of blue stain fungi associated with *Dendroctonus terebrans*. Plant Disease 71: 879–883.

Rao, G. S. R., M. A. Gerhart, R. T. Lee, L. A. Mitscher, and S. Drake. 1982. Antimicrobial agents from higher plants: dragon's blood resin. Journal of Natural Products 45: 646–648.

Rasnitsyn, A. P., and A. J. Ross. 2000. A preliminary list of arthropod families present in the Burmese amber collection at the Natural History Museum, London. Bulletin of the Natural History Museum, London (Geology) 56: 21–24.

Rausher, M. D. 1992. Natural selection and the evolution of plant–insect interactions. In: Insect Chemical Ecology, pp. 20–88, eds. B. D. Rostberg and M. B. Isman. Chapman and Hall, New York.

Raven, P. H., and R. M. Polhill. 1981. Biogeography of the Leguminosae. In: Advances in Legume Systematics, Part 1, pp. 27–34, eds. R. M. Polhill and P. H. Raven. Royal Botanic Gardens, Kew.

Ravetta, D. A., S. P. McLaughlin, and J. W. O'Leary. 1997. Evaluation of salt tolerance and resin production in coastal and Central Valley accessions of *Grindelia* species (Asteraceae). Madroño 44: 74–88.

Record, S. J. 1921. Further notes on intercellular canals in dicotyledonous wood. Journal of Forestry 19: 255–266.

Record, S. J., and R. W. Hess. 1943. Timbers of the New World. Yale University Press, New Haven, Connecticut.

Reed, A. H. 1972. The Gumdiggers: the Story of Kauri Gum. A. H. Reed and A. W. Reed, Wellington, New Zealand.

Rees, A. 1995. Frankincense and myrrh. New Plantsman 2: 55–59.

Reichardt, P. B., J. P. Bryant, T. P. Clausen, and G. Wieland. 1984. Defense of winter-dormant Alaska paper birch against snowshoe hare. Oecologia 65: 58–69.

Reichardt, P. B., J. P. Bryant, B. R. Mattes, T. P. Clausen, F. S. Chapin III, and M. Meyer. 1990. The winter chemical defense of balsam poplar against snowshoe hares. Journal of Chemical Ecology 16: 1941–1959.

Reichardt, P. B., F. S. Chapin III, J. P. Bryant, B. R. Mattes, and T. P. Clausen. 1991. Carbon / nutrient balance as a predictor of plant defense in Alaskan balsam poplar: potential importance of metabolite turnover. Oecologia 88: 401–406.

Reid, E. M., and M. E. Chandler. 1933. The Flora of the London Clay. British Museum (Natural History), London.

Reynolds, J. E. F. (ed.). 1996. Martindale: the Extra Pharmacopoeia. Royal Pharmaceutical Society, London.

Rhoades, D. G. 1977a. The antiherbivore chemistry of *Larrea*. In: Creosote Bush: Biology and Chemistry of *Larrea* in New World Deserts, pp. 135–175, eds. T. J. Mabry, J. H. Hunziker, and D. R. DiFeo. Hutchinson and Ross, Stroudsburg, Pennsylvania.

Rhoades, D. G. 1977b. Integrated antiherbivore, antidesiccant and ultraviolet screening properties of creosote bush resin. Biochemical Systematics and Ecology 5: 281–290.

Rhoades, D. G. 1979. Evolution of plant chemical defense against herbivores. In: Herbivores: Their Interactions with Plant Secondary Metabolites, pp. 4–48, eds. G. A. Rosenthal and D. H. Janzen. Academic Press, New York.

Rhoades, D. G. 1983. Herbivore population dynamics and plant chemistry. In: Variable Plants and Herbivores in Natural and Managed Systems, pp. 155–220, eds. R. F. Denno and M. S. McClure. Academic Press, New York.

Rhoades, D. G., and R. G. Cates. 1976. Toward a general theory of plant antiherbivore chemistry. In: Biochemical Interactions Between Plants and Insects, pp. 168–213, eds. J. W. Wallace and R. L. Mansell. Plenum Press, New York.

Rice, P. C. 1987. Amber. The Golden Gem of the Ages. Kosciusko Foundation, New York.

Richardson, D. M. (ed.). 1998. Ecology and Biogeography of *Pinus*. Cambridge University Press, New York.

Richardson, D. M., and W. T. Bond. 1991. Determinants of plant distribution: evidence from pine invasions. American Naturalist 137: 639–668.

Richardson, D. P., A. C. Messer, S. Greenberg, H. H. Hagedorn, and J. Meinwald. 1989. Defensive sesquiterpenoids from a dipterocarp *(Dipterocarpus kerrii)*. Journal of Chemical Ecology 15: 731–747.

Richmond, G. S., and Ghisalberti, E. L. 1994. The Australian desert shrub *Eremophila* (Myoporaceae): medicinal, cultural, horticultural and phytochemical uses. Economic Botany 48: 35–39.

Riddle, J. M. 1985. Dioscorides on Pharmacy and Medicine. University of Texas Press, Austin.

Riddle, J. M., and J. W. Estes. 1992. Oral contraceptives in ancient and medieval times. American Scientist 80: 226–233.

Riddle, J. M., J. W. Estes, and J. L. Russell. 1994. Ever since Eve . . . birth control in the ancient world. Archeology 47: 29–35.

Roberts, D. 1998. On the frankincense trail. Smithsonian 29: 120–135.

Roberts, D. R. 1973. Inducing lightwood in pine trees by paraquat treatment. U.S. Forest Service Research Note SE-191.

Roberts, D. R., N. M. Joyce, Jr., A. T. Proveaux, W. J. Peters, and R. V. Lawrence. 1973. Naval Stores Review 83: 4.

Robinson, C. 1999. Making forest biotechnology a commercial reality. Nature Biotechnology 17: 29–30.

Rodriguez, E. 1977. Ecogeographic distribution of secondary constituents in *Parthenium* (Compositae). Biochemical Systematics and Ecology 5: 207–218.

Rodriguez, E. 1985. Rubber and phytochemical specialties from desert plants of North America. In: Plants for Arid Lands, pp. 399–412, eds. G. E. Wickens, J. R. Goodin, and D. V. Field. George Allen and Unwin, London.

Rodriguez, E., P. L. Healey, and I. Mehta (eds.). 1984. Biology and Chemistry of Plant Trichomes. Plenum Press, New York.

Rohde, A. 1942. Das Bernsteinzimmer Friedrichs I in Königsberger Schloss. Pantheon 29: 200–203.

Romeo, J. T., J. A. Saunders, and P. Barbosa (eds.). 1996. Phytochemical Diversity and Redundancy in Ecological Interactions. Recent Advances in Phytochemistry 30.

Rosenthal, F. 1971. The Herb-Hashish versus Medieval Muslim Society. E. J. Brill, Leiden.

Ross, J. D., and C. Sombero. 1991. Environmental control of essential oil production in Mediterranean plants. In: Ecological Chemistry and Biochemistry of Plant Terpenoids, pp. 64–94, eds. J. B. Harborne and F. A. Tomás-Barberán. Clarendon Press, Oxford.

Rostenburg, A., Jr. 1955. An anecdotal biographical history of poison ivy. American Medical Association Archives of Dermatology 72: 438–445.

Rottländer, R. C. A. 1970. On the formation of amber from *Pinus* resin. Archaeometry 12: 35–52.

Rottländer, R. C. A. 1971. Über die Bildung des Bernsteins und sein chemisches Verhalten. Deutsche Farben-Zeitschrift 25: 66–69.

Roubik, D. W. 1983. Nest and colony characteristics of stingless bees from Panama (Hymenoptera: Apidae). Journal of the Kansas Entomological Society 56: 327–355.

Roubik, D. W. 1989. Ecology and Natural History of Tropical Bees. Cambridge University Press, Cambridge.

Roussi, M., J. Tahvanainen, and I. Uotila. 1989. Inter- and intraspecific variation in the resistance of winter-dormant birch (*Betula* spp.) against browsing by the mountain hare. Holarctic Ecology 12: 187–192.

Rudall, P., and M. W. Chase. 1996. Systematics of Xanthorrhoeaceae sensu lato: evidence for polyphyly. Telopea 6: 629–647.

Runeberg, J. 1960. The chemistry in the order Cupressales. 28. Heartwood constituents of *Juniperus virginiana* L. Acta Chemica Scandinavica 14: 1288–1294.

Rushforth, K. D. 1987. Conifers. Christopher Helm, London.

Rzedowski, J., and H. Kruse. 1979. Algunas tendencias evolutivas en *Bursera* (Burseraceae). Taxon 28: 103–116.

Safayhi, H., T. Mack, and H. P. T. Ammon. 1991. Protection by boswellic acids against galactosamine / endotoxin-induced hepatitis in mice. Biochemical Pharmacology 41: 1536–1537.

Sakakibara, M., D. DiFeo, N. Nakatani, B. Timmermann, and T. J. Mabry. 1976. Flavonoid methyl ethers in the external leaf surface of *Larrea tridentata* and *L. divaricata* (Zygophyllaceae). Phytochemistry 15: 727–731.

Sale, E. V. 1978. Quest for the Kauri. A. H. and A. W. Reed, Wellington, New Zealand.

Samini, M. N., and W. Unger. 1979. The gum resins of Afghan asafoetida-producing *Ferula* species. Observations on the provenance and quality of Afghan asafoetida. Planta Medica 36: 128–133.

Sánchez-Hidalgo, M. E., M. Martínez-Ramos, and F. J. Espinosa-García. 1999. Chemical differentiation between leaves of seedlings and spatially close adult trees from the tropical rainforest species *Nectandra ambigens* (Lauraceae): an alternate test of the Janzen-Connell model. Functional Ecology 13: 725–732.

Sánchez-Ramos, G., R. Dirzo, and F. J. Espinosa-García. 1999. Terpenoid variability in young and mature leaves of *Liquidambar styraciflua* from gap and closed forest in relation to herbivory, p. 498. Abstracts XVI International Botanical Congress.

Sanderson, M. J., and L. Hufford. 1996. Homoplasy: the Recurrence of Similarity in Evolution. Academic Press, San Diego.

San Feliciano, A., and J.-L. Lopez. 1991. Recent chemistry of conifer terpenoids. In: Ecological Chemistry and Biochemistry of Plant Terpenoids, pp. 1–27, eds. J. B. Harborne and F. A. Tomás-Barberán. Clarendon Press, Oxford.

Sanz, J. A. 1989. Naval stores status in South America with a focus on Brazil. Naval Stores Review 99: 6–9.

Saulei, S. M., and J. A. Aruga. 1994. The status and prospects of non-timber forest products development in Papua New Guinea. Commonwealth Forestry Review 73: 97–105.

Saunders, W. B., R. H. Mapes, F. M. Carpenter, and W. C. Elsik. 1974. Fossiliferous amber from the Eocene (Claiborne) of the Gulf Coast Plain. Geological Society of America Bulletin 85: 979–984.

Savage, T. J., M. W. Hatch, and R. Croteau. 1994. Monoterpene synthases of *Pinus contorta* and related conifers: a new class of terpenoid cyclase. Journal of Biological Chemistry 269: 4012–4020.

Savage, T. J., H. Ichii, S. D. Hume, D. B. Little, and R. Croteau. 1995. Monoterpene synthases from gymnosperms and angiosperms: enzyme stereospecificity and inactivation by cysteinyl- and arginyl-directed modifying reagents. Archives of Biochemistry and Biophysics 320: 257–265.

Sawadogo, M. 1985. Oleoresin of *Canarium schweinfurthii* Engl. Annales Pharmaceutiques Françaises 43: 89–96.

Schloman, W. W., Jr., R. A. Hively, A. Krishon, and A. M. Andrews. 1983. Guayule by product evaluation: extraction characterization. Journal of Agricultural and Food Chemistry 31: 873.

Schloman, W. W., Jr., O. J. Garrot, Jr., D. T. Ray, and D. J. Bennet. 1986. Seasonal effects on guayule resin composition. Journal of Agricultural Food Chemistry 34: 177–179.

Schlüter, T., and F. von Gnielinski. 1988. The East African copal. Its geologic, stratigraphic, paleontologic significance and comparison with fossil resins of similar age. National Museums of Tanzania Occasional Paper 8.

Schmincke, K. H. 1995. Non-wood forest products for rural income and sustainable forestry. Food and Agriculture Organization of the United Nations, Rome.

Schnepf, E. 1974. Gland cells. In: Dynamic Aspects of Plant Ultrastructure, pp. 331–357, ed. A. W. Roberts. McGraw Hill, Berkshire, United Kingdom.

Schubert, K. 1953. Mikroskopische Untersuchung pflanzlicher Einschlüsse des Bernsteins. Teil 2. Palaeontographica B 93: 103–119.

Schubert, K. 1961. Neue Untersuchungen über Bau und Leben der Bernsteinkiefern [*Pinus succinifera* (Conw.) emend]. Beihefte zum Geologischen Jahrbuch 45.

Schuck, H. J. 1982. Monoterpenes and resistance of conifers to fungi. In: Resistance to Diseases and Pests in Forest Trees, pp. 169–175, eds. H. M. Heybrock, B. M. Stephan, and K. Wissenberg. PUDOC, Wageningen.

Schultes, R. E. 1970. Random thoughts and queries on the botany of *Cannabis*. In: The Botany and Chemistry of *Cannabis*, pp. 11–38, eds. C. R. B. Joyce and S. H. Curry. Churchill, London.

Schultes, R. E., and R. F. Raffauf. 1990. The Healing Forest: Medicinal and Toxic Plants of the Northwest Amazonia. Historical, Ethno- & Economic Botany Series, Vol. 2. Dioscorides Press (Timber Press), Portland, Oregon.

Schultz, J. C., and T. Floyd. 1999. Desert survivor. Natural History 2–99: 24–29.

Schwartz, H. F. 1948. Stingless bees (Meliponidae) of the western hemisphere. Bulletin of the American Museum of Natural History 90: 1–546.

Scott, A. L., and T. N. Taylor. 1983. Plant / animal interactions during the Upper Carboniferous. Botanical Review 49: 259–307.

Sequeira, A. S., B. B. Normark, and B. D. Farrell. 2000. Evolutionary assembly of the conifer fauna: distinguishing ancient from recent associations in bark beetles. Proceedings of the Royal Society of London B267: 2359–2366.

Setia, R. C., M. P. Parthasarathy, and J. J. Shah. 1977. Development, histochemistry and ultrastructure of gum-resin ducts in *Commiphora mukul* Engl. Annals of Botany 41: 999–1004.

Setoguchi, H., T. A. Osawa, J.-C. Pintaud, T. Jaffré, and J.-M. Veillon. 1998. Phylogenetic relationships within Araucariaceae based on *rbc*L gene sequences. American Journal of Botany 85: 1509–1516.

Seybold, S. J., J. Bohlmann, and K. F. Raffa. 2000. The biosynthesis of coniferophagus bark beetle pheromones and conifer isoprenoids: an evolutionary perspective and synthesis. Canadian Entomologist 132: 1–57.

Shah, C. S., J. S. Qadry, and B. K. Shal. 1971. Evaluation of market samples of benzoin. Indian Journal of Pharmacy 33: 119–120.

Shanley, P., M. Cymerys, and J. Galvão. 1998. Frutiferas da Mata na Vida Amazônica. Patricia Shanley, Belém, Brazil.

Shao, Y., C.-t. Ho, C.-k. Chin, V. Badmaer, W. Ma, and M.-t. Huang. 1998. Inhibitory activity of boswellic acids from *Boswellia serrata* against human leukemia HL-60 cells in culture. Planta Medica 64: 328–331.

Sharma, G. K. 1979. Significance of eco-chemical studies of *Cannabis*. Science and Culture 45: 303–307.

Sharma, K. K., A. Bhattacharya, and S. N. Sushil. 1999. Indian lac insect, *Kerria lacca*, an important source of honeydew. Bee World 80: 115–118.

Sharma, M. L. 1994. Comparative chemical analysis of oleogum resin of *Commiphora wightii* Arnott. Bhandari. Current Science 60: 9–10.

Sharma, M. L., A. Kaul, A. Khajuria, S. Singh, and G. B. Singh. 1996. Immunodulatory activity of boswellic acids (pentacyclic triterpene acids) from *Boswellia serrata*. Phytotherapy Research 10: 107–112.

Sharma, P. K., and J. D. Sharma. 1996. Potent amoebicides from plant extracts—an in vitro assessment with the gum-oleo-resin of *Commiphora wightii*. Current Science 71: 68–70.

Shedrinsky, A. M., D. Grimaldi, T. P. Wampler, and N. S. Baer. 1989–91. Amber and copal: pyrolysis gas chromatographic (PyGC) studies of provenance. Wiener Berichte über Naturwissenschaft in der Kunst 6–8: 37–63.

Shedrinsky, A. M., D. A. Grimaldi, J. J. Boon, and N. S. Baer. 1993. Application of pyrolysis gas chromatography and pyrolysis gas chromatography / mass spectrometry to the unmasking of amber forgeries. Journal of Analytical and Applied Pyrolysis 25: 79–95.

Sheikh, M. I. 1981. Acid stimulation has no adverse effect on Chir pine *(Pinus roxburghii).* Pakistan Journal of Forestry 31: 26–28.

Sheikh, M. I. 1984. Face positions yield different quantities of resin from *Pinus roxburghii.* Pakistan Journal of Forestry 34: 97–100.

Shelef, L. A., O. A. Naglik, and D. W. Bogen. 1980. Sensitivity of some common food-borne bacteria to the spices sage, rosemary and allspice. Journal of Food Science 45: 1042–1044.

Shellard, E. J. 1961. Some convolvulaceous drugs: an historical survey. Chemist and Druggist 176: 219–222.

Shimada, I., K. B. Anderson, H. Haas, and J. H. Langenheim. 1997. Amber from 1000-year-old prehispanic tombs in northern Peru. In: Materials Issues in Art and Archeology V, pp. 3–18, eds. P. B. Vandiver, J. R. Druzik, J. J. Merkel, and J. Stewart. Symposium Proceedings 462. Materials Research Society, Pittsburgh.

Shivashankar, S., M. L. Shankaranarayana, and C. P. Natarajan. 1972. Asafoetida—varieties, chemical composition, standards and uses. Indian Food Packers 26: 36–44.

Shroeder, H. A. 1968. The *p*-hydroxycinnamyl compounds of Siam benzoin gum. Phytochemistry 7: 57–61.

Shukla, O. P., M. N. Moholay, and P. K. Bhattacharya. 1968. Microbial transformation of terpenes. Part X. Fermentation of α- and β-pinenes by a soil pseudomonad (PL strain). Indian Journal of Biochemistry and Biophysics 5: 79–91.

Siani, A. C., M. F. S. Ramos, O. Menezes de Lima Jr., R. Ribeiro dos Santos, E. Fernandez-Ferreira, R. O. A. Soares, E. C. Rosas, G. S. Susunaga, A. C. Guimarães, M. G. B. Zoghbi, and M. G. M. O. Henriques. 1999a. Evaluation of anti-inflammatory-related activity of essential oils from the leaves and resin of species of *Protium.* Journal of Ethnopharmacology 66: 57–69.

Siani, A. C., M. F. S. Ramos, A. C. Guimarães, G. S. Susunaga, and M. G. B. Zoghbi. 1999b. Volatile constituents from oleoresin of *Protium heptaphyllum* (Aubl.) March. Journal of Essential Oil Research 11: 72–74.

Singh, G. B., and C. K. Atal. 1986. Pharmacology of an extract of *salai-guggal* ex *Boswellia serrata,* a new nonsteroidal antiinflammatory agent. Agents and Action 1B: 407–412.

Singh, G. B., S. Singh, and S. Bani. 1993. Boswellic acids. Anti-inflammatory, anti-arthritic, anti-pyretic. Drugs of the Future 18: 307–309.

Singh, M. M., A. Agnihotri, S. W. Garg, D. N. Gupta, G. Keshri, and V. P. Kamboj. 1988. Antifertility and hormonal properties of certain carotane sesquiterpenes of *Ferula jaeschkeana.* Planta Medica 54: 492–494.

Singh, R. B., M. A. Niaz, and S. Ghosh. 1974. Hypolipidemic and antioxidant effects of *Commiphora mukul* as an adjunct to dietary therapy in patients with hypercholesterolemia. Cardiovascular Drugs and Therapy 8: 659–664.

Skutch, A. F. 1971. A Naturalist in Costa Rica. University of Florida Press, Gainesville.

Smiley, C. J. 1966. Cretaceous floras from Kuk River area, Alaska: stratigraphic and climatic interpretations. Geological Society of America Bulletin 77: 1–14.

Smith, N. J. H., D. L. P. Williams, and J. P. Talbot. 1992. Tropical Forests and Their Crops. Comstock Publishing Associates, Cornell University Press, Ithaca, New York.

Smith, R. H. 1964. Variation in the monoterpenes of *Pinus ponderosa* Laws. Science 143: 1337–1338.

Smith, R. H. 1966. Resin quality as a factor in resistance of pines to bark beetles. In: Breeding Pest Resistant Trees, pp. 189–196, eds. H. Gerhold, R. McDermott, E. Schreiner, and J. Winieksi. Pergamon Press, Oxford.

Smith, R. H. 1977. Monoterpenes of ponderosa pine xylem resin in the western United States. U.S. Department of Agriculture Technical Bulletin 1532: 1–48.

Smitinand, T., J. E. Vidal, and H. P. Hoang. 1990. Flore du Camboge du Laos et du Viêtnam. Muséum National d'Histoire Naturelle, Paris.

Snyder, M. A. 1992. Selective herbivory by Abert's squirrel mediated by chemical variability in ponderosa pine. Ecology 73: 1730–1741.

Snyder, M. A. 1993. Interactions between Abert's squirrel and ponderosa pine: the relationship between selective herbivory and host plant fitness. American Naturalist 141: 866–879.

Snyder, M. A. 1998. Abert's squirrel *(Sciurus aberti)* in ponderosa pine *(Pinus ponderosa)* forests: directional selection, diversifying selection. In: Ecology and Evolutionary Biology of Tree Squirrels, pp. 195–201, eds. M. A. Steele, J. F. Merritt, and D. A. Zegers. Special Publication 6. Virginia Museum of Natural History, Martinsville.

Snyder, M. A., and Y. B. Linhart. 1997. Porcupine feeding patterns: Selectivity by a generalist herbivore? Canadian Journal of Zoology 75: 2107–2111.

Snyder, M. A., and Y. B. Linhart. 1998. Subspecific selectivity by a mammalian herbivore: geographic interactions between two taxa[,] of *Sciurus aberti* and *Pinus ponderosa*. Evolutionary Ecology 12: 755–765.

Snyder, M. A., B. Finieschi, Y. B. Linhart, and R. H. Smith. 1996. Multivariate discrimination of host use by dwarf mistletoe *Arceuthobium vaginatum* subsp. *cryptopodum*: inter- and intraspecific comparisons. Journal of Chemical Ecology 22: 295–305.

Sochartono, T., and A. Mardiastuti. 1997. The current trade in *gaharu* in West Kalimantan. Biodiversitas Indonesia 1: 1–10.

Soenarno, M. M. I. 1987. Copal production on *Agathis* spp. of varying bark thicknesses, West Java. Duta Rimba 13: 3–6.

Solereder, H. 1908. Systematic Anatomy of the Dicotyledons, Vol. 1, transl. L. A. Boodle and F. E. Fritsch. Clarendon Press, Oxford.

Somerville, A. 1980. Resin pockets and related defects of *Pinus radiata* grown in New Zealand. New Zealand Forest Service 10: 439–444.

Soom, M. 1984. Bernstein vom Nordrand der schweizer Alpen. In: Bernstein-Neuigkeiten, ed. D. Schlee, Beiträge Naturkunde im Stuttgart, Serie C, 18: 15–20.

Sousa, S. M., and A. Delgado-S. 1993. Mexican Leguminosae: phytogeography, endemism, and origins. In: Biological Diversity of Mexico: Origins and Distribution, pp. 459–489, eds. T. Ramamoorthy, R. Bye, A. Lot, and J. Fa. Oxford University Press, Oxford.

Spekke, A. 1957. The Ancient Amber Routes and Geographic Discovery of the Eastern Baltic. M. Goppers, Stockholm.

Squillace, A. E. 1971. Inheritance of monoterpene composition in cortical oleoresin of slash pine. Forest Science 17: 381–387.

Squillace, A. F., O. Wells, and D. L. Rockwood. 1980. Inheritance of monoterpene composition cortical oleoresin in loblolly pine. Silvae Genetica 29: 141–152.

Stähelin, H. F., and A. V. Wartburg. 1991. The chemical and biological route from podophyllotoxin glycoside to etoposide. Ninth Cain Memorial Award Lecture. Cancer Research 51: 5–15.

Stanley, D. 1991. Demystifying the tragedy of the commons: the resin tappers of Honduras. Grassroots Development Journal of the Inter-American Foundation 15: 26–35.

Stauffer, D. F. 1989. Production, markets, and economics. In: Naval Stores: Production-Chemistry-Utilization, pp. 39 80, eds. D. E. Zinkel and J. Russell. Pulp Chemical Association, New York.

Steele, C. L., E. Lewinsohn, and R. Croteau. 1995. Induced oleoresin biosynthesis in grand fir as a defense against bark beetles. Proceedings of the National Academy of Sciences U.S.A. 92: 4146–4168

Steele, C. L., J. Crock, J. Bohlmann, and R. Croteau. 1998a. Sesquiterpene synthases from grand fir *(Abies grandis)*: comparison of constitutive and wound-induced activities, and cDNA isolation, characterization, and bacterial δ-selinene synthase and γ-humulene synthase. Journal of Biological Chemistry 273: 2078–2089.

Steele, C. L., S. Katoh, J. Bohlmann, and R. Croteau. 1998b. Regulation of oleoresinosis in grand fir *(Abies grandis)*. Differential transcriptional control of monoterpene, sesquiterpene, and diterpene synthase genes in response to wounding. Plant Physiology 116: 1497–1504.

Stefanoviac, S., J. Jager, J. Deutsch, J. Broutin, and M. Masselot., 1998. Phylogenetic relationships of conifers inferred from partial 28S rRNA gene sequences. American Journal of Botany 85: 688–697.

Stephan, G. 1990. Resin tapping in the German Democratic Republic. Naval Stores Review 100: 12.

Stewart, W. N. 1983. Paleobotany and the Evolution of plants. Cambridge University Press, Cambridge.

Stockey, R. A. 1978. Reproductive biology of Cerro Cuadrado fossil conifers: ontogeny and reproductive strategies in *Araucaria mirabilis*. Palaeontographica 166B: 1–15.

Stockey, R. A. 1982. The Araucariaceae: an evolutionary perspective. Review of Palaeobotany and Palynology 37: 133–154.

Stockey, R. A. 1989. Antarctic and Gondwana conifers. In: Antarctic Paleobiology—Its Role in Reconstruction of Gondwana, pp. 177–191, eds. T. N. Taylor and E. L. Taylor. Springer-Verlag, New York.

Stout, E. C., C. W. Beck, and B. Kosmowska-Ceranowicz. 1995. Gedanite and gedanosuccinite. In: Amber, Resinite and Fossil Resins, pp. 130–148, eds. K. B. Anderson and J. C. Crelling. Symposium Series 617. American Chemical Society, Washington, D.C.

Stout, S. A. 1995. Resin-derived hydrocarbons in fresh and fossil dammar resins and Miocene rocks and oils in the Mahakam Delta Indonesia. In: Amber, Resinite and Fossil Resins, pp. 43–75, eds. K. B. Anderson and J. C. Crelling. Symposium Series 617. American Chemical Society, Washington, D.C.

Strappaghetti, G. S., S. Cersano, A. Craveiro, and G. Proietti. 1982. Constituents of essential oil of *Boswellia frereana*. Phytochemistry 21: 2114–2115.

Strauss, S. H., K. F. Raffa, and P. C. List. 2000. Ethics and genetically engineered plantations. Journal of Forestry 98: 47–48.

Strobel, G. A., and F. Sugawara. 1986. The pathogenicity of *Ceratocystis montia* to lodgepole pine. Canadian Journal of Botany 64: 113–116.

Strong, D. E. 1966. Catalogue of the Carved Amber in the Department of Greek and Roman Antiquities. Trustees of the British Museum, London.

Stross, B. 1994–96. Mesoamerican copal resins. U Mut Maya 6: 177–196.

Stuart, G. A. 1911. Chinese Materia Medica: Vegetable Kingdom. Shanghai.

Stubblebine, W. H., and J. H. Langenheim. 1977. Effects of *Hymenaea courbaril* leaf resin on the generalist herbivore *Spodoptera exigua* (beet army worm). Journal of Chemical Ecology 3: 633–647.

Stubblebine, W. H., J. H. Langenheim, and D. E. Lincoln. 1975. Vegetative growth and leaf resin composition in *Hymenaea courbaril* under photoperiod extremes. Biochemical Systematics and Ecology 3: 219–228.

Stubblebine, W. H., D. E. Lincoln, and J. H. Langenheim. 1978. Vegetative responses to photoperiod in the tropical leguminous tree *Hymenaea courbaril*. Biotropica 10: 18–29.

Stubblefield, S. P., T. N. Taylor, and C. B. Beck. 1985. Studies of Paleozoic fungi. V. Wood-decaying fungi in *Callixylon newburyi* from the Upper Devonian. American Journal of Botany 72: 1765–1774.

Sturgeon, K. B. 1979. Monoterpene variation in ponderosa pine xylem resin related to western pine beetle predation. Evolution 33: 803–814.

Sturgeon, K. B., and J. B. Mitton. 1982. Evolution of bark beetle communities. In: Bark Beetles in North America: a System for Study of Evolutionary Biology, pp. 350–384, eds. J. B. Mitton and K. B. Sturgeon. University of Texas Press, Austin.

Sturgeon, K. B., and J. B. Mitton. 1986. Biochemical diversity of ponderosa pine and predation by bark beetles (Coleoptera: Scolytidae). Journal of Economic Entomology 79: 1064–1068.

Sturgeon, K. B., and J. L. Robertson. 1985. Microsomal polysubstrate monooxygenase activity in western and mountain pine beetles (Coleoptera: Scolytidae). Annals of the Entomological Society of America 78: 1–2.

Su, Zi-an. 1984. Naval stores industry in China. Naval Stores Review 94: 8–13.

Sukh Dev (ed.) 1985–86. CRC Handbook of Terpenoids-Diterpenoids, 4 vols. CRC Press, Boca Raton.

Sukh Dev. 1987. A modern look at an age old Ayurvedic drug—guggal. Science Age 5: 13–18.

Sukh Dev. 1988. Ayurveda and modern drug development. Proceedings of the Indian National Science Academy A54: 12–42.

Suter, A. F. 1929. East Indian copals and damars. Journal of the Royal Society of Arts 1929(Apr.): 576–598.

Suter, A. F. 1945. Natural resins of the British Empire. Crown Colonist 15: 711.

Swamy, A. V., S. Kalyanasundaram, and M. Balagopal. 1965. The plant that gives us jalap. Indian Farming 1965(Mar.): 30–31.

Syamasundar, K. V., G. R. Mallavarapu, and E. M. Krishna. 1991. Triterpenoids of the resin of *Bursera delpechiana*. Phytochemistry 30: 362–363.

Sykes, B. 1997. Really ancient DNA: lights turning red on amber. Nature 386: 764–765.

Tahvanainen, J., R. Julkunen-Tiitto, M. Roussi, and P. B. Reichardt. 1991. Chemical determinants of resistance in winter-dormant seedlings of European white birch *(Betula pendula)* to browsing by mountain hare. Chemoecology 2: 49–54.

Taipale, H. T., L. Harmala, M. Roussi, and S. P. Lapinjoki. 1994. Histological and chemical comparison of triterpene and phenolic deterrent contents of juvenile shoots of *Betula* species. Trees—Structure and Function 8: 232–236.

Takahashi, T., M. Ohsawa, S. Shimakoshi, H. Kishi, M. Kawahara, and N. Yoshikawa. 1993. Developmental cytology of the resin glands of hop *(Humulus lupulus* L.). Journal of Horticultural Science 68: 797–801.

Takhtajan, A. 1969. Flowering Plants: Origin and Dispersal. Smithsonian Institution Press, Washington, D.C.

Takhtajan, A. 1980. Outline of the classification of flowering plants (Magnoliophyta). Botanical Review 46: 225–359.

Takhtajan, A. 1997. Diversity and Classification of Flowering Plants. Columbia University Press, New York.

Tassou, C., and G. J. E. Nychas. 1995. Antimicrobial activity of the essential oil of mastic gum *(Pistacia lentiscus* var. *chia)* on gram positive and gram negative bacteria in broth and in model food system. International Biodeterioration and Biodegradation 36: 411–420.

Taylor, K. W. 1951. Guayule—an American source of rubber. Economic Botany 5: 255–273.

Taylor, T. N., and E. L. Taylor. 1993. The Biology and Evolution of Fossil Plants. Prentice Hall, Englewood Cliffs, New Jersey.

Teichmüller, M. 1958. Rekonstructionen verschiedener Moortypen des Hauptflöz der niederrheinischen Braunkohle. Fortschritte der Geologie von Rheinland und Westfalen 2: 599–612.

Theophrastus. Enquiry into Plants (and Minor Works in Odours and Weather Signs), 2 vols., 1961 transl. A. Hort. Loeb Classical Library, Harvard University Press, Cambridge.

Thevet, A. 1575. La Cosmographie Universelle. Guillaume Chaudière, Paris.

Thirgood, J. V. 1981. Man and the Mediterranean Forest. A History of Resource Depletion. Academic Press, London.

Thomas, B. R. 1969. Kauri resins—modern and fossil. In: Organic Geochemistry, pp. 599–618, eds. G. Eglington, and M. T. J. Murphy. Springer-Verlag, Berlin.

Thomas, B. R. 1970. Modern and fossil plant resins. In: Phytochemical Phylogeny, pp. 59–79, ed. J. B. Harborne. Academic Press, New York.

Thompson, J. W. 1994. The Coevolutionary Process. University of Chicago Press, Chicago.

Thomson, W. W., and P. L. Healey. 1984. Cellular basis of trichome secretion. In: Biology and Chemistry of Plant Trichomes, pp. 95–111, eds. E. Rodriguez, P. L. Healey, and I. Mehta. Plenum Press, New York.

Thomson, W. W., K. Platt-Aloia, and D. Koller. 1979. Ultrastructure and development of the trichomes of *Larrea* (creosote bush). Botanical Gazette 140: 249–260.

Thulin, M., and A. M. Warfa. 1987. The frankincense trees (*Boswellia* spp., Burseraceae) of North Somalia and southern Arabia. Kew Bulletin 42: 487–500.

Tillman, D. A. 1985. Forest Products: Advanced Technologies and Economic Analyses. Academic Press, Orlando, Florida.

Timmermann, B. N., and J. J. Hoffmann. 1985. Resins from *Grindelia:* a model for renewable resources in arid environments. In: Plants for Arid Lands, pp. 357–358, eds. G. E. Wickens, J. R. Goodin, and D. V. Field. Allen and Unwin, London.

Timmermann, B. N., D. J. Luzbetak, J. J. Hoffmann, S. D. Jolad, K. H. Schram, R. B. Bates, and R. E. Kleng. 1983. Grindelane diterpenoids from *Grindelia camporum* and *Chrysothamnus paniculatus*. Phytochemistry 22: 523–525.

Timmermann, B. N., S. P. McLaughlin, and J. J. Hoffmann. 1987. Quantitative variation of grindelane diterpene acids in 20 species of North American *Grindelia*. Biochemical Systematics and Ecology 15: 401–410.

Tippo, O., and W. L. Stern. 1977. Humanistic Botany. Norton and Company, New York.

Tisdale, R. A., and T. E. Nebek. 1992. Resin flow as a function of height along the bole of loblolly pine. Canadian Journal of Botany 70: 2509–2511.

Tollrian, R., and C. D. Harvell (eds.). 1999. The Ecology and Evolution of Inducible Defenses. Princeton University Press, Princeton, New Jersey.

Tomás-Barberán, F. A., C. García-Viquera, P. Vit-Olivier, F. Ferreres, and F. Tomás-Lorentz. 1993. Phytochemical evidence for the botanical origin of tropical propolis from Venezuela. Phytochemistry 34: 191–196.

Tomlin, E. S., R. I. Alfaro, J. H. Borden, and F. He. 1998. Histological response of resistant and susceptible white spruce to simulated white pine weevil damage. Tree Physiology 18: 21–28.

Tongacan, A. L., and F. F. Ordinario. 1974. Tapping of *almaciga* resin. Philippine Lumberman 20: 18–19, 22–23, 25.

Torquebiau, E. F. 1984. Man-made dipterocarp forest in Sumatra. Agroforestry Systems 2: 103–127.

Torquebiau, E. F. 1987. Multidisciplinary research on *Shorea javanica*. I. Introduction. Biotropica 1: 42–45.

Tŏth, G. 1985. Propolis: Medicine or fraud? American Bee Journal 125: 337–338.

Trapp, S., and R. Croteau. 2001. Defensive biosynthesis in conifers. Annual Review of Plant Molecular Biology 52: 689–724.

Trease, G. E., and W. C. Evans. 1972. Pharmacognosy, ed. 10. Baillière Tindall, London.

Tripplett, K. J. 1999. The ethnobotany of plant resins in the Maya cultural region of southern Mexico and Central America. Ph.D. dissertation, University of Texas, Austin.

Trlin, A. 1979. Now Respected, Once Despised. Dunsmare Press, Palmerton North, New Zealand.

Tschirch, A. 1906. Die Harze und die Harzebehälter. Borntraeger, Leipzig.

Tschirch, A. 1908–25. Handbuch der Pharmakognosie, Bande 1–3. Tauchnitz, Leipzig.

Tschirch, A., and E. Stock. 1933–36. Die Harze: die botanischen und chemischen Grundlagen unserer Kenntnisse über die Bildung, die Entwicklung und die Zusammensetzung der pflanzlichen Exkrete, ed. 3, 2 vols. Gebrüder Bornträger, Berlin.

Tsoumis, G. 1992. Harvesting Forest Products. Stobart Davies, Hertford, England.

Tucker, A. O. 1986. Frankincense and myrrh. Economic Botany 40: 425–433.

Turkel, H. S. 1968. Anatomical studies of the woods in the Chinle flora. Ph.D. Dissertation, Harvard University, Cambridge, Massachusetts.

Turner, G. W. 1999. A brief history of the lysigenous gland hypothesis. Botanical Review 65: 76–88.

Turner, G., J. Gershenzon, E. E. Nielson, J. E. Froehlich, and R. Croteau. 1999. Limonene synthase, the enzyme responsible for monoterpene biosynthesis in peppermint, is localized to leucoplasts of oil secretory cells. Plant Physiology 120: 879–886.

Turner, G. W., J. Gershenzon, and R. Croteau. 2000. Development of peltate glandular trichomes of peppermint. Plant Physiology 124: 665–679.

Tyler, V. E. 1982. The Honest Herbal. Stickley, Philadelphia.

Tyler, V. E., L. R. Brady, and J. E. Robbins. 1988. Pharmagcognosy, ed. 9. Lea and Febriger, Philadelphia.

Tyman, J. H. R. 1979. Non-isoprenoid long-chain phenols. Chemical Society Reviews 8: 499–537.

Uphof, T. C. T. 1968. Dictionary of Economic Plants, ed. 2, ed. J. C. Cramer. Stechert-Hafner, New York.

Urizar, N. L., A. B. Liverman, D. T. Dodds, F. V. Silva, P. Ordentlich, Y. Yan, F. J. Gonzalez, R. A. Heyman, D. J. Mangelsdorf, and D. D. Moore. 2002. A natural product that lowers cholesterol as an antagonist ligand for the FXR. Science 296: 1703–1706.

Urones, J. G., P. Basabe, I. S. Marcos, A. Jiminez, and others. 1994. Ring A functionalized neo-clerodane diterpenoids from *Cistus populifolius*. Tetrahedron 50: 10,791–10,802.

Usher, G. 1974. A Dictionary of Plants Used by Man. Macmillan, New York.

van Aarssen, B. G. K., H. C. Cox, P. Hoogendarn, and J. W. de Leeuw. 1990. A cadinene biopolymer present in fossil and extant dammar resins as a source for cadinenes and bicadinenes in crude oil from S.E. Asia. Geochimica et Cosmochimica Acta 54: 3021–3031.

van Aarssen, B. G. K., J. K. C. Hessels, O. Abbink, and J. W. de Leeuw. 1992. The occurrence of polycyclic sesqui-, tri- and oligoterpenoids derived from a resinous polymeric cadinene in crude oils from Southeast Asia. Geochimica et Cosmochimica Acta 56: 1231–1246.

van Aarssen, B. G. K., J. W. de Leeuw, M. Collinson, J. J. Boon, and K. Goth. 1994. Occurrence of polycadinene in fossil and recent resins. Geochimica et Cosmochimica Acta 58: 223–229.

van Beek, G. W. 1958. Frankincense and myrrh in ancient South Arabia. Journal of American Oriental Society 78: 141–152.

van Bergen, P. F., M. E. Collinson, A. C. Scott, and J. W. de Leeuw. 1995. Unusual resin chemistry from Upper Carboniferous pteridosperm resin rodlets. In: Amber, Resinite and Fossil Resins, pp. 149–169, eds. K. B. Anderson and J. C. Crelling. Symposium Series 617. American Chemical Society, Washington, D.C.

van den Berg, K. C., J. van der Horst, J. J. Boon, and O. O. Sudmeijer. 1998. *cis*-1,4-Poly-β-myrcene; the structure of the polymeric fraction of mastic resin (*Pistacia lentiscus* L.) elucidated. Tetrahedron Letters 39: 2645–2648.

van Steenis, C. G. G. J. 1932. The Styracaceae of Netherlands India. Bulletin du Jardin Botanique de Buitenzorg, III, 12: 212–272.

van Steenis-Kruseman, M. J. 1953. Select Indonesian medicinal plants. Organization for Scientific Research Indonesia, Bulletin 18: 1–90.

Vasek, F. C. 1966. The distribution and taxonomy of three western junipers. Brittonia 18: 350–371.

Vassilyev, A. E. 1970. On the localization of the synthesis of terpenoids in the plant cell. Rastitel'nye Resursy 6: 29–45.

Vassilyev, A. E. 1977. Functional Morphology of Plant Secretory Cells. Nauka Publishing House, Leningrad.

Vávra, N. 1984. Reich an armen Fundstellen Übersicht über die Fossilen Harze Österreichs, Stuttgarter Beiträge zur Naturkunde, Serie C, 18: 9–14.

Vávra, N. 1990. Gas liquid chromatography—an effective tool for chemical characterization of fossil resins. Prace Muzeum Ziemi 4: 3–14.

Vávra, N. 1993. A chemical characterization of fossil resins ("amber")—a critical review of methods, problems, and possibilities: determination of mineral species, botanical sources and geographic attribution. Abhandlungen der Geologische Bundesanstalt 49: 147–157.

Villanueva, M. A., R. C. Torres, K. H. C. Baser, T. Osek, and M. Kurkcuoglu. 1993. The composition of Manila elemi oil. Flavour and Fragrance Journal 8: 35–37.

Vité, J. P. 1961. The influence of water supply on oleoresin exudation pressure and resistance to bark beetle attack in *Pinus ponderosa*. Contributions of the Boyce Thompson Institute 21: 37–66.

Vogel, V. J. 1970. American Indian Medicine. University of Oklahoma Press, Norman.

Vogellehner, D. 1965. Untersuchungen zur Anatomie und Systematik der verkieselten Holzer aus dem frankischen sudthuringischen Keuper. Erlanger Geologische Abhandlungen 59: 3–76.

Vogt, E. 1976. Tortillas for the Gods: a Symbolic Analysis of Zinacanteco Rituals. Harvard University Press, Cambridge.

Vogt, T., P. Proksch, and P. G. Gultz. 1987. Epicuticular flavonoids in the genus *Cistus*, Cistaceae. Journal of Plant Physiology 131: 25–36.

von Reis, S. 1973. Drugs and Foods from Little-Known Plants: Notes in Harvard University Herbaria. Harvard University Press, Cambridge.

von Reis, S., and F. J. Lipp, Jr. 1982. New Plant Resources for Drugs and Foods from the New York Botanical Garden Herbarium. Harvard University Press, Cambridge.

von Rudloff, E. 1975. Volatile leaf oil analysis in chemosystematic studies of North American conifers. Biochemical Systematics and Ecology 2: 131–167.

von Rudloff, E., and G. Rehfelt. 1980. Chemosystematic studies of the genus *Pseudotsuga*. IV. Inheritance and geographic variation in the leaf oil terpenes of Douglas-fir from the Pacific Northwest. Canadian Journal of Botany 58: 546–556.

Wake, D. B. 1991. Homoplasy: the result of natural selection or evidence of design limitations. American Naturalist 138: 543–567.

Walberg, I., M. B. Hjelte, K. Karlsson, and C. R. Enzell. 1971. Constituents of commercial Tolu balsam. Acta Chemica Scandinavica 25: 3285.

Walker, D. W., and D. D. Jenkins. 1986. Influence of sweet potato plant residues on growth of sweet potato vine cuttings and cow pea plants. HortScience 21: 426–428.

Wallin, K. F., and K. F. Raffa. 1999. Altered constitutive and inducible phloem monoter-penes following natural defoliation of jack pine: implications to host mediated inter-guild interactions and plant defense theories. Journal of Chemical Ecology 25: 861–879.

Walter, C., D. S. Carson, M. I. Menzies, T. Richardson, and M. Carson. 1998. Applica-tion of biotechnology to forestry—molecular biology of conifers. World Journal of Microbiology and Biotechnology 14: 321–330.

Walter, J., J. Charon, A. Marpeau, and J. Launay. 1989. Effects of wounding on the ter-pene content of twigs on maritime pine (*Pinus pinaster* Ait.). Trees 4: 210–219.

Wang, J., R. Dechun, C. Zhiying, and Z. Ligang. 1999. Phytoalexins in *Dracaena cochinchinensis* resin. Chinese Journal of Applied Ecology 10: 255–256.

Wang, S., and J. B. Huffman. 1981. Botanochemicals: supplements to petrochemicals. Economic Botany 35: 369–382.

Wang, X. S., H. N. Poinar, G. O. Poinar, Jr., and J. L Bada. 1995. Amino acids in amber matrix and in entombed insects. In: Ambers, Resinite and Fossil Resins, pp. 255–262, eds. K. B. Anderson and J. C. Crelling. American Chemical Society Symposium Series 617, Washington, D.C.

Ward, B. B., K. J. Courtney, and J. H. Langenheim. 1997. Inhibition of *Nitrosomonas europea* by monoterpenes from coastal redwood *(Sequoia sempervirens)* in whole-cell studies. Journal of Chemical Ecology 23: 2583–2598.

Watanabe, K. 1944. Illustrations of Useful Plants in the Southern Regions, Vol. 1, Medic-inal Plants, transl. Mr. T. S. Wei.

Waterman, P. G. 1986. Resins and other exudates in the flora of the Kora National Reserve, Kenya. In: Kora, an Ecological Inventory of the Kora National Reserve, Kenya, pp. 137–156, eds. M. Coe and N. M. Collins. Royal Geographic Society, London.

Waterman, P. G., and S. Ampoto. 1985. Dammarane triterpenes from the stem bark of *Commiphora dalzielii*. Phytochemistry 24: 2925–2928.

Waterman, P. G., and M. F. Grandan (eds.). 1985. Chemistry and Chemotaxonomy of the Rutales. Academic Press, New York.

Watkins, R. W., J. A. Lumley, E. L. Gill, J. D. Bishop, S. D. Langton, A. D. MacNicoll, N. R. Price, and M. G. B. Drew. 1999. Quantitative structure-activity relationships (QSAR) of cinnamic-acid bird repellents. Journal of Chemical Ecology 25: 2825–2845.

Watson, J. G. 1927. A note on the exploitation of *damar penak* in the Federated Malay States. Indian Forester 53: 493–500, 501–560.

Watson, J. 1988. The Cheirolepidaceae. In: Origin and Evolution of Gymnosperms, pp. 382–447, ed. C. B. Beck. Columbia University Press, New York.

Wayne, R. K., J. A. Leonard, and A. Cooper. 1999. Full of sound and fury: the recent his-tory of ancient DNA. Annual Review of Ecology and Systematics 30: 457–477.

Webber, I. E. 1941. Systematic anatomy of the woods of the Burseraceae. Lilloa 6: 441–471.

Weigel, C. 1698. Künstler und Handwercher. Regensburg.

Weir, A., M. Dolan, D. Grimaldi, R. Guerrero, J. Wagensberg, and L. Margulis. 2002. Spirochete and protist symbionts of a termite *(Mastotermes electrodominicus)* in Mio-cene amber. Proceedings of the National Academy of Sciences U.S.A. 99: 1410–1413.

Weissman, G., and H. H. Dietrichs. 1975. The termicidal activity of the extractives of *Callitris* and their structural relations. Holzforschung 33: 54–56.

Weitschat, W., and W. Wichard. 1998. Atlas der Pflanzen und Tiere im Baltischen Bernstein. Verlag Dr Friedrich Pfeil, München.

Wells, F. V., and M. Billot. 1981. Perfumery Technology: Art, Science, Industry. E. Horwood, Halsted Press (distributor), New York.

Wenninger, J. A., R. L. Yates, and M. Dolinksy. 1967. Sesquiterpene hydrocarbons of commercial copaiba balsam and American cedar wood oils. Journal of AOAC (Association of Official Analytical Chemists) 50: 1304–1313.

Werker, E., and A. Fahn. 1969. Resin ducts of *Pinus halepensis* Mill.—their structure, development and pattern of arrangement. Botanical Journal of the Linnean Society 62: 379–411.

West, A. P., and W. H. Brown. 1920. Philippine Resins, Gums, Seed Oils, and Essential Oils. Department of Agriculture and Natural Resources, Bureau of Forestry of the Philippine Islands, Bulletin 20. Manila.

Wheatley, P. 1959. Geographical notes on some commodities involved in Sung maritime trade. Journal of the Malayan Branch of the Royal Asiatic Society 32: 5–139.

Wheatley, P. 1961. The Golden Khersonese: Studies in the Historical Geography of the Malay Peninsula Before A.D. 1500. Oxford University Press, Kuala Lumpur.

Wheeler, R. E. M. 1954. Rome Beyond the Imperial Frontiers. Bell, London.

Whistler, R. L. 1993. Exudate gums. In: Industrial Gums: Polysaccharides and Their Derivatives, ed. 3, pp. 309–339, eds. R. L. Whistler and J. N. BeMiller. Academic Press, New York.

Whistler, R. L., and J. N. BeMiller (eds.). 1993. Industrial Gums: Polysaccharides and Their Derivatives, ed. 3. Academic Press, New York.

White, C. S. 1994. Monoterpenes: their effects on ecosystem nutrient cycling. Journal of Chemical Ecology 20: 1381–1406.

White, D. 1914. Resins in Paleozoic plants and in coals of high rank. U.S. Geological Survey Professional Paper 85: 65–83.

White, E. E., and J.-E. Nilsson. 1984. Genetic variation in resin canal frequency and relationship to terpene production in foliage of *Pinus contorta*. Silvae Genetica 33: 79–84.

White, R., and J. Kirby. 2001. A survey of nineteenth- and early twentieth-century varnish compositions found on a selection of paintings in the National Gallery collection. National Gallery Technical Bulletin 22: 64–84.

Whitham, T. G. 1983. Host manipulation of parasites: within plant variation as a defense against rapidly evolving pests. In: Variable Plants and Herbivores in Natural and Managed Systems, pp. 15–41, eds. R. F. Denno and M. S. McClure. Academic Press, New York.

Whitmore, T. C. 1977. A first look at *Agathis*. Tropical Forest Papers 11.

Whitmore, T. C. 1980a. A monograph of *Agathis*. Plant Systematics and Evolution 135: 41–69.

Whitmore, T. C. 1980b. Utilization, potential and conservation of *Agathis*, a genus of tropical Asian conifers. Economic Botany 34: 1–12.

Whitmore, T. C. 1980c. Evolutionary implications of the distribution and ecology of the tropical conifer *Agathis*. New Phytology 84: 407–416.

Wild, H. 1959. A revised classification of the genus *Commiphora* Jacq. Boletim da Sociedade Broteriana, série 2, 33: 67–100.

Wilf, P., and C. C. Labandeira. 1999. Response of plant–insect associations to Paleocene-Eocene warming. Science 284: 2153–2156.

Williams, D. H., M. J. Stone, P. R. Hauck, and S. K. Rahman. 1989. Why are secondary metabolites (natural products) biosynthesized? Journal of Natural Products 52: 1189–1208.

Williams, K. S., D. E. Lincoln, and P. R. Ehrlich. 1983. The coevolution of *Euphydras chalcedona* butterflies and their larval host plants. I. Larval feeding behavior and host plant chemistry. Oecologia 56: 323–329.

Williams, L. O. 1989. The useful plants of Central America. Ceiba 24: 1–342.

Williams, M. 1989. Americans and Their Forests—a Historical Geography. Cambridge University Press, New York.

Williamson, G. C. 1932. The Book of Amber. Ernest Benn, London.

Wilson, B. J., and L. T. Burke. 1983. Sweet potato toxins and related toxic furans. In: Plant and Fungal Toxins, Vol. 1, pp. 3–41, eds. R. F. Keeler and A. T. Tu. Marcel Dekker, New York.

Winkel-Shirley, B. 1999. Evidence for enzyme complexes in the phenylpropanoid and flavonoid pathways. Physiologia Plantarum 107: 142–149.

Wisdom, C. 1941. The Chorti Indians of Guatemala. University of Chicago Press, Chicago, Illinois.

Wolda, H. 1978. Fluctuations in abundance of tropical insects. American Naturalist 112: 1017–1045.

Wolfe, J. A. 1972. An interpretation of Alaska floras. In: Floristics and Paleofloristics of Asia and Eastern North America, pp. 201–203, ed. A. Graham. Elsevier, Amsterdam.

Wolfe, J. A. 1975. Some aspects of plant geography of the northern hemisphere during late Cretaceous and Tertiary. Annals of the Missouri Botanical Garden 62: 264–279.

Wollenberg, E. K. 2001. Incentives for collecting *gaharu* (fungal infected wood of *Aquilaria* spp.; Thymelaceae) in East Kalimantan. Economic Botany 55: 444–456.

Wollenweber, E. 1975. Flavonoidmunster als systematisches Merkmal in der Gattung *Populus*. Biochemical Systematics and Ecology 3: 35–45.

Wollenweber, E. 1976. Rare methoxy flavonoids from buds of *Betula nigra*. Phytochemistry 15: 438–439.

Wollenweber, E. 1984. The systematic implication of flavonoids secreted by plants. In: Biology and Chemistry of Plant Trichomes, pp. 53–69, eds. E. Rodriguez, P. L. Healey, and I. Mehta, Plenum Press, New York.

Wollenweber, E., and S. L. Buchmann. 1997. Feral honeybees in the Sonoran desert: propolis sources other than poplar (*Populus* spp.). Zeitschrift für Naturforschung 52C: 530–535.

Wollenweber, E., and V. H. Dietz. 1981. Occurrence and distribution of free flavonoid aglycones in plants. Phytochemistry 20: 869–932.

Wollenweber, E., and M. Jay. 1988. Flavones and flavonols. In: The Flavonoids, pp. 233–328, ed. J. B. Harborne. Chapman and Hall, London.

Wollenweber, E., K. Egger, and E. Schnepf. 1971. Flavonoid-aglycone in *Alnus* Knospen und die Feinstruktur der Drüsenzellen. Biochemie und Physiologie der Pflanzen 162: 193–202.

Wollenweber, E., I. Schober, G. Schilling, F. J. Arriaga-Giner, and J. N. Roitman. 1989. A geranyl α-pyrone from the leaf resin of *Diplacus aurantiacus*. Phytochemistry 28: 3493–3496.

Wollenweber, E., M. Doerr, K. Siems, R. Faure, I. Bombarda, and E. M. Gayou. 1998. Triterpenoids in lipophilic leaf and stem coatings. Biochemical Systematics and Ecology 27: 103–105.

Wolters, O. W. 1967. Early Indonesian Commerce. Cornell University Press, Ithaca, New York.

Wood, D. L. 1982. The role of pheromones, kairomones and allomones on the host selection and colonization behavior of bark beetles. Annual Review of Entomology 27: 411–446.

Wood, S. L. 1982. The bark and ambrosia beetles of North and Central America (Coleoptera: Scolytidae), a taxonomic monograph. Great Basin Naturalist Memoirs 6.

Wooding, F. B. P., and D. H. Northcote. 1965. The fine structure of the mature canal cells of *Pinus pinea*. Journal of Ultrastructure Research 13: 233–244.

Wren, F. L. S. 1988. Potter's New Cyclopaedia of Botanical Drugs and Preparations. C. W. Daniel Company, Saffron Walden [Essex, England].

Wu, H., and Z.-h. Hu 1997. Comparative anatomy of resin ducts of the Pinaceae. Trees 11: 135–143.

Wyllie, S. G., and J. J. Brophy. 1989. The leaf oil of *Liquidambar styraciflua*. Planta Medica 55: 316–317.

Xena de Enrech, N., M. T. K. Arroyo, and J. H. Langenheim. 1983. Sistemática del genero *Copaifera* L. (Leguminosae: Caesalpinioideae, Detarieae) en Venezuela. Acta Botanica Venezuelica 14: 239–290.

Yazdani, R., D. Rudin, T. Alden, D. Lindgren, B. Harborn, and K. Ljung. 1982. Inheritance pattern of five monoterpenes in Scots pine (*Pinus sylvestris* L.). Hereditas 97: 261–272.

Zaddach, G. 1867. Das Tertiärgebirge Samlands. Schriften der Physikalishe-Ökonomische Gesellschaft Königsberg 8: 85–197.

Zahn, H., and R. Sapinkopf. 1947. The romance of incenses and spices. American Perfumer 49: 549–555.

Zavarin, E., and F. W. Cobbs, Jr. 1970. Oleoresin variability in *Pinus ponderosa*. Phytochemistry 9: 2509–2515.

Zavarin, E., and K. Snajberk. 1975. *Pseudotsuga menziesii* chemical races of California and Oregon. Biochemical Systematics and Ecology 2: 121–129.

Zavarin, E., K. Snajberk, T. Reichert, and E. Tsien. 1970. On the geographic variability of the monoterpenes from the cortical blister oleoresin of *Abies lasiocarpa*. Phytochemistry 9: 377–395.

Zavarin, E., K. Snajberk, and W. B. Critchfield. 1977. Terpenoid chemosystematic studies of *Abies grandis*. Biochemical Systematics and Ecology 5: 81–93.

Zeiss, H. H., and F. W. Grant. 1957. The constitution of cativo gum. Journal of the American Chemical Society 79: 1201–1205.

Zeybek, N. 1970. Liefert *Styrax officinalis* L. Ein Harz? Berichte der Schweizerischen Botanischen Gesellschaft 80: 189–193.

Zhang, X., and J. S. States. 1991. Selective herbivory of ponderosa pine by Abert squirrels: a re-examination of the role of terpenes. Biochemical Systematics and Ecology 19: 111–115.

Zhan-Qian, S. 1989. Production and market of naval stores in China. Naval Stores Review 99: 7–8.

Zherikhin, V. V., and K. Y. Eskov. 1999. Mesozoic and Lower Tertiary in former USSR. [In: Proceedings of the World Congress on Amber Inclusions, eds. J. Alonso, J. C. Corral, and R. López.] Estudios del Museo de Ciencias Naturales de Álava 14: 119–132.

Zherikhin, V. V., and A. J. Ross. 2000. A review of the history, geology and age of Burmese amber (Burmite). Bulletin of the Natural History Museum, London (Geology) 56: 1–3

Zias, J., H. Stark, J. Selgiman, et al. 1993. Early medical use of *Cannabis*. Nature 363: 215–215.

Zieck, J. 1975. Copal Industry in Papua New Guinea. Forest Products Research Centre, Boroko, Papua New Guinea.

Zinkel, D. E. 1975. Naval stores: silvichemicals from pine. Applied Polymer Symposium 28: 309–327.

Zinkel, D. E. 1977. Pine resin acids as chemotaxonomic and genetic indicators. TAPPI (Technical Association of the Pulp and Paper Industry) Conference Papers 53–56.

Zinkel, D. E. 1981. Turpentine, rosin and fatty acids from conifers. In: Organic Chemicals from Biomass, pp. 163–187, ed. I. S. Goldstein. CRC Press, Boca Raton.

Zinkel, D. E., and J. Russell (eds.). 1989. Naval Stores: Production-Chemistry-Utilization. Pulp Chemicals Association, New York.

Zobel, B. J., G. van Wyk, and P. Stahl. 1987. Growing Exotic Forests. John Wiley and Sons, New York.

Zoghbi, M. G. B., E. V. L. da Cunha, and W. W. Filho. 1993. Essential oil of *Protium unifoliatum* (Burseraceae). Acta Amazônica 23: 15–16.

Zoghbi, M. G. B., J. B. G. Siqueira, E. L. A. Walter, and O. L. P. Juniar. 1994. Constituintes químicos de *Protium paniculatum* (Burseraceae). Acta Amazônica 24: 59–62.

Zoghbi, M. G. B., J. G. S. Maia, and A. I. R. Luz. 1995. Volatile constituents from leaves and stems of *Protium heptaphyllum* (Aubl.) March. Journal of Essential Oil Research 7: 541–543.

Zumbühl, S., R. Knochenmuss, S. Wülfert, F. Dubois, M. J. Dale, and R. Zenobi. 1998. A graphite-assisted laser desorption / ionization study of light-induced aging in triterpene dammar and mastic varnishes. Analytical Chemistry 70: 707–715.

Zupanec, D. K. J., D. Vasili, S. Krali, and J. Pšeniĕnik. 1991. Variability of essential oils of hops, *Humulus lupulus* L. Journal of the Institute of Brewing 97: 197–206.

Plant Index

Subject Index